FORMULATIONS OF GENERAL RELATIVITY

This monograph describes the different formulations of Einstein's General Theory of Relativity. Unlike traditional treatments, Cartan's geometry of fibre bundles and differential forms is placed at the forefront, and a detailed review of the relevant differential geometry is presented. Particular emphasis is given to General Relativity in 4D space-time, in which the concepts of chirality and self-duality begin to play a key role. Associated chiral formulations are catalogued, and shown to lead to many practical simplifications. The book develops the chiral gravitational perturbation theory, in which the spinor formalism plays a central role. The book also presents in detail the twistor description of gravity, as well as its generalisation based on geometry of 3-forms in seven dimensions. Giving valuable insight into the very nature of gravity, this book joins our highly prestigious Cambridge Monographs in Mathematical Physics series. It will interest graduate students and researchers in the fields of Theoretical Physics and Differential Geometry.

KIRILL KRASNOV is Professor of Mathematical Physics at the University of Nottingham. Since receiving his PhD from Pennsylvania State University, he has worked at the University of California, Santa Barbara, and the Max Planck Institute for Gravitational Physics in Potsdam.

CAMBRIDGE MONOGRAPHS ON MATHEMATICAL PHYSICS

S. J. Aarseth *Gravitational N-Body Simulations: Tools and Algorithms*†
D. Ahluwalia *Mass Dimension One Fermions*
J. Ambjørn, B. Durhuus and T. Jonsson *Quantum Geometry: A Statistical Field Theory Approach*†
A. M. Anile *Relativistic Fluids and Magneto-fluids: With Applications in Astrophysics and Plasma Physics*
J. A. de Azcárraga and J. M. Izquierdo *Lie Groups, Lie Algebras, Cohomology and Some Applications in Physics*†
O. Babelon, D. Bernard and M. Talon *Introduction to Classical Integrable Systems*†
F. Bastianelli and P. van Nieuwenhuizen *Path Integrals and Anomalies in Curved Space*†
D. Baumann and L. McAllister *Inflation and String Theory*
V. Belinski and M. Henneaux *The Cosmological Singularity*†
V. Belinski and E. Verdaguer *Gravitational Solitons*†
J. Bernstein *Kinetic Theory in the Expanding Universe*†
G. F. Bertsch and R. A. Broglia *Oscillations in Finite Quantum Systems*†
N. D. Birrell and P. C. W. Davies *Quantum Fields in Curved Space*†
K. Bolejko, A. Krasiński, C. Hellaby and M-N. Célérier *Structures in the Universe by Exact Methods: Formation, Evolution, Interactions*
D. M. Brink *Semi-Classical Methods for Nucleus-Nucleus Scattering*†
M. Burgess *Classical Covariant Fields*†
E. A. Calzetta and B.-L. B. Hu *Nonequilibrium Quantum Field Theory*
S. Carlip *Quantum Gravity in 2+1 Dimensions*†
P. Cartier and C. DeWitt-Morette *Functional Integration: Action and Symmetries*†
J. C. Collins *Renormalization: An Introduction to Renormalization, the Renormalization Group and the Operator-Product Expansion*†
P. D. B. Collins *An Introduction to Regge Theory and High Energy Physics*†
M. Creutz *Quarks, Gluons and Lattices*†
P. D. D'Eath *Supersymmetric Quantum Cosmology*†
J. Dereziński and C. Gérard *Mathematics of Quantization and Quantum Fields*
F. de Felice and D. Bini *Classical Measurements in Curved Space-Times*
F. de Felice and C. J. S Clarke *Relativity on Curved Manifolds*†
B. DeWitt *Supermanifolds, 2nd edition*†
P. G. O. Freund *Introduction to Supersymmetry*†
F. G. Friedlander *The Wave Equation on a Curved Space-Time*†
J. L. Friedman and N. Stergioulas *Rotating Relativistic Stars*
Y. Frishman and J. Sonnenschein *Non-Perturbative Field Theory: From Two Dimensional Conformal Field Theory to QCD in Four Dimensions*
J. A. Fuchs *Affine Lie Algebras and Quantum Groups: An Introduction, with Applications in Conformal Field Theory*†
J. Fuchs and C. Schweigert *Symmetries, Lie Algebras and Representations: A Graduate Course for Physicists*†
Y. Fujii and K. Maeda *The Scalar-Tensor Theory of Gravitation*†
J. A. H. Futterman, F. A. Handler, R. A. Matzner *Scattering from Black Holes*†
A. S. Galperin, E. A. Ivanov, V. I. Ogievetsky and E. S. Sokatchev *Harmonic Superspace*†
R. Gambini and J. Pullin *Loops, Knots, Gauge Theories and Quantum Gravity*†
T. Gannon *Moonshine beyond the Monster: The Bridge Connecting Algebra, Modular Forms and Physics*†
A. García-Díaz *Exact Solutions in Three-Dimensional Gravity*
M. Göckeler and T. Schücker *Differential Geometry, Gauge Theories, and Gravity*†
C. Gómez, M. Ruiz-Altaba and G. Sierra *Quantum Groups in Two-Dimensional Physics*†
M. B. Green, J. H. Schwarz and E. Witten *Superstring Theory Volume 1: Introduction*
M. B. Green, J. H. Schwarz and E. Witten *Superstring Theory Volume 2: Loop Amplitudes, Anomalies and Phenomenology*
V. N. Gribov *The Theory of Complex Angular Momenta: Gribov Lectures on Theoretical Physics*†
J. B. Griffiths and J. Podolský *Exact Space-Times in Einstein's General Relativity*†
T. Harko and F. Lobo *Extensions of f(R) Gravity: Curvature-Matter Couplings and Hybrid Metric-Palatini Gravity*
S. W. Hawking and G. F. R. Ellis *The Large Scale Structure of Space-Time*†
B. B. Hu and E. Verdaguer *Semiclassical and Stochastic Gravity*
F. Iachello and A. Arima *The Interacting Boson Model*†
F. Iachello and P. van Isacker *The Interacting Boson-Fermion Model*†
C. Itzykson and J. M. Drouffe *Statistical Field Theory Volume 1: From Brownian Motion to Renormalization and Lattice Gauge Theory*†
C. Itzykson and J. M. Drouffe *Statistical Field Theory Volume 2: Strong Coupling, Monte Carlo Methods, Conformal Field Theory and Random Systems*†
G. Jaroszkiewicz *Principles of Discrete Time Mechanics*

G. Jaroszkiewicz *Quantized Detector Networks*
C. V. Johnson *D-Branes*†
P. S. Joshi *Gravitational Collapse and Spacetime Singularities*†
J. I. Kapusta and C. Gale *Finite-Temperature Field Theory: Principles and Applications, 2nd edition*†
V. E. Korepin, N. M. Bogoliubov and A. G. Izergin *Quantum Inverse Scattering Method and Correlation Functions*†
J. Kroon *Conformal Methods in General Relativity*
M. Le Bellac *Thermal Field Theory*†
L. Lusanna *Non-Inertial Frames and Dirac Observables in Relativity*
Y. Makeenko *Methods of Contemporary Gauge Theory*†
S. Mallik and S. Sarkar *Hadrons at Finite Temperature*
A. Malyarenko and M. Ostoja-Starzewski *Tensor-Valued Random Fields for Continuum Physics*
N. Manton and P. Sutcliffe *Topological Solitons*†
N. H. March *Liquid Metals: Concepts and Theory*†
I. Montvay and G. Münster *Quantum Fields on a Lattice*†
P. Nath *Supersymmetry, Supergravity, and Unification*
L. O'Raifeartaigh *Group Structure of Gauge Theories*†
T. Ortín *Gravity and Strings, 2nd edition*
A. M. Ozorio de Almeida *Hamiltonian Systems: Chaos and Quantization*†
M. Paranjape *The Theory and Applications of Instanton Calculations*
L. Parker and D. Toms *Quantum Field Theory in Curved Spacetime: Quantized Fields and Gravity*
R. Penrose and W. Rindler *Spinors and Space-Time Volume 1: Two-Spinor Calculus and Relativistic Fields*†
R. Penrose and W. Rindler *Spinors and Space-Time Volume 2: Spinor and Twistor Methods in Space-Time Geometry*†
S. Pokorski *Gauge Field Theories, 2nd edition*†
J. Polchinski *String Theory Volume 1: An Introduction to the Bosonic String*†
J. Polchinski *String Theory Volume 2: Superstring Theory and Beyond*†
J. C. Polkinghorne *Models of High Energy Processes*†
V. N. Popov *Functional Integrals and Collective Excitations*†
L. V. Prokhorov and S. V. Shabanov *Hamiltonian Mechanics of Gauge Systems*
S. Raychaudhuri and K. Sridhar *Particle Physics of Brane Worlds and Extra Dimensions*
A. Recknagel and V. Schiomerus *Boundary Conformal Field Theory and the Worldsheet Approach to D-Branes*
M. Reuter and F. Saueressig *Quantum Gravity and the Functional Renormalization Group*
R. J. Rivers *Path Integral Methods in Quantum Field Theory*†
R. G. Roberts *The Structure of the Proton: Deep Inelastic Scattering*†
P. Romatschke and U. Romatschke *Relativistic Fluid Dynamics In and Out of Equilibrium: And Applications to Relativistic Nuclear Collisions*
C. Rovelli *Quantum Gravity*†
W. C. Saslaw *Gravitational Physics of Stellar and Galactic Systems*†
R. N. Sen *Causality, Measurement Theory and the Differentiable Structure of Space-Time*
M. Shifman and A. Yung *Supersymmetric Solitons*
Y. M. Shnir *Topological and Non-Topological Solitons in Scalar Field Theories*
H. Stephani, D. Kramer, M. MacCallum, C. Hoenselaers and E. Herlt *Exact Solutions of Einstein's Field Equations, 2nd edition*†
J. Stewart *Advanced General Relativity*†
J. C. Taylor *Gauge Theories of Weak Interactions*†
T. Thiemann *Modern Canonical Quantum General Relativity*†
D. J. Toms *The Schwinger Action Principle and Effective Action*†
A. Vilenkin and E. P. S. Shellard *Cosmic Strings and Other Topological Defects*†
R. S. Ward and R. O. Wells, *Jr Twistor Geometry and Field Theory*†
E. J. Weinberg *Classical Solutions in Quantum Field Theory: Solitons and Instantons in High Energy Physics*
J. R. Wilson and G. J. Mathews *Relativistic Numerical Hydrodynamics*†

† Available in paperback

Formulations of General Relativity

Gravity, Spinors and Differential Forms

KIRILL KRASNOV
University of Nottingham

CAMBRIDGE
UNIVERSITY PRESS

CAMBRIDGE
UNIVERSITY PRESS

University Printing House, Cambridge CB2 8BS, United Kingdom

One Liberty Plaza, 20th Floor, New York, NY 10006, USA

477 Williamstown Road, Port Melbourne, VIC 3207, Australia

314–321, 3rd Floor, Plot 3, Splendor Forum, Jasola District Centre, New Delhi – 110025, India

79 Anson Road, #06–04/06, Singapore 079906

Cambridge University Press is part of the University of Cambridge.

It furthers the University's mission by disseminating knowledge in the pursuit of education, learning, and research at the highest international levels of excellence.

www.cambridge.org
Information on this title: www.cambridge.org/9781108481649
DOI: 10.1017/9781108674652

First published 2020

A catalogue record for this publication is available from the British Library.

ISBN 978-1-108-48164-9 Hardback

To the memory of the two Peter Ivanovichs in my life:
Peter Ivanovich Fomin, who got me interested in gravity,
and Peter Ivanovich Holod, who formed my taste for mathematics.

Contents

Preface

> Give thanks to God, who made necessary things simple, and complicated things unnecessary.
>
> Gregory Skovoroda, *Ukrainian Thinker,* 1722–1794

> There is always another way to say the same thing that doesn't look at all like the way you said it before. I don't know what the reason for this is. I think it is somehow a representation of the simplicity of nature? Perhaps a thing is simple if you can describe it fully in several different ways without immediately knowing that you are describing the same thing.
>
> Richard Feynman, *Nobel Lecture,* 1965

> Theories of the known, which are described by different physical ideas may be equivalent in all their predictions and are hence scientifically indistinguishable. However, they are not psychologically identical when trying to move from that base into the unknown. For different views suggest different kinds of modifications which might be made and hence are not equivalent in the hypotheses one generates from them in ones attempt to understand what is not yet understood. I, therefore, think that a good theoretical physicist today might find it useful to have a wide range of physical viewpoints and mathematical expressions of the same theory available to him.
>
> Richard Feynman, *Nobel Lecture,* 1965

Formulations of General Relativity. Facing this title the prospective reader should be thinking, what is there to formulate general relativity (GR)? GR can be formulated in one sentence: GR action functional is the integral of the scalar curvature over the manifold. Everything else that is there to say about GR is the consequence of the Euler–Lagrange equations one obtains by extremising this action, together with the action for matter fields. How can there be a book about 'formulations'? And why plural? Is there not just the usual Einstein–Hilbert formulation as stated previously?

A more sophisticated reader will know that there are several equivalent formulations of general relativity. There is the usual metric formulation, and then there is an equivalent formulation in terms of tetrads. But this is all well known. GR is about physical consequences of the dynamical postulate that fixes the theory. There may be several equivalent ways to define the dynamics. But this does not change the physics. So, one formulation is sufficient to unravel all the physics predicted by the theory. The usual metric formulation is by far the most studied and best understood. Why bother about developing any other equivalent formulation? And then why write a book about such unnecessary alternatives?

This is when the two quotes included previously from the Richard Feynman Nobel lecture become relevant. The first is about an empirical observation that theories that are relevant for describing the world around us tend to admit many different equivalent, but not obviously so, reformulations. The example Feynman has in mind is classical electrodynamics, not gravity. Feynman also notices that there is a deep link between the 'simplicity' of a theory, and the availability of many different, not manifestly equivalent, descriptions. He goes further and proposes this as the criterion of simplicity. This suggests that one can never fully appreciate the simplicity and beauty of GR without absorbing all the different available formulations of this theory.

The second quote is a different, but not unrelated, thought. There may be equivalent formulations of a theory, all leading to the same physical predictions. But such reformulations may be inequivalent if one decides to generalise. The example of most relevance for Feynman is the Hamiltonian and Lagrangian description of classical mechanics. The quantum generalisation of the Hamiltonian description leads to the usual operator formalism for quantum theory. The generalisation of the Lagrangian description leads to path integrals, which is arguably one of Feynman's main contributions to physics. These two equivalent formulations of classical mechanics are certainly not equivalent in terms of the new structures that can be generated from them. The same may well apply to gravity. We do not yet know which of the many available formulations of gravity will lead to the next big step in the quest for understanding the world around us.

So, the purpose of this book is to describe all the 'equivalent' formulations of general relativity that are known to the author, and that also put the geometry of differential forms and fibre bundles at the forefront of the description of gravity. What is meant by a 'formulation' here is a Lagrangian description, in which the dynamical equations are obtained by extremising the corresponding action. This gives us the most economic way of defining the theory.

Some of these equivalent formulations will likely be known to many readers. In particular, this is the already mentioned formulation in terms of tetrads. If this was the complete list, there would be no good reason to write this book.

What is much less known, and what really motivated this author to embark on the present project, is that there are some special features of GR in four spacetime dimensions. These special features are related to coincidences that occur precisely in four dimensions. Thus, in any dimension the Riemann curvature can be viewed as a matrix mapping antisymmetric rank-two tensors again into such tensors. And in four dimensions one also has the Hodge star operator that maps antisymmetric rank-two tensors into such tensors. One can ask how these two operations are related or compatible. It is then a simple to check but deep fact that a metric is Einstein if and only if these two operations commute. This fact leads to a whole series of *chiral* formulations of four-dimensional GR that have no analogs in higher spacetime dimensions. It is the development of these formulations, and contrasting them with the more known ones, that will occupy us for the large part of this book. There is no coherent account of these developments in the literature, certainly not in any book on GR. It is our desire to make such a coherent account available that was one of the main motivations for writing this monograph.

Another motivation for writing this exposition was our desire to promote the formalism(s) for GR that place the differential forms rather than metrics at the forefront. Differential forms are arguably the simplest and most natural geometric objects that can be placed on a smooth manifold, and are certainly simpler objects than a metric. It turns out to be possible to describe GR using the powerful calculus of differential forms and fibre bundles, which is largely due to Élie Cartan (see Chapter 1 for more on this). This book is in particular aimed at giving an exposition of the possible formalisms that achieve this.

A related theme is that of spinors and spinorial description. As is well known, and as we will also emphasise in the book, spinors and differential forms are essentially the same thing, with the Dirac operator being intimately related to the exterior derivative operator. This means that as soon as differential forms are being used as variables to describe the theory, the description has an interesting spinor translation. Viewed in this way, the kinetic operators arising in the field equations of formulations that use differential forms are various versions of the Dirac operator. This becomes especially pronounced in the so-called first-order formulations where field equations are first order in derivatives. This spinor aspect of gravity (and, as we shall see, Yang–Mills theory too), absent in the usual metric description, is another unifying theme of this book. In addition, the spinor description of gravity simplifies link to some recent developments in the field of scattering amplitudes, as we will touch on.

The more familiar of formalisms that use differential forms rather than metrics is that of tetrad (or vielbein, or moving frame or soldering form) introduced by Cartan. Historically, this formalism was first discovered in the context of two dimensions by the French mathematician Jean-Gaston Darboux (Cartan's PhD supervisor) in the late nineteenth century. It is particularly powerful in this context, as the two 1-forms that encode the metric information can be combined into a single complex-valued one-form on the manifold. This is related to the fact that any 2-manifold is a complex manifold. There is no direct analog of such a complexification trick in four dimensions because there is no longer a unique choice of an almost complex structure. But one gets a computationally powerful formalism in four dimensions via chiral formulations referred to above. These formulations, in the case of Lorentzian signature, bring into play complex-valued objects and in a certain sense provide the analog of the complexification trick that works so well in two dimensions. They also make a link to the twistor description of gravity, as we shall learn.

Our final introductory remark is about Einstein's cornerstone idea that gravity is geometry. At the time when Einstein formulated his theory, the only geometry available to him was the Riemannian geometry of metrics, described via the tensor calculus of Ricci and Levi-Civita. Einstein learned this mathematics guided by his friend and classmate Marcel Grossmann. It is thus no surprise that GR was formulated in the language of Riemannian geometry and tensor calculus. It is still being developed and also taught to graduate students in that way. However, already at the time of Einstein's formulation of GR, Élie Cartan was developing a very different type of geometry, the geometry in which the key role was played by differential forms and connections. His works, and works of those around him, strongly influenced the subject of differential geometry, and it is now far more rich and sophisticated than it was 100 years ago. The Riemannian geometry is now only its relatively small corner. This discussion is related to the theme of the present book because various different formulations of GR that we develop place various different geometric constructions at the forefront. In particular, the geometry of fibre bundles plays a much more important role than it does in the usual description of GR. It is thus certainly true that gravity continues to be geometry in the developments on this book, it is only that the word geometry is being understood more broadly than in the metric GR context. We do not yet know which of these 'geometries' is more fundamental than others, but a good researcher will certainly want to keep his/her mind open and learn all the available options.

The target audience for this book are postgraduate students interested in gravity, as well as already established researchers. To give encouraging words to the first audience, the author would like to recall his own experience as a student. This author remembers very distinctly that it was easiest to study, understand and prepare for exams on classical mechanics by reading Vladimir Arnold's book on the subject. And Paul Dirac's book played a similar role for quantum mechanics. Both books present their respective subjects in a beautiful and logical way, and both are inspired by mathematics. The moral here is that there are some students that learn best by understanding the overall logic of the formalism first, and only then embark on applications and problem solving. This is certainly not a universal way to learn, and most likely not the way to approach the subject for the first time. But it was important to the present author in his time as a student to have accounts of the usual subjects that concentrate more on the overall logic and the mathematical formalism, rather than on concrete problems that can be solved. The author hopes that there are similar minds out there, and that the present exposition will help such students to understand what GR is about.

In terms of the specialised knowledge that is required to understand this book, we do not assume any more than is usually assumed for graduate-level courses. Familiarity with concepts of differential geometry is desirable, but the aspects of this subject that are required to understand the present text are reviewed in the first chapter. So, a good graduate student should be able to follow this exposition without too much difficulty.

Thus, this book is mainly about different possible formalisms for doing calculations with GR, rather than about different possible physical consequences of this theory. So, this book certainly does not compete with the standard textbook expositions of GR, and the student must also study these more standard sources to understand the physics as predicted by general relativity. Excellent books on this subject that became the standard sources are *General Relativity* by R. Wald and *Spacetime and Geometry: An Introduction to General Relativity* by S. Carroll for GR in general and *Physical Foundations of Cosmology* by V. Mukhanov for applications to cosmology.

For the experienced researchers, the author suggests this book as a source on aspects of GR that are important about this theory, especially in four spacetime dimensions, but are not covered in any standard book on the subject. Thus, the book can be used as a compendium on different available formalisms for GR, as well as on some less standard aspects of geometry that are required to develop these formulations. Additional motivations for why different formulations of GR may lead to new developments and/or new generalisations are given in the concluding chapter.

We end by explaining why it is the quote from Gregory Skovoroda that we chose to be an epigraph for this whole exposition. First, the author is a Ukrainian, and it gives him a distinct pleasure to be able to quote Skovoroda, who was a deep thinker years ahead of his times, and who is still relevant today. He is almost unknown in the West, and maybe one of the readers will remember the name, and read his texts.

Second, we aim here to explain only simple, but in our view important things about GR in four spacetime dimensions. There is much more that can be said, and there is a great wealth of physical phenomena that the theory predicts and describes, and that we omit. Not because they are unimportant – on the contrary, they are the reason why physicists learn the subject. But rather because they are unnecessary to understand the overall logic of the theory. It is this overall logic and the facts likely needed to 'move to the unknown' that will concern us in the present book. And so we focus here only on things necessary to understand the overall logic of gravity, and hence only on things simple. We hope the reader will take this as a word of encouragement to follow the development of different formalisms described here.

Finally, I would like to thank my collaborators, from whom I learned a lot and without whose insights this book would not exist. Thanks in particular to Joel Fine, Yannick Herfray, Carlos Scarinci and Yuri Shtanov. Thanks also go to my family for their support to 'papa' working on his 'kniga'.

Introduction

[The tensor calculus] is the debauch of indices.

Élie Cartan, *from Introduction to 'Lecons sur la Geometrie des Espaces de Riemann', 1928*

In 1907, while still working as a clerk in a patent office in Bern, Albert Einstein had what he later referred to as 'the happiest thought' of his life. He realised that a freely falling observer does not experience gravity, and thus, effects of gravity are indistinguishable from those arising in an accelerating frame. These ideas were developed in two papers he published in 1908 and 1911. In these papers, Einstein argued that the rules of special relativity must continue to be applicable in an accelerated reference frame. This, in particular, led him to analyse experiences of an observer performing experiments on a rotating turntable. Einstein concluded that the ratio of the circumference of a circle to its diameter would be different from π. What this meant for Einstein was that if effects of gravity are those of a non-inertial coordinate system, and the geometry in the later is different from the Euclidean one, then gravity is geometry.

Einstein then searched for a mathematical description of this idea. On the return in 1912 to his *alma mater* ETH Zurich, he turned for help to his friend and classmate, now a professor of mathematics, Marcel Grossmann. Grossmann directed Einstein's attention to Riemannian geometry, the only one developed at that time, which had its origins in Gauss' work on the intrinsic geometry of two-surfaces in three-dimensional space. Bernhard Riemann lay the foundation of the subject in his famous 1854 Göttingen habilitation lecture, 'On the hypotheses that underlie geometry'. In this lecture he described the way to extend the Gauss' notion of curvature to an 'n-ply extended magnitude'. Thus, by the time Einstein studied this subject, it was far from being new. Einstein learned it in the form described in the 1900 exposition by Gregorio Ricci and Tulio Levi-Civita, 'Methods of the absolute differential calculus and their applications'. In a joint 1913 paper with Grossmann, Einstein described 'an outline' of a new gravity theory using precisely this language. The final version of the new theory of gravity was developed by late 1915, still using the language of tensor calculus. By this time Einstein was already in Berlin, and this work appeared single-authored. It is this 1915 theory that is now known as Einstein's general relativity (GR). Even to this day it is taught and applied using the nineteen-century language of tensor calculus.

Figure I.1 Bernhard Riemann

Bernhard Riemann was born on 17 September 1826 in Breselenz, a village in the kingdom of Hanover. His father was a poor Lutheran pastor. Riemann was the second of six children, shy and of not very strong health. His mother died when he was 20, and his brother and three of his sisters all died young, as eventually did he. Riemann exhibited exceptional mathematical skills, such as calculation abilities, from an early age, but suffered from a fear of speaking in public.

Even though Riemann was very gifted in mathematics, he planned to study theology and become a pastor, like his father. In 1846, his father gathered enough money to send him to Göttingen to study theology. However, once there, Riemann started attending mathematics lectures by Gauss. The latter recommended that Riemann give up his theological work and go into mathematics. After gaining his father's approval, Riemann transferred to Berlin in 1847, and returned to Göttingen in 1849. He defended his doctoral dissertation in 1851, on what we now call Riemann surfaces. He held his first lectures in 1854. His habilitation lecture founded the field of Riemannian geometry. In 1859, following Dirichlet's death, who had occupied Gauss' chair since 1855, Riemann became the head of mathematics at Göttingen.

In 1862, Riemann married Elise Koch and they had a daughter. He fled Göttingen in 1866 when the armies of Prussia and Hanover clashed there. He died in Italy the same year from tuberculosis. Riemann was a dedicated Christian, and saw his life as a mathematician as another way to serve God. During his life, he held closely to his Christian faith and considered it to be the most important aspect of his life. At the time of his death, he was reciting the Lord's Prayer with his wife and died before they finished saying the prayer.

Roughly around the same time, a French mathematician Élie Cartan was developing a very different type of geometry. In Cartan's work on differential geometry, the notions of differential forms and fibre bundles, both of which he to a large extent established, played a central role. Both of these will play a crucial role in this book too. Also, in 1913, constructing linear representations of Lie groups, Cartan discovered spinors. This will be important in our exposition as well. It was realised much later, in a 1954 book, *The Algebraic Theory of Spinors*, by another French mathematician (and one of the founding members of the Bourbaki group) Claude Chevalley, that spinors and differential forms are very closely related. We will explain this fact in due course.

Cartan was led to the notion of differential forms in his 1901 work developing a geometric approach to partial differential equations. What Cartan was after was a formalism that is invariant under arbitrary changes of variables. Cartan's main tool for this was the calculus of differential forms. Cartan then worked on problems of group theory, and in particular, as we already mentioned, discovered the spinor representations of the orthogonal groups in 1913.

Theory of Lie groups is intimately related to geometry. It is thus no surprise that Cartan turned to the latter. He was also motivated by Einstein's theory of gravity that came to prominence in 1919. It is in Cartan's works of the 1920s that his most important contributions to differential geometry were developed. Cartan's main realisation was that it is fruitful and necessary to consider other 'bundles' apart from the tangent bundle, and other 'connections' apart from the Levi–Civita connection. We put the words bundles and connections in quotes because these notions were only beginning to be understood in Cartan's works. In particular, Cartan himself, while working with different bundles extensively, never explicitly defined what is now known as a (principal) fibre bundle. Cartan was also responsible for a notion of what is now known as the (principal) connection, and in particular realised that such a connection is best described as a (Lie algebra valued) 1-form. Cartan was thus able to disassociate the notion of the connection and parallel transport from the very restricted form these take in the context of affine connections in the tangent bundle. This led him to the discovery of many new types of geometry, thus finding probably the most fruitful generalisation of Riemannian geometry. This was searched by many around the same time, in particular by Hermann Weyl, but it was Cartan who achieved this goal. As a bonus of his general programme on connections, Cartan was also able to give a very powerful and simple description of Riemannian geometry, in his 1925 paper, 'La géométrie des espaces de Riemann'. In the preface to his 1928 book, Leçons Sur la Géométrie des Espaces de Riemann, he stated his aim was that of bringing out the simple geometrical facts that have often been hidden under a debauch of indices. It is this description of Riemannian geometry that we will present under the name of 'tetrad' formalism for GR.

Élie Cartan was born on 9 April 1869 in Dolomieu (near Chambéry), a region Rhône-Alpes of France. His father was a blacksmith. The family was very poor, and it would be impossible for Élie to get good education if not for his talent for mathematics that was spotted early. Already at primary school, Élie impressed

Figure I.2 Élie Cartan

his teachers. One of them later said: 'Élie Cartan was a shy boy, but his eyes shone with an unusual light of great intelligence'. Still, Cartan could have never become a great mathematician if not for a young school inspector, and later important politician, Antonin Dubost. Dubost was visiting the school where the young Élie was taught and was impressed with young boy's talent. He encouraged Élie to participate in a competition for state funds that would enable him to study in a Lycée. Élie's school teacher M. Dupuis prepared him for the competitive examinations that were held in Grenoble. An excellent performance allowed Élie to study in good schools, and then later to study at the École Normale Supérieure (ENS) in Paris.

At ENS, Cartan became a student of Gaston Darboux, the inventor of the moving frame method, which Cartan later largely developed. Cartan's friend, Arthur Tresse, was studying under Sophus Lie in Leipzig, and told Cartan about the remarkable work of Wilhelm Killing on the classification of finite groups of continuous transformations. Cartan then set to complete Killing's work, and corrected some important mistakes and omissions in it. This became Cartan's doctoral dissertation. In one way or another, Cartan's whole scientific career revolved around the questions related to Lie groups and their geometry.

Cartan was a lecturer at the University at Montpellier during 1894–1896, and a lecturer at the University of Lyons, where he taught from 1896–1903. In 1903, he married Marie-Louise Bianconi (1880–1950), the daughter of a professor of chemistry there. The family moved to Paris in 1909, where Cartan was appointed professor first at the Sorbonne and later at ENS. The Cartans had four children. The eldest son, Henri, became a renowned mathematician. The second son, Jean, a composer of fine music, died of tuberculosis in 1932 at the age of 25. Their third son, Louis, became a physicist. He was a member of the Resistance fighting in France against the occupying German forces, and was arrested and executed by the Nazis in 1943. Cartan was 75 when he learned of his third son's fate, and this was a devastating blow for him. The fourth child of the family was a daughter, Hélène, who became a teacher of mathematics.

Cartan died in Paris in 1951 at the age of 82. Cartan's obituary by Chern and Chevalley opens with the words: 'Undoubtedly one of the greatest mathematicians of this century, his career was characterized by a rare harmony of genius and modesty'.

Cartan's more general connections were rediscovered by physicists only much later, in the 1954 work by Yang and Mills. Every known interaction in nature is now described by a gauge field or connection, of precisely the type that was first introduced by Cartan in his differential geometry work of the 1920s. Of course, Cartan did not write the Yang–Mills field equations, as his motivations were entirely different from those of particle physicists of the 1950s. It was thus Cartan who developed mathematics that is necessary to formulate gauge theories, and that can also be used to describe gravity. It is rather unfortunate that the theory of gravity is usually taught in the nineteenth-century language of tensor calculus and not in the twentieth-century language of principal connections in fibre bundles. Not only this second language is more clear – the debauch of indices is no longer there – but it is also more computationally efficient due to its usage of differential forms, and brings gravity closer in form to all the other interactions. We hope this book will serve to promote Cartan's language of differential forms and connections as the most appropriate one, not just for Yang–Mills theory, but also for gravity.

It must be admitted that for someone who was raised on notions of indices and tensor calculus, absorbing Cartan's geometric ideas is a rather difficult task. This is in particular manifested by the fact that Cartan's work on differential geometry was recognised to be of importance only late in his life. Quoting Cartan's obituary by Shiing-Shen Chern and Claude Chevalley, written in 1951, Cartan's 'death came at a time when his reputation and the influence of his ideas were in full ascent'. However, even in 1938, Hermann Weyl, in reviewing one of Cartan's books, wrote: 'Cartan is undoubtedly the greatest living master in differential geometry. ...I must admit that I found the book, like most of Cartan's papers, hard reading. ...' This sentiment was shared by many geometers at the time. The situation has changed however. Differential geometry is now taught, at least to mathematicians, in a way that incorporates Cartan's geometric ideas from the start. It is time that this powerful language is also taken on board by (gravitational) physicists.

Having given praise to Cartan's ideas, it should be said that the tetrad formalism *is* described in most standard textbooks on GR, often under the name of 'non-coordinate bases', see, e.g., Sean Carroll's 2019 book and/or *Geometry, Topology and Physics*, by Mikio Nakahara (2000). This formalism, however, is described only as secondary to the usual metric one. In particular, the spin connection, which is the central object that the tetrad formalism introduces, is considered to be only an object derived from the usual Christoffel connection. Also, the conceptual change that the tetrad formalism brings with itself, namely the fact that it works with a vector bundle different from the tangent bundle, is rarely emphasised, while this is the central point. Moreover, the presentation of the tetrad formalism in GR literature in fact avoids introducing any other bundle. The presentation of the tetrad formalism to be given in this book is different from the standard treatment in GR texts and is closer to the ones appearing in the mathematical literature.

Moreover, while a description of the tetrad formalism can often be found in the GR literature, it is rarely given any significance. Indeed, the usual attitude is that it is only a reformulation of GR, and, moreover, one that increases the number of field components that one has to work with, from 10 metric components in four dimensions in metric GR, to 16 tetrad components. This is clearly in the direction of loss of economy, and this appears to be a clear reason against using the tetrads. Furthermore, the tetrad formalism uses two different types of indices, the spacetime indices for vectors and forms on a manifold, and 'internal' indices for objects valued in the vector bundle on which the tetrad formalism is based. The usual attitude is that this leads to a notational nightmare. Why then use a formalism with two types of indices, if in the metric GR it is possible to work with only spacetime indices? Thus, the usual attitude to tetrads in the GR community is that this is a cumbersome formalism, which brings with it nothing new, and is therefore not worth the effort. It is nevertheless admitted that spinors can only be coupled to gravity by using the tetrads. But one is rarely interested in gravity effects caused by spinor matter, usually an effective description of matter using perfect fluids is completely sufficient to extract interesting physics. So, even though spinors do require tetrads, one rarely needs spinors in GR.

Yet another seemingly compelling reason to ignore tetrads is the description of the linearised excitations of the gravitational field. These carry spin two. As such, it appears to be natural to describe them by rank two tensors. The linearised dynamics is then readily available by either linearising the Einstein equations, or by looking for a second-order differential operator that is invariant under the linearised diffeomorphisms. Both procedures uniquely lead to the same linearised dynamics. The attitude of the particle physics community is then that Einstein's theory gives a nonlinear completion of this linearised description, which is moreover to a very large extent unique. This point of view has been advocated in Weinberg's 1972 book, *Gravitation and Cosmology*. From this point of view it appears to be unnatural to use any other object to describe gravity other than the metric.

Both arguments against the usage of tetrads actually underestimate the power of the formalism of differential forms. Yes, the tetrad carries more components, but the amount of gauge has also increased. And it is often the case in mathematics that a formalism that uses more independent functions allows for a simpler description. That this is the case with the tetrad formalism is manifested by the fact that the gravitational action in the tetrad formalism is just quartic in the basic fields, while the Einstein–Hilbert metric action is non-polynomial in the metric. Thus, the tetrad formalism gives an algebraically simpler description of the gravitational field. And working with objects with different types of indices is not a problem once an appropriate formalism is developed. Indeed, having fields with two different types of indices does not cause any problems in the treatment of the Yang–Mills theory. Finally, for the description of the linearised dynamics, it turns out that not only does the tetrad formalism not make things more complicated, on the contrary, the usage of differential forms brings with it simpler differential operators as compared to those that arise in the metric formalism. In fact, using differential forms, one achieves a description of the spin two linearised fields that is analogous to the description of Maxwell's theory, as we shall see in Chapter 8. There is no such analogy when one works with the metric variables. So, all in all, the formalism of differential forms does introduce simplifications in GR ranging from the full nonlinear dynamics to the linearised treatment. So, it is brushed aside in the usual GR texts for the wrong reasons, as we hope will become clear from the treatment in this book.

As we have already said in the preface, this book is more than just about the tetrad formalism. Its unifying theme is the formalisms for GR (in particular, GR in four spacetime dimensions) that are based on vector valued differential forms. Towards the end of the book, we will develop an even more exotic alternative, in which gravity in four dimensions will be seen to arise as the dimensional reduction of a theory of 'pure' differential forms, i.e., differential forms valued in $\mathbb{R}$, in seven dimensions. The development of all these different formulations would be impossible without Cartan's ideas and the example of the tetrad formalism, historically the first description of GR in terms of differential forms. This explains the considerable attention given to Cartan's type of differential geometry in this book. To put it provocatively, this book attempts to develop the theory of gravity using the twentieth-century differential geometry of Cartan, forgetting Einstein's theory of GR formulated using the nineteenth-century language of tensor calculus as much as possible.

1
Aspects of Differential Geometry

The purpose of this chapter is to review, in a concise manner, aspects of differential geometry that will be used in this book. It should be noted that the presentation here is more a list of things that are important rather than a pedagogical introduction to the subject. It is likely to be usable by those seeing this material for the first time only if accompanied by reading other texts. At the same time, the material here is standard and is covered in many books, so there is no shortage of more pedagogical sources. The books we like are *The Geometry of Physics* by Theodore Frankel (2012) and *Geometry, Topology and Physics* by Mikio Nakahara (2003). We have also taken some material from Dubrovin, Novikov and Fomenko's (1985) *Modern Geometry*, and some definitions are from the book by C. H. Taubes (2011) called *Differential Geometry.* Our presentation of differential forms is from R. Bott and L. W. Tu's (1982) *Differential Forms in Algebraic Topology.* Finally, an invaluable source on many aspects of Riemannian geometry is Besse's (1987) *Einstein Manifolds.* The only slightly original part in this chapter is our discussion of spinors in relation to differential forms. It is original just in the sense of not being covered in the standard books on differential geometry. Instead, it is standard in other books, in particular texts on Clifford algebras.

1.1 Manifolds

The arena of differential geometry is a differentiable (or smooth) manifold. The formalism to be described is important for two distinct reasons. First, it allows one to deal with 'topologically nontrivial' manifolds, which are, loosely speaking, manifolds that look like copies of $\mathbb{R}^n$ only locally, but not necessarily globally. Second, the formalism allows one to define objects to be placed on manifolds in a coordinate-independent manner. It is this second reason that is more important for the treatment of a theory like GR, rather than the first, because for physics purposes one is usually (but not always) happy to study the theory in a setting of trivial topology.

1.1.1 Cartography

The definition of a manifold is an abstraction of that originally described by Gauss' process of a cartographic representation of the Earth's surface. The process is as follows: the surface of the Earth is decomposed into sufficiently small regions. The regions are numbered, partially overlapping and each region is assigned to a group of cartographers. Each group produces a map of its region, with the map drawn on a paper. For each map, two coordinates can be used to identify every point. The collection of maps forms an atlas. And where regions are overlapping, there exists a clear rule that describes how points on one map correspond to points on another.

1.1.2 Topological Manifold

In this spirit, a *topological* n-dimensional manifold is a topological space[1] M such that every point has a neighbourhood U homeomorphic[2] to an open subset in $\mathbb{R}^n$. The neighbourhoods U with a map $\psi_U : U \to \mathbb{R}^n$ are called coordinate charts. Any topological manifold can be represented as a union of a finite, or countable, set of coordinate charts U, and a set of coordinate charts U that cover M is called an *atlas* on M. Two topological manifolds are said to be homeomorphic if there is a homeomorphism between them.

1.1.3 Smooth or Differentiable Manifold

A topological manifold M is called *differentiable* or *smooth*, if the transition function for overlapping regions $\psi_{U'} \circ \psi_U^{-1} : \psi_U(U' \cap U) \to \psi_{U'}(U' \cap U)$ is a map between open regions of $\mathbb{R}^n$ with partial derivatives of all orders. Two smooth

[1] A topological space is a set X with an additional notion of *neighbourhoods* defined on it. This is an assignment to each element (point) x of X a non-empty collection of subsets of X called neighbourhoods of x. These are required to satisfy the following axioms: (i) each point belongs to every one of its neighbourhoods; (ii) every subset of X containing a neighbourhood of x is also a neighbourhood of x; (iii) the intersection of two neighbourhoods of x is again a neighbourhood of x; (iv) every neighbourhood N of x contains a neighbourhood M of x such that N is a neighbourhood of every point in M. To this, one usually adds the *Hausdorff* assumption or axiom: for every two distinct points x and y of X there exist neighbourhoods of x and y that are disjoint from each other. Given the structure of neighbourhoods, there results the notion of *open* sets: A subset U of X is called open if it is the neighbourhood of all points in U.

[2] A homeomorphism is a continuous, one-to-one and onto (i.e., bijective) map between two topological spaces, whose inverse is also continuous. A map $f : X \to Y$ between two topological spaces is called continuous if for every $x \in X$ and every neighbourhood N of $f(x)$ there is a neighbourhood M of x such that $f(M) \subseteq N$.

manifolds are said to be diffeomorphic if there exists a diffeomorphism between them, i.e., a smooth homeomorphism with a smooth inverse.

1.1.4 Alternative Definition

The definition of smooth manifolds as given starts with the notion of a topological manifold, and adds a smooth structure. It is however possible to start directly with a smooth structure, and induce the topological structure from the smooth one. Thus, one can start directly in the world of smooth manifolds, and avoid talking about topological manifolds at all, which is what we will always do here.

This proceeds as follows. One starts by defining a notion of an atlas on M, which is a collection $\{U_\alpha, \psi_\alpha\}_{\alpha \in I}$ of coordinate charts such that (i) M is covered by the set of charts $\{U_\alpha\}_{\alpha \in I}$; (ii) for each $\alpha, \beta \in I$ the image $\psi_\alpha(U_\alpha \cap U_\beta)$ is open in $\mathbb{R}^n$, with an understanding that the empty set is open; and (iii) the map

$$\psi_\beta \circ \psi_\alpha^{-1} : \psi_\alpha(U_\alpha \cap U_\beta) \to \psi_\beta(U_\alpha \cap U_\beta)$$

is C^∞ with C^∞ inverse.

This gives M a topology by saying that a subset $V \subseteq M$ is open if, for each $\alpha, \psi_\alpha(V \cap U_\alpha)$ is an open subset of $\mathbb{R}^n$. This can be checked to give M topology in the sense that this equips M with the notion of open sets such that (i) M and empty set are open; (ii) an arbitrary union of open sets is open; and (iii) a finite intersection of open sets is open. These properties, as well as the fact that with this topology the maps ψ_α become homeomorphisms, are proven in lecture notes by N. Hitchin, entitled *Differentiable Manifolds.*

1.1.5 Constructions of Manifolds

The basic constructions of manifolds are submanifolds of $\mathbb{R}^n$, submanifolds of manifolds, products of manifolds, open subsets of manifolds, quotients of manifolds that are manifolds and the Grassmanians, see Taubes (2011, chapter 1) for more details.

We will only explain the submanifolds of $\mathbb{R}^n$ construction.

Theorem 1.1 *Let $F : U \to \mathbb{R}^m$ be a C^∞ function on an open set $U \subseteq \mathbb{R}^{n+m}$ and take $c \in \mathbb{R}^m$. Assume that for each $a \in F^{-1}(c)$ the derivative $DF_a : \mathbb{R}^{n+m} \to \mathbb{R}^m$ is surjective (which is the same as assuming that it has maximal rank m). Then $F^{-1}(c)$ has the structure of an n-dimensional smooth manifold.*

We note that the value c for which at every point of $F^{-1}(c)$ the matrix of partial derivatives of F has the maximal rank is called a **regular** value. A proof of this theorem is given, e.g., in Hitchin (2010) lectures.

Example 1.2 Consider the unit sphere in $\mathbb{R}^{n+1}$ given by

$$S^n = \{x \in \mathbb{R}^{n+1} : \sum_{a=1}^{n+1} (x^a)^2 = 1\}.$$

We define $F : \mathbb{R}^{n+1} \to \mathbb{R}$ by

$$F(x) = \sum_{i=a}^{n+1} (x^a)^2.$$

The previous theorem guarantees that $F^{-1}(1)$ is a manifold if 1 is a regular value of F. To check this we consider the matrix of partial derivatives

$$\frac{\partial F}{\partial x^a} = 2x^a.$$

This has rank one as long as not all x^a are identically zero, which is true for points on S^n. So, S^n is a manifold.

Example 1.3 It is instructive to also see an example when the conditions of the rank theorem are not satisfied. Let us consider the cone in $\mathbb{R}^{1,n}$ given by

$$C^n = \{x \in \mathbb{R}^{1,n} : -(x^0)^2 + \sum_{a=1}^{n} (x^a)^2 = 0\}.$$

We similarly construct the function

$$F(x) = -(x^0)^2 + \sum_{a=1}^{n} (x^a)^2.$$

Its matrix of derivatives is

$$\frac{\partial F}{\partial x^a} = (-2x^0, 2x^1, \ldots, 2x^n).$$

This has rank one as long as not all of x^a are zero. However, this is not satisfied for all of the points on $C^n = F^{-1}(0)$. Indeed, the tip of the cone is at the origin of $\mathbb{R}^{1,n}$, where all the coordinates vanish, and the conditions of the rank theorem are not satisfied. We can thus say that 0 is not a regular value of $F(x)$. Thus, the cone is not a smooth manifold because the conditions of the rank theorem fail at its tip. It is also intuitively clear that the tip is not a smooth point, and we see that this intuition is captured by the rank theorem.

Example 1.4 Here is an example of a more nontrivial submanifold in $\mathbb{R}^3$ – a torus. This is defined as a set of points

$$T^2 = \{x \in \mathbb{R}^3 : ((x_1^2 + x_2^2)^{1/2} - 1)^2 + x_3^2 = 1\}.$$

Again, the application of rank theorem shows this to be a manifold. This is a torus of revolution, which is made explicit by the following parametrisation by two angle coordinates

$$(\psi, \varphi) \to (x_1 = (1 + \cos\psi)\cos\varphi, x_2 = (1 + \cos\psi)\sin\varphi, x_3 = \sin\psi).$$

1.1.6 One More Manifold Example

Here is one more instructive example: The set of straight lines in the plane.

Example 1.5 Let X be the set of straight lines in the plane $\mathbb{R}^2$. Each such line can be described by an equation

$$Ax + By + C = 0,$$

with (A, B, C) and $(\lambda A, \lambda B, \lambda C), \lambda \neq 0$ describing the same line.

Let U_0 be the set of non-vertical lines. These are lines for which $B \neq 0$. Each such line has the equation of the form

$$y = mx + c,$$

where m, c are uniquely defined. This gives us the coordinate chart

$$\psi_0 : U_0 \to \mathbb{R}^2; \qquad \text{line} \to (m, c) \in \mathbb{R}^2.$$

Let U_1 be the set of non-horizontal lines. These are lines with $A \neq 0$. Every such line is described by an equation of the type

$$x = \tilde{m}y + \tilde{c}.$$

So, we have another coordinate chart and the coordinate map

$$\psi_1 : U_1 \to \mathbb{R}^2, \qquad \text{line} \to (\tilde{m}, \tilde{c}) \in \mathbb{R}^2.$$

Let us now consider the overlap $U_0 \cap U_1$. These are lines $y = mx + c$ that are not horizontal, i.e., $m \neq 0$. This gives

$$\psi(U_0 \cap U_1) = \{(m, c) \in \mathbb{R}^2 : m \neq 0\},$$

which is an open subset of $\mathbb{R}^2$, as is required. We can also describe explicitly the change of coordinates as one goes from U_0 to U_1. Indeed, when $m \neq 0$, the line $y = mx + c$ can be written as $x = m^{-1}y - cm^{-1}$. We thus have

$$\psi_1 \circ \psi_0^{-1}(m, c) = (m^{-1}, -cm^{-1}).$$

Away from $m = 0$ this is a smooth map with a smooth inverse. This gives the set of lines in $\mathbb{R}^2$ the structure of a smooth manifold.

1.2 Differential Forms

Differential forms are one of the most primitive objects that can be defined on a smooth manifold. These objects play a very important role in differential geometry. Our presentation here follows closely the book by R. Bott and L. W. Tu (1982) titled *Differential Forms in Algebraic Topology*, but with some differences in notation.

1.2.1 Differential Forms on $\mathbb{R}^n$

We start by defining differential forms on $\mathbb{R}^n$. Let $x^a, a = 1, \ldots, n$ be the Cartesian coordinates on $\mathbb{R}^n$. We define $\Lambda^\bullet$ to be the algebra over $\mathbb{R}$ generated by objects dx^a with the relations

$$dx^a dx^b = -dx^b dx^a. \tag{1.1}$$

Note that we omit the wedge product symbol, as being implied. In physics terminology, the objects dx^a are anti-commuting.

As a vector space $\Lambda^\bullet$ has basis

$$\begin{array}{cccccc} 1, & dx^a, & dx^a dx^b, & dx^a dx^b dx^c, & \ldots, & dx^1 dx^2 \ldots dx^n, \\ & & a<b & a<b<c & & \end{array} \tag{1.2}$$

and is of dimension

$$\dim \Lambda^\bullet = 2^n. \tag{1.3}$$

We define differential forms on $\mathbb{R}^n$ to be elements of

$$\Lambda^\bullet(\mathbb{R}^n) = \{C^\infty \text{ functions on } \mathbb{R}^n\} \otimes_{\mathbb{R}} \Lambda^\bullet. \tag{1.4}$$

Here $\otimes$ is the tensor product, whose definition is given in (1.16). This means that each such form can be uniquely written as

$$\omega = \sum_{a_1 < \cdots < a_q} f_{a_1 \ldots a_q} dx^{a_1} \ldots dx^{a_q},$$

where $f_{a_1 \ldots a_q}$ are smooth functions that are called components of the differential form ω. The algebra of differential forms is naturally graded

$$\Lambda^\bullet(\mathbb{R}^n) = \oplus_{q=0}^n \Lambda^q(\mathbb{R}^n), \tag{1.5}$$

where elements of $\Lambda^q(\mathbb{R}^n)$ are called q-forms on $\mathbb{R}^n$. We shall often omit the argument in $\Lambda^q(\mathbb{R}^n)$ if no confusion can arise as to what space the differential forms live on. An alternative expression for a degree q form is

$$\Lambda^q \ni \omega = \frac{1}{q!} f_{a_a \ldots a_q} dx^{a_1} \ldots dx^{a_q},$$

where the summation convention is implied.

1.2.2 Wedge Product of Forms

We can define the *wedge product* $\omega \wedge \tau$ or simply $\omega\tau$ of differential forms $\omega = \sum f_{a_1 \ldots a_q} dx^{a_1} \ldots dx^{a_q}$ and $\tau = \sum g_{b_1 \ldots b_p} dx^{b_1} \ldots dx^{b_p}$ to be

$$\omega\tau = \sum f_{a_1 \ldots a_q} g_{b_1 \ldots b_p} dx^{a_1} \ldots dx^{a_q} dx^{b_1} \ldots dx^{b_p}.$$

1.2.3 Exterior Derivative

The space $\Lambda^\bullet(\mathbb{R}^n)$ comes naturally equipped with a differential operator

$$d : \Lambda^q \to \Lambda^{q+1} \tag{1.6}$$

that is defined by two properties: (i) if $f \in \Lambda^0$ then $df = (\partial f/\partial x^a)dx^a$; (ii) if $\omega = \sum f_{a_1\dots a_q} dx^{a_1}\dots dx^{a_q}$ then $d\omega = \sum df_{a_1\dots a_q} dx^{a_1}\dots dx^{a_q}$.

Example 1.6 If $\omega = xdy$ then $d\omega = dxdy$.

The operator d is called *exterior differentiation*, or *exterior derivative*, and is the ultimate extension of the operators of gradient, curl and divergence of vector calculus, as the following example shows.

Example 1.7 On $\mathbb{R}^3$ the spaces Λ^0 and Λ^3 are 1-dimensional, and the spaces Λ^1, Λ^2 are 3-dimensional. Thus, the following identifications are possible

$$\begin{array}{ccccc} \{\text{functions}\} & \simeq & \{\text{0-forms}\} & \simeq & \{\text{3-forms}\} \\ f & \leftrightarrow & f & \leftrightarrow & f dxdydz \end{array}$$

and

$$\begin{array}{ccccc} \{\text{vector fields}\} & \simeq & \{\text{1-forms}\} & \simeq & \{\text{2-forms}\} \\ (f_1, f_2, f_3) & \leftrightarrow & f_1dx + f_2dy + f_3dz & \leftrightarrow & f_1dydz + f_2dzdx + f_3dxdy \end{array}$$

Then, on functions we have

$$df = \frac{\partial f}{\partial x}dx + \frac{\partial f}{\partial y}dy + \frac{\partial f}{\partial z}dz.$$

On 1-forms we have

$$\begin{aligned} &d(f_1dx + f_2dy + f_3dz) \\ &= \left(\frac{\partial f_3}{\partial y} - \frac{\partial f_2}{\partial z}\right)dydz + \left(\frac{\partial f_1}{\partial z} - \frac{\partial f_3}{\partial x}\right)dzdx + \left(\frac{\partial f_2}{\partial x} - \frac{\partial f_1}{\partial y}\right)dxdy. \end{aligned}$$

On 2-forms we have

$$d(f_1dydz + f_2dzdx + f_3dxdy) = \left(\frac{\partial f_1}{\partial x} + \frac{\partial f_2}{\partial y} + \frac{\partial f_3}{\partial z}\right)dxdydz.$$

This means that

$$\begin{aligned} d(\text{0-forms}) &= \text{gradient}, \\ d(\text{1-forms}) &= \text{curl}, \\ d(\text{2-forms}) &= \text{divergence}. \end{aligned}$$

It can be checked that d is an antiderivation, i.e.,

$$d(\omega\tau) = (d\omega)\tau + (-1)^{\deg(\omega)}\omega d\tau.$$

Note that this is the ordinary product rule at the level of functions. It can also be checked that, in view of the fact that the partial derivatives commute, the operator d squares to zero

$$d^2 = 0.$$

1.2.4 De Rham Complex

The space $\Lambda^\bullet(\mathbb{R}^n)$ equipped with the operator of exterior derivative is called the *de Rham complex* on $\mathbb{R}^n$. The kernel of d are *closed* forms, while the image of d are *exact* forms. On $\mathbb{R}^3$ the notion of closed forms subsumes the vector calculus terminology of irrotational and solenoidal vector fields, while the notion of exact forms generalises that of gradient and curl vector fields.

The de Rham complex may be viewed as a God-given set of differential equations, with solutions being closed forms. For example, on $\mathbb{R}^2$, finding a closed 1-form $fdx + gdy$ is equivalent to solving the differential equation $\partial g/\partial x - \partial f/\partial y = 0$. Since exact forms are automatically closed, they constitute 'trivial' or 'uninteresting' solutions. A measure of the size of the space of 'interesting' solutions is the definition of the de Rham cohomology: The q-th *de Rham cohomology* of $\mathbb{R}^n$ is the vector space

$$H^q(\mathbb{R}^n) = \{\text{closed } q\text{-forms}\}/\{\text{exact } q\text{-forms}\}.$$

For more on de Rham cohomology and the technology needed to compute it, see the Bott–Tu book (1982).

1.2.5 Pullback of Differential Forms

We would now like to extend the notion of differential forms and d from $\mathbb{R}^n$ to an arbitrary manifold. To this end, we need to understand how these notions are compatible with coordinate transformations between charts. To start with, let us introduce the notion of a *pullback* of a differential form. Thus, given a smooth map

$$f : \mathbb{R}^m \to \mathbb{R}^n$$

there is a natural notion of pullback on functions. Indeed, given $g \in \Lambda^0(\mathbb{R}^n)$ its pullback $f^*(g) \in \Lambda^0(\mathbb{R}^m)$ is defined as

$$f^*(g) = g \circ f.$$

We then extend this notion of pullback so that it commutes with the exterior differentiation. This defines f^* on forms uniquely

$$f^*(\sum g_{a_1\dots a_q} dy^{a_1} \dots dy^{a_q}) = \sum (g_{a_1\dots a_q} \circ f) d(y_{a_1} \circ f) \dots d(y_{a_q} \circ f),$$

so that

$$f^* : \Lambda^q(\mathbb{R}^n) \to \Lambda^q(\mathbb{R}^m).$$

The proof that with this definition the pullback commutes with the exterior differentiation is an exercise in chain rule. In practice, the pullback is computed by the simple change of coordinates.

Example 1.8 Let $\omega = xdy - ydx \in \Lambda^1(\mathbb{R}^2)$, and let the map $f : S^1 \to \mathbb{R}^2$ be given by $x = \cos(\varphi), y = \sin(\varphi)$. Then

$$f^*(\omega) = \cos(\varphi)d(\sin(\varphi)) - \sin(\varphi)d(\cos(\varphi)) = d\varphi.$$

1.2.6 Differential Forms on a Manifold

Having defined the notion of pullback of forms on $\mathbb{R}^n$, we are ready to define differential forms on an arbitrary smooth manifold M. Thus, a *differential form* on M is a collection of forms ω_U on $\mathbb{R}^n$ for each coordinate chart U, compatible in the following sense: If U, U' have a common overlap $U' \cap U$ then the pullback of $\omega_{U'}$ to the coordinate chart U coincides with ω_U

$$(\psi_{U'} \circ \psi_U^{-1})^*(\omega_{U'}) = \omega_U.$$

Example 1.9 Consider the space S^2. This is a manifold that can be covered by two coordinate charts each diffeomorphic to $\mathbb{R}^2$. Concretely, we take one coordinate chart to be given by the stereographic projection from the north pole from the sphere to the equatorial plane. We will identify the equatorial plane $\mathbb{R}^2$ with the complex plane $\mathbb{C}$. Then, if θ, φ are the usual spherical coordinates on the sphere, the complex coordinate of a point corresponding to the point θ, φ is

$$\psi_N : S^2/\{N\} \to \mathbb{C}, \qquad z = \frac{\cos(\theta/2)}{\sin(\theta/2)} e^{i\varphi}.$$

Let us now consider a 1-form on $\mathbb{C}$

$$\omega_N = 2\frac{zd\bar{z} + \bar{z}dz}{(1+|z|^2)^2} = d\frac{|z|^2 - 1}{|z|^2 + 1}.$$

Its pullback to S^2 is therefore given by

$$\psi_N^*(\omega_N) = d\cos\theta = dx_3,$$

where x_1, x_2, x_3 are the coordinates of the embedding of S^2 into $\mathbb{R}^3$. The form ω_N is a good 1-form in $U_N = S^2/\{N\}$, which moreover vanishes at $z = 0$ (south pole).

Let us now consider the second coordinate chart. It is taken to be the stereographic projection from the south pole to the equatorial plane. If we again identify $\mathbb{R}^2$ with $\mathbb{C}$, this coordinate map is

$$\psi_S : S^2/\{S\} \to \mathbb{C}, \qquad w = \frac{\sin(\theta/2)}{\cos(\theta/2)} e^{i\varphi}.$$

It is easy to see that $w = 1/\bar{z}$. This is a map $\psi_{N\to S}$ from $\mathbb{C}$ with origin removed to $\mathbb{C}$, and is the change of chart coordinate transformation in this case. Let us define the 1-form ω in the chart S to be given by

$$\omega_S = -2\frac{wd\bar{w} + \bar{w}dw}{(1+|w|^2)^2},$$

i.e., the same formula as that for ω_N but in terms of the coordinate w, and with an extra minus sign. The form ω_S is a good form on $U_S = S^2/\{S\}$, which moreover vanishes at $w = 0$ (north pole). It is now easy to check that

$$\psi^*_{N\to S}(\omega_S) = \omega_N,$$

and so the 1-form we have described is a globally defined 1-form on S^2.

Example 1.10 Let us now give an illustrative example of the fact that not every differential form that behaves well in some coordinate chart extends to a well-defined form on the whole manifold. Let us again consider the situation of S^2. Consider the north pole coordinate chart and the complex coordinate z on it as in the previous example. Consider the following 1-form

$$\omega_N = \frac{\mathrm{i}}{2}\frac{zd\bar{z} - \bar{z}dz}{1+|z|^2}. \tag{1.7}$$

This is a good form on $S^2/\{N\}$, vanishing at $z = 0$ (south pole). In order for this 1-form to be well-defined on the whole of S^2, there should be a well-defined 1-form ω_S on the coordinate chart $(S^2/\{S\}, \psi_S)$, whose pullback under the coordinate transformation $w = 1/\bar{z}$ should match ω_N everywhere apart from the south and north poles. Because ω_S should match ω_N we can guess an expression for it by performing the coordinate transformation $z = 1/\bar{w}$. This gives

$$\omega_S = \frac{\mathrm{i}}{2}\frac{(1/\bar{w})d(1/w) - (1/w)d(1/\bar{w})}{1+1/|w|^2} = \frac{\mathrm{i}}{2}\frac{wd\bar{w} - \bar{w}dw}{|w|^2(1+|w|^2)}.$$

We want this to be defined everywhere in the coordinate chart $(S^2/\{S\}, \psi_S)$, in particular at the north pole that corresponds to $w = 0$. However, it is clear that this 1-form is not defined at $w = 0$. So, there exists no globally defined 1-form on S^2 that agrees with (1.7) on the coordinate chart $(S^2/\{N\}, \psi_N)$. In fact, the object (1.7) does arise naturally, but corresponds to a connection in a certain line bundle over S^2 rather than a 1-form.

Now that we have the notion of differential forms on M, and given that the operator of exterior differentiation commutes with pullbacks, it is clear that d extends to a well-defined operator on differential forms on M. In particular, on overlaps $U' \cap U$, the operator d can be computed in either of the two coordinate systems and is coordinate-independent.

1.3 Integration of Differential Forms

One of the most important reasons to be interested in differential forms is the fact that they can be naturally integrated over (oriented) submanifolds. More precisely, a differential form of degree q can be naturally integrated over a submanifold of dimension q, closed or with boundary. A related reason why differential forms are important is the generalised Stokes' theorem stating that the integral of an exact form $d\omega$ is equal to the integral of ω over the boundary. This is the ultimate generalisation of the integral theorems of vector calculus.

1.3.1 Orientation

Differential forms can only be integrated over oriented submanifolds, so we first need to discuss the notion of an orientation. Our discussion follows closely the one in book by Theodore Frankel (2012) called *Geometry of Physics.*

Let us first discuss an orientation of a vector space V. Let $e_a \in V, a = 1, \ldots, n$ be a basis of vectors in V. Any other basis $e'_a \in V$ is obtained from e_a by a $\mathrm{GL}(n, \mathbb{R})$ transformation $e'_a = m_a{}^b e_b, m \in \mathrm{GL}(n, \mathbb{R})$. The determinant of m is either positive or negative. If it is positive one says that e'_a has the same orientation as e_a, if it is negative one says that the orientation is opposite. It is clear that the set of all possible bases is split into two subsets of opposite orientation. We can arbitrarily pick a basis from one of this two subsets and call it the positive orientation. To orient a vector space means to choose a basis that is said to have the positive orientation.

Let us now discuss orientation of manifolds. We can orient each tangent space $T_x M$ over a coordinate chart U by choosing a basis in the space of coordinate vector fields $\partial/\partial x^1, \ldots, \partial/\partial x^n$ and saying that it provides the positive orientation. We can do so over all the coordinate charts. The key issue is whether a global choice of orientation is possible, i.e., whether the Jacobians of the coordinate transformations over the overlaps are all of positive determinant. If it is possible to choose orientation of the charts in such a way, then we say that the manifold is orientable. If it's not possible, we say that the manifold is non-orientable. An example of a non-orientable manifold is the Möbius strip. It is clear that if M is connected and orientable, then there are just two different ways to orient it. The same discussion applies to submanifolds of M. Indeed, they are manifolds in themselves (possibly with a boundary but we will consider this later), and so they can be orientable or not, and if orientable, there are exactly two possible orientations to a connected submanifold.

Another situation arises if we have a submanifold $N \subset M$ of codimension exactly one, i.e., the situation of a hypersurface. In this case, there is a notion of a transverse vector field n along N. It is a vector field that is nowhere tangent to N, in particular it is nowhere zero. One then says that a hypersurface $N \subset M$ is two-sided in M it is possible to choose if there exists a transverse vector field along N. In general, if N is a two-sided hypersurface in an orientable manifold M then it is itself orientable.

1.3.2 Integration of a Form

One first defines the integral of a p-form $\omega = \omega(x)dx^1\wedge\cdots\wedge dx^p$ over an **oriented** region $(U,o) \subset \mathbb{R}^p$, where o is an orientation, i.e., a choice of a positively oriented basis. This is done as follows

$$\int_{(U,o)} \omega = \int_{(U,o)} \omega(x)dx^1 \wedge \cdots \wedge dx^p := o[x] \int_U \omega(x)dx^1 \dots dx^p, \tag{1.8}$$

where the integral on the right-hand side is the usual repeated integral in $\mathbb{R}^p$, and $o[x] = o[\partial_1, \dots, \partial_n] = \pm 1$ is the orientation of the basis of the coordinate vector fields $\partial_1, \dots, \partial_n$. We have reinstated the wedge product symbol in the previous formulas to make the passage from the integral of a differential form to the usual repeated integral clear. It is clear that the integral so defined changes the sign if the orientation is reversed

$$\int_{(U,-o)} \omega = -\int_{(U,o)} \omega. \tag{1.9}$$

Another property of the integral so defined is that it is independent of the coordinate system used to evaluate it.

We then define the integral of a differential form over an oriented **parametrised** subset of a manifold M. This is defined as follows. An oriented parametrised p-subset of M is a triple (U,o,F) consisting of an oriented region (U,o) of $\mathbb{R}^p$ together with a differentiable map $F: U \to M$. We then define

$$\int_{(U,o,F)} \omega := \int_{(U,o)} F^*\omega. \tag{1.10}$$

In other words, the differential form is pulled back to $\mathbb{R}^p$ via the parametrisation map, and then the integral is evaluated as a repeated integral over $\mathbb{R}^p$, taking the orientation into account. It is easy to check that, in the case of curves and surfaces in $\mathbb{R}^3$, this definition leads to the familiar vector calculus formulas for line and surface integrals. It can also be checked that the integral so defined is in fact independent of parametrisation.

Not every submanifold $N \subset M$ can be covered by a single parametrised subset, because the topology of N may be nontrivial. In this case, one defines the integral by covering N with patches each of which can be parametrised, and moreover patches that overlap only along edges and vertices. One then defines the integral of a form over N to be given by the sum of the integrals over the patches. This does not depend on how the surface is decomposed into a collection of parametrisable patches. We refer the reader to the book by Frankel (2012) for more details.

1.3.3 Stokes' Theorem

A manifold with boundary is a slight generalisation of the concept of a manifold. Every point of a manifold has an open neighbourhood diffeomorphic to a ball in $\mathbb{R}^n$. For a manifold with boundary this is true everywhere apart from points of the boundary. This is a special set of points denoted by ∂M that is required to be a manifold itself. In particular, ∂M does not have a boundary, or the boundary of a boundary is zero. Second, every point of the boundary has a neighbourhood that is diffeomorphic to a half-ball in $\mathbb{R}^n$, i.e., the set $|x| < \epsilon, x^n \leq 0$. It is clear that a manifold with empty boundary is just the concept of the manifold as we have previously defined it.

To state the Stokes' theorem for an integral of a p-form ω over a p-dimensional submanifold $N \subset M$ with boundary we need to define a canonical notion of the orientation of a boundary ∂N when an orientation of N is given. This is done as follows. Let $e_2, \ldots, e_n$ span the tangent space to ∂N at some point $x \in \partial N$. Let n be the tangent vector to N at x that is transverse to ∂N and points *out* of N. Then $e_2, \ldots, e_n$ is called positively oriented when $n, e_2, \ldots, e_n$ is a positively oriented basis of vectors in $T_x N$ according to the orientation of N chosen. With this choice of the orientation of ∂N we have

Theorem 1.11 *Let $N \subset M$ be a compact, i.e., having the property that every cover of N by open subsets has a finite subcover, oriented p-dimensional submanifold with boundary ∂N in a manifold M. Let ω be a continuously differentiable $(p-1)$-form on M. Then*

$$\int_N d\omega = \int_{\partial N} \omega. \tag{1.11}$$

1.4 Vector Fields

In differential geometry, vector fields get encoded into the operators of directional derivatives on functions. This gives a coordinate-free definition. However, in the spirit of the definition of the differential forms as given before, let us first state a definition in a coordinate chart.

1.4.1 Definition

One first defines the *tangent space* to M at a point p, denoted by $T_p M$, to be the vector space over $\mathbb{R}$ spanned by the operators of partial derivatives $\partial/\partial x^1, \ldots, \partial/\partial x^n$, where $x^1, \ldots, x^n$ are local coordinates in some coordinate chart U to which point p belongs. Then a smooth *vector field* on U is a linear combination

$$v_U = v^a \frac{\partial}{\partial x^a},$$

with summation convention implied. This way of defining vector fields encodes them into operators of directional derivatives on function.

In a different coordinate chart $y^a = y^a(x)$ the vector field is given by *push-forward*, with the relation between the coordinate vector fields given by the chain rule

$$\frac{\partial}{\partial x^a} = \frac{\partial y^b}{\partial x^a}\frac{\partial}{\partial y^b}.$$

A smooth vector field on M can then be viewed as a collection of vector fields on charts U that agree on overlaps $U' \cap U$. A vector field on M is an object in the tangent space to M denoted by TM. We will give a coordinate-free definition of vector fields in Section 1.4.4, after some examples are considered.

1.4.2 Push-Forward of Vector Fields

Let us develop the notion of push-forward of vector fields further. Thus, if $\phi : M \to N$ is a diffeomorphism from manifold M to manifold N, the push-forward maps vector fields on M to those on N, is defined by the chain rule, and is denoted by $\phi_* : TM \to TN$. Note, however, that the push-forward is a subtle notion because in general it is not possible to identify a vector field on N that is a push-forward of a given vector field on M. For example, the map $\phi : M \to N$ may not be surjective. In this case one can at most define the push-forward vector field on the image of the map. Another situation where the push-forward is not generally defined is when ϕ is not injective. In this case there is more than one choice of a push-forward at any given point in the image. In the situation of a surjective map ϕ, a vector field $X \in TM$ is called **projectable** if $\phi_*(X_x) \in T_yN$ is independent of a choice $x \in \phi^{-1}(y), y = \phi(x)$. This is precisely the condition that guarantees that the push-forward of $X \in TM$ as a vector field on N is well-defined. This discussion makes it clear that the notion of a push-forward of vector fields is much more complicated and subtle than the notion of the pullback of functions and forms. This is one more reason to say that differential forms are more fundamental objects than vector fields.

An equivalent way of stating the definition of the push-forward is to use the already available notion of the pullback on functions. Thus, if $f \in C^\infty(N)$ is a function on N, then $\phi^*(f) \in C^\infty(M)$ is its pullback. The push-forward vector field $\phi_*(v) \in TN$ is defined via the following relation

$$v(\phi^*(f)) = \phi_*(v)(f). \tag{1.12}$$

This is the same as the (implicit) definition by the chain rule. Indeed, if y^a are coordinates on N and $y^a = y^a(x)$ is the map $\phi : M \to N$, then $\phi_*(f)(x) = f(y(x))$, and we have

$$v(\phi^*(f)) = v^a \frac{\partial}{\partial x^a} f(y(x)) = v^a \frac{\partial y^b}{\partial x^a}\frac{\partial}{\partial y^b} f(y) = (\phi_*(v))^b \frac{\partial}{\partial y^b} f(y) = \phi_*(v)(f).$$

1.4.3 Example

Example 1.12 Let θ, ϕ be the usual spherical coordinates on S^2. Consider the following vector field on S^2

$$v = \frac{\partial}{\partial \phi}.$$

This is a globally defined vector field, as will be clear from considerations that follow later in this example. Consider the stereographic projection map

$$\psi_N : S^2/\{N\} \to \mathbb{C}, \qquad z = \frac{\cos(\theta/2)}{\sin(\theta/2)} e^{i\varphi}.$$

Let us find the push-forward of v with respect to this map. We have

$$v_N = \frac{\partial}{\partial \phi} = \frac{\partial z}{\partial \phi}\frac{\partial}{\partial z} + \frac{\partial \bar{z}}{\partial \phi}\frac{\partial}{\partial \bar{z}} = \mathrm{i}z\frac{\partial}{\partial z} - \mathrm{i}\bar{z}\frac{\partial}{\partial \bar{z}}. \tag{1.13}$$

Here $z, \bar{z}$ are interpreted as the local coordinates on $\mathbb{C}$.

Let us now take the other coordinate chart and consider its stereographic projection

$$\psi_S : S^2/\{S\} \to \mathbb{C}, \qquad w = \frac{\sin(\theta/2)}{\cos(\theta/2)} e^{i\varphi}.$$

In this coordinate chart the vector field v is given by the same expression $v = \partial/\partial\phi$, and its push-forward is

$$v_S = \mathrm{i}w\frac{\partial}{\partial w} - \mathrm{i}\bar{w}\frac{\partial}{\partial \bar{w}}, \tag{1.14}$$

which is the same expression as (1.13) with with w in place of z.

Now the overlap $U_N \cap U_S$ is all of the sphere without the north and south poles. On the overlap we have $w = 1/\bar{z}$. Let us find the push-forward of the vector field (1.13) under this map. We have

$$\psi_*^{N\to S} v_N = \mathrm{i}z\left(\frac{\partial w}{\partial z}\frac{\partial}{\partial w} + \frac{\partial \bar{w}}{\partial z}\frac{\partial}{\partial \bar{w}}\right) - \mathrm{i}\bar{z}\left(\frac{\partial w}{\partial \bar{z}}\frac{\partial}{\partial w} + \frac{\partial \bar{w}}{\partial \bar{z}}\frac{\partial}{\partial \bar{w}}\right).$$

The only nonvanishing derivatives here are $\partial w/\partial \bar{z}$ and its complex conjugate. We get

$$\psi_*^{N\to S} v_N = -\mathrm{i}z\frac{1}{z^2}\frac{\partial}{\partial \bar{w}} + \mathrm{i}\bar{z}\frac{1}{\bar{z}^2}\frac{\partial}{\partial w} = -\mathrm{i}\bar{w}\frac{\partial}{\partial \bar{w}} + \mathrm{i}w\frac{\partial}{\partial w} = v_S.$$

This coincides with the vector field in (1.14), which shows that the vector field given in two different coordinate charts by (1.13) and (1.14) defines a globally well-defined vector field on S^2.

1.4.4 Vector Fields as Derivations

Our definition of vector fields shows that they map smooth functions on M to smooth functions on M. Explicitly, in local coordinates

$$v(f)(x) = v^a(x)\frac{\partial f}{\partial x^a}(x).$$

It is clear that this map satisfies the Leibnitz property

$$v(fg) = fv(g) + gv(f).$$

In fact, any linear transformation with this property (called a derivation of the algebra $C^\infty(M)$) is a vector field:

Lemma 1.13 *Let $v : C^\infty(M) \to C^\infty(M)$ be a linear map that satisfies the property $v(fg) = fv(g) + gv(f)$. Then v is a vector field.*

A proof is simple and instructive, so we will spell it out. By linearity $v(cf) = cv(f)$, where c is a constant, and by Leibnitz property $v(cf) = cv(f) + fv(c)$, which means that $v(c) = 0$. Now, near a point with coordinates p^a any $f(x)$ can be written as

$$f(x) = (x^a - p^a)g_a(x) + c,$$

where $g_a(x)$ are some functions that satisfy

$$g_a(p) = \frac{\partial f}{\partial x^a}(p).$$

We now apply a derivation v to f written in this form. Using the Leibnitz property this gives

$$v(f) = v(x^a)g_a + (x^a - p^a)v(g_a).$$

Evaluating this at $x = p$ we get

$$v(f)(p) = v(x^a)(p)\frac{\partial f}{\partial x^a}(p).$$

Defining now $v(x^a)(p) := v^a$ we see that indeed any derivation is of the form

$$v = v^a\frac{\partial}{\partial x^a},$$

which coincides with our previous definition of vector fields.

The characterisation of vector fields as derivations can be used as an alternative way of defining them. The advantage of this definition is that it is clearly coordinate-independent.

1.4.5 Lie Bracket of Vector Fields

The previous characterisation of vector fields as derivations can be used to show that the commutator $[v, u]$ of two derivations is again a derivation and thus a vector field. Indeed, we have

$$\begin{aligned} uv(fg) &= u(v(f)g + fv(g)) = u(v(f))g + v(f)u(g) + u(f)v(g) + fu(v(g)), \\ vu(fg) &= v(u(f)g + fu(g)) = v(u(f))g + u(f)v(g) + v(f)u(g) + fv(u(g)), \end{aligned}$$

and so

$$(uv - vu)(fg) = (u(v(f)) - v(u(f)))g + f(u(v(g)) - v(u(g))),$$

which means that the Leibnitz property is satisfied and $uv - vu =: [u, v]$ is a vector field. The vector field $[u, v]$ is called the **Lie bracket** of u, v.

1.4.6 Interior Product

Vector fields are objects that can naturally be paired with the differential forms. This gives rise to the notion of interior product

$$i_v : \Lambda^q(M) \to \Lambda^{q-1}(M).$$

In components this is given by

$$(i_v\omega)_{a_1\ldots a_{q-1}} = v^a \omega_{aa_1\ldots a_{q-1}}. \tag{1.15}$$

In particular, the interior product of a vector field with a 1-form is a function $i_v\theta \equiv \theta(v)$, where $\theta \in \Lambda^1, v \in TM$.

The interior product can also be defined recursively. Thus, given a differential form

$$\omega = \frac{1}{q!}\omega_{a_1\ldots a_q} dx^{a_1} \ldots dx^{a_q}$$

we define

$$i_v\omega = \frac{1}{q!}\omega_{a_1\ldots a_q}(i_v dx^{a_1}) \ldots dx^{a_q} + \cdots + (-1)^{q-1}\frac{1}{q!}\omega_{a_1\ldots a_q} dx^{a_1} \ldots (i_v dx^{a_q}).$$

This means that we successively apply the operator i_v to all 1-form factors in ω, taking into account the arising signs. All terms in the previous expression are equal, and so the sum computes to

$$i_v\omega = \frac{1}{(q-1)!} v^{a_1}\omega_{a_1\ldots a_q} dx^{a_2} \ldots dx^{a_q}.$$

This clearly agrees with the component definition (1.15) stated previously.

Example 1.14 Considering again our previous example of S^2 manifold, let us compute the pairing of 1-form

$$\omega = \frac{\mathrm{i}}{2}\left(\frac{d\bar{z}}{\bar{z}} - \frac{dz}{z}\right)$$

with vector field

$$v = \mathrm{i}z\frac{\partial}{\partial z} - \mathrm{i}\bar{z}\frac{\partial}{\partial \bar{z}},$$

with both being given by their expressions on the coordinate chart $S^2/\{N\}$. It is computed using the rules

$$dz\left(\frac{\partial}{\partial z}\right) = 1, \quad d\bar{z}\left(\frac{\partial}{\partial \bar{z}}\right) = 1,$$

with the other two possible pairings being zero. This gives $\omega(v) = 1$, which is of course the expected pairing $d\phi(\partial/\partial\phi) = 1$.

Example 1.15 Here is a more nontrivial example of usage of the interior product. Let

$$\omega = dxdy, \qquad v = x\frac{\partial}{\partial x} + y\frac{\partial}{\partial y}.$$

Then

$$i_v\omega = xdy - ydx.$$

The way this is computed is that i_v is applied to every factor in $dxdy$, taking into account the arising signs. In other words $i_v(dxdy) = (i_v dx)dy - dx(i_v dy)$.

1.4.7 Coordinate-Free Definition of Forms

The fact that there is a natural pairing between vector fields and 1-forms means that the space of 1-forms can be identified with the space of linear functionals on vector fields

$$\theta : TM \to \mathbb{R}, \qquad v \to \theta(v) \equiv i_v\theta.$$

Given that we have a coordinate-independent definition of vector fields as derivations, this gives a coordinate-independent definition of 1-forms. In view of the possibility of this definition, the space of 1-forms is often denoted by T^*M, and referred to as the cotangent space to M.

1.5 Tensors

We have so far encountered differential forms as well as vector fields as analytical objects on manifolds. These are examples of more general objects called tensors.

1.5.1 Tensor Product

Let V, W be two finite-dimensional vector spaces over $\mathbb{R}$. We are going to define a new vector space $V \otimes W$ with the property that if $v \in V, w \in W$ then there is a product $v \otimes w \in V \otimes W$. The property of the tensor product $\otimes$ is that it is bilinear

$$\begin{aligned}(\lambda v_1 + \mu v_2) \otimes w &= \lambda v_1 \otimes w + \mu v_2 \otimes w, \\ v \otimes (\lambda w_1 + \mu w_2) &= \lambda v \otimes w_1 + \mu v \otimes w_2.\end{aligned} \tag{1.16}$$

So, the tensor product $V \otimes W$ is the vector space of all finite linear combinations of symbols like $v \otimes w$. Two such expressions are regarded as equal if they can be transformed one into another by a sequence of operations (1.16).

If $e_i \in V, i = 1, \ldots, n, f_j \in W, j = 1, \ldots, m$ are a basis for V, W, it is clear that the vectors $e_i \otimes f_j$ form a basis for $V \otimes W$, and so the dimension the tensor product space is $\dim(V \otimes W) = \dim(V)\dim(W)$. It is important to remember that a typical element of $V \otimes W$ can only be written as a sum

$$\sum_{i,j} a_{ij} e_i \otimes f_j,$$

and not as a pure product $v \otimes w$.

1.5.2 Tensors

At a point $p \in M$ of the manifold we have previously defined the vector spaces TM_p of vectors (derivations) and 1-forms (covectors) T^*M_p. We can take the tensor product of r copies of the tangent space and s copies of the cotangent space. An element t of this tensor product is called a tensor of type (r, s)

$$t \in TM_p \otimes TM_p \otimes T^*M_p \otimes T^*M_p \equiv T^{r,s}M_p.$$

If coordinates are chosen, then such a tensor can be expanded in coordinate bases in TM_p, T^*M_p

$$t = t^{a_1 \ldots a_r}_{b_1 \ldots b_r} \frac{\partial}{\partial x^{a_1}} \otimes \cdots \otimes \frac{\partial}{\partial x^{a_r}} \otimes dx^{b_1} \otimes \cdots \otimes dx^{b_s}, \tag{1.17}$$

where summation convention is implied.

Having defined tensors at a point, we can extend this definition to the whole of the manifold. As with differential forms and vectors, there are two possible definitions. One is the coordinate one, in which a tensor is a collection of objects like (1.17) in each coordinate chart that match on coordinate chart overlaps. Another, coordinate-free definition arises if we remember that vectors can be naturally paired with 1-forms, and 1-forms can be naturally paired with vectors. Then a tensor $t \in T^{r,s}M$ on M can be defined as a multi-linear functional

$$t \in T^{r,s}M : T^*M \times \cdots \times T^*M \times TM \times \cdots \times TM \to C^\infty(M). \tag{1.18}$$

In other words, a tensor of type (r, s) can be viewed as a machine into which one has to insert r 1-forms and s vector fields to get a function on M. This definition is clearly coordinate-independent.

It is worth remembering that type $(1, 0)$ tensors are vector fields, and type $(0, 1)$ tensors are 1-forms (covector fields). The type $(0, 0)$ tensors are defined to be functions.

Example 1.16 There exists a $(1, 1)$ tensor that in every coordinate system has components $1, 0$. Its components are denoted by δ^a_b, which is known as the Kronecker delta. The tensor itself is then

$$\delta = \delta^a_b \frac{\partial}{\partial x^a} \otimes dx^b.$$

The Jacobians arising under changes of coordinates cancel, and this tensor can be given by δ^a_b in any coordinate system.

Because vectors can be naturally paired with covectors, there exists on tensors a naturally defined operation of *contraction*, which maps a tensor of type (r, s) into a tensor of type $(r - 1, s - 1)$. This arises by pairing one of the vector slots with one of the covector slots. It is important to keep in mind that in general the position of the slots matter, and so, e.g., contraction of the first vector slot with the first covector slot gives a tensor different from the one that arises by contracting the second vector slot with the first covector slot. Given a tensor of type (r, s) there are in general nm different tensors of type $(r - 1, s - 1)$ that can be obtained by contraction.

Example 1.17 Taking the tensor δ of type $(1, 1)$ as an example, there is the only possible contraction. This contraction produces a function on M, whose value is constant and equal to the dimension of M.

1.5.3 Differential Forms as Tensor Fields

Having defined tensors we can see that the previously defined differential forms are just special type of tensors. Thus, a rank q differential form is a completely antisymmetric tensor of type $(0, q)$. Indeed, we have defined differential forms as elements of vector space generated by anti-commuting objects dx^a, with a general form of rank q given by

$$\Lambda^q \ni \omega = \frac{1}{n!} \omega_{a_1 \dots a_q} dx^{a_1} \dots dx^{a_q}, \tag{1.19}$$

with the summation convention implied. Given that the coordinate q-forms $dx^{a_1} \dots dx^{a_q}$ are completely antisymmetric in $a_1, \dots, a_1$, the form components $\omega_{a_1 \dots a_q}$ are also completely antisymmetric. There is then a natural correspondence under which q-forms go into tensors of type $(0, q)$

$$\frac{1}{n!}\omega_{a_1\ldots a_q}dx^{a_1}\ldots dx^{a_q} \to \frac{1}{n!}\omega_{a_1\ldots a_q}dx^{a_1}\otimes\cdots\otimes dx^{a_q},$$

where we just added the tensor product symbols. It is clear that this sends a form of rank q into a tensor of type $(0, q)$ that is completely antisymmetric. In the opposite direction, a tensor of type $(0, q)$ that is completely antisymmetric gives a q-form obtained by replacing every occurrence of the tensor product with the wedge product (or, as in this book, simply omitting the tensor product symbol with the wedge product symbol implied). With this in mind, we will not make any difference between the antisymmetric rank $(0, q)$ tensors and differential forms.

1.6 Lie Derivative

1.6.1 One-Parameter Groups of Diffeomorphisms

Definition 1.18 A one-parameter group of diffeomorphisms of a manifold M is a smooth map

$$\phi : M \times \mathbb{R} \to M$$

such that (introducing the notation $\phi_t(x) = \phi(x, t)$): (i) $\phi_t : M \to M$ is a diffeomorphism; (ii) $\phi_0 = id$; (iii) $\phi_{s+t} = \phi_s \circ \phi_t$.

It is important that this definition requires that the map $\phi_t : M \to M$ exists for all $t \in \mathbb{R}$. It is also possible to introduce a related notion of an *integral curve* of a vector field v though a point $p \in M$. This can also be described as a family of maps ϕ_t defined for some range of parameter t, with this range typically depending on the point p through which the integral curve is drawn. The maps satisfy $\phi_0 = id$ and $\phi_{t+s} = \phi_t \circ \phi_s$ wherever they are defined. On compact manifolds integral curves can be extended to one-parameter groups of diffeomorphisms (see Section 1.6.2 for the corresponding theorem).

We will see that one-parameter groups of diffeomorphisms generate vector fields (as their velocity vector field, see Section 1.6.2), and vice versa, at least on compact manifolds, vector fields generate one-parameter groups of diffeomorphisms. So, vector fields can be viewed as infinitesimal versions of one-parameter groups of diffeomorphisms.

1.6.2 Velocity Vector Field

Let ϕ_t be a one-parameter group of diffeomorphisms, and let f be a function on M. Then $f(\phi_t(x))$ is a smooth function of t. Differentiating with respect to t at $t = 0$ we get

$$\frac{\partial}{\partial t} f(\phi_t(x))\bigg|_{t=0} = v_x(f).$$

This is a tangent vector at x as can be seen from the fact that it satisfies the Leibnitz property. This follows from the Leibnitz rule for the derivative with respect to t and $\phi_0(x) = x$. In local coordinates we have

$$\phi_t(x^1, \dots, x^n) = \phi(y^1(x,t), \dots, y^n(x,t))$$

and

$$\frac{\partial}{\partial t} f(y^1, \dots, y^n)\Big|_{t=0} = \frac{\partial f}{\partial y^a}\Big|_{y=x} \frac{\partial y^a}{\partial t}\Big|_{t=0} = v^a(x) \frac{\partial f}{\partial x^a},$$

which corresponds to the vector field

$$v = v^a(x) \frac{\partial}{\partial x^a}.$$

Here

$$v^a(x) = \frac{\partial y^a}{\partial t}\Big|_{t=0}.$$

The vector we have obtained this way is the velocity vector of the curve $\phi_t(x)$ at point x. We, however, have the orbits of ϕ_t covering the whole of M, and we can compute this velocity vector at any point. This gives us the velocity vector field corresponding to a one-parameter group of diffeomorphisms of M.

It is clear that to define the notion of the velocity vector field of a map $\phi_t : M \to M$ we only need ϕ_t to be defined for small values of t. So, the velocity vector fields can be defined in situations more general than those of one-parameter groups of diffeomorphisms, as we will now see.

1.6.3 Integral Curves

We now explain how one can go in the opposite direction and, given a vector field, construct a map $\phi_t : M \to M$ (not necessarily defined for all $t \in \mathbb{R}$) that satisfies properties $\phi_t = id$ and $\phi_{t+s} = \phi_t \circ \phi_s$.

Definition 1.19 An integral curve of a vector field v is a smooth map $\phi : (\alpha, \beta) \subset \mathbb{R} \to M$ such that the velocity vector field of this curve coincides with v along this curve.

Example 1.20 Let $M = \mathbb{R}^2$ with coordinates (x, y) and let $v = \partial/\partial x$. We are looking for an integral curve for this vector field in the form $(x(t), y(t))$. Its velocity vector field is given by

$$\frac{dx}{dt}\frac{\partial}{\partial x} + \frac{dy}{dt}\frac{\partial}{\partial y}.$$

We thus get the following equations for the integral curves of $\partial/\partial x$:

$$\frac{dx}{dt} = 1, \qquad \frac{dy}{dt} = 0,$$

whose solution is

$$(x(t), y(t)) = (t + a_1, a_2).$$

Thus, the integral curves are horizontal lines.

We now have the following theorem.

Theorem 1.21 *Given a vector field v on M and a point $a \in M$ there exists a maximal integral curve of v through a.*

A proof is the generalisation of the previous example, see, e.g., N. Hitchin's lectures on differential geometry. We now allow the point a to vary. This produces the following theorem.

Theorem 1.22 *Let v be a vector field on M and for $(t, x) \in \mathbb{R} \times M$, let $\phi(t, x) = \phi_t(x)$ be the maximal integral curve of v through x. Then (i) the map $(t, x) \to \phi_t(x)$ is smooth; (ii) $\phi_t \circ \phi_s = \phi_{t+s}$ wherever the maps are defined; (iii) if M is compact then $\phi_t(x)$ is defined on $\mathbb{R} \times M$ and gives a one-parameter group of diffeomorphisms of M.*

A proof is given in N. Hitchin's lectures. This theorem can be rephrased by saying that vector fields generate integral curves viewed as maps $\phi_t : M \to M$ satisfying $\phi_t \circ \phi_s = \phi_{t+s}$ wherever the maps are defined. Further, on compact manifolds these extend to one-parameter groups of diffeomorphisms.

1.6.4 The Lie Bracket of Vector Fields Revisited

We have just seen that a vector field gives rise to a one-parameter family of maps $\phi_t : M \to M$ that have this vector field as its velocity. If we consider the natural action of this diffeomorphism on a function f, and evaluate its derivative at zero, we get the action of this vector field on f

$$\frac{\partial}{\partial t} f(\phi_t)\Big|_{t=0} = v(f).$$

An alternative way of stating this relation is to say that we are differentiating with respect to t the function $\phi_t^*(f)$ obtained as the pullback of f with respect to ϕ_t

$$\frac{\partial}{\partial t} \phi_t^*(f)\Big|_{t=0} = v(f). \tag{1.20}$$

We can also consider the action of the diffeomorphism ϕ_t on some other vector field u. In the context of Lie derivative, the convention is to always consider the pullback $\phi_t^*(u)$, which is defined as the push-forward with respect to the inverse map ϕ_t^{-1}. A simple calculation shows that this pullback satisfies

$$\phi_t^*(u(f)) = \phi_t^*(u)(\phi_t^* f).$$

This can be differentiated with respect to t at $t = 0$ keeping in mind (1.20). We get

$$v(u(f)) = \dot{u}(f) + u(v(f)),$$

which implies that

$$\mathcal{L}_v u \equiv \dot{u} = vu - uv = [v, u]. \tag{1.21}$$

This gives the Lie bracket $[v, u]$ the interpretation of the infinitesimal change of u with respect to the diffeomorphism generated by v. The operator $\mathcal{L}_v$ is called the Lie derivative with respect to a vector field v. In coordinates the components of the Lie bracket vector field are given by

$$[v, u]^a = v^b \frac{\partial u^a}{\partial x^b} - u^b \frac{\partial v^a}{\partial x^b} \equiv v^b \partial_b u^a - u^b \partial_b v^a. \tag{1.22}$$

1.6.5 Lie Derivative of 1-Forms

Let us now repeat a similar calculation, but for 1-forms. Thus, we again take the one-parameter group of diffeomorphisms ϕ_t that corresponds to a vector field v. We take a 1-form θ and consider its pullback $\phi_t^*(\theta)$ with respect to ϕ_t. We now pair the 1-form θ with an arbitrary vector field u. There is a simple relation between this pairing and the pairing of the pullback objects

$$\phi_t^*(\theta(u)) = \phi_t^*(\theta)(\phi_t^*(u)).$$

We then derive this with respect to t and set $t = 0$. This gives

$$v(\theta(u)) = \dot{\theta}(u) + \theta(\dot{u}). \tag{1.23}$$

Because we already know (1.21) we can get $\dot{\theta}$.

Let us derive a convenient formula for $\theta([u, v])$. We have, in coordinates

$$\theta([v, u]) = \theta_a(v^b \partial_b u^a - u^b \partial_b v^a) = v^b \partial_b(\theta_a u^a) - u^b \partial_b(\theta_a v^a) + u^b v^a (\partial_b \theta_a - \partial_a \theta_b).$$

This can be rewritten as

$$\theta([v, u]) = v(\theta(u)) - u(\theta(v)) - i_u i_v d\theta, \tag{1.24}$$

where we have used the interior product. We now combine this with (1.23) and notice that the terms $v(\theta(u))$ cancel on both sides. We also rewrite $u(\theta(v)) = i_u d i_v \theta$ and $\dot{\theta}(u) = i_u \dot{\theta}$, and cancel the i_u on all sides of this equation. We get

$$\mathcal{L}_v \theta \equiv \dot{\theta} = d i_v \theta + i_v d\theta. \tag{1.25}$$

The formula we have derived is known as the *Cartan's magic formula.*

It is a good exercise to rewrite this formula in components. We have

$$(\mathcal{L}_v \theta)_a = \partial_a(\theta_b v^b) + v^b(\partial_b \theta_a - \partial_a \theta_b) = v^b \partial_b \theta_a + \theta_b \partial_a v^b. \tag{1.26}$$

Let us derive another property of the Lie derivative $\mathcal{L}_v$. Let us apply it to a 1-form that is given by the exterior derivative of a function

$$\mathcal{L}_v df = di_v df + i_v ddf = di_v df = d\mathcal{L}_v f.$$

We thus see that the Lie derivative commutes with the operator of exterior differentiation, which is one of its most important properties. This property extends to Lie derivative of arbitrary degree forms, which will be discussed in Section 1.6.7.

1.6.6 The Practical Way of Computing the Lie Derivative

Let us again use the formula (1.26) for the components of the Lie derivative of a 1-form. Let us multiply this formula by dx^a on both sides. We have

$$(\mathcal{L}_v\theta)_a dx^a = (v^b\partial_b\theta_a)dx^a + (\theta_b\partial_a v^b)dx^a = (\mathcal{L}_v\theta_a)dx^a + \theta_b d(\mathcal{L}_v x^b).$$

Indeed, to write the last term on the right-hand side we have used $\mathcal{L}_v x^b = v^a\partial_a x^b = v^b$. We can thus write this formula as

$$\mathcal{L}_v\theta = \mathcal{L}_v(\theta_a dx^a) = (\mathcal{L}_v\theta_a)dx^a + \theta_b d(\mathcal{L}_v x^a). \tag{1.27}$$

This gives us an efficient practical rule for computing the Lie derivative of one forms: write 1-form in a coordinate basis as $\theta = \theta_a dx^a$, and then apply the Lie derivative first to the components θ_a viewed as functions, and then to the coordinate 1-forms dx^a. The latter can be computed using the fact that the Lie derivative commutes with the exterior derivative, and thus the Lie derivative must be applied again to just functions – the coordinate functions x^a. One can check that the similar rule can be used also for vector fields, and indeed for any tensors. This gives the most practical way of computing the Lie derivative of concretely specified tensors.

1.6.7 Lie Derivative of Differential Forms

Cartan's magic formula that expresses the Lie derivative of a 1-form in terms of the operations of exterior differentiation and interior product extends to forms of arbitrary degree.

Theorem 1.23 *The Lie derivative $\mathcal{L}_v\omega$ of a p-form ω is given by*

$$\mathcal{L}_v\omega = d(i_v\omega) + i_v d\omega. \tag{1.28}$$

A nice proof of this fact is given in N. Hitchin's differential geometry lecture notes. Cartan's formula (1.28) immediately shows that the Lie derivative on forms commutes with the exterior derivative.

1.7 Integrability Conditions

This section discusses the classical question of conditions of integrability of a given distribution of vector fields. We follow Theodore Frankel's (2012) book, *Geometry of Physics*, to which the reader is referred to for more details and proofs.

1.7.1 Distributions and Their Integrability Condition

Given a smooth nonvanishing vector field on $\mathbb{R}^n$ one can always, at least locally, find a smooth family of integral curves, which have the given vector field as their tangent. The classical question is if the same extends to more than one vector field, i.e., if, given a smooth family of k-planes in $\mathbb{R}^n$ it is possible to find an integral surface, which is a surface everywhere tangent to the planes. The answer is in general no.

Definition 1.24 A k-dimensional **distribution** Δ_k on M assigns in a smooth fashion to each $x \in M$ a k-dimensional subspace $\Delta_k(x)$ of the tangent space T_xM. An k-dimensional **integral manifold** of Δ_k is an k-dimensional submanifold of M that is everywhere tangent to the distribution. The distribution Δ_k is said to be **(completely) integrable** if locally there are coordinates $x^1, \ldots, x^k, y^1, \ldots, y^{n-k}$ for M of dimension n such that the coordinate slices $y^1 = \text{constant}, \ldots, y^{n-k} = \text{constant}$ are k-dimensional integral submanifolds of Δ_k. Such a coordinate system is called a **Frobenius chart** for M.

Definition 1.25 The distribution Δ is said to be in **involution** if $[\Delta, \Delta] \subset \Delta$, i.e., if the Lie bracket of any two vector fields from Δ is again in Δ.

Theorem 1.26 *The distribution Δ is integrable if and only if it is involutive, i.e., in involution.*

The proof in one direction is easy. If the distribution is integrable then the integral curve of any vector field in the distribution is in the integral manifold, and it is easy to show using the definition of the Lie derivative that uses pullback that the Lie bracket of two such vector fields is again tangent to the integral manifold. To prove the theorem in the other direction one uses a reformulation of the integrability condition in terms of differential forms, see the book, *Geometry of Physics*, by Frankel (2012).

If a distribution Δ is integrable, then the integral manifolds define a **foliation** of M, and each connected integral manifold is called a **leaf** of the foliation.

1.7.2 Distributions and 1-Forms

Let θ be a 1-form that does not vanish at a point $x \in M$. The **annihilator** or **null space** of θ at x is the $(n-1)$ dimensional subspace of T_xM of vectors

$v : \theta(v) = 0$. The classical literature on this subject writes $\theta = 0$ for this null space. It is also common to refer to θ as a Pfaffian, and $\theta = 0$ is called a Pfaffian equation. If $\theta_1, \ldots, \theta_r$ are $r = n - k$ linearly independent 1-forms $\theta_1 \wedge \cdots \wedge \theta_r \neq 0$ on some open subspace of M, then the intersection of their null spaces forms an $n - r = k$ dimensional distribution Δ_k. In other words, $v \in \Delta_k$ if and only if $\theta_1(v) = \cdots = \theta_r(v) = 0$. Note that there is no claim that every distribution can be *globally* defined by r Pfaffians. We now have the following theorem.

Theorem 1.27 *The following conditions are locally equivalent.*

1. *Δ is in involution, that is, $[\Delta, \Delta] \subset \Delta$.*
2. *$d\theta_i$ vanishes when restricted to Δ.*
3. *There are 1-forms λ_{ij} such that $d\theta_i = \lambda_{ij} \wedge \theta_j$.*
4. *$d\theta_i \wedge \Omega = 0$, where $\Omega = \theta_1 \wedge \cdots \wedge \theta_r$.*

The proof is easy, and is based on (1.24), see Frankel (2012, chapter 6). Thus, a distribution Δ_k can *locally* be described either by k linearly independent vector fields $v_1, \ldots, v_k$, that span Δ_k at each point, or by exhibiting $r = n - k$ linearly independent 1-forms $\theta_1, \ldots, \theta_r$ whose common null space is Δ_k. The distribution is involutive if either $[\Delta, \Delta] \subset \Delta$ or if $d\theta$ vanishes on vector fields from Δ. We know that an integrable distribution is involutive. The statement in the opposite direction, i.e., that an involutive distribution is locally completely integrable is known as the **Frobenius theorem**. For a proof, see Theodore Frankel's book (2012).

1.8 The Metric

A metric on a manifold M is a smoothly varying inner product on the tangent spaces TM_x. Because the inner product is a (symmetric) bilinear form, we want the metric to take values in $T^*M_x \otimes T^*M_x$ at each point, i.e., to be a $(0, 2)$ tensor. Moreover, this tensor must be symmetric. In local coordinates this can be written as

$$g \equiv ds^2 = g_{ab}\, dx^a \otimes dx^b, \tag{1.29}$$

where we introduced a new notation ds^2 (the squared interval), whose meaning will become clear in Section 1.8.1.

If there is such a tensor defined on M, the vector space TM of vectors becomes an inner product space. Indeed, using g we can define an inner product (a symmetric bilinear pairing) of two vectors $v, w \in TM_p$

$$(v, w) := g_{ab} v^a w^b.$$

This metric pairing of vectors v, w gives a real number per point of M. As with any inner product, this can be used to define the notions of norm and (when the inner product is positive definite) angle between vectors. Note that the metric does not in general need to be positive-definite, even though the cases when it are easier to analyse. This case is thus more studied.

1.8.1 Length of Curves

The metric plays a distinguished role in differential geometry because this is the object that is required to be able to compute length of curves on M. Thus, if $x^a(t)$ is any parametrised curve γ, then the length of any segment $\gamma_{[\alpha,\beta]} : t \in [\alpha, \beta]$ is defined to be

$$l(\gamma_{[\alpha,\beta]}) := \int dt \sqrt{g_{ab} \frac{dx^a}{dt} \frac{dx^b}{dt}}. \tag{1.30}$$

This is easily seen to be curve reparametrisation invariant. However, in order for the square root to be defined, one needs to make some assumption about the metric definiteness, e.g., assuming it to be positive definite makes the length of any curve well-defined. A metric g on M that is everywhere positive-definite is called a *Riemannian* metric on M. The fact that a metric g defines an infinitesimal squared interval (from which the length of any curve is obtained by integration) justifies the notation ds^2 for it in (6.223).

1.8.2 Pullback Metric

Given a map between two manifolds $f : M \to N$, and a metric in N, we can pull it back to get a metric in M. The pullback metric is easiest to derive working in some coordinate patch for both M, N. Thus, if x are coordinate on M and $y = f(x)$ are coordinates on N then

$$f^*(g_{ab} dy^a \otimes dy^b) = g_{ab} \frac{\partial f^a}{dx^c} \frac{\partial f^b}{dx^d} dx^c \otimes dx^d = g'_{cd} dx^c \otimes dx^d.$$

It is important that the map $f : M \to N$ does not need to be invertible. In particular, this map can be the inclusion map from M to a submanifold $M \subset N$. In this case the metric arising on M is called the *induced* metric.

Example 1.28 The length of a curve formula (1.30) can be understood in induced metric terms. Indeed, a parametrised curve $x^a(t)$ can be thought of as a map from $\mathbb{R}$ (or a segment thereof) to M. This gives rise to a pullback metric on $\mathbb{R}$ given by

$$f^*(g) = g_{ab} \frac{\partial x^a}{dt} \frac{\partial x^b}{dt} dt^2.$$

We can then compute the length of any segment of $\mathbb{R}$ in this metric by taking the square root of the squared interval and integrating, which is precisely what the formula (1.30) does.

Example 1.29 Let us consider $\mathbb{R}^3$ with its standard flat metric

$$ds^2_{\mathbb{R}^3} = dx^2 + dy^2 + dz^2.$$

Consider the unit sphere in $\mathbb{R}^3$ given by the surface $x^2+y^2+z^2=1$. This surface can be explicitly parametrised by the spherical coordinates

$$x = \sin\theta\cos\phi, \quad y = \sin\theta\sin\phi, \quad z = \cos\theta, \tag{1.31}$$

which gives the inclusion map $S^2 \to \mathbb{R}^3$. The pullback metric is then easily computed to be

$$ds^2_{S^2} = d\theta^2 + \sin^2\theta d\phi^2. \tag{1.32}$$

Example 1.30 Let us now consider the Minkowski space $\mathbb{R}^{1,2}$ in 2+1 dimensions, with the metric given by

$$ds^2_{\mathbb{R}^{1,2}} = dt^2 - dx^2 - dy^2.$$

Let us consider the surface $\mathbb{H}^2$, which is one of the two sheets (e.g., upper) of the two-sheeted hyperboloid $t^2 - x^2 - y^2 = 1$. This surface is known as the *hyperbolic plane*. Let us introduce an analog of the sphere inclusion map (1.31) by parametrising

$$x = \sinh\theta\cos\phi, \quad y = \sinh\theta\sin\phi, \quad t = \cosh\theta. \tag{1.33}$$

The metric induced on $\mathbb{H}^2$ is then

$$ds^2_{\mathbb{H}^2} = d\theta^2 + \sinh^2\theta d\phi^2, \tag{1.34}$$

which is just the metric (1.32) with the trigonometric functions replaced by the hyperbolic ones.

A different model for $\mathbb{H}^2$ is possible by introducing the stereographic projection from the point $(-1,0,0)$ coordinates. We project on the plane $t=0$, and the complex coordinate that on $\mathbb{R}^2$ that corresponds to a point on $\mathbb{H}^2$ is easily seen to be

$$z = \frac{\sinh(\theta/2)}{\cosh(\theta/2)} e^{\mathrm{i}\phi}. \tag{1.35}$$

It can then be checked by an explicit computation that in these coordinates

$$ds^2_{\mathbb{H}^2} = \frac{4|dz|^2}{(1-|z|^2)^2}. \tag{1.36}$$

1.8.3 Isometries

Definition 1.31 A diffeomorphism $f: M \to N$ between two manifolds equipped with a metric is called an *isometry* if $f^* g_N = g_M$.

Example 1.32 Let us consider the upper-half plane $M = \{(x, y) \in \mathbb{R}^2 : y > 0\}$ with the metric

$$ds^2 = \frac{dx^2 + dy^2}{y^2}. \tag{1.37}$$

Let us use the complex coordinate $w = x + \mathrm{i}y$. Consider the following transformations on M

$$f(w) = \frac{aw + b}{cw + d}, \tag{1.38}$$

where $a, b, c,$ and d are real and $ad - bc > 0$. These are known as Möbius transformations. We have

$$df(w) \equiv f^*(dw) = (ad - bc)\frac{dw}{(cw + d)^2}$$

and

$$f^*y = y \circ f = \frac{1}{2\mathrm{i}}\left(\frac{aw + b}{cw + d} - \frac{a\bar{w} + b}{c\bar{w} + d}\right) = \frac{ad - bc}{|cw + d|^2}y. \tag{1.39}$$

This means that

$$f^*g = (ad - bc)^2\frac{dwd\bar{w}}{|(cw + d)^2|^2}\frac{|cw + d|^4}{(ad - bc)^2y^2} = \frac{dwd\bar{w}}{y^2} = g. \tag{1.40}$$

So, Möbius transformations are isometries of this Riemannian metric on the upper-half plane.

Example 1.33 Our other example of an isometry is one linking the metric in (1.36) with (1.37). Consider a map

$$z = -\frac{w - \mathrm{i}}{w + \mathrm{i}}. \tag{1.41}$$

This can be checked to map the upper-half plane in the w-coordinate to the unit disc in the z coordinate. Let us then compute the pullback of the metric (1.36) under this map. We have

$$dz = -2\mathrm{i}\frac{dw}{(w + \mathrm{i})^2}$$

and

$$1 - |z|^2 = -\frac{2\mathrm{i}(w - \bar{w})}{|w + \mathrm{i}|^2} = \frac{4y}{|w + \mathrm{i}|^2}.$$

This means that

$$\frac{4|dz|^2}{(1 - |z|^2)^2} = \frac{16|dw|^2}{|(w + \mathrm{i})^2|^2}\frac{|w + \mathrm{i}|^4}{16y^2} = \frac{|dw|^2}{y^2}. \tag{1.42}$$

This shows that the pullback of (1.36) under the map in question is (1.37), and this is an isometry. The description (1.37) is known as the upper-half plane model of the hyperbolic plane.

1.9 Lie Groups and Lie Algebras

In this section we review basic facts about (matrix) Lie group and Lie algebras. In general, the notion of the group is important because symmetries form groups. Lie groups arise when symmetries are of continuous nature. We will follow the book by Taubes (2011) closely.

1.9.1 Definition of a Group

Definition 1.34 A **group** is a set G with a special element e (called the identity) and two operations:

- Multiplication $\mu : G \times G \to G, \quad \mu(g, h) = gh \in G.$
- Inverse $\sigma : g \to g^{-1}, \quad \sigma(g) = g^{-1}, \quad gg^{-1} = g^{-1}g = e.$

The multiplication is required to be associative $(gh)k = g(hk)$ and multiplication of any group element g by the identity element is required to return g, i.e., $ge = eg = g$.

Definition 1.35 A **Lie group** is a group with the structure of a smooth manifold such that both the multiplication and the inverse are smooth maps.

As we will discuss in section 1.9.6, Lie groups naturally act on themselves by left or right multiplication. Each of these actions preserves structures on G, and is thus a symmetry. So, Lie group is a manifold that is at the same time a set of transformations that acts on this manifold by symmetries. This is why a Lie group can be thought of as a manifold with a lot of symmetry.

1.9.2 The General Linear Group $\mathbf{GL}(n, \mathbb{R})$

This is the principal and most important example of a Lie group, as all other Lie groups can be realised as subgroups of a sufficiently large general linear group.

Let $M(n, \mathbb{R})$ denote the space of $n \times n$ matrices with real entries. Using the matrix elements as coordinates we can view this space as a copy of $\mathbb{R}^{n^2}$. Matrices can be multiplied

$$\mu : M(n, \mathbb{R}) \times M(n, \mathbb{R}) \to M(n, \mathbb{R}), \qquad \mu(m, m') = mm',$$

and the multiplication map is smooth (being linear in both arguments). There are two important special functions on $M(n, \mathbb{R})$:

- $\mathrm{Det}(m)$ – this is a polynomial of degree n.
- $\mathrm{Tr}(m)$ – this is a linear function.

Definition 1.36 Let $\mathrm{GL}(n, \mathbb{R})$ be the subset of *invertible* matrices in $M(n, \mathbb{R})$, i.e., $m \in \mathrm{GL}(n, \mathbb{R})$ if and only if $\mathrm{Det}(m) \neq 0$.

It is clear that $\mathrm{GL}(n, \mathbb{R})$ as defined is an open subset of $M(n, \mathbb{R})$, and is thus a smooth manifold of dimension $\dim(\mathrm{GL}(n, \mathbb{R})) = n^2$. The multiplication of matrices restricts to $\mathrm{GL}(n,\mathbb{R})$ as a smooth map. The inverse map is also smooth, because for any n it is given by the ratio of a polynomial by the determinant. So, we deduce that $\mathrm{GL}(n,\mathbb{R})$ forms a Lie group, known as the general linear group. This group plays the key role not only as a 'manifold with a lot of symmetry', but also as the group that naturally acts in the tangent space to every point of every n-dimensional manifold. This is because the (invertible) coordinate transformations act on vectors and covectors by the matrix of the Jacobian of the coordinate transformation. This matrix is nondegenerate and belongs to $\mathrm{GL}(n,\mathbb{R})$. For this reason one often refers to $\mathrm{GL}(n,\mathbb{R})$ as the *structure group* of an n-dimensional manifold. We will discuss this point when we describe fibre bundles.

1.9.3 Subgroups of $\mathbf{GL}(n,\mathbb{R})$

Definition 1.37 A subgroup H of group G is a subset that contains the identity e, is mapped into itself by the inverse map, and is closed under the multiplication.

We can then obtain Lie groups by taking subgroups of other Lie groups. Indeed, a subgroup of a Lie group that is also a submanifold is a Lie group. Let us apply this strategy to $\mathrm{GL}(n,\mathbb{R})$ and determine interesting Lie subgroups that this group contains.

The group $\mathrm{SL}(n,\mathbb{R}) \subset \mathrm{GL}(n,\mathbb{R})$ is defined to be the subgroup of the general linear group consisting of matrices of determinant one

$$\mathrm{SL}(n,\mathbb{R}) = \{m \in \mathrm{GL}(n,\mathbb{R}) : \mathrm{Det}(m) = 1\}.$$

This is a subgroup because $\mathrm{Det}(mm') = \mathrm{Det}(m)\mathrm{Det}(m')$. One proves that this is also a submanifold using the rank theorem, see Taubes (2011) for the proof.

The orthogonal group $\mathrm{O}(n,\mathbb{R}) \subset \mathrm{GL}(n,\mathbb{R})$ is defined to be

$$\mathrm{O}(n,\mathbb{R}) = \{m \in \mathrm{GL}(n,\mathbb{R}) : mm^T = 1\}.$$

This is a subgroup because, given $m, m' \in \mathrm{O}(n,\mathbb{R})$ we have $(mm')(mm')^T = mm'(m')^T m^T = 1$. One again shows that this is a submanifold of $\mathrm{GL}(n,\mathbb{R})$ using the rank theorem, see Taubes (2011). The dimension of $\mathrm{O}(n,\mathbb{R})$ is $n(n-1)/2$.

The determinant of any matrix in $\mathrm{O}(n,\mathbb{R})$ satisfies $(\mathrm{Det}(m))^2 = 1$ and so $\mathrm{Det}(m) = \pm 1$. One then defines the set of orthogonal matrices with unit determinant to be the special orthogonal group

$$\mathrm{SO}(n,\mathbb{R}) = \{m \in \mathrm{O}(n,\mathbb{R}) : \mathrm{Det}(m) = 1\}.$$

This is clearly a subgroup. As a manifold $\mathrm{O}(n,\mathbb{R})$ splits into two connected components, and the special orthogonal group is just one of these two connected components. So, it is also a Lie group. Its dimension is the same as that of $\mathrm{O}(n,\mathbb{R})$, and is equal to $n(n-1)/2$.

1.9.4 Complex Lie Groups

One can define Lie groups similar to those defined previously over $\mathbb{R}$ but working over $\mathbb{C}$ instead. One starts by defining the group $\mathrm{GL}(n,\mathbb{C})$, which is defined as the open subset of invertible $\mathrm{Det}(m) \neq 0$ matrices in the space $M(n,\mathbb{C})$ of $n \times n$ matrices with complex entries. Viewing $\mathbb{C}$ as $\mathbb{R}^2$, we can coordinatise $\mathrm{GL}(n,\mathbb{C})$ by $2n^2$ real coordinates.

There is, however, another definition of $\mathrm{GL}(n,\mathbb{C})$ that shows how this group actually arises naturally when working over $\mathbb{R}$. The idea of this definition is to start with what is called an **almost complex structure** on $\mathbb{R}^{2n}$, and then consider all matrices in $M(2n,\mathbb{R})$ that commute with the chosen almost complex structure. It is instructive to see how $\mathrm{GL}(n,\mathbb{C})$ arises in this 'real' fashion.

An almost complex structure is defined as an $2n \times 2n$ matrix that squares to minus the identity matrix

$$J \in M(2n,\mathbb{R}) : J^2 = -1.$$

We now define M_J to be the set of $2n \times 2n$ matrices that commute with J

$$M_J = \{m \in M(2n,\mathbb{R}) : mJ = Jm\}.$$

We then define G_J to be the set of invertible matrices with this property. This is a group by virtue of $mm'J = mJm' = Jmm'$ and $Jm^{-1} = m^{-1}J$.

To see how $n \times n$ complex matrices can arise in this setup we note that the eigenvalues of J are $\pm\mathrm{i}$. The matrices commuting with J preserve the eigenspaces. Because the matrix J we start from is real and acts on the real vector space of dimension $2n$, the eigenvectors come in complex conjugate pairs: if v is an eigenvector of eigenvalue $+\mathrm{i}$, i.e., $Jv = \mathrm{i}v$, then $\bar{v}$ is an eigenvector of eigenvalue $-\mathrm{i}$, i.e., $J\bar{v} = -\mathrm{i}\bar{v}$. Let $v_1, \ldots, v_n$ be a set of linearly independent eigenvectors of J of eigenvalue $+\mathrm{i}$. Then any m that commutes with J preserves the space spanned by $v_1, \ldots, v_n$. Its action is thus of the form

$$mv_i = m^{\mathbb{C}}_{ij} v_j,$$

where $m^{\mathbb{C}}_{ij} \in \mathbb{C}$ are a set of $n \times n$ complex numbers. So, we get an identification between real $2n \times 2n$ matrices m that commute with J and $n \times n$ complex matrices $m^{\mathbb{C}}$. Under this identification $\mathrm{Det}(m) = |\mathrm{Det}(m^{\mathbb{C}})|^2$. This means that invertible $\mathrm{Det}(m) \neq 0$ real matrices from M_J correspond to invertible complex matrices, i.e., elements of $\mathrm{GL}(n,\mathbb{C})$.

We will now define some naturally arising subgroups of $\mathrm{GL}(n,\mathbb{C})$ that are also submanifolds. But it should always be kept in mind that any of these complex groups can be viewed in a 'real' way, as a subgroup of the group of matrices that commute with an almost complex structure.

The group $\mathrm{SL}(n,\mathbb{C})$ is defined as the subgroup of matrices from $\mathrm{GL}(n,\mathbb{C})$ of determinant one. This gives a submanifold in R^{2n^2} because one is a regular value of the determinant, as can be checked. The real dimension of $\mathrm{SL}(n,\mathbb{C})$ is $2(n^2-1)$.

The unitary group $\mathrm{U}(n)$ is defined to be the subgroup of unitary matrices

$$\mathrm{U}(n) = \{m \in M(n,\mathbb{C}) : mm^\dagger = 1\},$$

where $m^\dagger = \bar{m}^T$ is the Hermitian conjugation (i.e., complex conjugate transposed). This is a group because $(mm')(mm')^\dagger = mm'(m')^\dagger m^\dagger = 1$ and $m^{-1}(m^{-1})^\dagger = 1$. This is a submanifold again by applying the rank theorem. The real dimension of this group is the dimension of $\mathrm{GL}(n,\mathbb{C})$ which is $2n^2$ minus the dimension of the space of Hermitian matrices, which is n^2. So, the dimension of the unitary group is n^2.

The determinant of any matrix in $\mathrm{U}(n)$ satisfies $|\mathrm{Det}(m)|^2 = 1$, which implies that it is a pure phase. The special unitary group is defined to be the group of matrices in $\mathrm{U}(n)$ of determinant one

$$\mathrm{SU}(n) = \{m \in \mathrm{U}(n) : \mathrm{Det}(m) = 1\}.$$

This is clearly a subgroup, and is a submanifold by a simple application of the rank theorem. Its dimension is $n^2 - 1$.

1.9.5 Classical Lie Groups

The Lie groups that we have just discussed are all examples of classical Lie groups. These arise as subgroups of $\mathrm{GL}(n,\mathbb{R})$ or $\mathrm{GL}(n,\mathbb{C})$ that preserves some structure on the space $\mathbb{R}^n$ or $\mathbb{C}^n$ where it naturally acts. Let us describe this structure case by case.

The groups $\mathrm{SL}(n,\mathbb{R})$ and $\mathrm{SL}(n,\mathbb{C})$ arise as those preserving the volume form $dx^1 \wedge \cdots \wedge dx^n$ on $\mathbb{R}^n$ or $dz^1 \wedge \cdots \wedge dz^n$ on $\mathbb{C}^n$.

The orthogonal group $\mathrm{O}(n,\mathbb{R})$ arises as the group of transformations from $\mathrm{GL}(n,\mathbb{R})$ that preserve the quadratic form

$$(x^1)^2 + \cdots + (x^n)^2$$

on $\mathbb{R}^n$. Indeed, introducing a column x with entries x^a we can write the previous quadratic form as $x^T x$. Then $\mathrm{GL}(n,\mathbb{R})$ acts on such columns by multiplication $x \to mx$, and the condition for the quadratic form to be invariant is $m^T m = 1$, which is what defines $\mathrm{O}(n,\mathbb{R})$.

The orthogonal group admits a generalisation called $\mathrm{O}(r,s)$ that preserves the indefinite quadratic form

$$(x^1)^2 + \cdots + (x^r)^2 - (x^{r+1})^2 - \cdots - (x^{r+s})^2.$$

The group $\mathrm{U}(n)$ arises as the subgroup of $\mathrm{GL}(n,\mathbb{C})$ that preserves the Hermitian form

$$\bar{z}^1 z^1 + \cdots + \bar{z}^n z^n$$

on $\mathbb{C}^n$. Indeed, we can write the previous Hermitian form as $z^\dagger z$, where z is a column with z^a as entries. Then $m \in \mathrm{GL}(n,\mathbb{C})$ acts on z as $z \to mz$ and the condition for the Hermitian form to be preserved is precisely $m^\dagger m = 1$, which defines the unitary group $\mathrm{U}(n)$.

This group admits a generalisation $\mathrm{U}(r,s)$ that replaces the previous Hermitian form with an indefinite one

$$\bar{z}^1 z^1 + \cdots + \bar{z}^r z^r - \bar{z}^{r+1} z^{r+1} - \cdots - \bar{z}^{r+s} z^{r+s}. \tag{1.43}$$

The last classical group that we have not yet encountered is the symplectic group $\mathrm{Sp}(n,\mathbb{R})$ and this is defined as the subgroup of $\mathrm{GL}(2n,\mathbb{R})$ that fixes the standard symplectic form

$$dx^1 \wedge dy^1 + \cdots + dx^n \wedge dy^n$$

on $\mathbb{R}^{2n}$, where $x^a, y^a, a = 1,\ldots,n$ are the coordinates on $\mathbb{R}^{2n}$. This 2-form can be encoded into an antisymmetric $2n \times 2n$ matrix I, and then $\mathrm{Sp}(n,\mathbb{R})$ arises as the subgroup of $\mathrm{GL}(2n,\mathbb{R})$ that preserves this matrix $m^T I m = I$.

1.9.6 Group Actions on Manifolds

Groups, and Lie groups in particular, can act on manifolds. If this action preserves some geometric structure on the manifold in question (e.g., metric), then one says that this action describes a **symmetry** (in the case when a metric is preserved one talks of isometries). Lie groups then give continuous symmetries. Lie groups naturally act on themselves by symmetry operations, and this is going to be important for what follows.

Definition 1.38 Let M be a manifold and G a Lie group. A left action of G on M is a map

$$\lambda : G \times M \to M$$

such that $\lambda(\mu(g,h),x) = \lambda(g,\lambda(h,x))$ and $\lambda(e,x) = x$.

An equivalent way of stating this definition is to say that a (left) action of G on M is a homomorphism from G to the group of diffeomorphisms of M. Then $\lambda_g : M \to M$ is a diffeomorphism satisfying $\lambda_{gh} = \lambda_g \circ \lambda_h$ and $\lambda_e = id$.

There are many examples of such actions that can be constructed. The simplest examples we can consider is the action of $\mathrm{GL}(n,\mathbb{R})$ on $\mathbb{R}^n$ or on $M(n,\mathbb{R})$. Both are examples of left actions. In the second case one can act with $\mathrm{GL}(n,\mathbb{R})$ matrices also on the right, which would give an example of a right action. Another basic example is the left action of G on itself.

1.9.7 Classification of Actions

The following is a standard terminology in relation to an action of G on M.

- The action of G on M is called **effective** if any nontrivial element acts nontrivially somewhere, i.e., if $\lambda_g = id$ implies $g = e$.

- The action of of G on M is called **free** if there are no fixed points, i.e., if $\lambda_g(x) = x$ for some x then $g = e$.
- of G on M is called **transitive** if any two points of M can be connected by the action of G, i.e., if $\forall x, y \in M$ there exists $g \in G : \lambda_g(x) = y$. This means that there is a single *orbit* of the action of G on M.

1.9.8 Group Homomorphisms

We have already encountered one example of a group homomorphism–that from G to the group of diffeomorphisms on M. Here we consider group homomorphisms in more generality and derive one important statement about the kernel of such homomorphisms.

Definition 1.39 A group homomorphism from group G to another group G' is a map $\psi : G \to G'$ that is compatible with the product on both G, G'. In other words, the map ψ is required to satisfy $\psi(gh) = \psi(g)\psi(h)$.

Definition 1.40 A subgroup $N \subset G$ is called **normal** if $\forall g \in G, gNg^{-1} \subset N$.

Definition 1.41 Given a group G and its subgroup $H \subset G$, the space G/H is defined to be the set of equivalence classes $g \sim gh, h \in H$. This set is called the (left) *coset* of H in G.

Cosets of the type G/H play a very important role in group theory, and when G is a Lie group, geometry. For now, let us consider the special case when $H \subset G$ is a normal subgroup. This situation is special because in this case, the coset G/N is a group itself. Indeed, consider two arbitrary elements of the coset G/N. They are of the form $gn, g'n'$, where $n, n' \in N$. We now want to see if their multiplication can be defined. To do so, we use the normal property of N to rewrite $ng' = g'n''$ for some $n'' \in N$. This means that $gng'n' = gg'n''n'$, which means that the product of two equivalence classes $gn, g'n'$ is the equivalence class of the product gg', which defines the product of equivalence classes and makes G/N into a group. We have thus proved a simple but important theorem.

Theorem 1.42 *When N is a normal subgroup G/N is a group.*

An important source of normal subgroups comes by considering group homomorphisms $\psi : G \to G'$. We then have

Theorem 1.43 *The kernel* Ker_ψ *of a group homomorphism $\psi : G \to G'$ is a normal subgroup of G, and $G' = G/\mathrm{Ker}_\psi$.*

1.9.9 Orbits of Group Action

Definition 1.44 Let G act on a manifold M (or, for purposes of this definition, any set M). The **orbit** of a point $x \in M$ under the action of G is a set of all points in M that can be obtained from x by the G action, i.e., $\mathcal{O}_x = \{y \in M : \lambda_g(x) = y\}$.

It is easy to see that the group action on any of its orbits is transitive, i.e., any point can be connected to any other point by the group action.

Definition 1.45 Let G act on M. The **stabiliser** of a point $x \in M$ is the set $H_x \subset G$ such that $\lambda_h(x) = x, \forall h \in H_x$.

It is easy to check that the set H_x is a subgroup of G, called the stabiliser subgroup of a point. We then have the following important statement:

Theorem 1.46 *Let G act on M, and let $\mathcal{O}_x$ be the orbit of a point $x \in M$, and let H_x be the stabiliser at that point. Then the coset G/H_x is canonically isomorphic to the orbit $\mathcal{O}_x$.*

This means that, given an action of G on some manifold M (or more generally a set M), the orbits of this action can be canonically identified with group cosets. This effectively means that if G acts on M then the orbits of this action can be thought of as sitting inside the group. All symmetric spaces, i.e., spaces where some Lie group acts by symmetries, are then coset spaces. The simplest examples are: the two-dimensional sphere $S^2 = \mathrm{SO}(3)/\mathrm{SO}(2)$ and the hyperbolic plane $H^2 = \mathrm{SL}(2,\mathbb{R})/\mathrm{SO}(2)$.

1.9.10 Lie Algebras

Definition 1.47 Lie algebra is a vector space L equipped with a bilinear map

$$L \times L \to L, \qquad (A, B) \to [A, B]$$

called Lie bracket. This is required to be antisymmetric $[A, B] = -[B, A]$, and satisfy Jacobi identity

$$[A, [B, C]] + [B, [C, A]] + [C, [A, B]] = 0.$$

The basic examples of Lie algebras are the following.

- Vector fields on a manifold with the Lie bracket of vector fields as the Lie bracket form a Lie algebra. Jacobi identity can be checked by an explicit calculation.
- The space $M(n, \mathbb{R})$ of $n \times n$ matrices forms Lie algebra with the Lie bracket given by the commutator $[A, B] = AB - BA$. The Jacobi identity is trivial to verify.

As we will see in Section 1.9.15, there is a relation between these two examples. Another useful definition is

Definition 1.48 A subspace of Lie algebra L that is closed under the Lie bracket is called a (Lie) subalgebra of L.

1.9.11 Homomorphism of Lie Algebras

Definition 1.49 A homomorphism between two Lie algebras L, L' is a linear map $\phi : L \to L'$ that is compatible with the Lie bracket on L, L', i.e.,

$$\phi([A, B]) = [\phi(A), \phi(B)].$$

Example 1.50 The vectors in $\mathbb{R}^3$ form a Lie algebra with Lie bracket given by the vector product

$$(\mathbf{x}, \mathbf{y}) \to [\mathbf{x}, \mathbf{y}] = \mathbf{x} \times \mathbf{y}.$$

Jacobi identity can be checked by an explicit calculation, or using $\mathbf{x} \times (\mathbf{y} \times \mathbf{z}) = \mathbf{y}(\mathbf{x} \cdot \mathbf{z}) - \mathbf{z}(\mathbf{x} \cdot \mathbf{y})$.

The other relevant Lie algebra is that of anti-Hermitian 2×2 matrices of zero trace. The matrix commutator of any two matrices has zero trace, so the zero trace condition is preserved by the commutator. The commutator of two Hermitian or anti-Hermitian matrices is anti-Hermitian. This is why it's anti-Hermitian matrices that make up a Lie algebra. Any such matrix can be written as

$$\phi(x) = -\frac{\mathrm{i}}{2}\sigma^i x^i,$$

where $x^i \in \mathbb{R}^3$. This also gives the map from $\mathbb{R}^3$ to the space of tracefree anti-Hermitian 2×2 matrices. Using $[\sigma^i, \sigma^j] = 2\mathrm{i}\epsilon^{ijk}\sigma^k$ we have

$$[\phi(x), \phi(y)] = \left(\frac{\mathrm{i}}{2}\right)^2 x^i y^j 2\mathrm{i}\epsilon^{ijk}\sigma^k = -\frac{\mathrm{i}}{2}\epsilon^{ijk}x^i y^j \sigma^k = \phi(\mathbf{x} \times \mathbf{y}).$$

This shows that the map ϕ is a homomorphism of Lie algebras.

1.9.12 Lie Algebra of a Lie Group: Left-Invariant Vector Fields

Every Lie group G is a manifold, and G itself acts on this manifold by the left action.

Definition 1.51 The left action of G on itself is a homomorphism $G \to \mathrm{Diff}(G)$ from the group into the group of diffeomorphisms of the manifold G. In other words, this is a map $\lambda : G \times G \to G$ given by the left multiplication $\lambda_g(g') = gg'$.

We note that this action is free and transitive, with the whole of G being one single orbit.

One can define a special subset of all the vector fields on the group manifold – the vector fields that are invariant with respect to the action of the group.

Definition 1.52 A vector field X on G is called left-invariant if its push-forward with respect to the left action of G on itself coincides with X, i.e., if $\forall g \in G$ we have $\lambda_{g*}(X) = X$.

We now define the **Lie algebra** Lie(G) of G to be the vector space of all left-invariant vector fields on G. The Lie bracket is given by the Lie bracket of vector fields, and it is clear that the Lie bracket of two left-invariant vector fields is also left-invariant. We will explicitly check this for matrix groups in Section 1.9.15. This characterisation of the Lie algebra is available for any Lie group. However, this characterisation is often not the most convenient to work with in practice. This is why we need the following alternative descriptions.

1.9.13 Description in Terms of the Tangent Space at the Identity

We note that every vector field on the group manifold can be restricted to the identity element, where it gives some vector in T_eG. In particular, left-invariant vector fields can be restricted in this way. In the opposite direction, given a vector in T_eG, we can use the left action to push-forward this vector to every point on the group manifold. What we obtain is by construction a left-invariant vector field on G.

Thus, there is a bijective correspondence between the space of left-invariant vector fields on G and the vector space T_eG. This also makes it clear that the Lie algebra of any Lie group has the same dimension as the group manifold. However, the Lie bracket on this description of the Lie algebra is not intrinsically defined. Indeed, the only way to compute it is to transport two given vectors in T_eG to the whole of the group, compute their Lie brackets, and then restrict the result to T_eG. An alternative description that allows to compute the Lie bracket directly is needed, and will be available for matrix groups.

1.9.14 Description in Terms of One-Parameter Subgroups

Definition 1.53 A one-parameter subgroup of G is a homomorphism $\mathbb{R} \to G$ from the real line to the group. Concretely, this is a one-parameter family $g_t \in G$ of group elements satisfying $g_0 = e$ and $g_{s+t} = g_s g_t$.

Now, using the left action, we see that every one-parameter subgroup of G gives rise to a one-parameter group of diffeomorphisms of G. In turn, we know that one-parameter groups of diffeomorphisms generate vector fields. It is easy to check that the vector field that results in this way from a one-parameter subgroup of G is left-invariant. We will do this check when we consider matrix groups.

In the opposite direction, given a left-invariant vector field, we already know that vector fields generate their integral curves. Every integral curve of a left-invariant vector field is obtained by the left action of some one-parameter subgroup of G. Again, we will explicitly check this for matrix groups in section 1.9.15.

So, we have a bijective correspondence between left-invariant vector fields and one-parameter subgroups of G. This provides the third description of the Lie algebra of G namely, as the set of one-parameter subgroups of G. Again, the Lie bracket is implicit in this description, as one needs to convert the two one-parameter subgroups into left-invariant vector fields, compute their Lie bracket, and then convert the result into a one-parameter subgroup. All these steps can be explicitly described for matrix groups, as we now discuss.

1.9.15 Matrix Groups

Various classical groups that we discussed previously are matrix groups, i.e., subgroups of $\mathrm{GL}(n, \mathbb{R})$. More generally, every finite-dimensional Lie group can be described as an appropriate subgroup of a matrix group of a sufficiently large dimension. To do this, one just needs to consider some representation of the group, for example the so-called adjoint representation, which arises by considering the action of the group on its Lie algebra. We will not consider representations here.

The first concept that we need to define for matrix groups is that of the exponential map. This is defined via the notion of the matrix exponent. Thus, for any matrix $A \in M(n, \mathbb{R})$, we define

$$e^A = \sum_{k=0}^{\infty} \frac{1}{k!} A^k = \mathbb{I} + A + \frac{1}{2} A^2 + \cdots .$$

The matrix exponent has some important properties. First of all we have

$$e^{tA} e^{sA} = e^{(t+s)A}, \qquad \text{and} \qquad e^{-A} = (e^A)^{-1}.$$

The second property means that e^A is an invertible matrix, i.e., $e^A \in \mathrm{GL}(n, \mathbb{R})$. The first property means that e^{tA} is a one-parameter subgroup of $\mathrm{GL}(n, \mathbb{R})$.

Theorem 1.54 *All one-parameter subgroups of* $\mathrm{GL}(n, \mathbb{R})$ *are of the type* e^{tA} *for some* $A \in M(n, \mathbb{R})$.

To prove this theorem, let $\phi_t \in \mathrm{GL}(n, \mathbb{R})$ be a one-parameter subgroup of $\mathrm{GL}(n, \mathbb{R})$. Denote by A the derivative of ϕ_t at $t = 0$, i.e.,

$$A := \frac{d}{dt} \phi_t \Big|_{t=0}.$$

The we have

$$\frac{d}{dt} \phi_t = \frac{d}{ds} \phi_{s+t} \Big|_{s=0} = \frac{d}{ds} \phi_s \Big|_{s=0} \phi_t = A \phi_t.$$

This means that ϕ_t satisfies the differential equation $d\phi_t/dt = A\phi_t$. The solution of this differential equation that passes through the identity at $t = 0$ is

$$\phi_t = e^{tA}.$$

Now that we have an explicit characterisation of the one-parameter subgroups of $\mathrm{GL}(n,\mathbb{R})$ as matrix exponents, we can provide also an explicit description of the Lie algebra of this group in terms of left-invariant vector fields, as well as the description in terms of the tangent space at the identity. We first need to compute explicitly the left-invariant vector field that arises as the velocity vector field of the one-parameter group of diffeomorphisms generated by e^{tA}. These are obtained as follows

$$\frac{d}{dt}f(ge^{tA})\Big|_{t=0} = \frac{\partial f}{\partial g^i{}_j}(gA)^i{}_j, \tag{1.44}$$

and so, the corresponding vector field is

$$\xi_A = (gA)^i{}_j \frac{\partial}{\partial g^i{}_j}. \tag{1.45}$$

The derivation (1.44) requires some explanation. What is done in (1.44) can be understood as the translation of the vector corresponding to A at the identity of the group manifold to an arbitrary point g using the left action of G on itself. This is why the resulting vector field is left-invariant.

We thus see that left-invariant vector fields on $\mathrm{GL}(n,\mathbb{R})$ are in correspondence with matrices $A \in M(n,\mathbb{R})$, via (1.45). We can now compute the Lie bracket of two left-invariant vector fields

$$\begin{aligned}[\xi_A,\xi_B] &= (gA)^i{}_j \frac{\partial g^k{}_m}{\partial g^i{}_j} B^m{}_l \frac{\partial}{\partial g^k{}_l} - (gB)^k{}_l \frac{\partial g^i{}_s}{\partial g^k{}_l} A^s{}_j \frac{\partial}{\partial g^i{}_j} \\ &= (gA)^i{}_j B^j{}_l \frac{\partial}{\partial g^i{}_l} - (gB)^k{}_l A^l{}_j \frac{\partial}{\partial g^k{}_j} = (gAB - gBA)^i{}_j \frac{\partial}{\partial g^i{}_j} = \xi_{[A,B]}.\end{aligned} \tag{1.46}$$

We thus see that we have a homomorphism of the Lie algebra of left-invariant vector fields on $\mathrm{GL}(n,\mathbb{R})$ into the matrix Lie algebra, with the Lie bracket of vector fields going into the commutator of matrices.

We can now summarise what we have learned about the three different descriptions of the Lie algebra of $\mathrm{GL}(n,\mathbb{R})$. The description in terms of left-invariant vector fields is that given by ξ_A given by (1.45). Evaluating these vector fields at the identity we get vectors $A^i{}_j(\partial/\partial g^i{}_j)$ that are in correspondence with matrices $A \in M(n,\mathbb{R})$. The description in terms of one-parameter subgroups is via $g = e^{tA}$. Again, every one-parameter subgroup is in correspondence with a matrix A. So, in all cases a Lie algebra element is in correspondence with a matrix A, and the Lie bracket can be computed as the matrix commutator (1.46).

1.9.16 Explicit Description of Lie Algebras of Some Classical Groups

Classical groups have beed previously defined as various subgroups of $\mathrm{GL}(n,\mathbb{R})$. We have just understood that the Lie algebra of $\mathrm{GL}(n,\mathbb{R})$ can be described as the Lie algebra of $n \times n$ real matrices, with the Lie bracket given by the matrix commutator. We can obtain a similar explicit description of the Lie algebras of the classical groups.

The group $\mathrm{SL}(n,\mathbb{R})$ consists of matrices of determinant one. The corresponding Lie algebra is that of matrices of trace zero. It is easy to see that the space of matrices of vanishing trace is closed under the operation of matrix commutator, because the commutator of any two matrices automatically has zero trace.

The group $\mathrm{O}(n,\mathbb{R})$ is the group of matrices satisfying $m^T m = id$. Taking $m = e^{tA}$ and differentiating the relation $m^T m = id$ at $t = 0$ we get $A^T + A = 0$. This means that the Lie algebra of the orthogonal group is that of antisymmetric matrices. The space of antisymmetric matrices is closed under the operator of taking the commutator.

The group $\mathrm{SU}(n)$ is the group of complex unitary matrices of unit determinant. We already know that the condition of unit determinant translates at the Lie algebra level to the condition that the matrices are tracefree. Let us see the consequences of the unitarity condition. This is the condition $m^\dagger m = id$. Taking $m = e^{tA}$ and differentiating $m^\dagger m = id$ at $t = 0$ we get $A^\dagger + A = 0$, which is the condition that the matrix A is anti-Hermitian. Thus, the Lie algebra of $\mathrm{SU}(n)$ consists of tracefree anti-Hermitian matrices.

1.10 Cartan's Isomorphisms

This section is just a quick look at this rather vast (and important) subject. For more details see the book, *Spinors and Calibrations*, by F. Reese Harvey (1990).

Orthogonal groups have spinor representations. Those come with various inner products. This means that the spin groups (which arise as covers of the orthogonal groups) are always subgroups of various classical groups preserving the relevant inner product on the space of spinors. In lower dimensions the spin groups coincide with various classical groups, and this is why the Cartan's isomorphisms arise. The most important of these isomorphisms (for physics) are that between the rotation group in three dimensions and the special unitary group in two dimensions, and that between the Lorentz group in four dimensions and the complex special linear group in two dimensions.

1.10.1 The Isomorphism $\mathrm{SO}(3) = \mathrm{SU}(2)/\mathbb{Z}_2$

Let us consider the space of anti-Hermitian 2×2 matrices with zero trace. Any such matrix is of the form

$$\mathbf{x} = \mathrm{i}\begin{pmatrix} x^3 & x^1 - \mathrm{i}x^2 \\ x^1 + \mathrm{i}x^2 & -x^3 \end{pmatrix}. \tag{1.47}$$

In other words, any such matrix is of the form $\mathbf{x} = \mathrm{i}\sigma^i x^i$, where σ^i are the Pauli matrices. The previous matrix has the property that

$$\det(\mathbf{x}) = (x^1)^2 + (x^2)^2 + (x^3)^2,$$

which is the squared interval in $\mathbb{R}^3$. So, we have an isomorphism between the space of anti-Hermitian 2×2 matrices of zero trace and $\mathbb{R}^3$, with the norm squared of an $\mathbb{R}^3$ vector (x^1, x^2, x^3) being represented by the determinant of the corresponding 2×2 matrix.

Now, the group SU(2) acts on the space of anti-Hermitian matrices of zero trace via

$$g \in \mathrm{SU}(2), \qquad \mathbf{x} \to g\mathbf{x}g^\dagger. \tag{1.48}$$

Because $g \in \mathrm{SU}(2)$ this action preserves the determinant. Remembering that there is an isomorphism between the space of anti-Hermitian matrices of zero trace and $\mathbb{R}^3$, we get a group homomorphism $\tau : \mathrm{SU}(2) \to \mathrm{O}(3)$. Because the arising transformations are orientation preserving, this is in fact a homomorphism into SO(3). This homomorphism has a nontrivial kernel consisting of $e, -e \in \mathrm{SU}(2)$. Thus, we get $\mathrm{SO}(3) = \mathrm{SU}(2)/\mathbb{Z}_2$.

1.10.2 Description of the Isomorphism $\mathbf{SO(1,3) = SL(2,}\boldsymbol{C}\mathbf{)/\mathbb{Z}_2}$

This is very similar to the previous section, except that the restriction to trace zero matrices is dropped. Thus, let us consider the space of anti-Hermitian 2×2 matrices. Any such matrix is of the form

$$\mathbf{x} = \mathrm{i}\begin{pmatrix} x^0 + x^3 & x^1 - \mathrm{i}x^2 \\ x^1 + \mathrm{i}x^2 & x^0 - x^3 \end{pmatrix}. \tag{1.49}$$

We also have

$$\det(\mathbf{x}) = -(x^0)^2 + (x^1)^2 + (x^2)^2 + (x^3)^2,$$

which is the squared interval in $\mathbb{R}^{1,3}$. This gives an isomorphism between the points in Minkowski space $\mathbb{R}^{1,3}$ and anti-Hermitian 2×2 matrices, with the squared interval being represented by the determinant.

Consider now the following action of $\mathrm{SL}(2, \mathbb{C})$ on anti-Hermitian matrices

$$g \in \mathrm{SL}(2, \mathbb{C}), \qquad \mathbf{x} \to g\mathbf{x}g^\dagger. \tag{1.50}$$

This maps anti-Hermitian matrices into themselves, and preserves the determinant. The transformations that one generates this way are orientation-preserving, and moreover, preserve the orientation of time. This gives a group homomorphism

$$\tau : \mathrm{SL}(2, \mathbb{C}) \to \mathrm{SO}^+(1,3) \tag{1.51}$$

from the group of special complex linear transformations in two dimensions to what is known as the restricted Lorentz group (consisting of orientation-preserving orthogonal transformations that also preserve the orientation of time). The kernel of τ is again a copy of $\mathbb{Z}_2$, the one generated by $\pm e \in \mathrm{SL}(2, \mathbb{C})$. The group $\mathrm{SL}(2, \mathbb{C})/\mathbb{Z}_2$ is known as $\mathrm{PSL}(2, \mathbb{C})$. So, we get the isomorphism

$$\mathrm{PSL}(2,\mathbb{C}) = \mathrm{SO}^+(1,3). \tag{1.52}$$

This isomorphism plays a very important role in physics, as it in particular provides a description of the spinor representations of the Lorentz group. Very concretely, these are the two-component columns of complex numbers on which the group $\mathrm{SL}(2,\mathbb{C})$ acts by matrix multiplication.

1.11 Fibre Bundles

The logic we follow in this section is to first define the most general notion of a fibre bundle without any Lie group action in the fibres. Already this setup allows a connection and curvature to be defined. Only then do we add the extra structure of a group acting in the fibres. We first define principal fibre bundles, and then discuss vector bundles, in particular vector bundles associated with principal bundles.

1.11.1 Definition

The notion of a fibre bundle is an abstractisation of the commonly encountered geometric setup where a manifold E is foliated by submanifolds, with each leaf of the foliation diffeomorphic to some given manifold F. One calls the set of leaves of the foliation the base space B.

In this geometric setup the base space B arises as the quotient space. This is not what is most convenient to have a workable definition. For this reason, the standard definition of the fibre bundle contains the base space built into it.

Definition 1.55 A fibre bundle (or simply bundle) is a triple (E, B, π), where E, B are topological manifolds and $\pi : E \to B$ is a map. The space E is called the **total space**, the space B is the **base**, and the map π is called the **projection** of the bundle. It is moreover required that for every $e \in E$ there is an open neighbourhood $U \subset B$ of $\pi(e)$ (which is called a trivialising neighbourhood) so that there is a homeomorphism $\phi : \pi^{-1}(U) \to U \times F$, where F is another smooth manifold, such that the projection π agrees with the projection on the first factor. This means that for every $b \in B$ the preimage $\pi^{-1}(b)$ is homeomorphic to F and is called the **fibre over** b. A fibre bundle is often denoted as

$$F \longrightarrow E \xrightarrow{\pi} B.$$

This mimics short exact sequences and represents the fact that the image of the first map, i.e., the fibres, are in the kernel of the second (projection) map. Another common notation for the bundle is $\pi : E \to B$. A **smooth** fibre bundle arises when all E, B, and F are smooth manifolds, and all the maps are smooth.

Example 1.56 The simplest (trivial) example of a fibre bundle is the product $(B \times F, B, \pi)$ where the projection π is that on the first factor.

Before we consider examples of nontrivial bundles, let us state two more important definitions.

Definition 1.57 A bundle (E', B', π') is a **subbundle** of (E, B, π) provided E' is a subspace of E, B' is a subspace of B, and $\pi' = \pi|_{E'} : E' \to B'$.

Definition 1.58 A **cross section** (or simply section) of a bundle (E, B, π) is a map $s : B \to E$ such that $s(b) \in \pi^{-1}(b)$.

1.11.2 Examples of Nontrivial Bundles

We give two examples, without proving that they really correspond to nontrivial bundles. Another basic example to have in mind is that of the Hopf fibration. It is treated in due course in Section 1.13.

Example 1.59 The tangent bundle over S^n, denoted (T, S^n, π) is a subbundle of the product bundle $(S^n \times \mathbb{R}^{n+1}, S^n, \pi)$, whose total space is defined by the relation $(p, x) \in T$ if and only if the Euclidean inner product $(p|x) = 0$. An element $(p, x) \in T$ is called a tangent vector to S^n at $p \in S^n \subset \mathbb{R}^{n+1}$. The fibre $\pi^{-1}(p)$ is of dimension n. A cross section of the tangent bundle is called a tangent vector field on S^n.

Example 1.60 The bundle of (orthonormal) k-frames over S^n denoted by (E, S^n, π) is a subbundle of the product bundle $(S^n \times (S^n)^k, S^n, \pi)$ where the total space E is the subspace of $S^n \times (S^n)^k$ consisting of (unit) vectors $(p, v_1, \ldots, v_k) \in R^{n+1} \times (\mathbb{R}^{n+1})^k$ such that $(p|v_i) = 0$ and $(v_i|v_i) = \delta_{ij}$. In other words, this is the subspace of k orthonormal tangent vectors to S^n. A cross section of this bundle is called a field of k-frames. The existence of a nowhere vanishing field of k-frames is a difficult problem, and in general, there is no such existence.

1.11.3 Restrictions and Pullbacks of Bundles

Definition 1.61 Let (E, B, π) be a fibre bundle, and let A be a subset of B. We can then define a new bundle $E|_A$ with A as the base, defined as (E', A, π'), where $E' = \pi^{-1}(A)$ and $\pi' = \pi|_{E'}$.

The other definition concerns a case when we have a map $f : B' \to B$, and a bundle over B. In this case we can define the pullback bundle f^*E over B'. The fibre of f^*E over $b' \in B'$ is just the fibre of $f(b') \in B$ in E.

Definition 1.62 Let (E, B, π) be a fibre bundle and $f : B' \to B$ be a map. The pullback bundle f^*E has B' as the base, the subspace of all pairs $(b', e) \in B' \times E$ such that $f(b') = \pi(e)$ as the total space, and the map $(b', e) \to b'$ as the projection π'.

1.11.4 Connections in Fibre Bundles: Ehresmann Connection

The setup of the fibre bundle is the minimal geometric setup that allows the notion of a connection to be defined. The connection that arises in this setting is known as the **Ehresmann** connection. When a bundle is given more structure (e.g., principal bundles or vector bundles that we consider in Sections 1.12 and 1.14, respectively), the Ehresmann connection becomes the more familiar connections that we have in those settings. It is, however, very important to understand that there is a rather minimal geometric structure that is required for a connection to be defined.

A fibre bundle (E, B, π) is a manifold that comes with a preferred subset in the set of all vector fields $X \in TE$. Indeed, vector fields can be pushed forward with the projection map. While this push-forward is in general ill-defined (because the projection map is not injective), the notion of vector fields that are in the kernel of the projection map is well-defined. One then calls such vector fields **vertical**. Alternatively, vertical vector fields are those tangent to the fibres. Let us denote the set of vertical vector fields by $V \subset TE$. We now note that, given an arbitrary vector field $X \in TE$, there is in general no way to represent this vector field as a sum of its vertical part and the remainder. A connection is then defined as a rule that provides such a decomposition. This motivates the following definition.

Definition 1.63 An **Ehresmann connection** on E is a smooth subbundle H of TE, called the **horizontal bundle** of the connection, which is complementary to V, in the sense that it defines a direct sum decomposition $TE = H \oplus V$. In other words, a connection is a rule that for each point $e \in E$ defines a vector subspace $H_e \subset T_e E$, called the horizontal subspace of the connection at e. The set of horizontal subspaces H_e is required to depend smoothly on e, and horizontal vectors should be complementary to vertical $H_e \cap V_e = \{0\}$. Any tangent vector $X \in T_e E$ should be representable as the sum of its vertical and horizontal parts $T_e E = V_e \oplus H_e, X = X_V + X_H$.

This notion of the connection is extremely useful, for it immediately allows several things to be defined. First, it allows us to define horizontal lifts of curves $\gamma(t) \in B$ on the base. Indeed, consider a curve on the base that passes through point $x = \gamma(0)$. Select a point $p \in \pi^{-1}(x)$ in the fibre over the point x. The horizontal lift of γ through p is a curve $\tilde{\gamma}(t)$ in the total space E such that for every t the tangent vector to the curve lies in H, i.e. $d\tilde{\gamma}/dt \in H_{\tilde{\gamma}(t)}$, and the curve $\tilde{\gamma}(t)$ projects to $\gamma(t)$ for every t. This should be compared to the notion of an integral curves of a vector field on a manifold. In general, the horizontal lift of a curve $\gamma(t)$ can be shown to exist for sufficiently small t.

1.11.5 Curvature

The other important and natural notion that can be defined for an Ehresmann connection is that of the **curvature**. This notion can be motivated by considering the horizontal lifts along two different paths that start and end at the same points on the base. In general, the horizontal lifts starting at the same point will fail to end at the same point. This failure of the horizontal lifts along different paths to agree is measured by the curvature.

Another way to motivate the definition of the curvature is to think about the Lie bracket of two horizontal vector fields. In general, the resulting vector field will fail to be horizontal. This would signal the fact that the horizontal distribution is not integrable, i.e., that the notion of the horizontal vector fields does not arise from some foliation of the total space E by submanifolds (diffeomorphic to B). The horizontal distribution is only integrable (and thus defines submanifolds diffeomorphic to B through every point of E) when the Lie bracket of two horizontal vector fields is again horizontal. This is precisely what is measured by the curvature of the connection, with curvature being zero being equivalent to the horizontal distribution being integrable.

With these remarks in mind we define the curvature of the Ehresmann connection as a two form on the total space E with values in the vertical subbundle V of TE. The curvature is given by

$$R(X,Y) = [X_H, Y_H]_V, \qquad R \in \Lambda^2(E,V),$$

where $R(X,Y)$ is the contraction of the 2-form R with the vector fields X, Y, X_H, Y_H are the horizontal parts of X, Y, $[X,Y]$ stands for the Lie bracket of vector fields, and the result of the Lie bracket is projected onto the vertical subspace.

1.11.6 Connection as a 1-Form

Given that a connection is an object that provides a decomposition of the tangent bundle TE into vertical and horizontal subbundles, we can encode a connection into a 1-form with values in the space of vertical vector fields. Indeed, let $v \in \mathrm{End}(TE)$ be an endomorphism of the tangent bundle that maps any vector field X into its vertical part

$$\omega : X \to \omega(X) := X_V,$$

and is the identity map on the space of vertical vector fields. We can encode such an endomorphism into a 1-form with values in V, i.e., $\omega \in \Lambda^1(E,V)$, so that the pairing of this 1-form with X is $\omega(X)$. This 1-form is required to satisfy $\omega(X_V) = X_V$. The horizontal vector fields are then those in the kernel of the map ω

$$H = \{X \in TE : \omega(X) = 0\}.$$

Concretely, the object ω is a linear map that acts in each T_eE, smoothly depends on the point e, and sends each tangent vector to its vertical part.

1.11.7 Metric in the Total Space Defines a Connection

Let us now consider a frequently encountered situation when there is a (Riemannian) metric in the total space of the bundle, i.e., a symmetric and (positive) definite element of $T^*E \otimes T^*E$. In this situation, there is a natural Ehresmann connection that is such that the horizontal vector fields are those metric orthogonal to the vertical ones

$$H = \{X \in TE : (X|Y_V) = 0, \quad \forall Y_V \in V \subset TE\}.$$

This is the situation encountered in Kaluza–Klein theory, where certain components of the metric in the total space of the bundle receive the interpretation of a connection. We will consider an example of the setup of this sort when we study the Hopf fibration in the next section.

1.12 Principal Bundles

The notion of a principal bundle arises when we have an additional structure of a Lie group acting in the fibres, and when, moreover, this action is free and transitive so that each fibre is diffeomorphic to the group.

1.12.1 Definition and Examples

A principal G-bundle is a bundle (E, B, π) with fibres copies of a Lie group G. The formal definition is as follows.

Definition 1.64 A principal G-bundle is a bundle (E, B, π) with extra structure of a smooth **right** action of G on E by diffeomorphisms, i.e., a map $R_g : E \to E$ with the property that the identity element in G acts as the identity map, and $R_h \circ R_g = R_{gh}$. It is usual in this context to write $R_g p = pg$. This map is required to leave the projection π invariant, i.e., $\pi(pg) = \pi(p)$, and thus act on fibres. Moreover, the action of G on the fibres is required to be free and transitive, so that each fibre is an orbit of this action, and is a copy of the group G. Concretely, this means that the trivialising maps can be chosen to commute with the G action. In other words, for every $p \in E$ there is an open neighbourhood U of $\pi(p) \in B$ such that there is a smooth map $\phi : \pi^{-1}(U) \to U \times G$ that has the property that if $\phi(p) = (\pi(p), h(p))$ then $\phi(pg) = (\pi(p), h(p)g)$.

Example 1.65 The canonical example of a principal H bundle is the bundle $(G, G/H, \pi)$, where the total space is the group manifold G, the base is the group coset $B = G/H$, and the projection is the map from G to the set of its right H cosets $g \sim gh$. The fibres of this bundle are copies of H, with H acting on G as $g \to gh$, and preserving the fibres.

Example 1.66 Another prototypical example of a principal bundle is the **frame bundle** of a smooth manifold M denoted by FM. The fibre over a point $x \in M$

is the set of all frames (i.e., ordered bases of the tangent space T_xM). The general linear group $\mathrm{GL}(n, \mathbb{R})$ acts freely and transitively on these frames, which makes this into a principal $\mathrm{GL}(n, \mathbb{R})$ bundle.

Example 1.67 Yet another example arises in the situation when there is a metric on M. In this case one can consider the bundle of orthonormal frames on M. This is a principal $\mathrm{O}(n, \mathbb{R})$ bundle over M. The orthogonal group acts freely and transitively on the space of orthonormal frames.

Example 1.68 An example that connects the orthonormal frame principal bundle with the coset $(G, G/H, \pi)$ bundle is the bundle of orthonormal frames on the sphere S^n. The sphere is the group coset $S^n = \mathrm{SO}(n+1)/\mathrm{SO}(n)$, and the bundle of oriented orthonormal frames on S^n is the principal $\mathrm{SO}(n)$ bundle whose total space is the group manifold $\mathrm{SO}(n+1)$. To see this explicitly, we view the sphere S^n as the space of unit vectors in $\mathbb{R}^{n+1}$. A point in the oriented, orthonormal frame bundle consists of an $(n+1)$-tuple of orthonormal vectors $(x, v_1, \ldots, v_n)$ in $\mathbb{R}^{n+1}$. The projection is that onto the first element. We can identify this $(n+1)$-tuple of vectors with a matrix in $\mathrm{SO}(n+1)$, with these vectors as columns.

1.12.2 Coordinate (Cocycles) Definition

It is possible to give another, equivalent definition of a principal bundle that views such a bundle as glued from copies of the trivial bundle over suitable cover of the base B by coordinate charts. This definition is sometimes more convenient in practice, as it allows some explicit constructions of principal bundles. It proceeds as follows.

Let us start with a finite cover of the base B by open coordinate charts U. The data that is needed for this construction of the bundle is a set of **principal bundle transition functions**. These are defined for any pair U, V of coordinate charts that have a nontrivial overlap. They are denoted by $g_{UV} \in G$ and have the following properties: (i) g_{UU} is the constant map to the identity in G; (ii) $g_{UV}^{-1} = g_{VU}$; and (iii) if U, V, W are any three coordinate charts with $U \cap V \cap W \neq 0$, then the condition $g_{UV} g_{VW} g_{WU} = 1$ must hold. This set of constraints on the transition functions is called the **cocycle constraints**. Given such transition functions, the principal bundle is constructed as the quotient of the set of trivial bundles $U \times G$ by the equivalence relation that puts $(x, g) \in U \times G$ equivalent to $(x', g') \in U' \times G$ if and only if $x = x'$ and $g = g_{UU'}(x) g'$.

This gluing construction clearly produces a principal bundle in the sense of our coordinate-free definition. Let us also see how to go from the coordinate-free construction to the construction with transition functions just described. For this purpose, one stars by selecting a cover of the base B by coordinate charts U. For every coordinate chart we have a trivialising map $\phi_U : E|_U \to U \times G$. Now, given an intersection of two coordinate charts, we get the composition of maps

$\phi_U \circ \phi_V^{-1}$ which maps $U \cap V$ to G. These are our transition functions g_{UV}. They satisfy all the cocycle constraints, and thus provide the coordinate description of the same bundle.

Example 1.69 Principal U(1) bundles over S^2 can be constructed using the cocycle definition as follows. We cover the base S^2 with two coordinate charts, the north N and south S hemisphere. Both of these are mapped into the complex plane by the stereographic projection. On the intersection of the two charts the complex coordinate z of, say, the north chart N is different from zero. Let us form the map $g : N \cap S \to \mathrm{U}(1)$ as $g : z \to z/|z|$. We then fix an integer m and take the transition function for our principal U(1) bundle over S^2 to be $g^m : N \cap S \to \mathrm{U}(1)$. There are no triple intersections in this case, and so there are no triple intersection constraint to satisfy.

1.12.3 Connections in Principal Bundles

A connection in an arbitrary fibre bundle (E, B, π) is a horizontal distribution $H_p \subset TE$ at every point of $p \in E$. In a principal bundle we have a group G acting on E by the right action and preserving the fibres. It is natural to demand that the horizontal distribution is invariant under this action, in the sense that the horizontal vectors at p pushed forward using the right action are horizontal vectors at $H_{\lambda_g p}$, i.e., $R_{g*}(H_p) = H_{pg}$. Such a connection is called a **principal (Ehresmann) connection**, or simply a connection in a principal bundle.

Now, since each fibre is a copy of the group manifold, the space of vertical vectors at each $p \in E$ can be identified with the Lie algebra $\mathfrak{g}$ of G. As we have previously noted, a connection can be described as a 1-form ω in the total space. If a tangent vector at p is inserted into ω, it returns a vertical vector. But we have identified vertical vectors at p with the Lie algebra of G. This means that we can encode the connection into a 1-form that is Lie algebra valued $\omega \in \Lambda^1(E, \mathfrak{g})$.

For the construction that follows, it will be convenient to identify the Lie algebra $\mathfrak{g}$ with the space of **left**-invariant vector fields on the group manifold. So, it will be convenient to have the group G act on the fibres from the right, while considering the realisation of $\mathfrak{g}$ by left-invariant vector fields. With this realisation of $\mathfrak{g}$ in mind, we can make the identification of the vertical tangent space V_p at any point $p \in E$ more concrete. Thus, each vertical tangent vector in V_p is the restriction of some left-invariant vector field on G, and in this way each tangent vector from V_p is identified with an element of $\mathfrak{g}$. The Ehresmann connection in general is a linear map from T_pE to V_p which is identity on V_p, and after the identification we just made, the corresponding 1-form ω has the property that ω with a left-invariant vector field inserted into it returns the same left-invariant vector field.

This 1-form ω representing the Ehresmann connection is not going to be invariant under the right action of the group G in the fibres, but is going to have certain simple transformation properties instead. To describe these, we note that the right action of the group on itself gives rise to a nontrivial action of G on its Lie algebra, realised as the space of left-invariant vector fields. This is the **adjoint** action. For example, for matrix groups, the left-invariant vector fields are those of the form

$$\xi_A = (gA)^i{}_j \frac{\partial}{\partial g^i{}_j}, \qquad A \in M(n, \mathbb{R}),$$

and pushing forward such a vector field using the right action $R_h g = gh$ produces a left-invariant vector field

$$R_{h*}\xi_A = \xi_{A^h}$$

with

$$A^h = \mathrm{Ad}_{h^{-1}} A = h^{-1}Ah.$$

We can now state the transformation property of the 1-form $\omega \in \Lambda^1(E, \mathfrak{g})$ that represents a connection. Given that the connection is a projector that maps an arbitrary vector field into its vertical part, and that this projector commutes with the right action of G on E, we form a linear map that is the composition of this projector with the map from the space of vertical vector fields to the Lie algebra realised as the space of left-invariant vector fields. This linear map is our 1-form ω. When acting on it with the right action, the projection onto the vertical vectors commutes with the right action, while the right action of the group on the space of left-invariant vector fields is the adjoint action. So, we get the following property that must be satisfied by ω

$$R_h^*\omega = \mathrm{Ad}_{h^{-1}}\omega.$$

Conversely, any 1-form $\omega \in \Lambda^1(E, \mathfrak{g})$ with this property and the property that when a left-invariant vector field is inserted it is returned defines an Ehresmann connection in a principal G-bundle.

1.12.4 Coordinate Description of Principal Connections

All this can be described very concretely in coordinates. Thus, we choose some coordinate chart $U \subset B$, and the corresponding trivialisation in which $E|_U = U \times G$. Then the 1-form on G with values in $\mathfrak{g}$ that is left-invariant and transforms under the right action by the adjoint representation is

$$g^{-1}dg.$$

This form maps left-invariant vector fields into the corresponding Lie algebra elements. Thus, for any connection ω, its restriction to the fibres must coincide with $g^{-1}dg$. Then any form that is equivariant with respect to the right action and coincides with $g^{-1}dg$ on the fibres can be written as

$$\omega = g^{-1}dg + g^{-1}Ag, \tag{1.53}$$

where $A \in \Lambda^1(B, \mathfrak{g})$ is a Lie algebra valued 1-form on the base. It is in order to be able to give this coordinate description of the connection that we have identified the Lie algebra with the left-invariant vector fields on G, while using the right action of G on E to define the principal bundle.

It is important to emphasise that while the description in terms of a Lie algebra valued 1-form on the base makes in general sense only locally, in a coordinate chart, the description of the connection as a Lie algebra valued form on the total space is global, and independent of any coordinates that can be chosen.

1.12.5 Coordinate Description of Horizontal and Vertical Vector Fields

Now that we wrote the connection 1-form ω as (1.53) in a trivialisation, we can work out explicitly what the horizontal subbundle of TE is. Let us consider the case of matrix groups, and let

$$X = a^i{}_j \frac{\partial}{\partial g^i{}_j} + a^\mu \frac{\partial}{\partial x^\mu} \tag{1.54}$$

be an arbitrary vector field in TE. Here both $a^i{}_j$ and a^μ are functions of $x^\mu, g^i{}_j$. When the Ehresmann connection is described by the corresponding 1-form ω, the horizontal vector fields arise as those in the kernel of ω. So, the component functions $a^i{}_j, a^\mu$ of a horizontal vector field satisfy

$$(g^{-1}a)^i{}_j + (g^{-1}a^\mu A_\mu g)^i{}_j = 0.$$

From this we see that the horizontal vector fields are those of the form

$$X_H = -(a^\mu A_\mu g)^i{}_j \frac{\partial}{\partial g^i{}_j} + a^\mu \frac{\partial}{\partial x^\mu}. \tag{1.55}$$

These horizontal vector fields are preserved by the right action of G on the fibres when a^μ is a set of functions on the base only, i.e., g-independent. The vector field obtained can also be referred to as the horizontal lift of an arbitrary vector field $a^\mu(\partial/\partial x^\mu)$ from the base into the total space. The horizontal lift explicitly depends on the connection components A_μ (that are matrix valued). Similarly, the vertical projection of X in (1.54) is given by

$$X_V = (a + a^\mu A_\mu g)^i{}_j \frac{\partial}{\partial g^i{}_j}. \tag{1.56}$$

1.12.6 Change of Trivialisation as Gauge Transformations

The given previously coordinate description of the connection is based on a trivialisation of the bundle over a coordinate chart on the base. Every trivialisation of a principle bundle comes from a cross section, and vice versa, if a principal bundle admits a global cross section, this implies that the bundle is trivial. Let us see this. If $s : B \to E$ is a section, then we can parametrise a point in E as the set of pairs $(x, s(x)g), x \in B, g \in G$. This provides a global identification between E and $B \times G$. In the opposite direction, if we have a trivialisation over U, which is a map $\phi : E|_U \to U \times G$, the preferred section is given by $s(x) = \phi^{-1}(x, 1)$.

Given a trivialisation and the corresponding section $s : B \to E$ of the bundle E, we can understand the 1-form on the base $A \in \Lambda^1(B, \mathfrak{g})$ in (1.53) as the pullback of the connection 1-form ω with respect to s, i.e.,

$$A = s^*(\omega).$$

This raises the question of what happens if a trivialisation is changed. As we shall see, this corresponds to what in physics is called gauge transformations.

We can go from one trivialisation corresponding to section $s(x)$ to another $s'(x)$ by the right action

$$s'(x) = s(x)h(x),$$

where $h(x) \in G$ is a function from the base into the group. Then the trivialisation that corresponds to $s'(x)$ is obtained by the parametrisation $(x, s'(x)g) = (x, s(x)h(x)g)$. The change of trivialisation thus corresponds to replacing $g \to h(x)g$, i.e., to the left local action of $h(x)$ on the group manifold. Let us see the effect of this transformation on the connection. We have

$$\omega^h = (h(x)g)^{-1}d(h(x)g) + (h(x)g)^{-1}A(h(x)g) = g^{-1}dg + g^{-1}A^h g,$$

where

$$A^h = h^{-1}(x)dh(x) + h^{-1}(x)Ah(x). \tag{1.57}$$

We thus see that ω^h preserves its form (1.53), but with the 1-form on the base A being replaced with the 1-form A^h. The 1-form A^h is said to be obtained from A by a **gauge transformation**.

1.12.7 Curvature as a 2-Form

The curvature of a connection was described as an antisymmetric map $R : H \times H \to V$ from the space of horizontal vector fields to the vertical ones. It can be given a further characterisation in the case of principal bundles. Thus, consider the 2-form

$$R = d\omega + \frac{1}{2}[\omega, \omega] \in \Lambda^2(E, \mathfrak{g}). \tag{1.58}$$

In the case of matrix groups this becomes simply

$$R = d\omega + \omega\omega, \tag{1.59}$$

where the wedge product in the second term is implied. This 2-form vanishes on vertical vector fields. This is easiest to see in a trivialisation. Indeed, for matrix groups we have

$$\begin{aligned} R &= d(g^{-1}dg + g^{-1}Ag) + (g^{-1}dg + g^{-1}Ag)(g^{-1}dg + g^{-1}Ag) \\ &= g^{-1}(dA + AA)g - g^{-1}dgg^{-1}dg - g^{-1}dgg^{-1}Ag - g^{-1}Adg \\ &\quad + g^{-1}dgg^{-1}dg + g^{-1}dgg^{-1}Ag + g^{-1}Adg = g^{-1}R(A)g, \end{aligned} \tag{1.60}$$

where

$$R(A) = dA + AA \tag{1.61}$$

is the curvature 2-form on the base. This form clearly vanishes when a vertical vector field is inserted into it.

1.12.8 Relation to Curvature Defined as Measure of Non-Integrability

It is an instructive exercise to relate the definition (1.61) of the curvature to the definition that was given earlier, where curvature appeared as a measure of non-integrability of the horizontal distribution. Thus, let us compute the quantity $[X_H, Y_H]_V$ for two vector fields X, Y of the form (1.54). Using (1.55) we have

$$X_H = -(a^\mu A_\mu g)^i{}_j \frac{\partial}{\partial g^i{}_j} + a^\mu \frac{\partial}{\partial x^\mu}, \qquad Y_H = -(b^\mu A_\mu g)^i{}_j \frac{\partial}{\partial g^i{}_j} + b^\mu \frac{\partial}{\partial x^\mu}.$$

The Lie bracket of these two vector fields computes to

$$\begin{aligned} [X_H, Y_H] &= (b^\nu A_\nu a^\mu A_\mu g)^i{}_j \frac{\partial}{\partial g^i{}_j} - (a^\nu(\partial_\nu b^\mu)A_\mu g + a^\mu b^\nu (\partial_\mu A_\nu) g)^i{}_j \frac{\partial}{\partial g^i{}_j} \\ &\quad + a^\mu \partial_\mu b^\nu \frac{\partial}{\partial x^\mu} - (a \leftrightarrow b). \end{aligned}$$

This expression contains derivatives of a^μ, b^μ. However, when we take the vertical projection using (1.56) these derivatives cancel out and we get

$$[X_H, Y_H]_V = -a^\mu b^\nu \left(R_{\mu\nu} g\right)^i{}_j \frac{\partial}{\partial g^i{}_j}, \tag{1.62}$$

where

$$R_{\mu\nu} = \partial_\mu A_\nu - \partial_\nu A_\mu + A_\mu A_\nu - A_\nu A_\mu. \tag{1.63}$$

These are just of course the components of the curvature 2-form

$$R = (1/2) R_{\mu\nu} dx^\mu dx^\nu,$$

with R given by (1.61). This shows that the curvature as computed using (1.61) is indeed the measure of non-integrability of the horizontal distribution as defined by the connection.

1.13 Hopf Fibration

The purpose of this section is to describe an example that illustrates most of the previous constructions. The Hopf fibration describes the three-sphere S^3 as a nontrivial S^1 bundle over S^2.

1.13.1 Construction of the Hopf Fibration

The Hopf fibration has a spinor origin. Consider the special unitary group SU(2) in two dimensions. This group naturally acts on columns

$$\psi = \begin{pmatrix} \alpha \\ \beta \end{pmatrix}, \qquad \alpha, \beta \in \mathbb{C}.$$

The action is by the matrix multiplication

$$\mathrm{SU}(2) \ni m : \psi \to m\psi.$$

We will refer to this as the **spinor** representation of SU(2).

We have the following Hermitian inner product on the space of spinors

$$|\psi|^2 = \psi^\dagger \psi = |\alpha|^2 + |\beta|^2.$$

This inner product is invariant under the SU(2) action. In fact, it is invariant under a larger group U(2), but we are after SU(2) here.

Because $\mathbb{C}^2 = \mathbb{R}^4$ the space of unit $|\psi|^2 = 1$ spinors is nothing else but the three-sphere $S^3 \subset \mathbb{R}^4$. If we take a unit spinor $|\psi|^2 = 1$ and act on it with SU(2) it will remain a unit spinor. This means that the space of unit spinors is an orbit of SU(2). It is not hard to see that the space of unit spinors with $\beta \neq 0$ can be parametrised as

$$\psi = \frac{e^{\mathrm{i}\phi}}{\sqrt{1+|z|^2}} \begin{pmatrix} z \\ 1 \end{pmatrix}, \qquad \phi \in [0, 2\pi), z \in \mathbb{C}. \tag{1.64}$$

One can then check that the SU(2) action on unit spinors is transitive and without fixed points. This means that that the points in this orbit are in one-to-one correspondence with SU(2) group elements. This also means that the group SU(2) is isomorphic to S^3 as the manifold. Thus, we can view (1.64) as providing a set of coordinates on S^3.

The three-sphere S^3 is going to be the total space of the bundle we are about to construct. The other elements that we need to define a bundle is the projection map to the base. This is designed as follows. Consider the map from the space of unit spinors to $\mathbb{R}^3$ given by

$$\mathbf{v}_\psi = \psi^\dagger \sigma \psi.$$

Here $\sigma = (\sigma^1, \sigma^2, \sigma^3)$ are the usual Pauli matrices. A simple computation gives

$$\mathbf{v}_\psi = \left(\frac{2\mathrm{Re}(z)}{1+|z|^2}, \frac{-2\mathrm{Im}(z)}{1+|z|^2}, \frac{|z|^2-1}{1+|z|^2} \right).$$

In particular, it is clear that $|\mathbf{v}|^2 = 1$, and so this is a vector lying on the unit $S^2 \subset \mathbb{R}^3$. Thus, we have constructed a map

$$\pi : S^3 \to S^2, \qquad \pi : \mathbb{R}^4 \supset S^3 \ni \psi \to \mathbf{v}_\psi \in S^2 \subset \mathbb{R}^3.$$

Explicitly, we see that the vector $\mathbf{v}$ is independent of the coordinate ϕ. The projection map $\pi : S^3 \to S^2$ is the map that 'forgets' about the ϕ coordinate and maps the z coordinate to the corresponding point on S^2, with ϕ, z viewed as coordinates on S^3. Here we can both view the base space S^2 of fibration as a set of one-dimensional submanifolds parametrised by ϕ, or as the target of the projection map π.

1.13.2 Metric in the Total Space and the Corresponding Ehresmann Connection

The sphere S^3 comes with its round metric, which is the restriction of the flat metric in $\mathbb{R}^4$ to S^3. Viewing $\mathbb{R}^4 = \mathbb{C}^2$ the flat metric takes the form

$$ds^2 = |d\alpha|^2 + |d\beta|^2.$$

Using the parametrisation (1.64) we write

$$\alpha = \frac{z e^{\mathrm{i}\phi}}{\sqrt{1+|z|^2}}, \qquad \beta = \frac{e^{\mathrm{i}\phi}}{\sqrt{1+|z|^2}}. \tag{1.65}$$

In this parametrisation, the metric on S^3 evaluates to

$$ds^2 = \left(d\phi + \frac{\mathrm{i}}{2} \frac{z d\bar{z} - \bar{z} dz}{1+|z|^2} \right)^2 + \frac{|dz|^2}{(1+|z|^2)^2}. \tag{1.66}$$

This is a metric in the total space of the bundle, which is S^3 in our case. As we know, a (Riemannian) metric in the total space defines a connection. The corresponding horizontal distribution is metric orthogonal to the vertical one.

Vertical vector fields in this bundle are those of the form

$$X_V = a(\phi, z) \frac{\partial}{\partial \phi}. \tag{1.67}$$

Let us find vector fields that are metric orthogonal to vertical. We look for horizontal vector fields in the form

$$X = a \frac{\partial}{\partial \phi} + b \frac{\partial}{\partial z} + \bar{b} \frac{\partial}{\partial \bar{z}}, \tag{1.68}$$

where a is a real valued function, and b is a complex valued function in the total space. Taking the metric pairing of a vertical vector field $a'(\partial/\partial\phi)$ with what we want to be a horizontal one and setting the result to zero gives

$$a' \left(a + \frac{\mathrm{i}}{2} \frac{z\bar{b} - \bar{z}b}{1+|z|^2} \right) = 0. \tag{1.69}$$

We thus see that horizontal vector fields are those of the form

$$X_H = -\frac{\mathrm{i}}{2}\frac{z\bar{b}-\bar{z}b}{1+|z|^2}\frac{\partial}{\partial\phi} + b\frac{\partial}{\partial z} + \bar{b}\frac{\partial}{\partial \bar{z}}. \tag{1.70}$$

If we want these vector fields to be invariant with respect to the action of S^1 on the fibres, the function b must be ϕ-independent, i.e., function on S^2 only.

We can also see that the 1-form that encodes this horizontal subbundle is given by

$$\tilde{\omega} = d\phi + \frac{\mathrm{i}}{2}\frac{zd\bar{z}-\bar{z}dz}{1+|z|^2}, \tag{1.71}$$

with the horizontal vector fields being those that satisfy $\tilde{\omega}(X_H) = 0$. We note that the metric on the three-sphere can be written as

$$ds^2_{S^3} = \tilde{\omega}^2 + \frac{1}{4}ds^2_{S^2}, \qquad ds^2_{S^2} = \frac{4|dz|^2}{(1+|z|^2)^2}, \tag{1.72}$$

where we also identified the standard metric on the two-sphere of radius one.

1.13.3 Hopf Fibration as a Principal Bundle

Hopf fibration is also an example of a principal bundle. Indeed, the fibres of the fibration (S^3, S^2, π) are copies of $S^1 \sim \mathrm{U}(1)$. Let us parametrise the group manifold $\mathrm{U}(1)$ as

$$\mathrm{U}(1) \ni g = e^{\mathrm{i}\phi}. \tag{1.73}$$

We have this group acting on the total space of the bundle by $\psi \to g\psi$. It is clear that the projection map π commutes with the $\mathrm{U}(1)$ action because $\mathbf{v}_{g\psi} = \mathbf{v}_{\psi}$.

Let us also understand what connections in this principal bundle are. First, we note that for $b = b(z,\bar{z})$ the horizontal distribution (1.70) is ϕ-independent, and so is $\mathrm{U}(1)$-invariant. Thus, it gives rise to a connection in the principal bundle. To describe this connection in coordinates, we note that the canonical 1-form $g^{-1}dg$ on the group manifold $\mathrm{U}(1)$ is

$$g^{-1}dg = \mathrm{i}d\phi. \tag{1.74}$$

This makes it natural to identify the Lie algebra of $\mathrm{U}(1)$ with the vector space $\mathrm{i}\mathbb{R}$ of imaginary numbers. A general connection in this principal bundle is then of the form

$$\omega = g^{-1}dg + g^{-1}Ag = \mathrm{i}d\phi + A, \tag{1.75}$$

where A is some Lie algebra valued form on the base. In our case A is a pure imaginary 1-form on S^2 (locally). Comparing to (1.71) we see that there is a canonical geometric connection in the Hopf bundle given by $\omega = \mathrm{i}\tilde{\omega}$, with $\tilde{\omega}$

being real and given by (1.71). In other words, the canonical U(1) connection in the Hopf bundle is given by the following pure imaginary 1-form

$$\omega = \mathrm{i}d\phi + \frac{1}{2}\frac{\bar{z}dz - zd\bar{z}}{1+|z|^2}. \tag{1.76}$$

1.14 Vector Bundles

A vector bundle is a fibre bundle with an additional structure of a vector space for fibres. We follow C. H. Taubes' 2011 book, *Differential Geometry,* in this section.

1.14.1 Definition

A vector bundle (E, B, π) is a fibre bundle with the structure of a finite-dimensional (real) vector space for each fibre $\pi^{-1}(x), x \in B$, such that the trivialisation maps can be chosen to be linear maps $\phi : \pi^{-1}(U) \to U \times \mathbb{R}^k$. In other words, it is required that the trivialisation map from a fibre $\pi^{-1}(x)$ can be chosen to be a linear isomorphism between the vector spaces $\pi^{-1}(x)$ and $\mathbb{R}^k$. The dimension k is called the **rank** of the vector bundle. The simplest example of a vector bundle is the trivial bundle $B \times \mathbb{R}^k$.

As for any fibre bundle, there is a notion of sections of a vector bundle. A section is a map $\psi : B \to E$ that respects the projection $\pi(\psi(x)) = x$. Given that fibres of a vector bundle have the structure of a vector space, every vector bundle has a preferred section, the zero section.

1.14.2 Cocycles Definition

A vector bundle can also be given a cocycles definition, similar to what we saw for the principal bundles. To see how this comes about, let us consider the trivialising maps ϕ_U, ϕ_V over two different but overlapping neighbourhoods U, V. Over the overlap $U \cup V$ we have

$$\phi_U \circ \phi_V^{-1} : (U \cap V) \times \mathbb{R}^k \to (U \cap V) \times \mathbb{R}^k.$$

This map is linear and thus satisfies

$$\phi_U \circ \phi_V^{-1}(x, v) = (x, g_{UV} v)$$

for some $\mathrm{GL}(k, \mathbb{R})$ valued function $g_{UV} : U \cap V \to \mathrm{GL}(n, \mathbb{R})$. These are called the **transition functions** for the vector bundle. Similarly to what we had for principal bundles, these transition functions satisfy

$$g_{UU} = \mathbb{I}, \quad g_{VU} = g_{UV}^{-1}, \quad g_{UV} g_{VW} g_{WU} = \mathbb{I}.$$

In the opposite direction, given a set of trivial bundles $U \times \mathbb{R}^k$ over the neighbourhoods U, and a set of transition functions satisfying the previous cocycle properties, one can construct a vector bundle by identifying the points of the trivial bundles with the cocycles, similar to what was done in the principal bundle case.

1.14.3 The Tangent and Cotangent Bundles as Vector Bundles

It is not hard to see that what we have previously defined as the tangent bundle (even prior to the definition of the notion of a bundle) is an example of a vector bundle. Indeed, let us recall how this was defined. Given a pair of coordinate neighbourhoods U, V, there are two maps $\psi_U : U \to \mathbb{R}^n, \psi_V : V \to \mathbb{R}^n$. The transition function $\psi_{UV} = \psi_U \circ \psi_V^{-1}$ maps $\psi_U(U \cup V) \to \psi_V(U \cup V)$, both subsets of $\mathbb{R}^n$. The differential of this map ψ_{UV*} is a map from $\psi_U(U \cup V)$ to $\mathrm{GL}(n, \mathbb{R})$, and is the matrix of the Jacobian of the corresponding coordinate transformation. So, we set $g_{UV} = \psi_{UV*}$. The cocycle conditions are satisfied by the chain rule. The tangent space above each point is spanned by the coordinate vector fields $\partial_1, \ldots, \partial_n$, and the corresponding trivial bundles $TU = U \times \mathbb{R}^n$ are glued together by the transition functions g_{UV}, which follow from the chain rule. This shows that the tangent bundle over a manifold, as it was previously defined, is an example of a vector bundle. In fact, the definition of a general vector bundle could be motivated by the example of the tangent bundle.

In a similar vein, the cotangent bundle T^*M is as well an example of a vector bundle, with transition functions given by the inverse of those for the tangent bundle.

A section of the tangent bundle TM is called a vector field, and a section of the cotangent bundle T^*M is called a 1-form.

1.14.4 Structure Group

There is an extra structure that can be added to a fibre bundle, and this structure brings general fibre bundles closer to the principal and vector bundles. This is the structure of a group that acts on the fibres so that the matching between overlapping local trivialisation charts is a group transformation. Often this group is a part of the definition of the fibre bundle, see e.g., Nakahara (2003), and is called the **structure group** of the bundle. But, as we have seen in the section on fibre bundles, it is not necessary to have this structure and one can develop a meaningful theory even in its absence. The structure group is naturally present in the case of principal bundles, where the fibre is a group itself. Structure group is also present in the case of vector bundles, where it is, most generally, the general linear group $\mathrm{GL}(n, \mathbb{R})$. However, the structure group of a vector bundle may also be a subgroup of $\mathrm{GL}(n, \mathbb{R})$, as in the case of associated vector bundles.

1.14.5 From Vector Bundles to Principal Bundles

There are two general constructions that relate vector bundles with principal bundles. One such construction is from a vector bundle E, to the principal frame bundle FE, which is also often denoted by $P_{\mathrm{Gl}(E)}$. Its fibre is the space of frames of E, and given that any two frames can be mapped one into another by a $\mathrm{GL}(n, \mathbb{R})$ transformation, this space can be identified with a copy of the general linear group. In case there is a metric in the fibres of E, one can consider the space of orthonormal frames of E. This is a principal bundle whose fibre is the orthogonal group.

1.14.6 From Principal Bundles to Associated Vector Bundles

It is also possible to go in the opposite direction, starting from a principal G-bundle and defining the notion of the associated vector bundle. This is a vector bundle whose fibre is any linear representation of G. We thus first need to define the notion of a representation. Let V denote the vector space $\mathbb{R}^n$ or $\mathbb{C}^n$. Denote by $Gl(V)$ either $\mathrm{GL}(n, \mathbb{R})$ of $\mathrm{GL}(n, \mathbb{C})$. A representation ρ of G is a group homomorphism from G to $Gl(V)$, i.e., a map that sends the identity element in G into the identity matrix in $Gl(V)$, and is compatible with the group multiplication $\rho(gg') = \rho(g)\rho(g')$.

Let $\pi : P \to B$ with the principal G-bundle over base B. Then one defines a new bundle, denoted by $P \times_\rho V$, which is the quotient of $P \times V$ by the equivalence relation

$$(p, v) \sim (pg, \rho(g^{-1})v).$$

In other words, the group G acts on $P \times V$, and the space $P \times_\rho V$ is the space of orbits of this action.

To see that this is indeed a vector bundle, we need to specify the projection, as well as check the vector space properties of the fibre. First, the projection sends the equivalence class of (p, v) to $\pi(p)$. This is well-defined, and it is clear that the base of the new bundle is the same as base of the principal bundle. The multiplication by real or complex number λ sends the equivalence class of (p, v) to that of $(p, \lambda v)$. The zero section is the equivalence class of $(p, 0) \in P \times V$. We can also add sections of $P \times_\rho V$ by adding the corresponding equivalence classes. The trivialisation of the new bundle is obtained as follows. First, let $\phi : P|_U \to U \times G$ be a trivialisation of P over neighbourhood U. Denote by $\psi : P \to G$ the function that is the composition of ϕ with the projection onto the second factor. We can then define $\phi^V : (P \times_\rho V)|_U \to U \times V$ as the map that sends the equivalence class of (p, v) to $(\pi(p), \rho(\psi(p))v)$. This is well-defined because $(pg, \rho(g^{-1})v)$ gets sent to $(\pi(pg), \rho(\psi(pg))\rho(g^{-1})v) = (\pi(p), \rho(\psi(p))v)$ because $\rho(\psi(pg))\rho(g^{-1}) = \rho(\psi(pg)g^{-1}) = \rho(\psi(p))$. Finally, if g_{UV} are the principal bundle transition function arising as $\phi_U \circ \phi_V^{-1}$, then the associated bundle transitions functions are $\rho(g_{UV})$.

Example 1.70 This example shows how to recover the original vector bundle as the bundle associated with its bundle of frames. Thus, let E be a vector bundle with fibre V, and FE be its principal bundle of frames. Let ρ be the defining representation of $GL(n,\mathbb{R})$. Then the associated bundle $FE \times_\rho V$ is canonically isomorphic to E. To see this, we view a section of FE as the set of elements $(e_1, \ldots, e_n)$ that span the fibre of E at every point, i.e., a frame. We can then define a function $f : FE \times V \to E$ that sends $e = (e_1, \ldots, e_n)$ and $v \in V$ to $f(e,v) = v^i e_i$, with summation convention implied. This map is invariant with respect to the G action on $FE \times V$, and this shows that the quotient $FE \times_\rho V$ is sent by f to E.

What is important about the associated bundle construction is that, given a vector bundle E, all bundles obtained from E via various algebraic operations, such as $\otimes_n E, \Lambda^p E, \mathrm{Sym}^p(E)$, can be viewed as arising from the frame bundle FE via the associated bundle construction. Thus, all these vector bundles can be studied at once by focusing on the one principal frame bundle. Because of this, all tensor bundles arise as vector bundles associated to the frame bundle of a Riemannian manifold.

1.14.7 Covariant Derivatives

Given a vector bundle E, a **covariant derivative** is an operation that maps sections of the bundle to 1-form valued sections

$$\nabla : C^\infty(M;E) \to C^\infty(M; E \otimes T^*M). \tag{1.77}$$

Moreover, this map satisfies several properties. First, it respects the vector structure of the fibres $\nabla(s+s') = \nabla s + \nabla s'$. Second, it obeys the analog of the Leibnitz's rule

$$\nabla(fs) = f\nabla s + s \otimes df, \qquad \forall f \in C^\infty(M).$$

The simplest covariant derivative is a version of the exterior derivative, as can be seen in a trivialisation. Indeed, let $x \to (x, f_1(x), \ldots, f_n(x))$ be a section of the bundle $M \times \mathbb{R}^n$. Then

$$x \to (x, df_1, \ldots, df_n)$$

is a covariant derivative. It is also easy to verify that the space of covariant derivatives is an affine space, as the difference $\nabla - \nabla'$ of two covariant derivatives is a linear map. Thus, if $a \in \mathrm{Hom}(E) \otimes T^*M$ is a 1-form valued map of E into itself, then $\nabla + a$ is also a covariant derivative.

As an example one can consider the situation when $E \to M$ is a subbundle of the trivial bundle $M \times \mathbb{R}^n$. Let $\Pi \in \mathrm{Hom}(M \times \mathbb{R}^n; E)$ be the fibrewise orthogonal projection in $\mathbb{R}^n$ onto E. Then $\nabla s = \Pi ds$ is a covariant derivative.

For instance, one can consider the tangent bundle to the unit sphere S^n in $\mathbb{R}^{n+1}$. This is the set of points $|x|^2 = 1$ in $\mathbb{R}^{n+1}$, and the tangent space at every point is the set of vectors $v \in \mathbb{R}^{n+1}$ orthogonal to x, i.e., $x^T v = 0$. The orthogonal projection to the tangent space is $\Pi = \mathbb{I} - xx^T$, which is a $(n+1) \times (n+1)$

matrix. The covariant derivative is then $\nabla s = \Pi ds$. For example, let us consider a constant vector $e \in \mathbb{R}^{n+1}$. A section of the tangent bundle is obtained by projecting this vector to lie in TS^n, i.e., $s = \Pi e = e - x(x^T e)$. Its covariant derivative is $\nabla s = (xx^T - \mathbb{I})dx(x^T e)$, because $(xx^T - \mathbb{I})x = 0$.

1.14.8 Coordinate Expression for a Covariant Derivative

Let $e_i \in V$ be a basis of the vector space V. Then any section s of the vector bundle V is of the form $s^i(x)e_i$, where $s^i(x)$ are functions on the base. Using the Leibnitz rule satisfied by the covariant derivative, the covariant derivative of this section is

$$\nabla s = ds^i e_i + s^i \nabla e_i = (ds^i + A^i{}_j s^j)e_i, \tag{1.78}$$

where we introduced the 1-form valued **connection coefficients** $A^i{}_j$ defined via

$$\nabla e_i := A^j{}_i e_j. \tag{1.79}$$

We will often write (1.78) as

$$\nabla s^i = ds^i + A^i{}_j s^j. \tag{1.80}$$

1.14.9 Covariant Exterior Derivative and Curvature

We have introduced the covariant derivative as a map (1.77). This derivative admits an extension that acts on differential forms with values in E, and maps $C^\infty(M; E \otimes \Lambda^p T^* M) \to C^\infty(M; E \otimes \Lambda^{p+1} T^* M)$. This extension is called the **exterior covariant derivative**, is denoted by the symbol d_∇, and will play an important role in what follows. It is defined to satisfy the following rules: (i) If ω is a p-form and s is a section of E, then $d_\nabla(s\omega) = \nabla s \wedge \omega + s d\omega$; (ii) it is linear $d_\nabla(\omega_1 + \omega_2) = d_\nabla \omega_1 + d_\nabla \omega_2$.

While $d^2 = 0$, it is not in general true that $d_\nabla^2 = 0$. However, d_∇^2 defines a section of $C^\infty(M; \mathrm{End}(E) \otimes \Lambda^2 T^* M)$, i.e., a 2-form on the base with values in the space of endomorphisms of the fibre. This is because $d_\nabla^2 \mathfrak{w} = F_\nabla \wedge \mathfrak{w}$ for every differential form $\mathfrak{w}$ with values in E. Let us verify this. Consider the result of application of d_∇^2 to the product of a section s and a differential form ω. We have

$$\begin{aligned} d_\nabla^2(s\omega) &= d_\nabla(d_\nabla s \wedge \omega) + d_\nabla(s d\omega) \\ &= d_\nabla^2 s \wedge \omega - d_\nabla s \wedge d\omega + d_\nabla s \wedge d_\omega = (d_\nabla^2 s) \wedge \omega. \end{aligned}$$

Here we have used $d_\nabla \omega = d\omega$ and $d^2 = 0$. When $\omega = f$ is a function we see that $d_\nabla^2(sf) = (d_\nabla^2 s)f$, which implies that d_∇^2 is an algebraic operator, and so $d_\nabla^2 \mathfrak{w} = F_\nabla \wedge \mathfrak{w}$.

As we have see in (1.78), in a trivialisation every exterior covariant derivative is of the form

$$d_\nabla = d + A, \qquad A \in \mathrm{End}(E) \otimes T^* M. \tag{1.81}$$

A simple computation then shows that the corresponding curvature 2-form is

$$F_\nabla = dA + A \wedge A \in \mathrm{End}(E) \otimes \Lambda^2 T^* M. \tag{1.82}$$

1.14.10 Principal Connections and Covariant Derivatives

Let E be a vector bundle with fibre V that is associated with a principal G-bundle P via the construction $E = P \times_\rho V$, where ρ is some representation of G. Let ω be a principal connection on P. Then it defines a covariant derivative on E as follows. Let us view a section s of E as a G equivariant map $\tilde{s} : P \to V$. This is the map that makes the following diagram commute

$$\begin{array}{ccc} P & \xrightarrow{(\mathrm{id},\tilde{s})} & P \times V \\ {\scriptstyle\pi}\big\downarrow & & \big\downarrow{\scriptstyle\sim} \\ M & \xrightarrow{\quad s \quad} & E \end{array}$$

Concretely, $\tilde{s}$ is a vector valued function on P with the property $R_g^* \tilde{s} = \rho(g^{-1})\tilde{s}$. Every such function defines a section of $E = P \times_\rho V$. The (exterior) covariant derivative ∇s is defined to be the horizontal projection $(d\tilde{s})_H$ of $d\tilde{s}$. Alternatively, for every vector field $v \in TM$, let v_H be the horizontal lift to TP. We then define the covariant derivative $\nabla_v s$ of s to be the section of E that corresponds to the equivariant vector field $v_H \tilde{s}$.

Let us work all this out in a trivialisation. For simplicity, let us only consider a $\mathrm{GL}(n, R)$ principal bundle of frames and the associated bundle corresponding to the defining representation. Let e_i be a basis of V and $s = v^i(x) e_i$ be a section of E. We then define the following equivariant vector valued function on P

$$\tilde{s}^i = (g^{-1})^i{}_j s^j,$$

where $g_i{}^j$ are the coordinates along the fibres of P. This function is indeed equivariant in the sense of satisfying $R_h^* \tilde{s} = \rho(h^{-1})\tilde{s}$. We then apply the horizontal vector field (1.55) to this 'function', getting again an equivariant function of the same type

$$X_H((g^{-1})^i{}_j v^j(x)) = a^\mu (g^{-1})^i{}_j (\partial_\mu v^j + A^j_{\mu\,k} v^k).$$

This defines the associated covariant derivative as

$$\nabla_\mu v^i = \partial_\mu v^i + A^i_{\mu\,j} v^j.$$

1.15 Riemannian Geometry

In this section, we follow the book by M. Nakahara (2003), *Geometry, Topology and Physics*. Our coordinates now carry Greek indices, as is standard in the physics literature.

1.15.1 Affine Connection, Connection Coefficients

A covariant derivative in the tangent bundle is called an **affine connection**. Thus, it is a map $\nabla : TM \to TM \otimes T^*M$, satisfying the Leibnitz rule appropriate for vector bundle covariant derivatives. We can also view this covariant derivative as a map $\nabla : TM \times TM \to TM$ satisfying

$$\nabla_X(Y+Z) = \nabla_X Y + \nabla_X Z, \tag{1.83}$$

$$\nabla_{(X+Y)} Z = \nabla_X Z + \nabla_Y Z, \tag{1.84}$$

$$\nabla_{(fX)} Y = f\nabla_X Y, \tag{1.85}$$

$$\nabla_X(fY) = X(f)Y + f\nabla_X Y. \tag{1.86}$$

Affine connections can be characterised in coordinates as follows. Let $e_\mu = \partial/\partial x^\mu$ be the coordinate vector fields. Then define symbols $\Gamma^\lambda{}_{\nu\mu}$ via

$$\nabla_\mu e_\nu \equiv \nabla_{e_\mu} e_\nu = e_\lambda \Gamma^\lambda{}_{\nu\mu}. \tag{1.87}$$

The covariant derivative of an arbitrary vector field is then

$$\nabla_\mu(X^\nu e_\nu) = (\partial_\mu X^\nu) e_\nu + X^\nu \nabla_\mu e_\nu = (\partial_\mu X^\lambda + \Gamma^\lambda{}_{\nu\mu} X^\nu) e_\lambda. \tag{1.88}$$

In the physics literature, this formula is usually written omitting the basis vectors as

$$(\nabla_\mu X)^\lambda \equiv \nabla_\mu X^\lambda = \partial_\mu X^\lambda + \Gamma^\lambda{}_{\nu\mu} X^\nu, \tag{1.89}$$

where the meaning of $\nabla_\mu X^\lambda$ is that of the objects $(\nabla_\mu X)^\lambda$ that arise as the components in $\nabla_\mu X = (\nabla_\mu X)^\lambda e_\lambda$.

The introduced covariant derivative can be extended to arbitrary tensors by requiring

$$\nabla_X(T_1 \otimes T_2) = (\nabla_X T_1) \otimes T_2 + T_1 \otimes (\nabla_X T_2). \tag{1.90}$$

Moreover, this relation must also hold when some of the indices are contracted. This immediately allows to extend ∇ to a connection in the cotangent bundle T^*M. Indeed, we must have

$$X(\omega(Y)) = \nabla_X(\omega(Y)) = (\nabla_X \omega)(Y) + \omega(\nabla_X Y). \tag{1.91}$$

Writing everything in coordinates, this gives

$$\nabla_\mu \omega_\nu = \partial_\mu \omega_\nu - \omega_\alpha \Gamma^\alpha{}_{\nu\mu}. \tag{1.92}$$

1.15.2 Curvature and Torsion

Given an affine connection in TM, there are two natural tensors that can be defined. One of them is the curvature that exists for a covariant derivative d_∇. This exists in any vector bundle. The other is the torsion, and this exists only for a covariant derivative in T^*M, which the derivative in TM produces. The torsion can be defined as the difference between the exterior covariant derivative $d_\nabla \omega$ and $d\omega$, for a 1-form $\omega \in T^*M$.

In terms of the operator $\nabla : T^M \times T^M \to T^M$ that we introduced previously, the curvature and torsion have the following definitions

$$\begin{aligned} T(X,Y) &:= \nabla_X Y - \nabla_Y X - [X,Y], \\ R(X,Y,Z) &:= \nabla_X \nabla_Y Z - \nabla_Y \nabla_X Z - \nabla_{[X,Y]} Z. \end{aligned} \tag{1.93}$$

It can be checked that both operations are multi-linear in all the entries, and are thus tensors. We also note that one often writes

$$R(X,Y,Z) := R(X,Y)Z. \tag{1.94}$$

The tensor R is called **Riemann curvature tensor**. In components, one defines the following objects

$$T^\lambda{}_{\mu\nu} := dx^\lambda(T(e_\mu, e_\nu)) = \Gamma^\lambda{}_{\nu\mu} - \Gamma^\lambda{}_{\mu\nu} = -2\Gamma^\lambda_{[\mu\nu]}, \tag{1.95}$$

and

$$\begin{aligned} R^\kappa{}_{\lambda\mu\nu} &:= dx^\kappa(R(e_\mu, e_\nu)e_\lambda) \\ &= \partial_\mu \Gamma^\kappa{}_{\lambda\nu} - \partial_\nu \Gamma^\kappa{}_{\lambda\mu} + \Gamma^\eta{}_{\lambda\nu}\Gamma^\kappa{}_{\eta\mu} - \Gamma^\eta{}_{\lambda\mu}\Gamma^\kappa{}_{\eta\nu}. \end{aligned} \tag{1.96}$$

1.15.3 The Ricci Tensor and the Scalar Curvature

Given the curvature tensor of type $(1,3)$ one can perform a contraction to produce the **Ricci tensor** given by

$$Ric(X,Y) := dx^\mu(R(e_\mu, Y)X). \tag{1.97}$$

In components

$$Ric_{\mu\nu} = Ric(e_\mu, e_\nu) = R^\lambda{}_{\mu\lambda\nu}. \tag{1.98}$$

Contracting further with a metric produces the **scalar curvature**

$$R := g^{\mu\nu} Ric(e_\mu, e_\nu) = g^{\mu\nu} R_{\mu\nu}. \tag{1.99}$$

1.15.4 Levi–Civita Connection

Theorem 1.71 The fundamental theorem of (psuedo-) Riemannian geometry. *On a (pseudo-) Riemannian manifold (M,g) there exists a unique torsion free affine connection that is compatible with g, i.e., one with $\nabla_\mu g_{\lambda\kappa} = 0$. This connection is called the* **Levi–Civita** *connection.*

A proof is by an explicit computation. The connection coefficients of the Levi–Civita connection are called the Christoffel symbols, and are given by

$$\Gamma^\kappa{}_{\mu\nu} = \frac{1}{2} g^{\kappa\lambda}(\partial_\mu g_{\lambda\nu} + \partial_\nu g_{\lambda\mu} - \partial_\lambda g_{\mu\nu}). \tag{1.100}$$

1.15.5 Bianchi Identities

Let R be the Riemann tensor defined with respect to the Levi–Civita connection. Then the following identities hold

$$R(X,Y)Z + R(Z,X)Y + R(Y,Z)X = 0, \tag{1.101}$$
$$(\nabla_X R)(Y,Z)V + (\nabla_Y R)(Z,X)V + (\nabla_Z R)(X,Y)V = 0.$$

In components, these read

$$R^{\kappa}{}_{\lambda\mu\nu} + R^{\kappa}{}_{\mu\nu\lambda} + R^{\kappa}{}_{\nu\lambda\mu} = 0, \tag{1.102}$$
$$\nabla_\kappa R^{\xi}{}_{\lambda\mu\nu} + \nabla_\mu R^{\xi}{}_{\lambda\nu\kappa} + \nabla_\nu R^{\xi}{}_{\lambda\kappa\mu} = 0.$$

1.16 Spinors and Differential Forms

In this section we describe some basic facts about Clifford algebras. In particular, we explicitly describe the isomorphism between a Clifford algebra of a vector space and its exterior algebra. This gives a concrete realisation of any Clifford algebra as the exterior algebra equipped with a special product. The product is given by the difference of the exterior and interior products. This allows for a concrete and efficient description of modules of the Clifford algebra – spinors. Spinors will be seen to be special types of elements of the exterior algebra, i.e., special types of differential forms. We follow the book by F. Reese Harvey (1990), called *Spinors and Calibrations*, closely in this section.

1.16.1 Clifford Algebras

Let V be a vector space (over $\mathbb{R}$), and let $(\cdot,\cdot)$ be a symmetric bilinear form on V, i.e., a metric. The **Clifford algebra** $Cl(V)$ associated with this bilinear form is defined as the quotient of the tensor algebra $\sum_{r=0}^{\infty} \otimes^r V$ by the ideal generated by all elements of the form $v \otimes v - (v,v)1$. In other words, this is the algebra generated by V subject to the relations

$$v \cdot v = -(v,v)1, \tag{1.103}$$

or, in polarised form

$$v \cdot w + w \cdot v = -2(v,w)1. \tag{1.104}$$

We note that Clifford algebra generalises the notion of the exterior algebra $\Lambda^\bullet V$. Indeed, when the bilinear form in question is zero, the Clifford algebra reduces to the exterior algebra.

Even when the bilinear form in question is not zero, there is a relation between the Clifford algebra and the exterior algebra, as is described by the following theorem.

Theorem 1.72 *Clifford algebra $Cl(V)$ is isomorphic (as a vector space, not as an algebra) to the exterior algebra $\Lambda^\bullet V$. Moreover, under this isomorphism the Clifford product is explicitly described as*

$$x \cdot u = x \wedge u - i_x u, \tag{1.105}$$

where $\cdot$ denotes the Clifford product, $x \in V$ and $u \in \Lambda^\bullet V \cong Cl(V)$. The operation i_x is the interior product $i_x : \Lambda^\bullet V \to \Lambda^\bullet V$ defined as the adjoint of the wedge product under the bilinear form in question, i.e.,

$$(x \wedge u, v) = (u, i_x v), \qquad x \in V, \quad \forall u, v \in \Lambda^\bullet V. \tag{1.106}$$

To prove the first statement of the theorem one notes that each tensor can be expressed as a skew tensor modulo relations (1.104), and skew tensors generate $\Lambda^\bullet V$. To prove the second statement one computes

$$x \cdot (x \cdot u) = x \cdot (x \wedge u - i_x u) = -x \wedge i_x u - i_x(x \wedge u) = -(x, x)u.$$

This shows that indeed, V viewed as sitting inside $\Lambda^\bullet V$ gives rise to operators on $\Lambda^\bullet V$ that satisfy the Clifford algebra defining relations. The Clifford product is extended to arbitrary elements of $\Lambda^\bullet V$ by linearity. Indeed, let $v = v_1 \wedge \cdots \wedge v_p$ be an element of $\Lambda^\bullet V$. We can write this as a tensor

$$v = \frac{1}{p!} \sum_\sigma \mathrm{sign}_\sigma v_{\sigma(1)} \otimes \cdots \otimes v_{\sigma(p)}. \tag{1.107}$$

Then, if we rewrite the Clifford product (1.105) as

$$x \cdot u = (E_x - I_x)u, \tag{1.108}$$

where $E_x := x\wedge$ and $I_x := i_x$ are the exterior and interior products respectively, the Clifford algebra element corresponding to v becomes the following operator on $\Lambda^\bullet V$

$$v = \frac{1}{p!} \sum_\sigma \mathrm{sign}_\sigma (E_{v_{\sigma(1)}} - I_{v_{\sigma(1)}}) \circ \cdots \circ (E_{v_{\sigma(n)}} - I_{v_{\sigma(p)}}). \tag{1.109}$$

This gives an explicit description of the Clifford algebra $Cl(V)$ as the exterior algebra $\Lambda^\bullet V$ equipped with the product (1.109). In other words, if we denote the isomorphism $Cl(V) \cong \Lambda^\bullet V$ by ϕ, then an element $v \in \Lambda^\bullet V$ corresponds in $Cl(V)$ to an operator $\phi(v)$ given by the right-hand side of (1.109), and the Clifford product is $v \cdot u = \phi(v)u$.

1.16.2 The Groups Spin and Pin

Let $Cl^*(V)$ denote the multiplicative group of invertible elements of the Clifford algebra $Cl(V)$. This is the group generated by non-null vectors $u \in V$. For such vectors, the inverse is given by $u^{-1} = -u/(u, u)$, as follows from the Clifford algebra defining relation $u \cdot u = -(u, u)1$.

Definition 1.73 The Pin group is the subgroup of $Cl^*(V)$ generated by unit vectors in V.

Definition 1.74 The spin group is defined as the subgroup of even elements in Pin, i.e., Spin $=$ Pin $\cup\, Cl^{even}(V)$. Concretely

$$\text{Spin} = \{a \in Cl^*(V) : a = u_1 \dots u_{2r}, \quad u_j \in V, \quad |u_j| = 1\}.$$

The following construction relates the groups Pin and Spin to the orthogonal transformations of V. First, let us introduce an involution of $Cl(V)$ that is an identity on the space of even elements, and reverses the sign of odd elements. It is generated by $\tilde{x} = -x$ on V. With this in mind, we define the twisted adjoint representation of $Cl^*(V)$ on $Cl(V)$ via

$$\widetilde{\text{Ad}}_a x := \tilde{a} x a^{-1}. \tag{1.110}$$

We then have the following theorem.

Theorem 1.75 *The twisted adjoint action* (1.110) *of* $Cl^*(V)$ *on* $Cl(V)$ *is orthogonal. Moreover, this action gives rise to two isomorphisms*

$$\text{O}(V) = \text{Pin}/\mathbb{Z}_2, \qquad \text{SO}(V) = \text{Spin}/\mathbb{Z}_2. \tag{1.111}$$

A proof is given in Harvey's 1990 book, in chapter 10. In particular, this theorem says that the Spin group is a double cover of the special orthogonal group. A more explicit description of the Spin group is possible in terms of reflections, but we will not need it. See the book by Harvey (1990) for more details.

1.16.3 The Split Case $Cl(p,p)$

There is a general theory of classification of Clifford algebras depending on the dimension and the signature. See, e.g., the book by Harvey (1990), chapter 11. However, this theory can be circumvented in the split case, when an explicit description is possible in terms of the exterior algebra. Furthermore, this explicit description can be used to describe other signatures, by realising $Cl(r,s)$ with $r+s=2p$ as subalgebras of the complexification $Cl(p,p) \otimes_{\mathbb{R}} \mathbb{C}$. It is here that we will explicitly see that spinors are differential forms.

The representation of Clifford algebra acting on itself by the Clifford multiplication is not irreducible. An explicit and beautiful description of irreducible representations is available in the split signature case. Consider the space $\mathbb{R}^p$ of half the dimension. Let x^a be the Cartesian coordinates on this space. Consider the exterior algebra $\Lambda^\bullet(\mathbb{R}^p)$. The operation of multiplication by objects dx^a increases the degree of a differential form, while the operation of insertion of a vector field $\partial/\partial x^a$ lowers the degree. This suggests that we define

$$(a^a)^\dagger := dx^a, \qquad a_a := i_{\partial/\partial x^a}. \tag{1.112}$$

These **creation-annihilation** operators satisfy the following relations

$$\begin{aligned} (a^a)^\dagger (a^b)^\dagger = -(a^b)^\dagger (a^a)^\dagger, \qquad a_a a_b = -a_b a_a, \\ a_a (a^b)^\dagger + (a^b)^\dagger a_a = \delta_a^b. \end{aligned} \tag{1.113}$$

This is the Clifford algebra in dimension $2p$, which corresponds to the metric of spit signature (p,p). Indeed, it is generated by objects of the form $X = v + \theta, v \in \mathbb{R}^p, \theta \in \mathbb{R}^{p*}$, with the defining relations being

$$X_1X_2 + X_2X_1 = 2g(X_1, X_2)\mathbb{I}, \tag{1.114}$$

where the metric is

$$g(v_1 + \theta_1, v_2 + \theta_2) = \frac{1}{2}(\theta_1(v_2) + \theta_2(v_1)). \tag{1.115}$$

Introducing the linear combinations $V = (v + \theta)/2, U = (v - \theta)/2$ diagonalises the metric and gives $|v + \theta|^2 = |V|^2 - |U|^2$, which shows that this is a metric of signature (p, p).

This Clifford algebra admits a representation on differential forms from $\Lambda^\bullet(\mathbb{R}^{p*})$ defined by

$$(v + \theta)\omega = i_v\omega + \theta\omega. \tag{1.116}$$

This gives a representation of the Clifford algebra because

$$\begin{aligned}(v + \theta)(v + \theta)\omega &= i_v(i_v\omega + \theta\omega) + \theta(i_v\omega + \theta\omega) \\ &= i_v(\theta\omega) + \theta i_v\omega = (i_v\theta)\omega = g(v + \theta, v + \theta)\omega.\end{aligned} \tag{1.117}$$

This means that differential forms in dimension p, i.e., the space $\Lambda^\bullet(\mathbb{R}^p)$ is the space of *spinors* of the pseudo-orthogonal group $\mathrm{SO}(p,p)$. This is a fact of fundamental importance, and in particular gives one of the easiest ways to explicitly construct the spinor representations of many orthogonal groups. In particular, the Weyl representations of $\mathrm{SO}(p,p)$ are the spaces of even and odd degree differential forms. The Lie algebra $\mathfrak{so}(p,p)$ is realised in this formalism as the span of all operators quadratic in the creation-annihilation operators (1.112).

Example 1.76 Let us see explicitly how the spinor representations of $\mathrm{SO}(2,2)$ are differential forms in $\mathbb{R}^2$. To this end, we introduce a pair of creating annihilation operators $a_1, a_1^\dagger$ and $a_2, a_2^\dagger$, with the usual anti-commutation relations $a_i a_j^\dagger + a_j^\dagger a_i = \delta_{ij}$ and all other pairs anti-commuting. The Lie algebra of $\mathrm{SO}(2,2)$ is spanned by all elements of degree two in the Clifford algebra (commutators of γ-matrices in the physics terminology), and these are all the quadratic operators one can construct from $a_1, a_1^\dagger$ and $a_2, a_2^\dagger$.

Let us consider the following operators

$$H = a_1a_1^\dagger - a_2a_2^\dagger, \qquad E_+ = a_1a_2^\dagger, \qquad E_- = a_2a_1^\dagger. \tag{1.118}$$

It is easy to check that the following $\mathfrak{sl}(2)$ commutation relations hold

$$[E_+, E_-] = H, \qquad [H, E_\pm] = \pm 2E_\pm. \tag{1.119}$$

This gives us one copy of $\mathfrak{sl}(2)$ Lie algebra. One can form the second copy of $\mathfrak{sl}(2)$ in the following way

$$\bar{H} = a_1a_1^\dagger + a_2a_2^\dagger - 1 \equiv a_1a_1^\dagger - a_2^\dagger a_2, \qquad \bar{E}_+ = a_1a_2, \qquad \bar{E}_- = a_2^\dagger a_1^\dagger.$$

Again we get the usual $\mathfrak{sl}(2)$ commutation relations

$$[\bar{E}_+, \bar{E}_-] = \bar{H}, \qquad [\bar{H}, \bar{E}_\pm] = \pm 2\bar{E}_\pm. \tag{1.120}$$

Together, the six operators we have constructed span the Lie algebra $\mathfrak{so}(2,2)$. And it is not hard to check that all barred operators commute with unbarred ones, so we have two commuting copies of $\mathfrak{sl}(2)$. So we get an explicit realisation of the Lie algebra $\mathfrak{so}(2,2)$ as two commuting Lie algebras $\mathfrak{sl}(2,\mathbb{R})$.

Let us now discuss its action on spinors. The Weyl representations are formed by forms of even and odd degrees. The forms of odd degree are spanned by dx^1, dx^2. The action of the first copy of $\mathfrak{sl}(2)$ is as follows:

$$\begin{aligned} H dx^2 &= (a_1 a_1^\dagger - a_2 a_2^\dagger) dx^2 = dx^2, & H dx^1 &= (a_1 a_1^\dagger - a_2 a_2^\dagger) dx^1 = -dx^1, \\ E_- dx^2 &= a_2 a_1^\dagger dx^2 = -dx^1, & E_+ dx^1 &= a_1 a_2^\dagger dx^1 = -dx^2, \end{aligned}$$

while the second copy acts trivially on these states. So, the state dx^2 is the spin up, and dx^1 is the spin down state for the first copy of $\mathfrak{sl}(2)$.

The even degree forms are spanned by 1 and $dx^1 dx^2$. The first copy of $\mathfrak{sl}(2)$ acts trivially, while the action of the second copy is

$$\begin{aligned} \bar{H}\,1 &= (a_1 a_1^\dagger - a_2^\dagger a_2)\,1 = 1, & \bar{H} dx^1 dx^2 &= (a_1 a_1^\dagger - a_2^\dagger a_2) dx^1 dx^2 = -dx^1 dx^2, \\ \bar{E}_-\,1 &= a_2^\dagger a_1^\dagger\,1 = -dx^1 dx^2, & \bar{E}_+ dx^1 dx^2 &= a_1 a_2 dx^1 dx^2 = -1. \end{aligned}$$

Thus, the state 1 is the spin up, and $dx^1 dx^2$ is spin down for the second copy of $\mathfrak{sl}(2)$.

2

Metric and Related Formulations

In this chapter we describe the standard metric formulation of general relativity (GR). We attempt to be as concise as possible, covering in detail aspects of the formalism that are not already available in the standard sources.

2.1 Einstein–Hilbert Metric Formulation

This section covers the standard Einstein–Hilbert formulation.

2.1.1 Affine Connection and Riemann Curvature

For the convenience of the reader, we start by collecting all the useful formulas related to Riemannian geometry from the previous chapter. This fixes our conventions. For the covariant derivatives in TM, T^*M we have

$$\nabla_\mu v^\nu = \partial_\mu v^\nu + \Gamma^\nu{}_{\rho\mu} v^\rho, \qquad \nabla_\mu v_\nu = \partial_\mu v_\nu - \Gamma^\alpha{}_{\nu\mu} v_\alpha. \tag{2.1}$$

The torsion-free metric Christoffel connection components are given by

$$\Gamma^\rho{}_{\mu\nu} = \frac{1}{2} g^{\rho\sigma} \left(\partial_\mu g_{\nu\sigma} + \partial_\nu g_{\mu\sigma} - \partial_\sigma g_{\mu\nu} \right). \tag{2.2}$$

A simple, but very useful, consequence of this formula is obtained by contraction

$$\Gamma^\rho{}_{\mu\rho} = \frac{1}{2} g^{\rho\sigma} \left(\partial_\mu g_{\rho\sigma} + \partial_\rho g_{\mu\sigma} - \partial_\sigma g_{\mu\rho} \right) = \frac{1}{2} g^{\rho\sigma} \partial_\mu g_{\rho\sigma} = \frac{1}{\sqrt{-g}} \partial_\mu \sqrt{-g}. \tag{2.3}$$

The Riemann curvature tensor components are

$$R^\sigma{}_{\rho\mu\nu} = \partial_\mu \Gamma^\sigma{}_{\rho\nu} - \partial_\nu \Gamma^\sigma{}_{\rho\mu} + \Gamma^\alpha{}_{\rho\nu} \Gamma^\sigma{}_{\alpha\mu} - \Gamma^\alpha{}_{\rho\mu} \Gamma^\sigma{}_{\alpha\nu}. \tag{2.4}$$

We have

$$\begin{aligned} (\nabla_\mu \nabla_\nu - \nabla_\nu \nabla_\mu) v^\rho &= R^\rho{}_{\alpha\mu\nu} v^\alpha, \\ (\nabla_\mu \nabla_\nu - \nabla_\nu \nabla_\mu) v_\rho &= -R^\alpha{}_{\rho\mu\nu} v_\alpha. \end{aligned} \tag{2.5}$$

The contractions of the curvature tensor are

$$R_{\mu\nu} = R^{\lambda}{}_{\mu\lambda\nu}, \qquad R = g^{\mu\nu} R_{\mu\nu}. \tag{2.6}$$

The two Bianchi identities read

$$\begin{aligned} R^{\sigma}{}_{\rho\mu\nu} + R^{\sigma}{}_{\mu\nu\rho} + R^{\sigma}{}_{\nu\rho\mu} &= 0, \\ \nabla_{\alpha} R^{\sigma}{}_{\rho\mu\nu} + \nabla_{\mu} R^{\sigma}{}_{\rho\nu\alpha} + \nabla_{\nu} R^{\sigma}{}_{\rho\alpha\mu} &= 0. \end{aligned} \tag{2.7}$$

The differential Bianchi identity (2.7) can be contracted to produce another very useful consequence. Contracting the indices $\alpha\rho$ we get

$$\nabla^{\alpha} R_{\alpha\sigma\mu\nu} = \nabla_{\mu} R_{\nu\sigma} - \nabla_{\nu} R_{\mu\sigma}. \tag{2.8}$$

A further contraction produces

$$\nabla^{\alpha} R_{\alpha\mu} = \frac{1}{2} \nabla_{\mu} R. \tag{2.9}$$

2.1.2 Einstein–Hilbert Action

GR differs from all other physical theories in the fact that from the metric and its first derivatives it is impossible to built a (covariantly transforming under diffeomorphisms) scalar whose square could play the role of the Lagrangian density. Indeed, the components $\Gamma^{\rho}{}_{\mu\nu}$ of the affine connection that are built from the first derivatives of the metric can be made to vanish (at a point) by a choice of a coordinate system, and so no covariant scalar of the schematic form $\Gamma\Gamma$ can be constructed. The simplest scalar that arises in Riemannian geometry is the Ricci scalar, and this involves second derivatives of the metric. A Lagrangian linear in the Ricci scalar is then possible, and can lead to second-order field equations, as will be explicitly verified in Section 2.2.1. So, we write

$$S_{\text{EH}}[g] = \frac{1}{16\pi G} \int \sqrt{-g}(R - 2\Lambda), \tag{2.10}$$

where the coordinate volume element is omitted, g is the determinant of the metric, which we assume to have Lorentzian signature. The quantity G is the Newton's constant, and Λ is the cosmological constant. The latter can be set to zero if desired. The sign in front of the action is signature-dependent, and is fixed by an argument in Section 2.2.2. We use conventions in which metric signature is mostly plus.

2.1.3 Einstein Equations

Taking into account

$$\delta\sqrt{-g} = -\frac{1}{2} \sqrt{-g} g_{\mu\nu} \delta g^{\mu\nu} \tag{2.11}$$

we have the following expression for the first variation of the Einstein–Hilbert action

$$\delta S_{\mathrm{EH}} = \frac{1}{16\pi G}\int \sqrt{-g}\left(R_{\mu\nu} - \frac{1}{2}g_{\mu\nu}(R-2\Lambda)\right)\delta g^{\mu\nu}. \tag{2.12}$$

Vanishing of the expression in the brackets here is the vacuum Einstein equations (with nonzero Λ)

$$R_{\mu\nu} - \frac{1}{2}g_{\mu\nu}(R-2\Lambda) = 0. \tag{2.13}$$

Taking the trace of this equation, and assuming we are in four dimensions gives $R = 4\Lambda$ and $R_{\mu\nu} = \Lambda g_{\mu\nu}$. Metrics satisfying the latter equation are referred to as Einstein.

2.2 Gamma–Gamma Formulation

The following discussion follows Landau–Lifschitz's book (1987), *Field Theory*, chapter 93, closely.

2.2.1 Action Rewritten in Terms of Christoffel Symbols

To convince oneself that the previous variational principle leads to sensible second order field equations, one can rewrite the Einstein–Hilbert Lagrangian as a quantity of the type $\Gamma\Gamma$, plus a surface term. Schematically

$$\int \sqrt{-g}R = \int \sqrt{-g}\,\Gamma\Gamma + \int \partial_\mu(\sqrt{-g}w^\mu), \tag{2.14}$$

where w^μ is a vector field constructed from the metric and its first derivatives. The last term is a surface term, does not affect the extremisation problem that leads to the field equations, and can thus be ignored for the problem of deriving the latter. And the quantity $\Gamma\Gamma$ contains only the first derivatives of the metric. It will be explicitly obtained later in this subsection. This argument shows that the field equations obtained from the Einstein–Hilbert Lagrangian are sensible equations of the second order in derivatives.

Let us derive this $\Gamma\Gamma$ formulation explicitly. We have

$$\sqrt{-g}R = \sqrt{-g}g^{\mu\rho}\left(\partial_\nu\Gamma^\nu{}_{\mu\rho} - \partial_\mu\Gamma^\nu{}_{\nu\rho} + \Gamma^\alpha{}_{\mu\rho}\Gamma^\nu{}_{\alpha\nu} - \Gamma^\alpha{}_{\nu\rho}\Gamma^\nu{}_{\alpha\mu}\right). \tag{2.15}$$

Integrating by parts in the first two terms, and omitting the surface terms, we get

$$\begin{aligned}\sqrt{-g}g^{\mu\rho}\partial_\nu\Gamma^\nu{}_{\mu\rho} &\hat{=} -\partial_\nu(\sqrt{-g}g^{\mu\rho})\Gamma^\nu{}_{\mu\rho},\\ \sqrt{-g}g^{\mu\rho}\partial_\mu\Gamma^\nu{}_{\nu\rho} &\hat{=} -\partial_\mu(\sqrt{-g}g^{\mu\rho})\Gamma^\nu{}_{\nu\rho},\end{aligned} \tag{2.16}$$

where $\hat{=}$ means modulo surface terms. The quantity $\partial_\nu(\sqrt{-g}g^{\mu\rho})$ can be rewritten in terms of the Christoffel symbol and the metric. Indeed, we have

$$\partial_\nu\sqrt{-g} = \sqrt{-g}\Gamma^\rho{}_{\nu\rho}, \quad \partial_\nu g^{\mu\rho} = -\Gamma^\mu{}_{\nu\sigma}g^{\sigma\rho} - \Gamma^\rho{}_{\nu\sigma}g^{\mu\sigma}. \tag{2.17}$$

The last relation is just a rewriting of the fact that the covariant derivative of $g^{\mu\rho}$ vanishes. Overall, we have

$$\sqrt{-g}R\hat{=}\sqrt{-g}\Big(-\Gamma^{\sigma}{}_{\nu\sigma}g^{\mu\rho}\Gamma^{\nu}{}_{\mu\rho}+(\Gamma^{\mu}{}_{\nu\sigma}g^{\sigma\rho}+\Gamma^{\rho}{}_{\nu\sigma}g^{\mu\sigma})\Gamma^{\nu}{}_{\mu\rho}$$
$$+\Gamma^{\sigma}{}_{\mu\sigma}g^{\mu\rho}\Gamma^{\nu}{}_{\nu\rho}-(\Gamma^{\mu}{}_{\mu\sigma}g^{\sigma\rho}+\Gamma^{\rho}{}_{\mu\sigma}g^{\mu\sigma})\Gamma^{\nu}{}_{\nu\rho}+g^{\mu\rho}(\Gamma^{\alpha}{}_{\mu\rho}\Gamma^{\nu}{}_{\alpha\nu}-\Gamma^{\alpha}{}_{\nu\rho}\Gamma^{\nu}{}_{\alpha\mu})\Big).$$

Taking into account the arising cancellations we have

$$\sqrt{-g}R\hat{=}\sqrt{-g}g^{\rho\sigma}(\Gamma^{\mu}{}_{\nu\rho}\Gamma^{\nu}{}_{\mu\sigma}-\Gamma^{\mu}{}_{\rho\sigma}\Gamma^{\nu}{}_{\nu\mu}). \tag{2.18}$$

This shows that an action that is manifestly quadratic in first derivatives of the metric is possible, at the expense of this action not having manifest transformation properties as far as diffeomorphisms are concerned. The action reads

$$S_{\Gamma\Gamma}[g]=\frac{1}{16\pi G}\int\sqrt{-g}\,(g^{\rho\sigma}(\Gamma^{\mu}{}_{\nu\rho}\Gamma^{\nu}{}_{\mu\sigma}-\Gamma^{\mu}{}_{\rho\sigma}\Gamma^{\nu}{}_{\nu\mu})-2\Lambda)\,. \tag{2.19}$$

2.2.2 Fixing the Sign in Front of the Action

The formulation (2.19) is a convenient starting point for the analysis that fixes the sign in front of the action. Our desire is to have the Lagrangian given by kinetic minus potential energy. The kinetic energy term is the one involving the time derivatives. It is most convenient to perform the analysis by fixing a gauge. We thus set the mixed temporal-spatial components of the metric to zero $g_{0i}=0$. We only keep the time derivatives. The only terms surviving in the kinetic term are then

$$-\frac{1}{4}g^{00}\dot{g}_{ij}g^{ik}g^{jl}\dot{g}_{kl}+\frac{1}{4}g^{00}(g^{ij}\dot{g}_{ij})^2. \tag{2.20}$$

The last term here can be set to zero by fixing the gauge in which the determinant of the spatial metric is constant. This leaves the first term, which we write as

$$-\frac{1}{4}g^{00}(\dot{g}_{ij})^2. \tag{2.21}$$

This is nonnegative in the mostly plus signature $(-,+,+,+)$, which fixes the sign in front of the Einstein–Hilbert action for this signature. The action would require a minus sign in front in the mostly minus signature.

2.2.3 Lagrangian in Terms of the Metric

We can further rewrite the Lagrangian (2.19) by substituting in it the explicit expression for the Christoffel symbol in terms of the derivatives of the metric. An explicit calculation gives the following identities

$$\sqrt{-g}\,g^{\rho\sigma}\Gamma^{\mu}{}_{\nu\rho}\Gamma^{\nu}{}_{\mu\sigma}=\sqrt{-g}\,\partial_{\mu}g^{\rho\alpha}\partial_{\nu}g_{\sigma\alpha}\left(\frac{1}{4}g^{\mu\nu}\delta^{\sigma}_{\rho}-\frac{1}{2}g^{\mu\sigma}\delta^{\nu}_{\rho}\right),$$

where we have used $g^{\mu\alpha}g^{\nu\beta}\partial_\rho g_{\alpha\beta} = -\partial_\rho g^{\mu\nu}$, and

$$\sqrt{-g}\, g^{\rho\sigma}\Gamma^\mu{}_{\rho\sigma}\Gamma^\nu{}_{\nu\mu} = -\sqrt{-g}\left(\partial_\nu g^{\mu\nu}\partial_\mu \ln(\sqrt{-g}) + g^{\mu\nu}\partial_\mu \ln(\sqrt{-g})\partial_\nu \ln(\sqrt{-g})\right),$$

where we used $\Gamma^\nu{}_{\nu\mu} = \partial_\mu(\ln\sqrt{-g})$. Integrating by parts in the first term in the previous expression we have

$$\int \sqrt{-g}\, g^{\rho\sigma}\Gamma^\mu{}_{\rho\sigma}\Gamma^\nu{}_{\nu\mu} = \int \sqrt{-g} g^{\mu\nu}\partial_\mu\partial_\nu(\ln\sqrt{-g}).$$

Overall, neglecting a surface term, the action (2.19) written in terms of the metric becomes

$$S[g] = \frac{1}{16\pi G}\int\sqrt{-g}\left[\partial_\mu g^{\rho\alpha}\partial_\nu g_{\sigma\alpha}\left(\frac{1}{4}g^{\mu\nu}\delta^\sigma_\rho - \frac{1}{2}g^{\mu\sigma}\delta^\nu_\rho\right) - g^{\mu\nu}\partial_\mu\partial_\nu(\ln\sqrt{-g}) - 2\Lambda\right]. \tag{2.22}$$

2.2.4 Linearisation on the Minkowski Background

The action (2.22) is a possible starting point for the gravitational perturbation theory. Thus, let us set $\Lambda = 0$ and consider $g_{\mu\nu} = \eta_{\mu\nu} + h_{\mu\nu}$, where $\eta_{\mu\nu}$ is the (constant) Minkowski metric. If we are to keep only the terms of second order in $h_{\mu\nu}$ in the action, then in the first term in (2.22) only the terms involving the derivative can depend on $h_{\mu\nu}$. All other occurrences of the metric must be replaced with the Minkowski metric. In the second term in (2.22), we integrate by parts and then use

$$\delta\ln\sqrt{-g} = \frac{h}{2}, \qquad \delta\partial_\mu(\sqrt{-g}g^{\mu\nu}) = \frac{1}{2}\partial_\mu h\,\eta^{\mu\nu} - \partial_\mu h^{\mu\nu}. \tag{2.23}$$

Here $\delta g_{\mu\nu} = h_{\mu\nu}, h = \eta^{\mu\nu}h_{\mu\nu}$, and we used $\delta g^{\mu\nu} = -h^{\mu\nu}$.

The Einstein–Hilbert action to second order in the expansion in $\eta_{\mu\nu}$ is then

$$S^{(2)}_{\rm EH}[h] = \frac{1}{32\pi G}\int\left[\frac{1}{2}h^{\mu\nu}\Box h_{\mu\nu} + \partial_\mu h^{\mu\rho}\partial^\nu h_{\nu\rho} - \partial_\mu h^{\mu\nu}\partial_\nu h - \frac{1}{2}h\Box h\right], \tag{2.24}$$

where and $\Box := \partial^\mu\partial_\mu$. The first term here is the kinetic term for the gravitons, while all other terms can be set to zero by a choice of gauge. This computation of the linearised action can also serve to fix the sign in front of the action. Indeed, in the mostly plus signature the kinetic term for a scalar ϕ is $-(\partial_\mu\phi)^2$, which is $\phi\Box\phi$ neglecting a surface term. It can be checked that $S^{(2)}_{\rm EH}[h]$ is invariant (modulo surface terms) under the linearised diffeomorphisms

$$\delta_\xi h_{\mu\nu} = \partial_\mu\xi_\nu + \partial_\nu\xi_\mu. \tag{2.25}$$

2.2.5 Inverse Densitiesed Metric as an Independent Variable

A further rewriting of the action (2.22) is possible in the important case $\Lambda = 0$. When this is the case we can introduce the combinations

$$\sqrt{-g}g^{\mu\nu} = \sigma^{\mu\nu}, \qquad \sigma_{\mu\nu} = \frac{1}{\sqrt{-g}}g_{\mu\nu}, \tag{2.26}$$

so that $\sigma^{\mu\rho}\sigma_{\nu\rho} = \delta^\mu_\nu$. The fact that this is a natural variable for GR directly follows from the first-order Palatini formalism that will be described in Section 2.4. In terms of these variables the Lagrangian becomes

$$S[\sigma] = \frac{1}{16\pi G}\int \Big[\partial_\mu\sigma^{\rho\alpha}\partial_\nu\sigma_{\sigma\alpha}\left(\frac{1}{4}\sigma^{\mu\nu}\delta^\sigma_\rho - \frac{1}{2}\sigma^{\mu\sigma}\delta^\nu_\rho\right) + \frac{1}{2}\sigma^{\mu\nu}\omega_\mu\omega_\nu\Big], \tag{2.27}$$

where

$$\omega_\mu = \partial_\mu(\ln\sqrt{-g}) = \frac{1}{2}\sigma_{\alpha\beta}\partial_\mu\sigma^{\alpha\beta}. \tag{2.28}$$

All the formulas are valid in four dimensions. The action (2.27) was in particular derived in Cheung and Remmen (2017), where it is argued to be a convenient starting point for the flat space gravitational perturbation theory. If one sets $\sigma^{\mu\nu} = \eta^{\mu\nu} - h^{\mu\nu}$ then its inverse is given by the simple geometric series $\sigma_{\mu\nu} = \eta_{\mu\nu} + h_{\mu\nu} + h^2_{\mu\nu} + \cdots$. This produces the most economic known perturbative expansion of the $\Lambda = 0$ Einstein–Hilbert action. Every order of this expansion contains the same number of terms, after the last term in (2.27) is taken care of by a gauge-fixing, see Cheung and Remmen (2017). This should be compared to the perturbative expansion of the Einstein–Hilbert action in the usual variables $g_{\mu\nu} = \eta_{\mu\nu} + h_{\mu\nu}$. The expansion in the Einstein–Hilbert case contains a rapidly increasing number of terms at every order, see, e.g., the appendix of the paper Goroff and Sagnotti (1986) for the expansion up to quartic order in $h_{\mu\nu}$. The arising quartic order Lagrangian occupies half a page.

2.3 Linearisation

The purpose of this section is to obtain formulas for linearisation of the Einstein–Hilbert action around an arbitrary background. This helps to understand what type of kinetic operator for gravitational perturbations arises in the metric formulation. In perturbation theory, the GR Lagrangian expands to produce a series of Lagrangians, one for every order in the perturbation. We want the terms in the Lagrangian at each order to involve covariant derivatives with respect to the background connection. For this purpose the covariant Einstein–Hilbert action is a better starting point for perturbation theory than (2.22).

2.3.1 Linearisation of the Connection and Curvature

We have the following formula for the linearisation of the Christoffel symbol

$$\delta\Gamma^\rho{}_{\mu\nu} = \frac{1}{2}g^{\rho\sigma}\left(\nabla_\mu\delta g_{\nu\sigma} + \nabla_\nu\delta g_{\mu\sigma} - \nabla_\sigma\delta g_{\mu\nu}\right). \tag{2.29}$$

Here $g_{\mu\nu}$ is the background metric, and ∇ is the covariant derivative with respect to the background metric. This formula is checked by explicit verification.

For the linearisation of the Riemann curvature we have

$$\delta R^\sigma{}_{\rho\mu\nu} = \nabla_\mu\delta\Gamma^\sigma_{\rho\nu} - \nabla_\nu\delta\Gamma^\sigma_{\rho\mu}. \tag{2.30}$$

Together with (2.29) this implies

$$\delta R_{\mu\nu} = \frac{1}{2} g^{\rho\sigma} \left(\nabla_\rho \nabla_\mu \delta g_{\nu\sigma} + \nabla_\rho \nabla_\nu \delta g_{\mu\sigma} - \nabla_\rho \nabla_\sigma \delta g_{\mu\nu} - \nabla_\nu \nabla_\mu \delta g_{\rho\sigma} \right).$$

The last term here is not manifestly $\mu\nu$ symmetric, but its antisymmetric part is a multiple of $R^{\rho\sigma}{}_{\mu\nu} \delta g_{\rho\sigma}$ and so vanishes.

2.3.2 Lichnerowicz Laplacian

The Lichnerowicz Laplacian $\Delta^{(2)}$ on $(0,2)$ tensors is defined as

$$\Delta^{(2)} h_{\mu\nu} := -\nabla^\alpha \nabla_\alpha h_{\mu\nu} - 2R_{\mu\rho\nu\sigma} h^{\rho\sigma} + R_\mu{}^\rho h_{\rho\nu} + R_\nu{}^\rho h_{\rho\mu}. \tag{2.31}$$

The reason for this particular combination of the usual Laplacian corrected with curvature dependent terms becomes clear if we introduce two further Laplacians, one on vectors and one on functions

$$\begin{aligned} \Delta^{(1)} \xi_\mu &:= -\nabla^\alpha \nabla_\alpha \xi_\mu + R_\mu{}^\alpha \xi_\alpha, \\ \Delta^{(0)} \phi &:= -\nabla^\alpha \nabla_\alpha \phi. \end{aligned} \tag{2.32}$$

We then have

$$\begin{aligned} \Delta^{(1)} \nabla_\mu \phi &= -\nabla^\alpha \nabla_\alpha \nabla_\mu \phi + R_{\mu\alpha} \nabla^\alpha \phi \\ &= -\nabla_\mu \nabla^\alpha \nabla_\alpha \phi + R^\beta{}_\alpha{}^\alpha{}_\mu \nabla_\beta \phi + R_{\mu\alpha} \nabla^\alpha \phi = -\nabla_\mu \nabla^\alpha \nabla_\alpha \phi = \nabla_\mu \Delta^{(0)} \phi. \end{aligned}$$

Thus, one can either first take the gradient of a function to produce a 1-form and then apply the Laplacian $\Delta^{(1)}$, or first apply the Laplacian to the scalar and then take the gradient. These operations commute. One can rephrase this by saying that the Laplacians $\Delta^{(0)}$, $\Delta^{(1)}$ are intertwined by the operator ∇_μ mapping scalars into $(0,1)$ tensors.

The Lichnerowicz Laplacian on $(0,2)$ tensors is introduced with similar idea in mind. Thus, we can apply the Laplacian $\Delta^{(1)}$ to a vector ξ_ν, and then take the covariant derivative, and symmetrise to produce a symmetric $(0,2)$ tensor. This is the same as applying the Laplacian $\Delta^{(2)}$ to $\nabla_\mu \xi_\nu + \nabla_\nu \xi_\mu$, modulo some curvature dependent terms that vanish when the background is Einstein. Thus, we have the following identity

$$\Delta^{(2)}(\nabla_\mu \xi_\nu + \nabla_\nu \xi_\mu) = \nabla_\mu \Delta^{(1)} \xi_\nu + \nabla_\nu \Delta^{(1)} \xi_\mu + 2(\nabla_\alpha R_{\mu\nu} - \nabla_\mu R_{\nu\alpha} - \nabla_\nu R_{\mu\alpha}) \xi^\alpha. \tag{2.33}$$

It is proved by explicit verification. The terms involving the curvature vanish when the background is Einstein $R_{\mu\nu} = \Lambda g_{\mu\nu}$, and so we have the desired intertwining property of $\Delta^{(1)}$, $\Delta^{(2)}$. It should be kept in mind, however, that this now holds only on Einstein backgrounds, unlike the intertwining property of $\Delta^{(0)}$, $\Delta^{(1)}$ that is true in general.

2.3.3 Linearised Ricci and Lichnerowicz Operator

Let us now rewrite the Linearised Ricci tensor in terms of the Lichnerowicz operator just introduced. Denoting $\delta g_{\mu\nu} := h_{\mu\nu}, g^{\mu\nu}h_{\mu\nu} = h$ and $\delta R_{\mu\nu} = R_{\mu\nu}(h)$ we have

$$2R_{\mu\nu}(h) = -\nabla^\alpha\nabla_\alpha h_{\mu\nu} + \nabla^\alpha\nabla_\mu h_{\nu\alpha} + \nabla^\alpha\nabla_\nu h_{\mu\alpha} - \nabla_\mu\nabla_\nu h. \tag{2.34}$$

In the second term here we can commute the covariant derivatives to write it as

$$\begin{aligned}\nabla^\alpha\nabla_\mu h_{\nu\alpha} &= \nabla_\mu\nabla^\alpha h_{\nu\alpha} - R^\beta{}_\nu{}^\alpha{}_\mu h_{\beta\alpha} - R^\beta{}_\alpha{}^\alpha{}_\mu h_{\nu\beta} \\ &= \nabla_\mu(\delta h)_\nu - R_{\alpha\mu\beta\nu}h^{\alpha\beta} + R_\mu{}^\alpha h_{\alpha\nu},\end{aligned} \tag{2.35}$$

where we have introduced an operator

$$(\delta h)_\mu := \nabla^\nu h_{\mu\nu} \tag{2.36}$$

that maps symmetric $(0,2)$ tensors into $(0,1)$ tensors. Thus, we have

$$\nabla^\alpha\nabla_\mu h_{\nu\alpha} + \nabla^\alpha\nabla_\nu h_{\mu\alpha} = \nabla_\mu(\delta h)_\nu + \nabla_\nu(\delta h)_\mu - 2R_{\alpha\mu\beta\nu}h^{\alpha\beta} + R_\mu{}^\alpha h_{\alpha\nu} + R_\nu{}^\alpha h_{\alpha\mu}. \tag{2.37}$$

We already recognise the curvature terms appearing in the Lichnerowicz Laplacian. Overall, we have

$$2R_{\mu\nu}(h) = \Delta^{(2)}h_{\mu\nu} + \nabla_\mu(\delta h)_\nu + \nabla_\nu(\delta h)_\mu - \nabla_\mu\nabla_\nu h. \tag{2.38}$$

Thus, on traceless $h = 0$ and transverse $(\delta h)_\mu = 0$ tensors the linearised Ricci is (half) of the Lichnerowicz operator.

Now using

$$\nabla^\alpha(\nabla_\alpha\xi_\mu + \nabla_\mu\xi_\alpha) = \nabla^\alpha\nabla_\alpha\xi_\mu + R_\mu{}^\alpha\xi_\alpha = -\Delta^{(1)}\xi_\mu + 2R_\mu{}^\alpha\xi_\alpha \tag{2.39}$$

and the intertwining property (2.33) together with the assumption that the background is Einstein $R_{\mu\nu} = \Lambda g_{\mu\nu}$ we see that

$$R_{\mu\nu}(\nabla\xi + \nabla\xi) = \Lambda(\nabla_\mu\xi_\nu + \nabla_\nu\xi_\mu), \tag{2.40}$$

which verifies that the linearised Einstein equations are automatically satisfied by metric perturbations that are pure diffeomorphisms.

2.3.4 Second Variation of the Einstein–Hilbert Action

Applying to (2.12) the variation the second time we get

$$\begin{aligned}\delta^2 S_{\text{EH}} = \frac{1}{16\pi G}\int & \sqrt{-g}\,\delta g^{\mu\nu}\left(\delta R_{\mu\nu} - \frac{1}{2}g_{\mu\nu}g^{\rho\sigma}\delta R_{\rho\sigma}\right) \\ & - \frac{\sqrt{-g}}{2}\delta g^{\mu\nu}\delta g^{\rho\sigma}\left(g_{\mu\nu}R_{\rho\sigma} + R_{\mu\nu}g_{\rho\sigma} - \frac{1}{2}(R-2\Lambda)(2g_{\mu\rho}g_{\nu\sigma} + g_{\mu\nu}g_{\rho\sigma})\right).\end{aligned}$$

On an Einstein background in four dimensions $R_{\mu\nu} = \Lambda g_{\mu\nu}$. Let us also divide the second variation by two to get the second-order action, and replace $\delta g_{\mu\nu} = h_{\mu\nu}, g^{\mu\nu}h_{\mu\nu} = h$. We also use $\delta g^{\mu\nu} = -h^{\mu\nu}$. We get

$$S^{(2)}[h] = \frac{1}{32\pi G}\int \sqrt{-g}\left[-h^{\mu\nu}R_{\mu\nu}(h) + \frac{1}{2}hR(h) + \Lambda(h^{\mu\nu}h_{\mu\nu} - \frac{h^2}{2})\right],$$

where $R(h) = g^{\mu\nu}R_{\mu\nu}(h)$. Let us rewrite this action in terms of the metric. We have

$$R(h) = \Delta^{(0)}h + \nabla^\mu\nabla^\nu h_{\mu\nu}. \tag{2.41}$$

This gives

$$S^{(2)}[h] = \frac{1}{32\pi G}\int \sqrt{-g}\left[-\frac{1}{2}h^{\mu\nu}\Delta^{(2)}h_{\mu\nu} + (\delta h)^\mu(\delta h)_\mu + h\nabla^\mu\nabla^\nu h_{\mu\nu}\right. \tag{2.42}$$
$$\left. + \frac{1}{2}h\Delta^{(0)}h + \Lambda(h^{\mu\nu}h_{\mu\nu} - \frac{h^2}{2})\right].$$

We note that the kinetic terms here are just the covariantisations of those appearing in the flat space second-order Lagrangian (2.24), with, importantly, the flat space Laplacian on rank two tensors being replaced by the Lichnerowicz Laplacian.

2.4 First-Order Palatini Formulation

In the first-order formulation one introduces an independent connection field into the game, to convert the Lagrangian into first order in derivatives form. The Lagrangian is

$$S_{\text{Palatini}}[g,\Gamma] = \frac{1}{16\pi G}\int d^Dx\sqrt{-g}\,(g^{\mu\nu}R_{\mu\nu}(\Gamma) - 2\Lambda)\,. \tag{2.43}$$

The Ricci tensor present in (2.43) is formed out of the Riemann curvature $R_{\mu\nu}(\Gamma) := R^\sigma{}_{\mu\sigma\nu}$. In Palatini formalism the affine connection is assumed to be torsion-free, i.e., to satisfy the symmetry

$$\Gamma^\mu_{\nu\rho} = \Gamma^\mu_{\rho\nu}. \tag{2.44}$$

The Ricci curvature $R_{\mu\nu}(\Gamma)$ is not automatically symmetric, but the symmetric part is selected in (2.43) when $R_{\mu\nu}$ gets contracted with the symmetric metric.

Variation of (2.43) with respect to the affine connection gives an equation that implies that $\nabla_\rho g^{\mu\nu} = 0$, i.e., that the connection is metric-compatible. The solution to this equation is the usual expression (2.2) for Γ in terms of the derivatives of the metric. Substituting this solution into the action one gets back the second-order Einstein–Hilbert action.

We also note that in the case $\Lambda = 0$, if one views $\sqrt{-g}g^{\mu\nu}$ as the basic variable of the theory, the action (2.43) is cubic in the fields. This has been emphasised by Deser (1970), who used this cubic formulation to reconstruct GR from the linear Fierz–Pauli theory and hence prove its uniqueness. The inverse densitiesed metric has already been used in (2.27) to rewrite the action of GR in a way that is minimally nonlinear. The action (2.27) can be obtained from (2.43) by integrating out the Γ field, as was explicitly shown in Cheung and Remmen (2017). This reference also derives Feynman rules for GR in the cubic first-order formulation (2.43).

2.5 Eddington–Schrödinger Affine Formulation

Instead of "integrating out" from (2.43) the affine connection to get back the Einstein–Hilbert action one can integrate out the metric field. Indeed, varying the Palatini action with respect to the metric one gets an equation that is trivially solved

$$g_{\mu\nu} = \frac{1}{\Lambda} R_{(\mu\nu)}(\Gamma). \tag{2.45}$$

This can then be substituted into the action to get a second-order pure affine formulation

$$S_{\mathrm{ES}}[\Gamma] = \frac{1}{8\pi G\Lambda} \int d^4x \sqrt{-\det(R_{(\mu\nu)}(\Gamma))}. \tag{2.46}$$

The trick of integrating out the metric is possible in any dimension, and we wrote the four-dimensional version here. The field equation that results by varying this action with respect to the connection implies that the metric defined in (2.45) is compatible with the connection. The definition of the metric (2.45) then becomes the Einstein equation. We note that this purely affine formulation is only available with a nonzero cosmological constant. Note also that (in any dimension) the coefficient in front of the Eddington–Schrödinger action is dimensionless. In four dimensions we have $(G\Lambda)^{-1} \sim 10^{120}$, a very large number.

While the action (2.46) appears to be a natural construct, the pure affine formalism brings with it arbitrariness that is not present in the metric formalism. This has been emphasised in particular by Pauli, see Goenner (2014), section 8.2. Thus, the tensor $R_{\mu\nu}$ is not automatically symmetric even for a symmetric affine connection. It can be split into its symmetric and antisymmetric parts, and these can be separately used in constructing the Lagrangian. The elementary building blocks are then

$$L_0 = \sqrt{-\det(R_{(\mu\nu)}(\Gamma))}, \quad L_1 = \sqrt{\tilde{\epsilon}^{\mu\nu\rho\sigma}\tilde{\epsilon}^{\alpha\beta\gamma\delta} R_{(\mu\alpha)} R_{(\nu\beta)} R_{[\rho\gamma]} R_{[\sigma\delta]}},$$
$$L_2 = \tilde{\epsilon}^{\mu\nu\rho\sigma} R_{[\mu\nu]} R_{[\rho\sigma]},$$

where $\tilde{\epsilon}^{\mu\nu\rho\sigma}$ is the densitized antisymmetric tensor that exists without any background structure on the manifold. The previous blocks are all densities of weight one, and can be integrated over the manifold. However, one can also consider their ratios. The most general Lagrangian is then

$$L = L_0 f\left(\frac{L_1}{L_0}, \frac{L_2}{L_0}\right) \tag{2.47}$$

for an arbitrary function f of two variables. The case $f = 1$ gives GR, but other choices are possible. A general theory from this class has been studied in Hejna (2006), where it was shown that it is equivalent to a nonlinear Einstein–Proca system. This ambiguity that arises in writing down the most general Lagrangian is a drawback of all "pure connection" formulations, as we will see in the following chapters.

Another drawback of the pure affine formulation is the very large number of field components one has to deal with. Indeed, in four dimensions we have $4 \times 10 = 40$ components in $\Gamma_\mu{}^\rho{}_\nu$ as compared to only 10 components in $g_{\mu\nu}$. This makes the pure affine formalism not very useful in practice.

2.6 Unification: Kaluza–Klein Theory

One can consider Einstein's theory in spacetime dimensions higher than four, and dimensionally reduce it to 4D by e.g., assuming that the fields are independent of all but four spacetime coordinates. This is the Kaluza–Klein idea. The resulting theory contains 4D Einstein's gravity, but also contains other fields. Importantly, if we interpret the original higher-dimensional space as the total space of the fibre bundle, with the dimensional reduction giving the projection map, then the fact that there is a metric in the total space of the bundle implies that there is a natural connection that gets induced, by the requirement that the horizontal vector fields are those metric orthogonal to vertical ones. This means that the dimensionally reduced theory naturally contains gauge fields. This is one of the most attractive ideas towards unifying gravity with other forces in nature. However, we refrain its quantitative discussion until the next chapter, where the frame formalism will allow us to simplify computations that must be made to see what kind of dimensionally reduced Lagrangian arises.

3
Cartan's Tetrad Formulation

We now come to the first key chapter of this book, where the customary metric geometry of a typical exposition of general relativity (GR) gives in to a more powerful geometric description, with differential forms and fibre bundles playing the key role.

The book, *Einstein Gravity in a Nutshell*, by Anthony Zee quotes (p. 787) Einstein, a year before his death, speaking to a group of John Wheeler's students. Einstein in particular made a comment: 'There is much reason to be attracted to a theory with no space, and no time. But nobody has any idea how to build it up'.

The question of why there is a nontrivial metric field apparently filling all of the universe is perhaps one of the most interesting questions to which physics currently gives no answer. General relativity describes the dynamics of such a metric field, but it does not explain why this field is nonzero rather than zero. In fact, GR starts by postulating that this field is nondegenerate, and thus cannot even be formulated for zero metrics. Neither can it be formulated perturbatively around the zero metric. The metric is thus essentially nonzero in GR. This fact has often been pointed out as the reason for various theoretical problems with GR, in particular, its worse than desired behaviour as a quantum theory (non-renormalisability).

It should be clear that a metric description of geometry has no chance of answering the question as to why a metric exists in the first place. In such a description the metric is simply postulated from the very beginning. Hence, developing a formalism where some other geometrical objects plays central role, and metric arises only as a secondary objects is a necessary step in the direction of explaining why it exists. This is probably the strongest theoretical motivation for developing alternative geometric viewpoints on gravity.

The point of view that we take on gravity in this book is that it is a dynamical theory of a collection of differential forms rather than a dynamical theory of metrics. The possibility of a metric interpretation then appears to be somewhat of an accident. To study gravity not coupled to any matter fields and to solve its field equations one never needs the metric interpretation. It is only when matter coupled to gravity is introduced, that the matter is seen to follow the geodesics of a certain metric constructed from the collection of differential forms that the theory is about. This is how the metric interpretation arises.

This is, of course, profoundly different from the usual viewpoint on GR. The metric GR is based on the equivalence principle that is built into its formalism by requiring that all matter couples to the same metric. It can then be seen to be natural that this very metric is the main dynamical variable of the theory. In the viewpoint advocated in the previous paragraph, the equivalence principle would not be automatic, because in principle, different species of matter may couple differently to the collection of differential forms that describe the gravitational field. This means that there can exist some theories of gravity and matter that are mathematically consistent, but which would violate the equivalence principle and thus would not be of any physical interest. We will not need to worry about such issues for quite a while because in most of the theories we consider in what follows the metric is apparent, and the most natural matter coupling is the coupling to this metric. But there will be examples where the questions of matter coupling become quite nontrivial. This is when the previous remarks will become relevant.

While gravity is seen as a dynamical theory of a collection of differential forms in all the formulations to be developed in the rest of this book, most of them (if not all) still fall short of answering the question of why the metric is nonzero. Thus, in most of these formulations the postulate of nonvanishing (nondegenerate) metric is replaced by a similar nondegeneracy assumption. So, most of the formulations we develop, while providing a new geometric viewpoint on GR and leading to simplifications in the structure of Einstein equations, do not answer the fundamental 'Why nonzero metric' question. But these formulations do suggest even more exotic alternatives to which we will turn in the last chapter, and which may have the potential to answer this question.

In addition, the formulations that are based on differential forms make gravity look much more similar to Yang-Mills gauge theories, where the dynamical variables are also collections of differential forms. In fact, we will see that gravity can be viewed as the ordinary sort of gauge theory, with fibre bundles and gauge fields playing central roles. The main difference between gravity and gauge theories of the more familiar sort is the existence of an object that links the geometry of the fibres with that of the tangent space. It is this object that encodes the metric, and it is this object whose presence makes the gravity theory so profoundly different from the usual gauge theory. In this chapter we will develop the series of formulations that all become possible thanks to Cartan's idea of a frame field.

3.1 Tetrad, Spin Connection

There are two conceptually very different points of view on the tetrad. One that is followed in many textbooks of GR goes under the name of 'non-coordinate bases', see, e.g., the book by Sean Carroll (2019) entitled *Spacetime and Geometry: An Introduction to General Relativity*, appendix J. We will briefly describe this viewpoint first, mainly to make it clear that this is not what is later adopted. We then proceed to develop Cartan's point of view.

3.1.1 Non-Coordinate Bases

In this viewpoint one never introduces any bundles apart from the tangent and cotangent (and more generally tensor) bundles over the manifold. This is what makes this approach so logically different from that of Cartan. We present the material of this subsection for completeness only. It can be skipped by readers not interested in this viewpoint on the frame.

In Riemannian geometry, one usually works with coordinate bases in the space of vector fields and 1-forms. A coordinate basis of vector fields is rarely orthonormal. The idea is then that one can choose a not necessarily coordinate basis of vectors $e_I \in TM$, where I is simply an index that enumerates them. These vectors can be chosen to be orthonormal in the sense that

$$g(e_I, e_J) = \eta_{IJ}, \tag{3.1}$$

where η_{IJ} is the flat metric of desired signature. As we discuss later, there may in fact be a difficulty in choosing an everywhere nonvanishing set of vectors that satisfy the orthonormality condition (3.1), on some manifolds there simply do not exist everywhere nonvanishing vector fields (S^2 is an example). But we will ignore this subtlety for now, working in a single coordinate chart. One can then expand the coordinate basis ∂_μ in terms of the new basis $\partial_\mu = e^I_\mu e_I$, where e^I_μ is a collection of 4×4 coefficient functions. We can rewrite the metric in terms of the objects e^I_μ. Indeed,

$$g_{\mu\nu} = g(\partial_\mu, \partial_\nu) = g(e^I_\mu e_I, e^J_\nu e_J) = e^I_\mu e^J_\nu g(e_I, e_J) = e^I_\mu e^J_\nu \eta_{IJ}, \tag{3.2}$$

where in the last step we have used (3.1). This shows that e^I_μ is the 'square root' of the metric, and completely encodes the latter.

One can then encode the operation of covariant differentiation in non-coordinate bases. This proceeds as follows. We write any vector as $TM \ni v = v^I e_I$. We then have

$$\nabla(v^I e_I) = (\partial v^I) e_I + v^I \nabla e_I. \tag{3.3}$$

If we introduce the connection coefficients

$$\nabla e_I = \omega^J{}_I e_J, \tag{3.4}$$

we have $\nabla v = (\partial v^I + \omega^I{}_J v^J)e_I$, which we can write as

$$\nabla v^I = \partial v^I + \omega^I{}_J v^J. \tag{3.5}$$

On the other hand we have $v^I e_I = v^\mu \partial_\mu = v^\mu e^I_\mu e_I$, from which $v^I = v^\mu e^I_\mu$. We also have $\nabla_\rho(v^\mu \partial_\mu) = (\partial_\rho v^\mu + \Gamma^\mu{}_{\nu\rho} v^\nu)\partial_\mu$. Comparing these two expressions we see that $\Gamma^\mu{}_{\nu\rho} e^I_\mu = \partial_\rho e^I_\nu + \omega^I_{\rho\,J} e^J_\nu$. This can be rewritten as

$$\Gamma^\mu{}_{\nu\rho} = e^\mu_I \partial_\rho e^I_\nu + e^\mu_I \omega^I_{\rho\,J} e^J_\nu, \tag{3.6}$$

where we introduced the inverse objects $e^\mu_I : e^\mu_I e^J_\mu = \delta^J_I, e^\mu_I e^I_\nu = \delta^\mu_\nu$. This relates the non-coordinate base connection coefficients $\omega^I_{\rho J}$ to the Christoffel symbols. This relation can also be rewritten as

$$\partial_\rho e^I_\nu - \Gamma^\mu{}_{\nu\rho} e^I_\mu \equiv \nabla_\rho e^I_\nu = -\omega^I_{\rho\,J} e^J_\nu. \tag{3.7}$$

This means that

$$0 = \nabla_\rho g_{\mu\nu} = \nabla_\rho(e^I_\mu e^J_\nu \eta_{IJ}) = (\omega^I_{\rho\,K} e^K_\mu e^J_\nu + e^I_\mu \omega^J_{\rho\,K} e^K_\nu)\eta_{IJ}. \tag{3.8}$$

We now contract this with $e^\mu_L e^\nu_M$ to get

$$\omega^I_{\rho\,L}\eta_{IM} + \omega^I_{\rho\,M}\eta_{IL} = 0, \tag{3.9}$$

which means that the connection coefficients $\omega^I{}_J$ take values in the Lie algebra of the Lorentz group of the appropriate signature. From (3.6) we also see that the vanishing of the torsion $\Gamma^\mu{}_{[\nu\rho]} = 0$ is equivalent to

$$de^I + \omega^I{}_J \wedge e^J = 0, \tag{3.10}$$

which is an equation written in terms of the wedge product and the exterior derivative. Here we think of the coefficients e^I_μ as components of a differential form $e^I := e^I_\mu dx^\mu$.

The viewpoint described does not necessitate introducing any other bundles apart from the tangent/cotangent bundles. However, the fact that the connection coefficients $\omega^I{}_J$ are naturally Lie algebra valued strongly suggests that we are in fact dealing with a principal connection here, even if this fact is not apparent from our description. The viewpoint of Cartan, which we are to develop now, puts a certain associated vector bundle at the forefront of the description. Importantly, it also leads to a set of natural generalisations of Riemannian geometry that are hard to come up with in the usual tensorial formalism.

3.1.2 Geometric Structures

One of Cartan's key ideas is that a 'geometric structure' on a manifold (e.g., metric, but other examples are possible, see what follows) can be efficiently encoded in a collection of differential forms. This proceeds as follows. First, we need a 'local model' for our desired geometric structure. Thus, let $V = \mathbb{R}^n$ and $\mathrm{GL}(V) = \mathrm{GL}(n, \mathbb{R})$. Let T be the space of certain type tensors for V.

For example, it can be the space $T = S^2V^*$ of symmetric rank $(0,2)$ tensors, or $T = \Lambda^2V^*$ the space of antisymmetric $(0,2)$ tensors. Another important example is $T = V \otimes V^* = \mathrm{End}(V)$, the space of linear maps on V. Let $\psi \in T$ be a tensor and $G_\psi \subset \mathrm{GL}(V)$ be the subgroup preserving ψ. For example, if $Q \in S^2V^*$ is a (nondegenerate) symmetric tensor then $G_Q = O(V,Q)$, the group of orthogonal transformations preserving Q. When $\omega \in \Lambda^2V^*$ is a (nondegenerate) antisymmetric tensor, the stabiliser subgroup is $G_\omega = \mathrm{Sp}(V,\omega)$, the symplectic group. For $J \in V\otimes V^*$ such that $J^2 = -\mathrm{Id}$, the group $G_J = \mathrm{GL}(m,\mathbb{C})$ where $n = 2m$.

Let us now fix, in each case, a canonical tensor of the corresponding type. In the case of symmetric tensors Q can be chosen to be a flat metric of the desired signature. For the case of antisymmetric tensors this can be chosen to be

$$\omega = \begin{pmatrix} 0 & I_m \\ -I_m & 0 \end{pmatrix}, \tag{3.11}$$

where I_m is the $n\times n$ identity matrix. For the case of an almost complex structure $J \in V \otimes V^*$, it can be chosen to be given again by (9.79), but now interpreted as an element of $V \otimes V^*$ rather than Λ^2V^*.

Let us now consider a manifold M of dimension n. Our desire is to encode a geometric structure on M, modelled on one of the previously described tensors, into a collection of differential forms. To this end we first define a notion of **coframe or soldering form**. A coframe at $x \in M$ on M with values in V is an isomorphism

$$e : T_xM \to V. \tag{3.12}$$

Concretely, a soldering form is (locally, in a coordinate chart) a 1-form on M with values in V, i.e., the object $e^I = e^I_\mu dx^\mu$ already encountered in the previous subsection.

The general linear group $\mathrm{GL}(V)$ acts transitively on the space of coframes at a point. Concretely, let us define a right action of $\mathrm{GL}(V)$ on a coframe via $e \to g^{-1}e$, $g \in \mathrm{GL}(V)$. Let F^*M be the principal $\mathrm{GL}(V)$ bundle of coframes on M. Let us now consider some geometric structure on M, e.g., a metric or a nondegenerate 2-form, or an almost complex structure. In each case, let us fix in V a canonical object of the corresponding type, as previously discussed. There arises a notion of a coframe **adapted** to the geometric structure chosen, which is a coframe such that the geometric structure in question gets mapped into the canonical one by the coframe mapping. For example, in the case of a metric, an adapted coframe is the one for which the map (3.12) is an isometry between the metric in M chosen and the canonical fixed metric in V. For the example of a 2-form, an adapted frame is one for which the map (3.12) is a symplectic map, so that the original 2-form in M is the pullback of the canonical 2-form in V. For the example of an almost complex structure, an adapted frame is the one that satisfies $e(Jv) = J_Ve(v)$, where J_V is the canonical almost complex structure in V.

What makes this construction interesting is that there are in general many coframes that are adapted to a given geometric structure. Indeed, a coframe mapping (3.12) can be followed by a transformation in V that preserves the canonical object in V. There is then the notion of a bundle of adapted coframes, which is a principal $G_\psi \subset \mathrm{GL}(n,\mathbb{R})$ bundle for one of the G_ψ groups discussed previously. Thus, in the case of a metric we have the principal $\mathrm{O}(V)$ bundle of orthonormal coframes. In the case of a 2-form we get the principal $\mathrm{Sp}(V)$ bundle of symplectic coframes. In the case of an almost complex structure we get an $\mathrm{GL}(m,\mathbb{C}), n = 2m$ bundle of coframes that preserve the eigenspace decompositions of J and J_V. Thus, a geometric structure reduces the principal $\mathrm{GL}(V)$ coframe bundle F^*M to one of the principal bundles of the adapted coframes. This motivates the following definition

Definition 3.1 A G-structure on M is a reduction of the principal $\mathrm{GL}(V)$ bundle F^*M of coframes to a principal G-bundle, $G \subset \mathrm{GL}(V)$.

As we have seen, concretely a G-structure is encoded in a collection of V-valued differential forms on M. The original geometric structure on M in many cases arises as simply the pullback of a canonical tensor in V under the coframe map (3.12). This viewpoint on geometric structures unifies many different types of geometries. In particular, as we have seen, geometry of metrics, symplectic geometry, and complex geometry are all treated from a uniform viewpoint.

3.1.3 Tetrad as the Soldering Form

We see that there are two conceptually very different perspectives on the tetrad formulation of gravity. One of them brings to forefront the **frame field**, which is a non-coordinate basis of vectors in TM that are orthonormal with respect to the given metric. One can then define the dual coframe which is a collection of 1-forms that are orthonormal. All tensors can then be decomposed in such frames and/or coframe and one can set up the operation of covariant differentiation that introduces the connection coefficients $\omega^I{}_J$. This point of view does not bring into play any other bundles apart from the already available tangent bundle. And this is the point of view that is most often described in relation to the tetrad formalism. This is not the point of view that will be adopted here.

One problem with the point of view just discussed is that it brings with itself a difficulty that many manifolds do not admit nowhere vanishing vector fields. The simplest example where there are no such vector fields is S^2. Manifolds admitting a global section of the frame bundle are called **parallelisable**, and this property is rare. All Lie groups are parallelisable, but not all manifolds of interest are Lie groups.

This difficulty is avoided if one takes Cartan's viewpoint on the frame. This other viewpoint introduces a new vector bundle E over space(time) manifold M, with fibres copies of $V = \mathbb{R}^n$. This vector bundle is a priori unrelated to the tangent or cotangent bundles. The tetrad, soldering form, or a coframe is then an object that ties this bundle E to the tangent bundle. It is for this reason that the tetrad is referred to as the soldering form, because it solders an abstract vector bundle over the spacetime to the tangent bundle. Vector bundles such as E can be equipped with a connection, and it is this connection that will play the central role in our description of gravity, rather than the Levi–Civita connection on TM. All in all, the emphasis shifts from objects intrinsically defined on the manifold (tensors) to certain differential forms with values in E. This does make gravity look more similar to Yang–Mills theory, because the latter also starts by introducing a bundle over spacetime, with dynamical objects being those naturally living in this bundle. The difference between Yang–Mills theory and gravity is then simply in the fact that the latter comes with an object that ties the vector bundle in question to the tangent bundle–the soldering form.

From now on we will only develop the description of (pseudo-) Riemannian geometry, leaving the other examples (symplectic, complex) behind. But it should be kept in mind that they can be developed in parallel with the Riemannian geometry case.

With the previous remarks in mind, we introduce a vector bundle $V \hookrightarrow E \to M$, whose fibres V have a metric $\langle \cdot, \cdot \rangle$, are of dimension n, and are copies of $\mathbb{R}^{p,q}, p+q = n$, depending on the desired signature. One requires this new bundle to be isomorphic to the tangent bundle TM. The tetrad, soldering form, vielbein, or coframe[1] is then the object that provides this isomorphism

$$e : TM \to E. \tag{3.13}$$

Locally, the tetrad is a collection of n linearly independent 1-forms

$$e^I = e^I_\mu dx^\mu, \qquad I = 1, \dots, n,$$

with the map from TM to E being $e : v \to e^I(v) = e^I_\mu v^\mu$. We will refer to indices $I, J, \dots$ as 'internal' indices, to signify the fact that they are indices that refer to a vector bundle E that has in principle nothing to do with the natural bundles that are defined on M, such as the tangent and cotangent bundle.

Given a tetrad, the metric on M is defined to be the pullback of the metric on E

$$g(v, u) := \langle e(v), e(u) \rangle, \qquad \text{or} \qquad g_{\mu\nu} = e^I_\mu e^J_\nu \eta_{IJ}, \tag{3.14}$$

[1] The terms 'tetrad' and 'veilbein' both have the drawback that they refer explicitly to four dimensions. In three dimensions the same fields are usually referred to as triads or dreibeins; in higher dimensions, the term 'veilbein' is used. We will ignore this and use the same term in all dimensions.

where η_{IJ} is the metric on fibres V. We note that there are many different choices of tetrad that lead to the same metric. Indeed, the tetrad can be locally Lorentz rotated $e^I \to \Lambda^I{}_J e^J, \Lambda^I{}_J \in \mathrm{O}(p,q)$ without any change in the metric. This is why the (co-) frame bundle is an $\mathrm{O}(p,q)$ principal bundle.

3.1.4 Spin Connection

Having defined the tetrad, we can introduce the spin connection. It starts its life as a metric connection on the vector bundle E, or as a connection that is associated to a connection in the principal $\mathrm{O}(p,q)$ bundle. That is, locally, this is an object $\omega^I{}_J$ that is 1-form valued and defines the covariant derivative on sections of E

$$d^\omega V^I := dV^I + \omega^I{}_J V^J. \tag{3.15}$$

The metric property is

$$0 = d^\omega \eta^{IJ} = \omega^I{}_K \eta^{KJ} + \omega^J{}_K \eta^{IK}, \tag{3.16}$$

where we assumed that the components of the inverse metric η^{IJ} are constants. This is just the statement that the object $\omega^I{}_J$ is valued in the Lie algebra of $\mathrm{O}(p,q)$, i.e., valued in the Lie algebra of the corresponding Lorenz group. In the simplest case of Riemannian signature $\eta^{IJ} = \delta^{IJ}$ and the spin connection is a 1-form with values in antisymmetric $n \times n$ matrices.

3.1.5 Torsion-Free Spin Connection

When one has a tetrad e^I at one's disposal, one can introduce the following $\Lambda^2(M) \otimes E$ valued object called **torsion**

$$T^I := d^\omega e^I \equiv de^I + \omega^I{}_J e^J. \tag{3.17}$$

Note that d here is the exterior derivative and the wedge product of forms is implied in the second term.

Lemma 3.2 *Given a tetrad, there exists a unique metric torsion-free connection. Explicitly, it is given by*

$$\omega^I_{\mu\, J} = e^{\rho I} e^\sigma_J (-C_{\mu\rho\sigma} + C_{\rho\sigma\mu} + C_{\sigma\mu\rho}), \qquad C_{\mu\rho\sigma} := e_{\mu I} \partial_{[\rho} e^I_{\sigma]}. \tag{3.18}$$

The proof is by explicit verification. The object e^μ_I is the inverse tetrad defined via $e^\mu_I e^J_\mu = \delta^J_I$ and $e^\mu_I e^I_\nu = \delta^\mu_\nu$. The internal indices are raised and lowered with the internal metric η_{IJ} and its inverse. To convince oneself that such a statement can be true one can first count the number of equations in $d^\omega e^I = 0$ versus the number of unknowns. The number of equation is the dimension of the space of 2-forms $n(n-1)/2$ times n, while this is also the number of components in the connection $\omega_{\mu}^{I}{}_{J}$. The formula (3.18) is an analog of the Christoffel formula in Riemannian geometry.

3.1.6 Relation to the Christoffel Connection

In the tetrad formalism the Christoffel connection arises simply as the pullback of the spin connection from E to TM. This is expressed as follows. Given $u^I = e_\mu^I u^\mu$ we define the connection on TM via

$$e_\mu^I \nabla u^\mu = d^\omega u^I. \tag{3.19}$$

A quick calculation then gives

$$e^I{}_\sigma \Gamma^\sigma{}_{\nu\mu} = \partial_\mu e_\nu^I + \omega_\mu^I{}_J e_\nu^J. \tag{3.20}$$

Note that we can rewrite this relation as

$$0 = \nabla_\mu^\omega e_\nu^I := \partial_\mu e_\nu^I + \omega_\mu^I{}_J e_\nu^J - \Gamma^\sigma{}_{\nu\mu} e_\sigma^I. \tag{3.21}$$

Here we have introduced a new 'total' covariant derivative ∇^ω that acts on both the spacetime index and internal index of e_ν^I. The relation (3.20) is then interpreted as the statement that the total covariant derivative of the tetrad is zero. Note that (3.21) immediately implies that the connection on TM that appears in ∇^ω is the Christoffel connection for the metric $g_{\mu\nu} = e_\mu^I e_\nu^J \eta_{IJ}$. Indeed, we can act on $g_{\mu\nu}$ so defined with ∇^ω and the result is zero because both e_μ^I and η_{IJ} are killed by ∇^ω.

Let us also note that the equation $\nabla_\mu^\omega e_\nu^I = 0$ can be taken as defining both the spin connection and the Levi–Civita connection. Indeed, we can take the antisymmetric in $\mu\nu$ part of this equation and recover the torsion-free condition. Having solved it, we then get the Levi–Civita connection algebraically from the spin connection and the partial derivatives of the tetrad. An alternative way of reaching the same conclusion is by counting equations. There are n^3 equations in $\nabla_\mu^\omega e_\nu^I = 0$, and we can use them to find $n \times n(n-1)/2$ components of the spin connection and $n \times n(n+1)/2$ components of $\Gamma^\mu_{\nu\rho}$.

3.1.7 Curvature of the Spin Connection vs. Riemann Curvature

Having established a relation between the torsion-free spin and Levi–Civita connections, we are ready to establish a link between their respective curvatures. The curvature of the spin connection is defined via

$$2 d_{[\mu}^\omega d_{\nu]}^\omega u^I := R_{\mu\nu}{}^I{}_J u^J. \tag{3.22}$$

This gives, in form notations

$$R^I{}_J = d\omega^I{}_J + \omega^I{}_K \omega^K{}_J. \tag{3.23}$$

We can now obtain a relation to the Riemann curvature as follows. Since the 'total' covariant derivative ∇^ω kills the tetrad we have

$$0 = 2\nabla^\omega_{[\mu}\nabla^\omega_{\nu]}e^I_\rho = R_{\mu\nu}{}^I{}_J e^J_\rho - R^\alpha{}_{\rho\mu\nu}e^I_\alpha, \tag{3.24}$$

and thus

$$R^\alpha{}_{\rho\mu\nu} = R_{\mu\nu}{}^I{}_J e^\alpha_I e^J_\rho. \tag{3.25}$$

This gives a very efficient way of computing the Riemann curvature! Indeed, given a metric, we only need to choose the corresponding tetrad, and then compute the spin connection from the zero-torsion condition. There are only 24 components of the spin connection to compute in four dimensions, which are all compactly stored in six 1-forms. This should be compared to 40 components of the Christoffel symbol, for which no good storing device exists. The curvature of the spin connection is then obtained by simple operations of exterior differentiation and wedge product, and this leads directly to the components of the Riemann curvature tensor, as the relation (3.25) tells us.

The easiest way to find the Ricci curvature and Ricci scalar using the tetrad formalism is to convert all indices of $R_{\mu\nu}{}^I{}_J$ not to the spacetime indices as in (3.25), but to internal indices. Thus, introducing the object

$$R_{MN}{}^I{}_J := R_{\mu\nu}{}^I{}_J e^\mu_M e^\nu_N, \tag{3.26}$$

we can get the internal indices Ricci tensor as

$$R_{IJ} = R_{MI}{}^M{}_J, \tag{3.27}$$

so that $R_{\mu\nu} = R_{IJ}e^I_\mu e^J_\nu$ and the Ricci scalar as

$$R = R_{IJ}\eta^{IJ}. \tag{3.28}$$

3.1.8 Examples

We now spell out some examples of Riemann curvature computations using the frame formalism.

Example 3.3 Let us consider one of the simplest possible applications of the frame formalism, which is the computation of the curvature of the two-sphere. The metric is given by

$$ds^2 = R^2(d\theta^2 + \sin^2\theta d\phi^2), \tag{3.29}$$

which leads to the following frame 1-forms, or a 'dyad'

$$e^\theta = Rd\theta, \qquad e^\phi = R\sin\theta d\phi. \tag{3.30}$$

The two torsion-free equations to solve are $de^\theta + \omega^\theta{}_\phi e^\phi = 0, de^\phi + \omega^\phi{}_\theta e^\theta = 0$, which become

$$\omega^\theta{}_\phi d\phi = 0, \qquad \cos\theta d\phi d\theta = \omega^\phi{}_\theta d\theta, \tag{3.31}$$

and immediately give

$$\omega^{\phi}{}_{\theta} = \cos\theta d\phi. \tag{3.32}$$

Since the Lie algebra of SO(2) is one-dimensional, there is only this 1-form that constitutes the full spin connection. Its curvature is given by

$$R^{\phi}{}_{\theta} = -\sin\theta d\theta d\phi = \frac{1}{R^2} e^{\phi} e_{\theta}. \tag{3.33}$$

This immediately gives the only independent component of the Riemann curvature $R_{\phi\theta}{}^{\phi}{}_{\theta} = 1/R^2$, and thus $R_{\theta\theta} = 1/R^2$ and $R_{\phi\phi} = 1/R^2$. The Ricci scalar is then

$$R = \frac{2}{R^2}. \tag{3.34}$$

Example 3.4 Let us now redo the previous example using the complexified formalism. To this end, we first map the 1-forms (3.30) to 1-forms on $\mathbb{R}^2$ using the stereographic projection (from the north pole). The map from S^2 to $\mathbb{R}^2 = \mathbb{C}$ is explicitly given by

$$x + iy \equiv z = \frac{\cos(\theta/2)}{\sin(\theta/2)} e^{i\phi}. \tag{3.35}$$

The inverse map $\mathbb{C} \to S^2$ is given by

$$\cos\theta = \frac{|z|^2 - 1}{|z|^2 + 1}, \qquad e^{2i\phi} = \frac{z}{\bar{z}}. \tag{3.36}$$

We can now pull back the 1-forms (3.30) with this map. We get

$$d\theta = -\frac{d|z|^2}{|z|(|z|^2+1)}, \qquad \sin\theta d\phi = \frac{i(zd\bar{z} - \bar{z}dz)}{|z|(|z|^2+1)}. \tag{3.37}$$

This gives for the unit $R = 1$ sphere metric

$$ds^2 = \frac{4dzd\bar{z}}{(1+|z|^2)^2}. \tag{3.38}$$

To find the spin connection for this metric, it is convenient to work with the complexified frame. To introduce this, let us write the torsion-free conditions

$$de^x + \omega^x{}_y e^y = 0, \qquad de^y + \omega^y{}_x e^x = 0. \tag{3.39}$$

Let us introduce

$$e := e^x + ie^y. \tag{3.40}$$

Then the complex linear combination of the two equations in (3.39) gives $de + i\omega^y{}_x e = 0$. Thus, if we now introduce the notation $\omega := \omega^y{}_x$ there is one complex equation

$$de + i\omega e = 0 \tag{3.41}$$

to solve instead of two real equations (3.39).

For the metric (3.38) the complexified 1-form e is

$$ds^2 = e\bar{e}, \qquad e = \frac{2dz}{1+|z|^2}. \tag{3.42}$$

We can now easily solve (3.41) for the spin connection

$$\omega = -\frac{\mathrm{i}(zd\bar{z} - \bar{z}dz)}{1+|z|^2}. \tag{3.43}$$

Note that if we pull it back to S^2 we get $\omega = -(1+\cos\theta)d\phi$, which does not coincide with (3.32), nor should it.

We can now compute the curvature using this complex formalism. We have

$$d\omega = \frac{1}{2\mathrm{i}} e\bar{e}, \tag{3.44}$$

where now of course the wedge product of forms appears on the right-hand side. Let us see how to extract the scalar curvature from here. We have $(1/2\mathrm{i})e\bar{e} = e^y e^x$, and so what the equation (3.44) says is that $R^y{}_x = e^y e_x$, which immediately implies that $R_{xx} = R_{yy} = 1$ and the scalar curvature is equal to $R = 2$, as we have previously determined without using the complexified frame.

It is clear that the described complex version of the tetrad formalism in two dimensions, whose main formulas are $ds^2 = e\bar{e}$ and (3.41) as well as the rule

$$d\omega = \frac{R/2}{2\mathrm{i}} e\bar{e} \tag{3.45}$$

for reading the scalar curvature R from the curvature of ω is more efficient than the real formalism because it halves the number of equations that need to be written down. We will describe a similar formalism in four dimensions in the chapter on chiral descriptions of GR.

Example 3.5 We now note that the two sets of equations, namely (3.41) and (3.44) can be put together into a single matrix-valued equation. For this purpose, let us introduce a 1-form with values in 2×2 anti-Hermitian matrices

$$\theta := \frac{\mathrm{i}}{2} \begin{pmatrix} \omega & e \\ \bar{e} & -\omega \end{pmatrix}. \tag{3.46}$$

One then easily checks that both equations in question arise as the components of a single matrix-valued equation

$$d\theta + \theta\theta = 0. \tag{3.47}$$

Thus, the U(1) spin connection together with the frame field on S^2 combines into a flat SU(2) connection. This construction has its origin in the fact that $S^2 = \mathrm{SU}(2)/\mathrm{U}(1)$, and thus there is a natural flat $\mathfrak{su}(2)$-valued 1-form on S^2 given by the corresponding Maurer–Cartan 1-form $g^{-1}dg$, with g being a representative of a point $x \in S^2$ realised as the coset $S^2 = \mathrm{SU}(2)/\mathrm{U}(1)$.

This construction can be trivially generalised to constant scalar curvature 2-manifolds other than S^2. For example, there exists a similar $\mathrm{SL}(2,\mathbb{R})$ connection on the two-dimensional hyperbolic space H^2. The construction of combining the two Cartan equations (3.41) and (3.44) into a single equation requiring some connection to be flat has an analog in 3D, as we will see in the next chapter.

Example 3.6 Let us now consider a much more involved example of a static, spherically symmetric Lorentzian signature metric in four dimensions

$$ds^2 = -f^2 dt^2 + g^2 dr^2 + r^2(d\theta^2 + \sin^2\theta d\phi^2). \tag{3.48}$$

Here functions f, g depend on the radial coordinate only. The most natural tetrad is then

$$e^t = f dt, \quad e^r = g dr, \quad e^\theta = r d\theta, \quad e^\phi = r \sin\theta d\phi. \tag{3.49}$$

Let us start the process of determining the spin connection. We have $de^t = f' dr dt$, and the equation to solve is

$$de^t + \omega^t{}_r e^r + \omega^t{}_\theta e^\theta + \omega^t{}_\phi e^\phi = 0. \tag{3.50}$$

The last two terms here will involve de and $d\phi$, which are not present in the first term. So, the simplest possibility for this equation to be satisfied is to assume that $\omega^t{}_\theta$ and $\omega^t{}_\phi$ are actually zero, and that $\omega^t{}_r$ only has the dt component. This gives

$$\omega^t{}_r = \frac{f'}{g} dt. \tag{3.51}$$

The next equation to consider is

$$de^r + \omega^r{}_t e^t + \omega^r{}_\theta e^\theta + \omega^r{}_\phi e^\phi = 0. \tag{3.52}$$

The first two terms here are zero in view of the already known solution for $\omega^t{}_r$. The simplest possibility to have this equation satisfied is to assume that $\omega^r{}_\theta \sim d\theta$ and $\omega^r{}_\phi \sim d\phi$.

The next equation is

$$de^\theta + \omega^\theta{}_t e^t + \omega^\theta{}_r e^r + \omega^\theta{}_\phi e^\phi = 0. \tag{3.53}$$

The first term here is $de^\theta = dr d\theta$. We have already assumed that $\omega^t{}_\theta = 0$, and so there is no second term. We have also assumed that $\omega^r{}_\theta \sim d\theta$, and this is precisely the structure needed for the third term to cancel the first. Thus, the simplest option is to assume that $\omega^\theta{}_\phi \sim d\phi$. We then find

$$\omega^\theta{}_r = \frac{1}{g} d\theta. \tag{3.54}$$

The last equation reads

$$de^\phi + \omega^\phi{}_t e^t + \omega^\phi{}_r e^r + \omega^\phi{}_\theta e^\theta = 0. \tag{3.55}$$

The first term is a sum of two $de^\phi = \sin\theta dr d\phi + r\cos\theta d\theta d\phi$. We have already assumed previously that $\omega^t{}_\phi = 0$, so the second term does not contribute. And we can then read off the last two connection 1-forms

$$\omega^\phi{}_r = \frac{1}{g}\sin\theta d\phi, \qquad \omega^\phi{}_\theta = \cos\theta d\phi. \tag{3.56}$$

This finishes the most laborious process of any curvature computation – the determination of the connection.

We can now find the six components of the curvature 2-form. We have

$$R^t{}_r = d\omega^t{}_r + \omega^t{}_\theta\omega^\theta{}_r + \omega^t{}_\phi\omega^\phi{}_r = \left(\frac{f'}{g}\right)' dr dt,$$

$$R^t{}_\theta = \omega^t{}_r\omega^r_\theta = -\frac{f'}{g^2}dt d\theta, \qquad R^t{}_\phi = \omega^t{}_r\omega^r{}_\phi = -\frac{f'}{g^2}\sin\theta dt d\phi,$$

where we have used $\omega^t{}_\theta = 0, \omega^t{}_\phi$ in all three equations, and also used the fact that $\omega^r{}_\theta = -\omega^\theta{}_r$, and $\omega^r{}_\phi = -\omega^\phi{}_r$ to write the last two formulas. The other three curvature components are

$$\begin{aligned}
R^\theta{}_r &= d\omega^\theta{}_r + \omega^\theta{}_t\omega^t{}_r + \omega^\theta{}_\phi\omega^\phi{}_r = \left(\frac{1}{g}\right)' dr d\theta,\\
R^\phi{}_r &= de^\phi{}_r + \omega^\phi{}_t\omega^t{}_r + \omega^\phi{}_\theta\omega^\theta{}_r\\
&= \left(\frac{1}{g}\right)' \sin\theta dr d\phi + \frac{1}{g}\cos\theta d\theta d\phi + \cos\theta d\phi\frac{1}{g}d\theta = \left(\frac{1}{g}\right)' \sin\theta dr d\phi,\\
R^\phi{}_\theta &= de^\phi{}_\theta + \omega^\phi{}_t\omega^t{}_\theta + \omega^\phi{}_r\omega^r{}_\theta\\
&= -\sin\theta d\theta d\phi - \frac{1}{g^2}\sin\theta d\phi d\theta = -\sin\theta\left(1 - \frac{1}{g^2}\right) d\theta d\phi.
\end{aligned}$$

We can rewrite what we have found as follows

$$R^t{}_r = -\frac{1}{fg}\left(\frac{f'}{g}\right)' e^t e_r, \quad R^t{}_\theta = -\frac{f'}{rfg^2}e^t e_\theta, \quad R^t{}_\phi = -\frac{f'}{rfg^2}e^t e_\phi,$$

$$R^\theta{}_r = -\frac{1}{gr}\left(\frac{1}{g}\right)' e^\theta e_r, \quad R^\phi{}_r = -\frac{1}{gr}\left(\frac{1}{g}\right)' e^\phi e_r, \quad R^\phi{}_\theta = \frac{1}{r^2}\left(1 - \frac{1}{g^2}\right) e^\phi e_\theta.$$

We can now form components of the Ricci tensor. It is this state where most care is needed in order not to commit a sign mistake coming from raising-lowering the indices. One should take into account that raising-lowering the index t gives a minus sign. This in particular means that $R^t{}_r = R^r{}_t$, and similarly for the other components of the curvature involving t. We then have

$$R_{tt} = R_{rt}{}^r{}_t + R_{\theta t}{}^\theta{}_t + R_{\phi t}{}^\phi{}_t = \frac{1}{fg}\left(\frac{f'}{g}\right)' + \frac{2f'}{rfg^2},$$

$$R_{rr} = R_{tr}{}^{t}{}_{r} + R_{\theta r}{}^{\theta}{}_{r} + R_{\phi r}{}^{\phi}{}_{r} = -\frac{1}{fg}\left(\frac{f'}{g}\right)' - \frac{2}{gr}\left(\frac{1}{g}\right)',$$

$$R_{\theta\theta} = R_{t\theta}{}^{t}{}_{\theta} + R_{r\theta}{}^{r}{}_{\theta} + R_{\phi\theta}{}^{\phi}{}_{\theta} = -\frac{f'}{rfg^2} - \frac{1}{gr}\left(\frac{1}{g}\right)' + \frac{1}{r^2}\left(1 - \frac{1}{g^2}\right).$$

The $R_{\phi\phi}$ component of Ricci is the same as $R_{\theta\theta}$ in view of the spherical symmetry, and so carries no new information.

If we now want the Ricci flat Schwarzschild solution we set all the three obtained components of Ricci to zero. The sum of the first two equations thus gives immediately

$$\frac{f'}{f} + \frac{g'}{g} = 0, \tag{3.57}$$

which means that $fg = const.$ This constant must be unity if we demand that the metric approaches the Minkowski metric at $r \to \infty$. Then the last equation gives

$$\frac{d(1-f^2)}{1-f^2} + \frac{dr}{r} = 0, \tag{3.58}$$

which implies

$$1 - f^2 = \frac{r_+}{r}, \tag{3.59}$$

where r_+ is a constant of integration. We get the Schwarzschild solution. The difference of the first two equations, which is the equation that we did not yet use, is satisfied automatically.

3.1.9 Spin Connection vs. Levi–Civita Connection

It is worth emphasising the principal difference between the spin and Levi–Civita connections. The latter is a connection in the tangent bundle. Connections in a tangent bundle can be viewed as those associated with principal $\mathrm{GL}(n,\mathbb{R})$ connections. However, as soon as the torsion-free condition $\Gamma^{\mu}{}_{[\nu\rho]} = 0$ is imposed this connection can no longer be interpreted as a Lie algebra valued 1-form. This is the technical reason why it is rather difficult to work with the Levi–Civita connection, at least as compared to the spin connection. The latter is a principal connection, and is a 1-form with values in the Lie algebra of the corresponding Lorentz group. The former is not a principal connection, and is not a 1-form in any natural way. The powerful machinery of exterior differentiation and differential forms is only available in the case of the principal $\mathrm{O}(p,q)$ bundle spin connection.

However, having only a connection in a vector bundle E over M does not say anything about the geometry of M. It is for this reason that an additional object is introduced, which is the soldering form, viewed as a map $e : TM \to E$. This object ties objects in E to objects in TM. In particular, both the metric and the Levi–Civita connections on TM arise this way, as pullbacks of objects from E. The presence of this object is the main distinguishing feature of a theory of geometry as compared to a gauge theory.

3.2 Einstein–Cartan First-Order Formulation

We now have all the necessary ingredients to describe the Einstein–Cartan tetrad formulation. It has different versions, and here we describe the version in which the action is written in terms of differential forms. We will only give this action in four dimensions. The story in $2+1$ dimensions will be described in the following chapter. A generalisation to $n > 4$ dimensions is straightforward, but will not be needed. The action is a functional of tetrad e and spin connection ω that are treated as independent variables. It reads

$$S_{\text{EC}}[e,\omega] = \frac{1}{32\pi G}\int \epsilon_{IJKL}\, e^I e^J \left(R^{KL}(\omega) - \frac{\Lambda}{6} e^K e^L\right). \tag{3.60}$$

Here $R^{IJ}(\omega) = R^K{}_M(\omega)\eta^{ML}$ is the curvature 2-form of the Lorentz (spin) connection with one of its indices raised using the internal metric η^{IJ}. The wedge product of forms is implied in (3.60). The integrand is a top form, and thus to evaluate the integral a choice of orientation of M needs to be made. The object ϵ_{IJKL} is a completely antisymmetric tensor in $\Lambda^4 V^*$. It takes values ± 1, and an orientation of $V = \mathbb{R}^4$ needs to be chosen to fix this tensor, by requiring that it takes value $+1$ in the orientation chosen. A convenient choice of this orientation that ties it to the orientation of M is described in Section 3.3.1.

When one varies (3.60) with respect to the connection, one obtains the equation $d^\omega(e^I e^J) = 0$, which implies $d^\omega e^I = 0$, i.e., the zero-torsion condition. As we already know, this is an algebraic equation for the spin connection, and can be solved uniquely in terms of the derivatives of e^I, see (3.18). Substituting this solution into the action (3.60) brings us back to the Einstein–Hilbert action.

Varying the action (3.60) with respect to the frame one gets

$$\epsilon_{IJKL} e^J R^{KL} = \frac{\Lambda}{3}\epsilon_{IJKL} e^J e^K e^L, \tag{3.61}$$

which is the Einstein equation in the tetrad formalism.

We note that the Einstein–Cartan action (3.60) is polynomial in the fields, and contains just up to quartic terms. This is true even for $\Lambda \neq 0$, in contrast to the case of the Palatini action (2.43), which is only polynomial (with the choice of the inverse densitiesed metric as the main variable) for $\Lambda = 0$. This, as well as the necessity of tetrads when spinors are present, are the two reasons why the tetrad formulation can be considered superior to the formulation in terms of the metric.

However, one drawback of the Einstein–Cartan formulation as compared to the metric description is more complicated character of its Hamiltonian formulation obtained via the 3 + 1 split. It is known that in this case there are second-class constraints, see, e.g., Holst (1996) for the Hamiltonian analysis. This should be contrasted with the Arnowitt-Deser-Misner (ADM) formalism, shown in Arnowitt et al. (1960) where no second-class constraints appear. The appearance of second-class constraints in the Einstein–Cartan formalism is not surprising because 24 'momentum' variables of the connection have been introduced in addition to the 16 'configuration' variables. The extra variables are then eliminated by second-class constraints. A formalism that shares all the good features of Einstein–Cartan but does not suffer from the problem of second-class constraints is the chiral first-order formalism to be described in the next chapter.

3.3 Teleparallel Formulation

In the previous chapter we have seen that the GR Lagrangian can be rewritten in the $\Gamma\Gamma$ form, modulo surface terms, see (2.19). A similar rewriting is possible for the Einstein–Cartan Lagrangian (3.60). Let us carry out this exercise. It will suggest a different way of thinking about GR that goes under the name of teleparallel gravity. The action we are going to derive is not written in terms of wedge product of differential forms. Thus, the geometry of fibre bundles and differential forms that we previously emphasised as important for the interpretation of the tetrad formalism plays no role in this section. Our description here is very brief, and the reader is directed to Heisenberg (2018) for a more thorough discussion on encoding gravity into torsion or non-metricity.

3.3.1 Torsion-Squared Form of the Gravitational Lagrangian

Recalling the definition of the curvature, and integrating by parts, we can rewrite the action (3.60) as

$$S[e,\omega] = \frac{1}{32\pi G}\int \epsilon_{IJKL}\left(-2de^I e^J \omega^{KL} + e^I e^J \omega^K{}_M \omega^{ML} - \frac{\Lambda}{6} e^I e^J e^K e^L\right).$$

The idea now is to substitute here the explicit solution (3.18) for the spin connection in terms of derivatives of the tetrad. But for this solution $de^I = -\omega^I{}_J e^J$. Substituting this we get

$$S[e] = \frac{1}{32\pi G}\int \epsilon_{IJKL}\left(2\omega(e)^I{}_M e^M e^J \omega(e)^{KL} + e^I e^J \omega(e)^K{}_M \omega(e)^{ML} - \frac{\Lambda}{6} e^I e^J e^K e^L\right),$$

where $\omega(e)$ is the torsion-free spin connection given by (3.18).

Let us now rewrite the previous action in index notations, making the spacetime indices of all the forms explicit. We will be using the following identity

$$dx^\mu dx^\nu dx^\rho dx^\sigma = \tilde{\epsilon}^{\mu\nu\rho\sigma} d^4x.$$

Here $\tilde{\epsilon}^{\mu\nu\rho\sigma}$ is a densitiesed completely antisymmetric tensor that in any coordinate system has components ± 1. Its value is $+1$ for the ordering of $\mu\nu\rho\sigma$ that coincides with the orientation of M chosen. The quantity d^4x is the coordinate volume element. We also have the identities

$$\epsilon_{IJKL}\tilde{\epsilon}^{\mu\nu\rho\sigma} e^I_\mu e^J_\nu e^K_\rho e^L_\sigma := 24e, \tag{3.62}$$

which is essentially the volume element of the metric defined by the tetrad

$$e = \sqrt{-g(e)}.$$

Here $g(e)$ is the determinant of the metric (3.14) defined by the tetrad. In writing the identity (3.62) we assumed that the orientation that defines ϵ_{IJKL} is chosen so that this identity is true. Another useful identity is

$$\frac{1}{2}\epsilon_{IJKL}\tilde{\epsilon}^{\mu\nu\rho\sigma} e^I_\mu e^J_\nu = 2e\, e^{[\rho}_K e^{\sigma]}_L. \tag{3.63}$$

It is proved by multiplying both sides with $e^K_\rho e^L_\sigma$.

We now rewrite the first term in the action using

$$e^M e^J = -\frac{1}{4}\epsilon^{MJRS}\epsilon_{RSPQ} e^P e^Q. \tag{3.64}$$

This gives

$$\begin{aligned} 2\epsilon_{IJKL}\omega^I{}_M e^M e^J \omega^{KL} &= -\frac{1}{2}\epsilon_{IJKL}\omega^I{}_M \epsilon^{MJRS}\epsilon_{RSPQ} e^P e^Q \omega^{KL} \\ &= -2e\epsilon^{MJRS}\epsilon_{IJKL} e^{[\mu}_R e^{\nu]}_S \omega^I_{\mu M}\omega^{KL}_\nu . \end{aligned}$$

Now using

$$\epsilon^{MJRS}\epsilon_{IJKL} = -\left(\delta^M_I(\delta^R_K\delta^S_L - \delta^R_L\delta^S_K) + \delta^M_K(\delta^R_L\delta^S_I - \delta^R_I\delta^S_L) + \delta^M_L(\delta^R_I\delta^S_K - \delta^R_K\delta^S_I)\right)$$

Using these identities the action becomes

$$S[e] = \frac{1}{8\pi G}\int e\left(-e^{[\mu}_K e^{\nu]}_L \omega^K_{\mu\ M}\omega^{ML}_\nu - \Lambda\right), \tag{3.65}$$

where we have used $\omega^M_{\mu M} = 0$. The next step is to substitute the solution (3.18).

To make the next step, let us introduce the torsion

$$\mathcal{T}^I_{\mu\nu} = 2\partial_{[\mu} e^I_{\nu]}. \tag{3.66}$$

This is the torsion of the zero connection, and should be contrasted with the torsion of the frame-compatible spin connection, which is zero. If we define the version of the torsion $\mathcal{T}^I{}_{\mu\nu}$ with all indices replaced by internal indices $\mathcal{T}^I{}_{JK} = e^\mu_J e^\nu_K \mathcal{T}^I_{\mu\nu}$ we can rewrite the torsion-free spin connection in terms of the objects $\mathcal{T}^I{}_{JK}$. We have

$$2e^\mu_S \omega^I_{\mu\ J} = \mathcal{T}_{SJ}{}^I + \mathcal{T}^I{}_{JS} + \mathcal{T}_{JS}{}^I. \tag{3.67}$$

This gives $\omega_I{}^I{}_J = \mathcal{T}^I{}_{JI}$ and

$$\omega_L{}^K{}_M \omega_K{}^{ML} = \frac{1}{4}\left(-\mathcal{T}_{LMK}\mathcal{T}^{LMK} + 2\mathcal{T}_{MLK}\mathcal{T}^{LKM}\right).$$

This gives the final expression for the GR Lagrangian in terms of the derivatives of the tetrad

$$S[e] = \frac{1}{16\pi G}\int e\left(-\frac{1}{4}\mathcal{T}_{LMK}\mathcal{T}^{LMK} + \frac{1}{2}\mathcal{T}_{MLK}\mathcal{T}^{LKM} + \mathcal{T}^I{}_{KI}\mathcal{T}^{JK}{}_J - 2\Lambda\right),$$

where $\mathcal{T}$ is the torsion given by (3.66). This form of the Lagrangian should be compared to the conceptually similar rewriting (2.19) of the Lagrangian in the metric formulation.

3.3.2 Weitzenbock Connection

Given that we can rewrite the gravitational Lagrangian in the torsion-squared form, there arises the possibility of trading torsion for the curvature. Thus, one introduces the so-called **Weitzenbock** connection defined via

$$\partial_\mu e^I_\nu - W^\rho{}_{\nu\mu} e^I_\rho = 0. \tag{3.68}$$

As this relation shows, the connection $W^\rho_{\mu\nu}$ is designed to parallel transport the frame field. This is an affine connection that has nonzero torsion $\mathcal{T}^I_{\mu\nu} = 2e^I_\rho W^\rho{}_{\nu\mu}$, but zero curvature $R^\rho{}_{\sigma\mu\nu}(W) = 0$. Thus, the curvature has been traded for torsion in this formulation. Teleparallel gravity allows for a two-parameter family of modifications in which the relative coefficients in front of the different torsion squared terms are changed as compared to the Lagrangian that describes GR. More details on this and other aspects of teleparallel gravity is available in Aldrovandi and Pereira (2013) and Heisenberg (2018).

3.4 Pure Connection Formulation

Given that it is possible to 'integrate out' the metric variable from Palatini Lagrangian (2.43) to obtain the pure affine formulation (2.46), one can ask whether a similar trick is possible with the Einstein–Cartan formulation. The field equations one gets for the tetrad are algebraic in any dimension, so this is always possible in principle. In 3D it is possible to obtain a closed-form expression for the corresponding pure connection Lagrangian, see the next chapter. In 4D the equation one needs to solve is (3.61). At present it is not known how to solve this equation for e^I in a closed form. However, a perturbative solution (around constant curvature background) is possible, see Zinoviev (2005) and Basile et al. (2016). It is also possible to 'integrate out' the frame field in a closed form using a trick with Lagrange multiplier fields, see Section 3.7.2.

We now describe this solution. For simplicity, we only treat the Riemannian case so that there are no subtle signs coming from the internal metric η_{IJ}. The constant curvature background corresponds to

$$R^{IJ}(\omega) = \frac{\Lambda}{3} e^I e^J. \tag{3.69}$$

Denoting by e^I, ω^{IJ} the background and by $\delta e^I, a^{IJ}$ the perturbations we have the following linearisation of (3.61)

$$\epsilon_{IJKL} e^J d^\omega a^{KL} = \frac{2\Lambda}{3} \epsilon_{IJKL} e^J e^K \delta e^L, \tag{3.70}$$

whose solution is

$$\delta e^I = \frac{3}{2\Lambda} \hat{f}^I_J e^J, \qquad \hat{f}^I_J := f^I_J - \frac{1}{6} \delta^I_J f, \tag{3.71}$$

where we introduced the linearised curvature $f^{IJ}_{KL} := 2 d^\omega_{[\mu} a^{IJ}_{\nu]} e^\mu_K e^\nu_L$ and $f^I_J = f^{IK}_{JK}, f = f^I_I$. Note that the linearised 'Ricci' tensor f^I_J does not need to be symmetric.

The linearisation of the action (3.60), evaluated on the solution (3.71) gives

$$S^{(2)}[a] = \frac{3}{32\pi G\Lambda} \int e(\delta^I_K \delta^J_M - \delta^I_M \delta^J_K) \hat{f}^K_I \hat{f}^M_J + \frac{\Lambda}{3} \epsilon_{IJKL} e^I e^J a^K{}_M a^{ML}, \tag{3.72}$$

where $e := (1/24)\epsilon_{IJKL} e^I e^J e^K e^L$ is the volume form for e^I. The last term here can be rewritten in a convenient form. Thus, one uses the background condition (3.69) to replace the wedge product of two e's with the curvature. The term $\epsilon_{IJKL} R^{IJ} a^K{}_M a^{ML}$ is then rewritten by replacing $a^{ML} = (1/4)\epsilon^{MLPQ}\epsilon_{PQRS} a^{RS}$, and decomposing the product of two of the ϵ's. We get

$$\epsilon_{IJKL} R^{IJ} a^K{}_M a^{ML} = R^{IM} a_M{}^J \epsilon_{IJKL} a^{KL} = (1/2)(d^\omega d^\omega) a^{IJ} \epsilon_{IJKL} a^{KL}. \tag{3.73}$$

Integrating by parts we can then replace the last term in (3.72) with

$$-\epsilon_{IJKL} d^\omega a^{IJ} d^\omega a^{KL} = -(e/4)\epsilon_{IJKL} \epsilon^{MNPQ} f^{IJ}_{MN} f^{KL}_{PQ}.$$

Thus, the last term in (3.72) can also be rewritten in the form curvature squared. The final result for the linearised action can be written very compactly as Basile et al. (2016)

$$S^{(2)}[a] = -\frac{3}{64\pi G\Lambda} \int e\, C^{KL}_{IJ}[a] C^{IJ}_{KL}[a], \tag{3.74}$$

where the Weyl-like tensor is defined as

$$C^{IJ}_{KL}[a] := f^{IJ}_{KL} - (\delta^I_{[K} f^J_{L]} - \delta^J_{[K} f^I_{L]}) + \frac{f}{3} \delta^I_{[K} \delta^J_{L]}. \tag{3.75}$$

Note that in Euclidean signature the action (3.74) has a definite sign. This is similar to Eddington–Schrödinger action (2.46), but in contrast to the Einstein–Hilbert action. The previous manipulations can be simplified by starting with the MacDowell–Mansouri action instead, as in Basile et al. (2016). In that case there is no need for integration by parts manipulations, and the linearised action (3.74) results immediately. We will describe this Section 3.5.1.

3.5 MacDowell–Mansouri Formulation

The idea in MacDowell and Mansouri (1977) is to combine the spin connection ω^{IJ} of the Einstein–Cartan formalism together with the tetrad e^I into a connection for the gauge group $SO(1,4)$ or $SO(2,3)$, depending on the sign of the cosmological constant. The Lie algebra of these groups splits as the sum of the Lorentz subalgebra plus an additional four-dimensional part. The frame receives the interpretation of the component of the connection in this four-dimensional part. A similar idea can be put to use in 3D gravity, where it leads to its Chern–Simons formulation in Witten (1988) and, when the cosmological constant is zero, in Poincaré gauge theories of gravity, see, e.g., Hehl (2012).

The connection that combines the spin connection and the tetrad is an example of the Cartan connection, as is explained in Wise (2010). Cartan geometry changes the nature of the object that ties together an abstract fibre bundle and the manifold. In the previous discussion this role has been played by the soldering form. In the Cartan's case this role is played by a 1-form in a principal H bundle P over M that is valued not in the Lie algebra of H (as would be appropriate for a principal connection), but rather in the Lie algebra of a bigger Lie group G of which H is a Lie subgroup. Moreover, the mapping $T_pP \to \mathfrak{g}$ is required to be an isomorphism. Thus, this connection provides a local identification of the total space of an H bundle over M with the group manifold G; see Wise (2010) for more details.

There are two versions of this formulation. In the original formulation of MacDowell and Mansouri (1977), the basic field is an $SO(1,4)$ or $SO(2,3)$ connection, but the Lagrangian is only invariant under the four-dimensional Lorentz group.[2] Invariance under $SO(1,4)$ or $SO(2,3)$ is explicitly broken. In the version of Stelle and West (1980) the symmetry breaking from $SO(1,4)$ or $SO(2,3)$ to $SO(1,3)$ is dynamical, due to an auxiliary vector field, often referred to as the compensator in the literature.

3.5.1 MacDowell–Mansouri Version

The curvature of an $SO(1,4)$ or $SO(2,3)$ connection has two parts. First, there is the part valued in the Lie algebra of the Lorentz group $SO(1,3)$. It is given by

$$\mathcal{F}^{IJ} = R^{IJ}(\omega) - \frac{\Lambda}{3} e^I e^J. \tag{3.76}$$

[2] Supergravity can also be described along the same lines, by replacing the gauge group that gives pure gravity with a supergroup, see MacDowell and Mansouri (1977).

Second, there is the remaining part, which is just a multiple of the torsion tensor $d^\omega e^I$. The four-dimensional MacDowell–Mansouri action is

$$S_{\rm MM}[e,\omega] = -\frac{3}{64\pi G\Lambda}\int \epsilon_{IJKL}\mathcal{F}^{IJ}\mathcal{F}^{KL}. \tag{3.77}$$

Using (3.76) we get the Einstein–Cartan action (3.60) plus a topological term.

The action (3.77) thus differs from (3.60) by a total derivative term, and leads to the same field equations. However, it has many advantages over the Einstein–Cartan action. First, its value on maximally symmetric backgrounds $\mathcal{F}^{IJ} = 0$ is zero. Second, in relation to the problem of evaluating the gravitational action on, e.g., asymptotically anti-de Sitter (AdS) spaces, the usual Einstein–Hilbert or Einstein-Cartan actions diverge on such backgrounds and require renormalisation. This is usually done by adding to the action certain boundary terms that also diverge as one approaches the AdS boundary. The difference between the divergent bulk and boundary actions is then the renormalised action, see, e.g., de Haro et al. (2001). The action (3.77) vanishes on exact AdS and is finite on asymptotically AdS solutions. Moreover, the difference between the Einstein–Cartan and MacDowell–Mansouri actions is a total derivative, or equivalently a boundary term. Thus, the boundary terms needed for the renormalisation on asymptotically AdS backgrounds are automatically included in (3.77).

Another advantage of (3.77) over (3.60) is that it is very easy to linearise this action on maximally symmetric backgrounds. Indeed, we have

$$S^{(2)}_{\rm MM}[\delta e, a] = -\frac{3}{64\pi G\Lambda}\int \epsilon_{IJKL}\left(d^\omega a^{IJ} - \frac{2\Lambda}{3}e^I\delta e^J\right)\left(d^\omega a^{KL} - \frac{2\Lambda}{3}e^K\delta e^L\right),$$

where, as in the previous subsection, $\delta e^I, a^{IJ}$ are the perturbations of the tetrad and the spin connection respectively. Substituting here the solution (3.71) gives the pure connection linearised action (3.74) with very little work. Indeed, the combination that appears in the previous linearised action evaluates to

$$d^\omega a^{IJ} - \frac{2\Lambda}{3}e^{[I}\wedge\delta e^{J]} = \frac{1}{2}\left(f^{IJ}_{KL} - 2\delta^{[I}_M \hat{f}^{J]}_N\right)e^M\wedge e^N = \frac{1}{2}C^{IJ}_{MN}[a]e^M\wedge e^N,$$

and the result (3.74) follows immediately.

In the MacDowell–Mansouri formulation the fields of the first-order formulation (3.60) have been unified into a single connection field, but now the Lagrangian (3.77) is no longer manifestly of the first order. Schematically, it is of the type F^2. However, the two-derivative term in (3.77) is, modulo total derivative terms, a term with no derivatives. This is why (3.77) is equivalent to the first-order Einstein–Cartan Lagrangian.

A final remark is that it is possible to put (3.77) into a manifestly first-order form by 'integrating in' a 2-form field, as in BF-type formulations that we consider in Section 3.7. This manifestly first-order form of the MacDowell–Mansouri theory has been studied by Freidel and Starodubtsev (2005).

3.5.2 Stelle–West Version

The action (3.77) can be rewritten in manifestly $SO(1,4)$ or $SO(2,3)$ invariant form by introducing an extra field. Let us denote the five-dimensional indices by lowercase latin letters, so that $SO(1,4)$ or $SO(2,3)$ Lie algebra-valued objects are of the form $v^{ab} = v^{[ab]}$. Let us introduce a new field v^a. This field is required to have unit norm $|v|^2 = \pm 1$, depending on the sign of the cosmological constant. Let us consider the following action

$$S[A,v] = -\frac{3}{64\pi G\Lambda}\int \epsilon_{abcde}\mathcal{F}^{ab}(A)\mathcal{F}^{cd}(A)v^e. \tag{3.78}$$

Here A^{ab} is a $SO(1,4)$ or $SO(2,3)$ connection, and $\mathcal{F}^{ab}(A)$ is its curvature. The action is manifestly invariant under the large group. Choosing v^a to point in a particular direction breaks the symmetry down to the Lorentz group, and reproduces (3.77). The unit norm constraint can be explicitly added to the action with a Lagrange multiplier, see Section 3.5.3.

To couple gravity in this form to matter one just has to note that the frame is readily recovered as the covariant derivative $d^A v^a$ (with respect to the connection A^{ab}) of the vector v^a. This allows to convert, e.g., the Dirac Lagrangian to an explicitly $SO(1,4)$ or $SO(2,3)$ invariant form by replacing all occurrences of e^I with ∇v^a.

3.5.3 Pure $SO(1,4)$ or $SO(2,3)$ Connection Formulation

The idea of this formulation is to integrate out the vector field v^a of the Stelle–West formulation. The corresponding Lagrangian has been described in West (1978). A similar procedure has been considered in Freidel and Starodubtsev (2005) in a related context, but with the curvature squared action (3.78) replaced by a BF-type action containing an additional auxiliary 2-form field B^{ab}.

Let us add to (3.78) a Lagrange multiplier term to enforce the constraint. For definiteness, we consider the case of positive Λ so that the relevant constraint is $|v|^2 = 1$. The action is

$$S[A,v,\mu] = -\frac{3}{64\pi G\Lambda}\int \epsilon_{abcde}\mathcal{F}^{ab}(A)\mathcal{F}^{cd}(A)v^e - \frac{\mu}{2}(|v|^2 - 1). \tag{3.79}$$

Varying this action with respect to v gives

$$\frac{1}{4}\tilde{\epsilon}^{\mu\nu\rho\sigma}\epsilon_{abcde}\mathcal{F}^{ab}_{\mu\nu}\mathcal{F}^{cd}_{\rho\sigma} \equiv \tilde{X}_a = \tilde{\mu}v_a, \tag{3.80}$$

where we introduced a convenient notation, and $\tilde{\mu}\, d^4x = \mu$. The Lagrange multiplier can now be solved from the constraint and reads

$$\tilde{\mu} = \sqrt{|\tilde{X}|^2}. \tag{3.81}$$

The resulting pure connection action West (1978) is the integral of the Lagrange multiplier

$$S[A] = -\frac{3}{64\pi G\Lambda}\int \sqrt{|\tilde{X}|^2}. \tag{3.82}$$

This action, however, is not very useful for a perturbative expansion. Indeed, one typically wants to expand around a maximally symmetric background, which in this case corresponds to $\mathcal{F}^{ab} = 0$. We cannot expand the square root around zero, and so (3.82) is not useful as a starting point for gravitational perturbation theory. But the action (3.79) one step before the pure connection action, and especially its MacDowell–Mansouri version (3.77) in which the de Sitter symmetry is explicitly broken to Lorentz is very convenient for developing perturbation theory, as we saw previously.

3.6 Dimensional Reduction

We postponed the treatment of the Kaluza–Klein dimensional reduction to this chapter because it is much easier to perform the required connection computations using the frame formalism. We will only consider here the original case of $4+1$ dimensional space and reduction to four spacetime dimensions. We follow C. Pope's Lectures on Kaluza–Klein in this section.

3.6.1 Metric Parametrisation and the Spin Connection

With anticipation that the metric should be parametrised in such a way as to make the corresponding frame as simple as possible, we choose the $4+1$ dimensional metric in the following form

$$d\hat{s}^2 = e^{2\alpha\phi}ds^2 + e^{2\beta\phi}(dz + A)^2. \tag{3.83}$$

Here α, β are constants that will be chosen later, the quantity ds^2 is the four-dimensional metric, z is the coordinate along the fifth dimension, and A is a 1-form on the four-dimensional manifold. The hatted quantities refer to the five-dimensional spacetime. The corresponding (co-)frame is then given by

$$\hat{e}^I = e^{\alpha\phi}e^I, \qquad \hat{e}^z = e^{\beta\phi}(dz + A). \tag{3.84}$$

We now assume that all fields depend just on the four coordinates of M. We then solve for the components of the spin connection from the equations

$$d\hat{e}^I + \hat{\omega}^I{}_J\hat{e}^J + \hat{\omega}^I{}_z\hat{e}^z = 0, \qquad d\hat{e}^z + \hat{\omega}^z{}_I\hat{e}^I = 0. \tag{3.85}$$

It is easiest to solve the second equation first. We get

$$\hat{\omega}^z{}_I = \beta e^{-\alpha\phi}\partial_I\phi\hat{e}^z + \frac{1}{2}e^{(\beta-2\alpha)\phi}F_{IJ}\hat{e}^J. \tag{3.86}$$

Here $\partial_I \phi = e^\mu_I \partial_\mu \phi$, and $F_{IJ} = e^\mu_I e^\nu_J F_{\mu\nu}$, where $F_{\mu\nu} = 2\partial_{[\mu} A_{\nu]}$ is the field strength. One then substitutes this solution into the first equation and obtains

$$\hat{\omega}^{IJ} = \omega^{IJ} + \alpha e^{-\alpha\phi}(\partial^J \phi\, \hat{e}^I - \partial^I \phi\, \hat{e}^J) - \frac{1}{2} e^{(\beta - 2\alpha)\phi} F^{IJ} \hat{e}^z. \tag{3.87}$$

3.6.2 Curvature, Action, and Field Equations

Let us now compute the components of the curvature, the dimensionally reduced action and finally the field equations.

It is convenient to make some choices regarding the constants α, β first. We would like the dimensionally reduced Lagrangian to reproduce the Einstein–Hilbert Lagrangian plus terms for the other fields. The volume element for the 5D metric $d\hat{s}^2$ is

$$\sqrt{-\hat{g}} = e^{(4\alpha+\beta)\phi} \sqrt{-g}. \tag{3.88}$$

On the other hand, the 5D Ricci scalar is a multiple of two copies of the inverse frame times the curvature of the 5D spin connection. The curvature of the 5D spin connection is a sum of the curvature of the 4D spin connection and other terms. The two copies of the inverse frame give a factor of $e^{-2\alpha\phi}$ times the 4D inverse frame. This means that we will have a factor of $e^{(2\alpha+\beta)}$ multiplying the 4D Einstein–Hilbert action, and so we want to set $\beta = -2\alpha$. The other choice one makes is to make sure that the coefficient in front of the kinetic term for the scalar field to be canonical, namely $-(1/2)\sqrt{-g}(\partial_\mu \phi)^2$. This can be shown to require $\alpha^2 = 1/12$.

The computation of the curvature 2-forms is quite technical and we just quote the result from C. Pope's lectures. The components of the Ricci tensor in the tetrad basis are

$$\begin{aligned} \hat{R}_{IJ} &= e^{-2\alpha\phi}\left(R_{IJ} - \frac{1}{2}\partial_I \phi \partial_J \phi - \alpha \eta_{IJ} \Box \phi\right) - \frac{1}{2} e^{-8\alpha\phi} F_I{}^K F_{JK}, \\ \hat{R}_{zI} &= \frac{1}{2} e^{\alpha\phi} \nabla^J \left(e^{-6\alpha\phi} F_{IJ}\right), \\ \hat{R}_{zz} &= 2\alpha e^{-2\alpha\phi} \Box \phi + \frac{1}{4} e^{-8\alpha\phi} F^2, \end{aligned} \tag{3.89}$$

where $F^2 = F_{IJ} F^{IJ}$. Therefore the 5D Ricci scalar is $\hat{R} = \eta^{IJ} \hat{R}_{IJ} + \hat{R}_{zz}$ is given by

$$\hat{R} = e^{-2\alpha\phi}\left(R - \frac{1}{2}(\partial\phi)^2 - 2\alpha \Box \phi\right) - \frac{1}{4} e^{-8\alpha\phi} F^2, \tag{3.90}$$

and so the dimensionally reduced Lagrangian, modulo a surface term, is

$$\mathcal{L} = \sqrt{-\hat{g}} \hat{R} = \sqrt{-g}\left(R - \frac{1}{2}(\partial\phi)^2 - \frac{1}{4} e^{-6\alpha\phi} F^2\right). \tag{3.91}$$

3.6.3 Consistent Truncation

If we could set $\phi = const$ we would have obtained 4D GR coupled to Maxwell theory. However, one of the 5D field equations prevents us from doing this. Indeed, assuming the 5D cosmological constant to be zero we have the equation $\hat{R}_{zz} = 0$, which reads

$$\Box\phi = -\frac{1}{8\alpha} e^{-6\alpha\phi} F^2. \tag{3.92}$$

This means that the electromagnetic field strength serves as the source for the scalar field, and so it is not consistent to set $\phi = const$ when $F^2 \neq 0$. Thus, the dimensionally reduced theory is not just GR coupled to Maxwell, it is necessarily a scalar tensor theory of gravity coupled to electromagnetism. It is a general feature of dimensional reduction that parameters determining the 'volume' or more generally 'shape' of the space one reduces on become (typically massless) fields in the dimensionally reduced theory. One does not see such fields in nature, which is one of the most serious problems of this approach to gravity/gauge theory unification.

3.7 BF Formulation

We now describe another formulation that is related to the tetrad formalism. A formulation of this type is also possible in the chiral context of Chapter 5, and will play an important role. As in the case of MacDowell–Mansouri formalism, there is a conceptual change occurring in this description, which is in the nature of the object that ties the bundle and manifold geometries. Thus, there is no longer the soldering form $e : TM \to E$ that ties the vector bundle to the tangent bundle. Instead, its role is played by a 2-form field B^{IJ} that maps antisymmetric rank $(2, 0)$ tensors (also known as bivectors) into the Lie algebra of the orthogonal group, i.e., objects in $\Lambda^2 V^*$. The vector bundle E with a connection $\omega^I{}_J$ on it is still present in this formalism.

3.7.1 Formulation with Lagrange Multiplier Fields

The idea of BF-type formulations is to replace the wedge product $\epsilon_{IJKL} e^K \wedge e^L$ of two tetrads in the Einstein–Cartan action with a new 2-form field B_{IJ}. The kinetic term of the Einstein–Cartan action then takes the form $B_{IJ} R^{IJ}$, where $R^{IJ} = R^{IJ}(\omega)$ is the curvature. If this was the only term in the action, the theory would coincide with the so-called BF theory, where the acronym stands for the fact that it is usual to label the 2-form field of this theory using letter B and the curvature of the connection using letter F. However, this is not the only term in the action, in particular because in 4D not every 2-form field B^{IJ} is of the required form. So, one adds a set of constraints on the 2-form field to guarantee that it 'comes from a tetrad'. In 4D this has been first considered by

Pietri and Freidel (1999), and so we will refer to the corresponding model by the initials of these authors.[3] The higher dimensional version has been developed in Freidel et al. (1999).

Consider the following action

$$S_{\text{FdP}}[B,\omega,\Psi] = \frac{1}{16\pi G}\int B_{IJ}R^{IJ}(\omega) - \frac{1}{2}\left(\Psi^{IJKL} + \frac{\Lambda}{6}\epsilon^{IJKL}\right)B_{IJ}B_{KL}. \quad (3.93)$$

The Lagrange multiplier field Ψ^{IJKL} is required to be tracefree $\Psi^{IJKL}\epsilon_{IJKL} = 0$. When $B_{IJ} = (1/2)\epsilon_{IJKL}e^K \wedge e^L$ the action (3.93) reduces to (3.60).

Varying (3.93) with respect to the Lagrange multiplier field Ψ^{IJKL} we get the constraint

$$B^{IJ} \wedge B^{KL} \sim \epsilon^{IJKL}. \quad (3.94)$$

As is shown in Freidel et al. (1999), Theorem 1, this equation implies that B^{IJ} is either the wedge product of two frame fields, or the dual of such a wedge product

$$B^{IJ} = \pm e_I \wedge e_J \qquad \text{or} \qquad B^{IJ} = \pm\frac{1}{2}\epsilon_{IJKL}e^K \wedge e^L. \quad (3.95)$$

The second set of solutions to the constraints (3.94) is what gives GR, because the action then reduces to (3.60). The first set of solutions gives the so-called Holst term; see Holst (1996). After integrating out the spin connection it becomes a total derivative.

The Lorentz group $SO(1,3)$, in whose Lie algebra the 2-forms fields B^{IJ} are valued, is not simple. The general invariant metric on the Lie algebra is an arbitrary linear combination of two metrics $\delta^{[I}_K\delta^{J]}_L$ and $\epsilon^{IJ}{}_{KL}$. In (3.94) we have imposed the tracelessness of Ψ^{IJKL} with respect to a particular metric from this class. It is also possible to consider a more general tracefree constraint, as was first studied in Capovilla et al. (2001). This removes the degeneracy present in (3.95) and gives a single solution, which is a linear combination of the two solutions in (3.95). The action evaluated on the solution is then the Einstein–Cartan action with the addition of the Holst term.

Thus, classically, the theory (3.93), or its version Capovilla et al. (2001) where one imposes a more general tracefree condition on Ψ^{IJKL}, describes GR in the sense that all solutions of GR are also solutions of this theory. It is also interesting to note that the theory (3.93) with nonzero Λ actually contains not more solutions than those of GR. Indeed, the field equations arising from (3.93) are as follows

$$R^{IJ}(\omega) = \left(\Psi^{IJKL} + \frac{\Lambda}{6}\epsilon^{IJKL}\right)B_{KL}, \quad (3.96)$$

$$d^\omega B^{IJ} = 0, \quad (3.97)$$

$$B_{IJ}B_{KL} \sim \epsilon_{IJKL}. \quad (3.98)$$

[3] Plebánski (1977) has considered essentially the same model before, as his paper also contains an action that includes both the self-dual and anti–self-dual sectors.

We then note that while the second and third of these equations are solved by $B^{IJ} = \pm e^I e^J$ and ω being the torsion-free connection, the first of the equations is in conflict with the Bianchi identity. Indeed, it would say that $\pm R^{KLIJ} = \Psi^{IJKL} + (\Lambda/6)\epsilon^{IJKL}$, where $R^{KLIJ} := (1/2)e^{\mu K}e^{\nu L}R_{\mu\nu}{}^{IJ}(\omega)$. The antisymmetrisation of the left-hand side of this relation vanishes, while that of the right-hand side is nonzero. Thus, there are no solutions to the theory (3.93) coming from the unwanted first set of solutions (3.95) of the constraints (3.94). All solutions of (3.93) with $\Lambda \neq 0$ are also solutions of GR.

The formulation (3.93) is the starting point of the so-called spin foam model quantisation of gravity; see Perez (2013).

3.7.2 Pure Connection Formulation Revisited

While it appears to be difficult to 'integrate out' the tetrad field from the Einstein–Cartan action (3.60), it is trivial to integrate out the 2-form field B^{IJ} from (3.93). Indeed, let us introduce a notation for the matrix that appears in front of the BB-term, and add a Lagrange multiplier that fixes the trace of this matrix

$$S[B,\omega,M,\mu] = \frac{1}{16\pi G}\int B_{IJ}R^{IJ}(\omega) - \frac{1}{2}M^{IJKL}B_{IJ}B_{KL} + \frac{\mu}{2}(M^{IJKL}\epsilon_{IJKL} - 4\Lambda), \tag{3.99}$$

where μ is a new Lagrange multiplier field. For simplicity, we carry out all manipulations in the case of Euclidean signature where $\epsilon^{IJKL}\epsilon_{IJKL} = +24$, but similar considerations apply to the Lorentzian signature case. It is trivial to integrate out the B-field. The resulting Lagrangian is

$$S[\omega, M, \mu] = \frac{1}{32\pi G}\int (M^{-1})_{IJKL}R^{IJ}R^{KL} + \mu(M^{IJKL}\epsilon_{IJKL} - 4\Lambda).$$

We now proceed to integrate out the matrix M. Its field equation is

$$M^{-1}XM^{-1} = \mu\epsilon. \tag{3.100}$$

Here we suppressed the indices and introduced a 4-form valued matrix

$$X^{IJKL} := R^{IJ} \wedge R^{KL}. \tag{3.101}$$

The equation (3.100) is interpreted as an equation in the space of 6×6 symmetric matrices. Its solution has first been spelled out in Mitsou (2019) and reads

$$M^{-1} = \pm\sqrt{\mu}\sqrt{\epsilon}\left(\sqrt{\epsilon}X\sqrt{\epsilon}\right)^{-1/2}\sqrt{\epsilon}. \tag{3.102}$$

Both X and μ here are 4-form valued, with X being a 4-form with additional values in the space of 6×6 matrices. To make sense of a relation like (3.102) one can introduce an arbitrary volume form on the manifold, and obtain an actual matrix by dividing X^{IJKL} by the introduced volume form. Similarly, μ

can be written as a function times the same volume form. One then sees that the relation (3.102) is homogeneity degree zero in this volume form, and so it does not matter which volume form is used to make sense of it, as all 4-forms are related by multiplication by a nowhere zero function.

The Lagrange multiplier field μ is then obtained from the constraint $\mathrm{Tr}(M\epsilon) = 4\Lambda$. Taking the plus branch in (3.102) this gives

$$\sqrt{\mu} = \frac{1}{4\Lambda}\mathrm{Tr}\sqrt{\sqrt{\epsilon}X\sqrt{\epsilon}} = \frac{1}{4\Lambda}\mathrm{Tr}\sqrt{\epsilon X}, \tag{3.103}$$

which fixes M in terms of X completely. In the last expression we took into account that the trace is cyclic even in the presence of the square root. The latter is proved as follows

$$\mathrm{Tr}(\sqrt{NM}) = \mathrm{Tr}(N^{-1}\sqrt{NM}N) = \mathrm{Tr}(\sqrt{N^{-1}NMN}) = \mathrm{Tr}(\sqrt{MN}),$$

where we used the fact that the similarity transformation commutes with the matrix square root $N^{-1}\sqrt{MN} = \sqrt{N^{-1}MN}$. Substituting this back into the action we get the closed form pure connection action

$$S[\omega] = \frac{1}{128\pi G\Lambda}\int \left(\mathrm{Tr}\sqrt{\epsilon X}\right)^2. \tag{3.104}$$

Some comments are in order. First, we note that, interestingly, the object $\sqrt{\epsilon}$ that appears in the intermediate stages of the previous derivation properly exists as a real matrix only in the Lorentzian signature. Indeed, we have in general

$$\epsilon^2 = 4\sigma\,\mathbb{I}, \tag{3.105}$$

where $\sigma = \pm 1$ and equals minus one in the Lorentzian case. We have introduced

$$\mathbb{I}^{IJ}{}_{KL} := \delta^{[I}_K \delta^{J]}_L, \tag{3.106}$$

which is the identity operator on the space of antisymmetric 4×4 tensors. We can search for the square root of ϵ in the space of linear combinations of the matrix ϵ and the identity matrix

$$\sqrt{\epsilon} = \alpha\,\mathbb{I} + \beta\,\epsilon. \tag{3.107}$$

Squaring both sides gives that α, β must satisfy

$$\alpha^2 + 4\sigma\beta^2 = 0, \qquad 2\alpha\beta = 1, \tag{3.108}$$

which is only possible for real α, β for $\sigma = -1$. So, in the case of the Euclidean signature the object $\sqrt{\sigma}$ is complex. However, this does not cause any problems because in all the final expressions only the matrix ϵ itself always appears.

Our second comment is about the value of the action (3.104) on the background (3.69). We have

$$X^{IJKL} = \left(\frac{\Lambda}{3}\right)^2 e^I e^J e^K e^L = \left(\frac{\Lambda}{3}\right)^2 e\,\epsilon^{IJKL}, \tag{3.109}$$

which means that the background value of X is a multiple of the ϵ. This gives

$$\sqrt{\sqrt{\epsilon}X\sqrt{\epsilon}} = \frac{2\Lambda}{3}\sqrt{e}\,\mathbb{I}. \tag{3.110}$$

We will need the result in this form Section 3.7.3. Substituting all this into (3.104) gives the correct $(\Lambda/8\pi G)\int e$ for the action on the maximally symmetric background.

Our final comment here is that the action (3.104) is only defined perturbatively around the maximally symmetric background $X \sim \epsilon$. In this case the matrix that appears under the square root is close to the identity matrix, and the square root can be defined in the sense of a perturbative expansion in powers of deviations of the matrix from the identity. For a general matrix X the square root has many different branches. These can be seen by diagonalising the matrix ϵX and then taking the square roots of the eigenvalues. The first problem that one can encounter is that some of the eigenvalues may be negative. One will obtain a complex action in this case. The second problem is that even if the eigenvalues are all nonnegative, the square root of each of the eigenvalues has two branches. While it does not matter which branch is chosen if the same choice is followed for all the eigenvalues (because one takes the square of the trace of the square root in the action), one can also take the positive branch for some of the eigenvalues and the negative branch for some others. The resulting action is of course very different from the one where one only takes say the positive branch for all the eigenvalues. The similar problems occur in the chiral pure connection action to be written down in Chapter 5, except that in that case it can be guaranteed that the eigenvalues of the matrix of the square root are all nonnegative and the action is real. We will discuss this in due course. But the problem of different individual branches is present also in the chiral formulation. Moreover, analysing some explicit solutions one can convince oneself that it is not consistent to restrict one's attention only to the uniformly positive branch. The situation is more complex, and we will return it in the chapter on the chiral pure connection formalism. This discussion illustrates that, as it stands, the action (3.104) is only perturbatively defined.

3.7.3 Linearised Pure Connection Action

In this subsection we perform the exercise of linearising the action (3.104) around the maximally symmetric background. Our goal is to compare the result with (3.74).

We need some preparation. First, we have

$$\delta X^{IJKL} = d^\omega a^{IJ} R^{KL} + R^{IJ} d^\omega a^{KL}. \tag{3.111}$$

Using

$$d^\omega_{[\mu} a^{IJ}_{\nu]} = \frac{1}{2} f^{IJ}_{KL} e^K_\mu e^L_\nu \tag{3.112}$$

the first variation of X can be rewritten as

$$\delta X = \frac{\Lambda}{6} e(\epsilon f + f\epsilon). \tag{3.113}$$

We also need an expression for the second variation of this matrix

$$\delta^2 X^{IJKL} = 2(aa)^{IJ} R^{KL} + 2R^{IJ}(aa)^{KL} + 2d^\omega a^{IJ} d^\omega a^{KL}. \tag{3.114}$$

Here $(aa)^{IJ} = a^I{}_M a^{MJ}$. We will not attempt to transform this expression any further for now.

Let us now write

$$\sqrt{\sqrt{\epsilon} X \sqrt{\epsilon}} \sqrt{\sqrt{\epsilon} X \sqrt{\epsilon}} = \sqrt{\epsilon} X \sqrt{\epsilon}, \tag{3.115}$$

and vary this expression. We use the fact (3.110) that the background value of the matrices on the left-hand side is a multiple of the identity. This gives

$$\delta \sqrt{\sqrt{\epsilon} X \sqrt{\epsilon}} = \frac{3}{4\Lambda\sqrt{e}} \sqrt{\epsilon}\, \delta X \sqrt{\epsilon}, \tag{3.116}$$

from where we get an expression for the trace

$$\mathrm{Tr}\left(\delta \sqrt{\epsilon X}\right) = \sqrt{e}\, \mathrm{Tr}(f). \tag{3.117}$$

We now vary the expression (3.115) the second time to get

$$\frac{4\Lambda}{3} \sqrt{e}\, \delta^2 \sqrt{\sqrt{\epsilon} X \sqrt{\epsilon}} + 2\delta \sqrt{\sqrt{\epsilon} X \sqrt{\epsilon}}\, \delta \sqrt{\sqrt{\epsilon} X \sqrt{\epsilon}} = \sqrt{\epsilon}\, \delta^2 X \sqrt{\epsilon}. \tag{3.118}$$

We need the integral of the first term on the left-hand side. Let us consider the integral of the trace of the right-hand side

$$\int \mathrm{Tr}\left(\sqrt{\epsilon}\, \delta^2 X \sqrt{\epsilon}\right) = \int \epsilon_{IJKL}(4R^{IJ}(aa)^{KL} + 2d^\omega a^{IJ} d^\omega a^{KL}). \tag{3.119}$$

Integrating by parts in the second term we have

$$\begin{aligned} 2\int \epsilon_{IJKL} d^\omega a^{IJ} d^\omega a^{KL} &= -2\int \epsilon_{IJKL} a^{IJ} d^\omega d^\omega a^{KL} \\ &= -4\int \epsilon_{IJKL} a^{IJ} R^K{}_M a^{ML}. \end{aligned}$$

But we have already see in (3.73) that the quantities $\epsilon_{IJKL} a^{IJ} R^K{}_M a^{ML}$ and $\epsilon_{IJKL} R^{IJ}(aa)^{KL}$ are equal. This means that the integral in (3.119) vanishes. Thus, under the integral sign we have

$$\begin{aligned} \int \sqrt{e}\, \mathrm{Tr}\left(\delta^2 \sqrt{\sqrt{\epsilon} X \sqrt{\epsilon}}\right) &= -\frac{3}{2\Lambda} \int \mathrm{Tr}\left(\delta \sqrt{\sqrt{\epsilon} X \sqrt{\epsilon}}\, \delta \sqrt{\sqrt{\epsilon} X \sqrt{\epsilon}}\right) \\ = -2\left(\frac{3}{4\Lambda}\right)^3 \int \frac{1}{e} \mathrm{Tr}\left(\epsilon\, \delta X \epsilon\, \delta X\right) &= -\frac{3}{16\Lambda} \int e\, \mathrm{Tr}(4f^2 + \epsilon f \epsilon f), \end{aligned}$$

where we have used (3.116) to get the first expression in the second line and (3.113) to get the last expression. The quantity $\mathrm{Tr}(\epsilon f \epsilon f)$ can be computed by expanding the product of the ϵ tensors. This gives

$$\mathrm{Tr}(\epsilon f \epsilon f) = 4 f^{IJ}_{KL} f^{KL}_{IJ} - 16 f_{IJ} f^{IJ} + 4 f^2, \tag{3.120}$$

where $f^J_I := f^{KJ}_{KI}, f = f^I_I$. Overall, this gives

$$\int \sqrt{e}\,\mathrm{Tr}\left(\delta^2\sqrt{\sqrt{\epsilon}X\sqrt{\epsilon}}\right) = -\frac{3}{2\Lambda}\int e\left(f^{IJ}_{KL}f^{KL}_{IJ} - 2f_{IJ}f^{IJ} + \frac{1}{2}f^2\right).$$

We are now ready to assemble the pieces of the linearised action. The second-order action is half the second variation and we get

$$\begin{aligned} S^2[a] &= \frac{1}{128\pi G\Lambda}\int \mathrm{Tr}\left(\delta\sqrt{\epsilon X}\right)\mathrm{Tr}\left(\delta\sqrt{\epsilon X}\right) + \mathrm{Tr}\sqrt{\epsilon X}\,\mathrm{Tr}\left(\delta^2\sqrt{\epsilon X}\right) \\ &= -\frac{3}{64\pi G\Lambda}\int e\left(f^{IJ}_{KL}f^{KL}_{IJ} - 2f_{IJ}f^{IJ} + \frac{1}{3}f^2\right), \end{aligned}$$

which coincides with (3.74). This proves that (3.104) reproduces the linearised pure connection action that can be obtained from the Einstein–Cartan formalism. It is thus is a good starting point for gravitational perturbation theory around $\Lambda \neq 0$ background.

3.7.4 Modifications of GR in BF Formalism

An interesting class of modifications of GR can be obtained by changing the constraint on the second line in (3.99). The constraint present in the GR action (3.99) fixes a certain specific gauge-invariant function of the matrix M to be constant. However, there are many gauge-invariant functions of 6×6 symmetric matrices M that can be written down, e.g., $\mathrm{Tr}(M^2)$ or traces of higher powers, and, e.g., $\det(M)$. Such a change causes the theory to be modified rather dramatically, a generic modified theory of this type turns out to be a bi-metric theory of gravity. Such modifications have first been studied in Smolin (2009), and then in papers by Speziale (2010), Lisi et al. (2010), Beke (2011), Beke et al. (2012), and Alexander et al. (2014). It was shown in Alexandrov and Krasnov (2009) that such theories in general propagate $2 + 6$ degrees of freedom, which is the generic propagating content of a bi-metric theory of gravity.

3.7.5 Field Redefinitions

In view of the discussion in the previous section, it is perhaps surprising that one can modify the constraint on the second line in (3.99) nontrivially without changing the theory Krasnov (2018). To see how this becomes possible, let us first rewrite the action (3.99) in an index-free notation. We have

$$S[B,\omega,M,\mu] = \frac{1}{16\pi G}\int B^t R - \frac{1}{2}B^t M B + \frac{\mu}{2}(\mathrm{Tr}(\epsilon M) - 4\Lambda). \tag{3.121}$$

Here we think about objects B, R as columns on which matrices such as M, ϵ can act, and t is the transpose. Let us now carry out a field redefinition that replaces B by a mixture of B and R

$$B = G\tilde{B} + HR. \tag{3.122}$$

The action in terms of the new 2-form field $\tilde{B}$ will contain the R^2 terms. The idea is to choose the (not necessarily symmetric) matrices G, H in such a way that these terms are just multiples of the two topological invariants that can be formed of R, namely

$$\int R^t R, \qquad \int R^t \epsilon R. \tag{3.123}$$

Thus, we will require the R^2 term of the action in terms of $\tilde{B}$ to be

$$\int R^t T R, \qquad T = t_1 \mathbb{I} + t_2 \star . \tag{3.124}$$

where $\star = \frac{1}{2}\epsilon$. This is a convenient numerical factor because $\star^2 = \mathbb{I}$. Equating the matrix arising in the R^2 term of the new Lagrangian to T we get the following equation

$$H^t - \frac{1}{2} H^t M H = T. \tag{3.125}$$

We will also require that the BR term of the new Lagrangian keeps its canonical form. This gives another equation

$$G^t - G^t M H = \mathbb{I}. \tag{3.126}$$

We now note that the equation (3.125) tells us that H is a symmetric matrix. So, we drop the transpose symbol from it from now on. We also note that one can rewrite the second equation as $\mathbb{I} - MH = (G^t)^{-1}$, and the first as $\mathbb{I} - (1/2)MH = H^{-1}T$. We assumed that all matrices are invertible. Taking the difference of these equations gives $1 + (G^t)^{-1} = 2H^{-1}T$, from which we can write

$$H = 2T(1 + (G^t)^{-1})^{-1}. \tag{3.127}$$

Substituting this into the first equation we get the solution for G^t, and then the solution for H. So, the solutions of equations (3.125) and (3.126) are given by

$$H = 2T(\mathbb{I} + (G^t)^{-1})^{-1}, \qquad (G^t)^{-2} = \mathbb{I} - 2MT. \tag{3.128}$$

We now have to be careful again because the matrix $(G^t)^{-2}$ on the left-hand side of the second equation is not necessarily symmetric, and so it is not clear how to take the square root to get G itself. However, we can rewrite the right-hand side of the second equation as

$$\mathbb{I} - 2MT = T^{-1/2}(\mathbb{I} - 2T^{1/2} M T^{1/2}) T^{1/2}, \tag{3.129}$$

where we made some choice of a symmetric square root $T^{1/2}$ of the symmetric matrix T, and also assumed that T is invertible. We now note that the matrix that appears in the above expression (3.129) in brackets is symmetric, and so the notion of its square root makes sense. Thus, we can take the square root of $\mathbb{I} - 2MT$ as

$$(\mathbb{I} - 2MT)^{1/2} = T^{-1/2}(\mathbb{I} - 2T^{1/2}MT^{1/2})^{1/2}T^{1/2}. \tag{3.130}$$

This gives the final solution for G^t, H

$$\begin{aligned} G^t &= T^{-1/2}(\mathbb{I} - 2T^{1/2}MT^{1/2})^{-1/2}T^{1/2}, \\ H &= T^{1/2}(\mathbb{I} + T^{1/2}MT^{1/2})^{-1}T^{1/2}. \end{aligned} \tag{3.131}$$

We note that the expression for H is symmetric, as it should be. Finally, the coefficient matrix of the $\tilde{B}^2$ term, given by $\tilde{M} = G^{tr}MG$ is given by

$$\tilde{M} = T^{-1/2}(\mathbb{I} - 2T^{1/2}MT^{1/2})^{-1/2}(T^{1/2}MT^{1/2})(\mathbb{I} - 2T^{1/2}MT^{1/2})^{-1/2}T^{-1/2},$$

which is manifestly symmetric as it should be. But now the three terms in the middle only contain the matrix $T^{1/2}MT^{1/2}$ and the identity matrix, and so they commute. Therefore, we can also write

$$\tilde{M} = T^{-1/2}(T^{1/2}MT^{1/2})(\mathbb{I} - 2T^{1/2}MT^{1/2})^{-1}T^{-1/2}, \tag{3.132}$$

or more compactly

$$\mathbb{I} + 2T^{1/2}\tilde{M}T^{1/2} = (\mathbb{I} - 2T^{1/2}MT^{1/2})^{-1}, \tag{3.133}$$

from which the matrix M in terms of $\tilde{M}$ can be explicitly expressed as

$$M = \tilde{M}(\mathbb{I} + 2T\tilde{M})^{-1}, \tag{3.134}$$

which finally eliminates the square root of T from the expressions. Formally expanding the inverse on the right-hand side in powers of $T\tilde{M}$ it can be seen that the right-hand side is symmetric, as it should be, in spite of T and $\tilde{M}$ not commuting.

All in all, we learn that there is a two-parameter $t_{1,2}$ family of Lagrangians all giving a classically equivalent description of GR. They are all of BF-type and can be written as

$$S[B, A, M] = \frac{1}{16\pi G}\int B^tF - \frac{1}{2}B^tMB + R^tTR + \frac{\mu}{2}\left(\mathrm{Tr}\left[\epsilon M(\mathbb{I} + 2TM)^{-1}\right] - 4\Lambda\right), \tag{3.135}$$

with $T = 0$ corresponding to the original Lagrangian (3.121). Note that we have omitted the tildes from all the quantities in the above Lagrangian (3.135). The action (3.135) is still of the BF type with a constraint for the matrix appearing in front of the BB term. However, this constraint has changed considerably as compared to what it is in the standard action (3.99). In particular, as we shall see in the next subsection, while it is impossible to 'integrate out' the matrix M from (3.99), it is possible to obtain a Lagrange multiplier-free formulation starting with (3.135).

3.7.6 Formulation with a Potential for the 2-Form Field

One of the purposes of the manipulations of the previous subsection is that it turns out to be possible to 'integrate out' the matrix M from the action (3.135) to obtain a pure BF-type action with a potential for the 2-form field. It is perhaps surprising that such a formulation is at all possible, because in the original Lagrangian (3.121) there is a Lagrange multiplier field that imposes a constraint on the 2-form field, so some components of the B-field are non-dynamical. And it is not possible to integrate out the Lagrange multiplier fields from (3.121). However, after the field redefinitions, it will be possible to eliminate all the Lagrange multiplier fields from (3.135), with all the components of the redefined 2-form field becoming dynamical. It is far from obvious that such a formulation of GR should be possible.

Let us carry out this exercise. The field equation that arises by varying (3.135) with respect to M is

$$(\mathbb{I} + 2TM)X_B(\mathbb{I} + 2MT) = \mu\epsilon. \tag{3.136}$$

We introduced a 4-form valued matrix $X_B := BB^t$ constructed from the 2-form field. This equation can be rewritten as

$$(T + 2TMT)T^{-1}X_B T^{-1}(T + 2TMT) = \mu\epsilon. \tag{3.137}$$

Following Mitsou (2019), the solution for the symmetric matrix $T + 2TMT$ is given by

$$T + 2TMT = \pm\sqrt{\mu}\,\sqrt{\epsilon}\,(\sqrt{\epsilon}T^{-1}X_B T^{-1}\sqrt{\epsilon})^{-1/2}\sqrt{\epsilon}. \tag{3.138}$$

Taking the plus-branch solution, this gives

$$2M = \sqrt{\mu}T^{-1}\sqrt{\epsilon}(\sqrt{\epsilon}T^{-1}X_B T^{-1}\sqrt{\epsilon})^{-1/2}\sqrt{\epsilon}T^{-1} - T^{-1}, \tag{3.139}$$

and

$$2M(1 + 2TM)^{-1} = T^{-1} - \frac{1}{\sqrt{\mu}}\frac{1}{\sqrt{\epsilon}}(\sqrt{\epsilon}T^{-1}X_B T^{-1}\sqrt{\epsilon})^{1/2}\frac{1}{\sqrt{\epsilon}}. \tag{3.140}$$

Taking the trace of the product of this matrix with ϵ we determine μ

$$\sqrt{\mu} = \frac{\mathrm{Tr}\sqrt{\sqrt{\epsilon}T^{-1}X_B T^{-1}\sqrt{\epsilon}}}{\mathrm{Tr}(T^{-1}\epsilon) - 8\Lambda}, \tag{3.141}$$

which determines M completely. We now substitute the result into the action to obtain

$$S[B,\omega] = \frac{1}{16\pi G}\int B^t F + \frac{1}{4}B^t T^{-1}B - (4\text{Tr}(T^{-1}\epsilon) - 32\Lambda)^{-1}\left(\text{Tr}\sqrt{\epsilon T^{-1}X_B T^{-1}}\right)^2. \tag{3.142}$$

We have used the cyclicity of the trace to convert two factors of $\sqrt{\epsilon}$ into a single copy of ϵ. It is clear that this action depends on the existence of T^{-1}, where T is given by (3.124). Explicitly, this tensor is given by

$$T^{-1} = \frac{t_1}{t_1^2 - \sigma t_2^2} - \frac{t_2 \star}{t_1^2 - \sigma t_2^2}, \tag{3.143}$$

where again $\star = \frac{1}{2}\epsilon$ and $\sigma = \pm 1$ is the signature-dependent sign. Thus, T^{-1} exists provided $t_1^2 \neq \sigma t_2^2$. This is always the case for real t_1, t_2 and Lorentzian signature.

The action (3.142) is of the BF-type plus a potential for the B-field. The potential consists of two terms. One is quadratic in the B field. In fact, the first two terms in the action (3.142) correspond to a topological theory with no propagating degrees of freedom. The last term is what breaks the topological symmetry and gives rise to a theory with degrees of freedom. This potential is non-polynomial, and of the same type as arises in considerations (3.104) of the pure connection formulation. As we have already remarked, it is far from obvious that such a formulation of GR is at all possible.

As the last step in this line of developments, one can solve the equation $d^\omega B^{IJ} = 0$, which is the Euler–Lagrange equation arising by varying (3.142) with respect to the spin connection. The number of equations here is 4×6, which matches the number of unknowns in $\omega_\mu^I{}_J$. Thus, when B^{IJ} is suitably non-degenerate, one expects the solution for $\omega(B)$ to exist. One can then substitute this solution back into (3.142) and obtain a second order in derivatives theory for the 2-form field B^{IJ}. So far, nobody has carried out this exercise, but a similar procedure is possible and has been done in detail in the chiral case to be considered in Chapter 5.

4

General Relativity in 2+1 Dimensions

The purpose of this section is to describe the subtleties of the tetrad formalism for general relativity (GR) in the case of gravity in 2+1 spacetime dimensions. We describe both the Einstein–Cartan formalism in this case, as well as the Chern–Simons formulation. We then describe the pure connection formalism, which in this number of dimensions can be obtained in closed form directly from the tetrad formulation. As is well-known, and as will also be apparent from our considerations of this chapter, there are no propagating degrees of freedom in 2+1 gravity. This follows, e.g., from the fact that the Weyl tensor is identically zero in this number of dimensions, or from the fact that all solutions of Einstein equations in 2+1 have constant curvature. We also develop a very convenient index-free notation that is possible by identifying the Lie algebra of the Lorentz (orthogonal) group with 2×2 anti-hermitian matrices.

4.1 Einstein–Cartan and Chern–Simons Formulations

Let us start by reviewing some basic facts about 3D gravity.

4.1.1 Einstein–Cartan Frame Formalism in 3D

Let $e^i, i = 1, 2, 3$ be a frame field so that the 3D metric is

$$ds^2 = e^i \otimes e^j \eta_{ij}, \tag{4.1}$$

where η_{ij} is either $\eta_{ij} = \text{diag}(1, 1, 1)$ or $\eta_{ij} = \text{diag}(-1, 1, 1)$ depending on the desired signature. There are subtle differences between the two signature cases. For definiteness, let us consider the all-plus signature case.

For the Riemannian signature we raise and lower indices with the metric δ_{ij}, and the SO(3) spin connection is the set of 1-forms $w^{ij} = w^{[ij]}$. The antisymmetry is the statement that the connection is δ_{ij}-metric compatible. Let f^{ij} be the curvature

$$f^{ij} = dw^{ij} + w^{ik} w_k{}^j. \tag{4.2}$$

We then write the following action

$$S[e,w] = -\frac{1}{4}\int_M \left(e^i f^{jk} - \frac{\Lambda}{3} e^i e^j e^k \right) \epsilon_{ijk}. \tag{4.3}$$

The orientation implied here is that of the 3-form $e^i e^j e^k \epsilon_{ijk}$. The minus sign in front of the action is the usual choice for the all-plus signature. We work in units in which the 3D Newton's constant satisfies $4\pi G = 1$. Varying this action with respect to w we get the torsion-free condition

$$d_w e^i \equiv de^i + w^i{}_j e^j = 0. \tag{4.4}$$

It says that the connection w is the unique e-compatible connection. Substituting this connection into (4.3) we find

$$S[e, w(e)] = -\frac{1}{4}\int_M (R - 2\Lambda) v_g, \tag{4.5}$$

where R is the Ricci scalar of the metric, and the integration is carried out with respect to the metric volume element v_g.

Varying the action with respect to the frame field we get

$$f^{ij} = \Lambda e^i e^j, \tag{4.6}$$

which says that the curvature of an Einstein metric in three dimensions is constant. Thus, there are no propagating degrees of freedom in 2+1 dimensional gravity.

The connection matrix w^{ij} being antisymmetric, we can write

$$w^{ij} = \epsilon^{ikj} w^k, \tag{4.7}$$

which defines the new connection 1-forms w^i. We then have for the curvature

$$f^{ij} = \epsilon^{ikj} f^k, \qquad f^i = dw^i + \frac{1}{2}\epsilon^{ijk} w^j w^k. \tag{4.8}$$

Thus, the exceptional feature of this number of dimensions is that the spin connection and the tetrad can both be thought of as $\mathbb{R}^3$ valued 1-forms. In particular, this allows them to be mixed in a Cartan connection in a particularly simple way, see the next section.

4.1.2 Matrix Notations

It is very convenient to get rid of the internal $i, j, \ldots$ indices at the expense of making all objects 2×2 matrix valued. To this end, we use the isomorphism of the Lie algebras $\mathfrak{so}(3) = \mathfrak{su}(2)$. The Lie algebra generators are

$$\tau_i = -\frac{\mathrm{i}}{2}\sigma_i, \tag{4.9}$$

where σ_i are the usual Pauli matrices. We have

$$\mathrm{Tr}(\tau_i \tau_j) = -\frac{1}{2}\delta_{ij}, \qquad [\tau_i, \tau_j] = \epsilon_{ij}{}^k \tau_k. \tag{4.10}$$

The index of ϵ here is raised with the δ^{ij} metric.

We then form a matrix-valued connection

$$\mathbf{w} := w^i \tau_i. \tag{4.11}$$

In what follows we will always denote a matrix-valued object by a bold-face letter. The matrix valued curvature $\mathbf{f} := f^i \tau_i$ is computed as

$$\mathbf{f} = d\mathbf{w} + \mathbf{w}\mathbf{w}. \tag{4.12}$$

We also form anti-hermitian frame field 1-forms

$$\mathbf{e} := e^i \tau_i, \tag{4.13}$$

in terms of which the metric is

$$ds^2 = -2\,\mathrm{Tr}(\mathbf{e} \otimes \mathbf{e}). \tag{4.14}$$

In terms of the matrix-valued fields the torsion-free condition (4.4) takes the form

$$d_{\mathbf{w}}\mathbf{e} \equiv d\mathbf{e} + \mathbf{w}\mathbf{e} + \mathbf{e}\mathbf{w} = 0. \tag{4.15}$$

The field equation obtained by varying the action (4.3) with respect to $\mathbf{e}$ takes the form

$$\mathbf{f} = -\Lambda \mathbf{e}\mathbf{e}. \tag{4.16}$$

In the described index-free notation the action takes the form

$$S[\mathbf{e}, \mathbf{w}] = -\int_M \mathrm{Tr}\left(\mathbf{e}\mathbf{f} + \frac{\Lambda}{3}\mathbf{e}\mathbf{e}\mathbf{e}\right). \tag{4.17}$$

In what follows, we will mainly consider the $\Lambda < 0$ case. We set for simplicity

$$\Lambda = -1. \tag{4.18}$$

A different value of $|\Lambda|$ can always be reinstalled by rescaling the frame field.

We note that in case of Lorentzian signature metrics the relevant Lorentz group $\mathrm{SO}(1,2)$ is isomorphic to $\mathrm{SL}(2,\mathbb{R})/\mathbb{Z}_2$. This means that a similar index-free notation is also possible in this case, except that one has to work with real 2×2 tracefree matrices instead. We will leave details of the corresponding formalism as an exercise to the reader.

4.1.3 Chern–Simons Formulation

The two sets of equations $\nabla \mathbf{e} = 0, \mathbf{f} = \mathbf{ee}$ can be combined as the real and imaginary parts of a single complex-valued equation by introducing the complex tracefree 2×2 matrix-valued field

$$\mathbf{a} := \mathbf{w} + \mathrm{i}\mathbf{e}. \tag{4.19}$$

The field equations of 3D gravity then combine into the statement that the curvature of the $\mathrm{SL}(2,\mathbb{C})$ connection $\mathbf{a}$ is zero

$$0 = \mathbf{f}(\mathbf{a}) \equiv d\mathbf{a} + \mathbf{aa}. \tag{4.20}$$

These are the field equations following from the so-called Chern–Simons Lagrangian. Alternatively, we can write the Einstein–Cartan Lagrangian (4.17) (with $\Lambda = -1$), modulo a surface term, as

$$S[\mathbf{e}, \mathbf{w}] = -\frac{1}{2}\mathrm{Im} \int_M CS[\mathbf{a}], \tag{4.21}$$

where

$$CS[\mathbf{a}] := \mathrm{Tr}\left(\mathbf{a}d\mathbf{a} + \frac{2}{3}\mathbf{aaa}\right) \tag{4.22}$$

is the Chern–Simons 3-form for $\mathbf{a}$.

4.1.4 Topological Term

It is possible to add to (4.17) also the real part of the Chern–Simons functional of $\mathbf{a}$ with an arbitrary coefficient, see Witten (1988), section 2.3. When written in terms of $\mathbf{e}, \mathbf{w}$ this reads

$$\mathrm{Re} \int_M CS[\mathbf{a}] = \int_M \left(CS[\mathbf{w}] - \mathrm{Tr}(\mathbf{e}\, d_{\mathbf{w}}\mathbf{e})\right). \tag{4.23}$$

It is not hard to check that this term does not affect the field equations, in the sense that a linear combination of the two resulting field equations still says that the connection is metric-compatible.

4.1.5 Quantum Theory

Even though this book is about classical theory, it is appropriate to give some comments on the quantum case. Because the theory of 3D gravity is topological, one expects to be able to construct the corresponding quantum theory.

This is because one is dealing with a problem in quantum mechanics rather than in quantum field theory, and so the theory should exist. This is in contrast with the situation in higher dimensions where problems with (non-) renormalisability signal that the quantum theory is not well-defined. The easiest case is that of $\Lambda = 0$. In this case, for Riemannian signature, the theory becomes what is known as SU(2) BF theory. It is one-loop exact, and the partition function can be explicitly computed. It reduces to the Ray–Singer torsion for the operator ∇, see, e.g., Birmingham et al. (1991).

The case of nonzero Λ is much harder, as the theory is no longer one-loop exact. In spite of this, the quantum version of the Riemannian signature $\Lambda > 0$ gravity is known. It is based on the quantum group $\text{SU}_q(2)$ (at root of unity); see Reshetikhin and Turaev (1991). The partition function on a given 3-manifold M is constructed by choosing a simplicial decomposition of M, and then decorating the arising simplicial complex with certain combinatorial data. The arising state sum is independent of the chosen simplicial decomposition and is a topological invariant of M. At least in part of the literature, this construction is referred to as the Turaev–Viro model. Another way of seeing why the $\Lambda > 0$ case is understood is by noticing that in this case the Lagrangian can be represented as the difference of two Chern–Simons Lagrangians for $\mathbf{w} \pm \mathbf{e}$. The quantum Chern–Simons theory for the gauge group SU(2) is understood, and in a precise sense the $\Lambda > 0$ 3D gravity partition function is the product of two CS partition functions; see, e.g., Roberts (1997) for a nice proof.

As far as we are aware, there is no complete construction of the much more difficult $\Lambda < 0$ quantum theory, even though there is some recent progress in this direction; see, e.g., Blau and Thompson (2016) and references therein.

4.2 The Pure Connection Formulation

In this section we review the pure connection description of 3D gravity. As far as we are aware, the pure connection formulation of 3D gravity was first worked out in Peldan (1992), starting from the Hamiltonian point of view. A simpler description, directly at the level of the Lagrangian, appears in section 3.4 of Peldan (1994). We will only give the Lagrangian description.

We consider the case of negative cosmological constant $\Lambda = -1$, and consider pure gravity. The idea is to start with the first-order Einstein–Cartan action (4.17), and solve the equation $\mathbf{f} = \mathbf{ee}$ for $\mathbf{e}$ as a function of $\mathbf{f}$, substituting the result back into the action. To describe the solution, we introduce the notion of definiteness and sign of a connection. We follow Herfray et al. (2017) in this section.

4.2.1 Definite Connections

Let $\mathbf{w}$ be a set of 2×2 anti-hermitian matrix-valued 1-forms on M, i.e., an SU(2) connection. Let $\mathbf{f} = d\mathbf{w} + \mathbf{ww}$ be the curvature 2-forms. Let us pick an orientation on M. Then, for any volume form v in the fixed orientation class, we define a map from the set of 1-forms to the Lie algebra

$$\phi_{\mathbf{f}} : T^*M \to \mathfrak{su}(2), \qquad \phi_{\mathbf{f}}(\alpha) := \alpha \wedge \mathbf{f}/v. \tag{4.24}$$

This is a map from the three-dimensional space T^*M to the three-dimensional Lie algebra $\mathfrak{su}(2)$. We call a connection $\mathbf{w}$ **definite** or nondegenerate if this map is an isomorphism.

For a definite connection, we can construct a certain invariant from its curvature. Thus, consider

$$\lambda(\mathbf{f}) := \frac{4}{3}\mathrm{Tr}\left(\phi_{\mathbf{f}} \otimes \phi_{\mathbf{f}}(\mathbf{f})\right). \tag{4.25}$$

The notation here is that $\phi_{\mathbf{f}}$ acts on both form indices of $\mathbf{f}$, and we have a product of three Lie algebra elements under the trace. Note that the sign of $\lambda(\mathbf{f})$ is invariantly defined. Indeed, if we change the orientation by sending $v \to -v$, the sign of (4.25) does not change. A connection $\mathbf{w}$ is definite if and only if its curvature satisfies $\lambda(\mathbf{f}) \neq 0$.

In this book we are mainly interested in the case when $\lambda(\mathbf{f}) < 0$. This corresponds to the negative cosmological constant case $\Lambda < 0$. We will refer to such connections as *negative definite*. In this case, the connection defines a frame field $\mathbf{e}_{\mathbf{f}}$ such that

$$\mathbf{f} = \mathbf{e}_{\mathbf{f}} \wedge \mathbf{e}_{\mathbf{f}}. \tag{4.26}$$

In order to see that $\mathbf{f}$ satisfying this equation indeed corresponds to $\lambda(\mathbf{f}) < 0$, we can substitute (4.26) into (4.25) and compute the sign. This computation is easy if we first compute the action of $\phi_{\mathbf{f}}$ with $\mathbf{f}$ given by (4.26) on the frame fields. Thus, let us write $\alpha = \alpha^i e^i$ for some choice of the coefficients α^i. Using $\mathbf{e} = e^i \tau^i$ and the algebra of τ^i we can write the curvature as $\mathbf{f} = (1/2)\epsilon^{ijk} e^i_{\mathbf{f}} e^j_{\mathbf{f}} \tau^k$. To compute $\phi_{\mathbf{f}}$ let us divide in (4.24) by the volume form for the frame field $e^i_{\mathbf{f}}$, which is given by $v_{\mathbf{e}} = (1/6)\epsilon^{ijk} e^i e^j e^k$. We get the following result for the map $\phi_{\mathbf{f}}$

$$\phi_{\mathbf{f}}(\alpha) = \alpha^i \tau^i. \tag{4.27}$$

In other words, if the curvature is as in (4.26) and we divide by the frame volume form in (4.24), then the map $\phi_{\mathbf{f}}$ takes the frame $e^i_{\mathbf{f}}$ into the generator τ^i. It is then easy to see that $\lambda(\mathbf{f})$ for $\mathbf{f}$ as in (4.26) and with the volume form for $\mathbf{e}_{\mathbf{f}}$ used in the definition of $\phi_{\mathbf{f}}$ equals to minus one $\lambda(\mathbf{f}) = -1$.

To prove that for $\lambda(\mathbf{f}) < 0$ the curvature can be written in the form (4.26) we will describe the corresponding frame field explicitly in the next subsection.

4.2.2 The Pure Connection Formulation

Consider a negative definite connection $\mathbf{w}$. The volume form

$$v_{\mathbf{f}} := \sqrt{-\lambda(\mathbf{f})}\, v, \tag{4.28}$$

which is in the fixed orientation class, does not depend on the choice of the volume form v used in its construction. It is thus invariantly defined by the negative definite connection $\mathbf{w}$ and the fixed orientation of M. The pure connection formulation gravity action is just the total volume

$$S[\mathbf{w}] = \int v_{\mathbf{f}}. \tag{4.29}$$

We can now describe $\mathbf{e}_{\mathbf{f}}$ that solves (4.26). It is obtained via the following construction

$$i_\xi \mathbf{e}_{\mathbf{f}} = (\mathbf{f} \wedge i_\xi \mathbf{f} - i_\xi \mathbf{f} \wedge \mathbf{f})/2v_{\mathbf{f}}, \qquad \forall \xi \in TM. \tag{4.30}$$

The matrix on the right-hand side is anti-hermitian, as the commutator of two anti-hermitian matrices. The frame $\mathbf{e}_{\mathbf{f}}$ defines the metric $ds^2_{\mathbf{f}} := -2\,\mathrm{Tr}(\mathbf{e}_{\mathbf{f}} \otimes \mathbf{e}_{\mathbf{f}})$, which is of Riemannian signature. The frame $\mathbf{e}_{\mathbf{f}}$ has the property that

$$v_{\mathbf{f}} = -\frac{2}{3}\mathrm{Tr}(\mathbf{e}_{\mathbf{f}}\mathbf{e}_{\mathbf{f}}\mathbf{e}_{\mathbf{f}}). \tag{4.31}$$

Note that the action (4.29) is just the value of the first-order action (4.17) on the solution (4.30) of (4.26).

4.2.3 The First Variation and Euler-Lagrange Equations

The expression (4.31) makes it clear that the first variation of the pure connection action is given by

$$\begin{aligned} \delta S[\mathbf{w}] = -\int 2\,\mathrm{Tr}(\delta \mathbf{e}_{\mathbf{f}}\mathbf{e}_{\mathbf{f}}\mathbf{e}_{\mathbf{f}}) &= -\int \mathrm{Tr}(\delta(\mathbf{e}_{\mathbf{f}}\mathbf{e}_{\mathbf{f}})\mathbf{e}_{\mathbf{f}}) \\ &= -\int \mathrm{Tr}(\delta \mathbf{f}\, \mathbf{e}_{\mathbf{f}}). \end{aligned} \tag{4.32}$$

This shows that the critical points of the pure connection action are connections satisfying the following second-order partial differential equation (PDE)

$$d_{\mathbf{w}} \mathbf{e}_{\mathbf{f}} = 0, \tag{4.33}$$

with $d_{\mathbf{w}}$ given by (4.15). This equation says that the connection $\mathbf{w}$ is the unique torsion-free metric connection compatible with the frame $\mathbf{e}_{\mathbf{f}}$. The equation (4.26) that defines $\mathbf{e}_{\mathbf{f}}$ then becomes the statement that the metric constructed from $\mathbf{e}_{\mathbf{f}}$ is of constant negative curvature. This shows that (4.29) is indeed the pure connection formulation of 3D gravity (with negative Λ).

5

The 'Chiral' Formulation of General Relativity

We now come to what is possibly the least familiar description. The original idea was proposed on the physics side in a paper by Plebánski (1977). Related structures were discovered about the same time by mathematicians in Atiyah et al. (1978). This description of gravity was later rediscovered in Capovilla et al. (1991), in the authors' search for a Lagrangian formulation for Ashtekar's new Hamiltonian formulation of general relativity (GR); see Ashtekar (1987). There were also much earlier related ideas, as we review in the historical remarks section at the end of this chapter.

The fundamental reason for the existence of 'chiral' formulations of 4D GR is the fact that the Lie algebra of the four-dimensional 'Lorentz' groups[1] is not simple.[2] It is interesting to note that this is the only dimension when this phenomenon occurs.[3] We have the following 'accidental' isomorphisms

$$\begin{aligned} \mathfrak{so}(4) &= \mathfrak{su}(2) \oplus \mathfrak{su}(2), \\ \mathfrak{so}(1,3) &= \mathfrak{sl}(2,\mathbb{C}) \oplus \overline{\mathfrak{sl}(2,\mathbb{C})}, \\ \mathfrak{so}(2,2) &= \mathfrak{sl}(2,\mathbb{R}) \oplus \mathfrak{sl}(2,\mathbb{R}). \end{aligned} \tag{5.1}$$

In turn, these isomorphisms are related to the fact that the Hodge operator in four dimensions maps 2-forms into 2-forms, and defines the decomposition of the space of 2-forms into its eigenspaces of self-dual (SD) and anti–self-dual (ASD) forms. Indeed, the Lie algebra $\mathfrak{so}(n)$ of the orthogonal group can be realised as the matrix algebra of antisymmetric matrices. In four dimensions, antisymmetric matrices can be split into their SD and ASD parts, and this is why the first of relations in (5.1) arises. A similar mechanism is at play for other signatures.

[1] When we refer to 'Lorentz' group in quotes we always mean one of the appropriate (pseudo-) orthogonal groups, considering all possible signatures at the same time.
[2] In the Lorentzian case $\mathfrak{so}(1,3)$, strictly speaking, this is only true at the level of the complexified Lie algebra.
[3] The dimension two is also special because in it the 'Lorentz' groups are abelian.

The accidental isomorphisms (5.1) mean that the decomposition of the Riemann curvature tensor into its 'Lorentz' irreducible pieces (which is possible in any dimension) in four dimensions is related to the SD/ASD decomposition. This is what is ultimately responsible for the formulations that we describe in this chapter.

5.1 Hodge Star and Self-Duality in Four Dimensions

We start with a quick reminder of the Hodge operator in four dimensions. Then, after describing the Riemann curvature decomposition that becomes possible in 4D, we return to the Hodge star and discuss some of its important properties in much more detail.

Given a metric and choosing an orientation, we can form the Hodge star operator that sends 2-forms into 2-forms

$$* : B_{\mu\nu} \to {}^*B_{\mu\nu} = \frac{1}{2}\epsilon_{\mu\nu}{}^{\rho\sigma} B_{\rho\sigma}. \tag{5.2}$$

This operator squares to a multiple of the identity

$$(*)^2 = \frac{1}{4}\epsilon_{\mu\nu}{}^{\rho\sigma}\epsilon_{\rho\sigma}{}^{\alpha\beta} = \sigma\delta^{\alpha}_{[\mu}\delta^{\beta}_{\nu]}, \tag{5.3}$$

where $\sigma = \pm 1 = (-1)^p$ is the sign depending on the signature of $\mathbb{R}^{p,q}, p+q=4$. Note that the object $\delta^{\alpha}_{[\mu}\delta^{\beta}_{\nu]}$ is the identity operator on the space of 2-forms. This means that the eigenvalues of $*$ in the case of Euclidean and split signatures are ± 1, and in the case of Lorentzian signature $\pm \mathrm{i}$. Correspondingly, the space of 2-forms can be decomposed into eigenspaces of the Hodge star

$$\Lambda^2 = \Lambda^+ \oplus \Lambda^-, \tag{5.4}$$

where in the case of Lorentzian signature it is the space $\Lambda^2_{\mathbb{C}}$ of complexified 2-forms that admits such a decomposition. To fix our conventions, the eigenvectors of Hodge are 2-forms satisfying

$$\Lambda^{\pm} \ni B_{\mu\nu} : \frac{1}{2}\epsilon_{\mu\nu}{}^{\rho\sigma} B_{\rho\sigma} = \pm\sqrt{\sigma} B_{\mu\nu}, \tag{5.5}$$

where $\sqrt{\sigma}$ is defined to be either 1 or i depending on the signature, and the projectors on the spaces $\Lambda^{\pm}$ are

$$P^{\pm}{}_{\mu\nu}{}^{\rho\sigma} = \frac{1}{2}\left(\delta^{\rho}_{[\mu}\delta^{\sigma}_{\nu]} \pm \frac{1}{2\sqrt{\sigma}}\epsilon_{\mu\nu}{}^{\rho\sigma}\right). \tag{5.6}$$

5.2 Decomposition of the Riemann Curvature

In any dimension, the Riemann curvature tensor can be decomposed into pieces that take values in spaces of (finite-dimensional) irreducible representations of the Lorenz group. These pieces are the scalar curvature, the tracefree part of Ricci curvature, and the Weyl curvature. The metric is called Einstein if the tracefree part of its Ricci curvature vanishes, in other words if $R_{\mu\nu} \sim g_{\mu\nu}$. The proportionality coefficient is called by physicists the cosmological constant, and by mathematicians the scalar curvature. It must be a constant by one of the Bianchi identities, so its constancy is not an independent Einstein equation. But if one wants to fix this constant to a particular value, this constitutes an independent equation.

In four dimensions, something very special happens. As we have already said, the Lorentz group is not simple, and this is related to self-duality. So, there exists yet another decomposition of the Riemann curvature, specific only to four dimensions and related to self-duality, and this decomposition is related to the scalar/Ricci/Weyl decomposition. This makes it possible to impose the Einstein condition in a particularly elegant and efficient manner.

As we already discussed, the special property of 4D is that the Hodge star maps 2-forms into 2-forms $* : \Lambda^2 \to \Lambda^2$, and introduces the decomposition of the space of 2-forms into SD and ASD parts (5.4).

When the Levi–Civita connection is metric and torsion-free, the Riemann curvature $R_{\mu\nu\rho\sigma}$ is symmetric $R_{\mu\nu\rho\sigma} = R_{\rho\sigma\mu\nu}$, and can be viewed as a symmetric $\Lambda^2 \otimes \Lambda^2$–valued matrix. Decomposing this matrix into its $\Lambda^\pm$ components we get the following block form

$$\text{Riemann} = \begin{pmatrix} A & B \\ B^T & C \end{pmatrix}. \tag{5.7}$$

Here A is the SD–SD part, C is the ASD–ASD part, and both are symmetric as $\Lambda^2 \otimes \Lambda^2$ matrices, while B is the SD–ASD part

$$A := P_+ \text{Riemann}\, P_+, \quad C := P_- \text{Riemann}\, P_-, \quad B := P_+ \text{Riemann}\, P_-,$$

where $P_\pm$ are the SD/ASD projectors. These parts satisfy a set of properties that are signature-dependent and that we summarise as

Theorem 5.1 *In the case of Euclidean and split signatures the tensors A, B, and C are real. For Lorentzian signature the tensors A and C are complex and complex conjugates of each other $\overline{C} = A$, and B is Hermitian $\overline{B^T} = B$. In all cases the Bianchi identity $\epsilon^{\mu\nu\rho\sigma} R_{\mu\nu\rho\sigma} = 0$ implies that traces of A and C are equal, and equal to the scalar curvature $\text{Tr}(A) = \text{Tr}(C) = R/2$. The tracefree parts of A and C encode the self- and anti–self-dual parts of the Weyl curvature*

$$A = P_+ \text{Riemann}\, P_+ = P_+ \left(\text{Weyl} + \frac{R}{6} \mathbb{I} \right) P_+, \tag{5.8}$$
$$C = P_- \text{Riemann}\, P_- = P_- \left(\text{Weyl} + \frac{R}{6} \mathbb{I} \right) P_-.$$

Here $\mathbb{I}$ is the identity tensor in $\Lambda^2 \otimes \Lambda^2$. The SD–ASD part B encodes the tracefree part of Ricci curvature in the sense that $B = 0$ if and only if the tracefree part of Ricci is zero.

Let us prove these statements. First, let us show that the condition $B = 0$ is equivalent to the Einstein condition, and also equivalent to the statement that Riemann commutes with the Hodge operator. To this end, it is convenient to do the calculation of the SD–ASD projection of Riemann in an index-free way, using the SD/ASD projectors (5.6) that can be written as

$$P^{\pm} = \frac{1}{2}\left(\mathbb{I} \pm \frac{1}{\sqrt{\sigma}}*\right). \tag{5.9}$$

We then have

$$\begin{aligned} 4B = 4P^{+}\,\text{Riemann}\,P^{-} &= \text{Riemann} - \frac{1}{\sigma} * \text{Riemann}* \\ &+ \frac{1}{\sqrt{\sigma}} * \text{Riemann} - \frac{1}{\sqrt{\sigma}}\text{Riemann} * . \end{aligned} \tag{5.10}$$

The first line of the right-hand side here is a symmetric 6×6 matrix, and the second line is antisymmetric. In case of the Lorentzian signature the second line is also purely imaginary. Thus, in the Lorentzian case the matrix on the right-hand side is Hermitian, which is what we stated previously about B block of (5.7). In any case, the condition that $P^{+}\,\text{Riemann}\,P^{-}$ vanishes is equivalent to

$$\text{Riemann} = \frac{1}{\sigma} * \text{Riemann} * \quad \text{and} \quad * \text{Riemann} = \text{Riemann} * . \tag{5.11}$$

But it is clear that these two equations are equivalent. Indeed, by taking the left Hodge dual of the first equation one obtains the second equation and vice versa. So, we see that $B = 0$ is equivalent to the second equation in (5.11). This equation can be rephrased as follows. Let us think about the Riemann curvature as a map $R_{\mu\nu}{}^{\rho\sigma} : \Lambda^2 \to \Lambda^2$. Then the second equation in (5.11) is the statement that this map commutes with the Hodge star. This proves

Lemma 5.2 *The SD–ASD part B of the Riemann curvature vanishes if and only if the Riemann curvature commutes with the Hodge star.*

Let us now prove that this commutativity of Riemann and Hodge is equivalent to the Einstein condition. To this end we need the formula

$$\epsilon_{\mu\nu\rho\sigma}\epsilon^{\alpha\beta\gamma\delta} = 24\sigma\, \delta^{\alpha}_{[\mu}\delta^{\beta}_{\nu}\delta^{\gamma}_{\rho}\delta^{\delta}_{\sigma]}. \tag{5.12}$$

Here $\epsilon_{\mu\nu\rho\sigma}$ is the volume form, and indices are raised with the metric. Using this we get

$$\frac{1}{4\sigma}\epsilon_{\mu\nu}{}^{\rho\sigma}R_{\rho\sigma\gamma\delta}\epsilon^{\gamma\delta\alpha\beta} = R_{\mu\nu}{}^{\alpha\beta} + R\delta^{\alpha}_{[\mu}\delta^{\beta}_{\nu]} - 2\delta^{\alpha}_{[\mu}R^{\beta}_{\nu]} + 2\delta^{\beta}_{[\mu}R^{\alpha}_{\nu]}. \tag{5.13}$$

Thus, the first equation in (5.11) is equivalent to

$$R\delta^{\alpha}_{[\mu}\delta^{\beta}_{\nu]} - 2\delta^{\alpha}_{[\mu}R^{\beta}_{\nu]} + 2\delta^{\beta}_{[\mu}R^{\alpha}_{\nu]} = 0. \tag{5.14}$$

This, on the other hand, is equivalent to the tracefree part of Ricci being zero. Indeed, if the tracefree part of Ricci is zero, then this equation is satisfied by inspection. On the other hand, taking the contraction of say $\mu\alpha$ we get $R\delta_\nu^\beta - 4R_\nu^\beta = 0$, which is just the tracefree Ricci condition. Thus, we have proved

Theorem 5.3 *In four dimensions a metric is Einstein if and only if the Riemann tensor viewed as an endomorphism of Λ^2 commutes with the Hodge star.*

Together with Lemma 5.2 this implies that the Einstein condition is equivalent to vanishing of the SD–ASD part of the Riemann curvature $B = 0$.

It is also easy to prove that the traces of the SD–SD and ASD–ASD parts of Riemann are equal. Indeed, we have

$$\begin{aligned}\mathrm{Tr}\,(P^\pm \mathrm{Riemann}\, P^\pm) &= \mathrm{Tr}\,(P^\pm \mathrm{Riemann}) \\ &= \frac{1}{2}\left(\delta^\rho_{[\mu}\delta^\sigma_{\nu]} \pm \frac{1}{2\sqrt{\sigma}}\epsilon_{\mu\nu}{}^{\rho\sigma}\right) R^{\mu\nu}{}_{\rho\sigma} = \frac{1}{2}R,\end{aligned} \tag{5.15}$$

where we have used the Bianchi identity $R_{[\mu\nu\rho\sigma]} = 0$.

In a similar way, it is easy to prove that the SD–SD and ASD–ASD parts of Riemann are composed of just the Weyl curvature and the scalar part. Indeed, we have

$$\begin{aligned}4A = 4P^+\,\mathrm{Riemann}\,P^+ &= \mathrm{Riemann} + \frac{1}{\sigma} * \mathrm{Riemann} * \\ &\quad + \frac{1}{\sqrt{\sigma}} * \mathrm{Riemann} + \frac{1}{\sqrt{\sigma}}\mathrm{Riemann} * \\ &= \left(\mathbb{I} + \frac{1}{\sqrt{\sigma}}*\right)\left(\mathrm{Riemann} + \frac{1}{\sigma} * \mathrm{Riemann} *\right).\end{aligned} \tag{5.16}$$

We note that while the left-hand side is clearly SD with respect to both pair of indices, the last expression on the right-hand side is only explicitly SD with respect to the first pair. It can of course be equivalently rewritten with the SD projector on the second pair of indices instead of the first, or on both sides. On the other hand, we have

$$R_{\mu\nu}{}^{\alpha\beta} + \frac{1}{4\sigma}\epsilon_{\mu\nu}{}^{\rho\sigma}R_{\rho\sigma\gamma\delta}\epsilon^{\gamma\delta\alpha\beta} = 2R_{\mu\nu}{}^{\alpha\beta} + R\delta^\alpha_{[\mu}\delta^\beta_{\nu]} - 2\delta^\alpha_{[\mu}R^\beta_{\nu]} + 2\delta^\beta_{[\mu}R^\alpha_{\nu]} \tag{5.17}$$

Further, in four dimensions, the Weyl curvature tensor is

$$C_{\mu\nu\rho\sigma} = R_{\mu\nu\rho\sigma} - (g_{\mu[\rho}R_{\sigma]\nu} - g_{\nu[\rho}R_{\sigma]\mu}) + \frac{R}{3}g_{\mu[\rho}g_{\sigma]\nu}, \tag{5.18}$$

and so

$$R_{\mu\nu}{}^{\alpha\beta} + \frac{1}{4\sigma}\epsilon_{\mu\nu}{}^{\rho\sigma}R_{\rho\sigma\gamma\delta}\epsilon^{\gamma\delta\alpha\beta} = 2C_{\mu\nu}{}^{\alpha\beta} + \frac{R}{3}\delta^\alpha_{[\mu}\delta^\beta_{\nu]}. \tag{5.19}$$

These manipulations prove the first formula in (5.8). Similar transformations are used to prove the second formula. This concludes the proof of Theorem 5.1.

While in the previous proof it may appear that all matrices A, B, and C are $\Lambda^2 \otimes \Lambda^2$ valued and thus 6×6, the spaces $\Lambda^\pm$ are in fact three-dimensional, and so A, B, and C are in fact 3×3 matrices. This will be made explicit in Section 5.5 when we discuss the SD/ASD decomposition of the Lie algebra of the 'Lorentz' groups.

The idea of the chiral formulation is then that it is sufficient to have access to only one row of the matrix (5.7) to impose the Einstein condition. We will later show that the two rows of (5.7) can be given the interpretation of the curvatures of the SD and ASD parts of the spin connection. It then becomes possible to impose the Einstein condition working with only one of the chiral parts of the spin connection.

5.3 Chiral Version of Cartan's Theory

As we have seen in the previous section, in four dimensions it is enough to have access only to the SD part of the Riemann curvature with respect to a pair of indices, rather than to the full Riemann curvature, to impose the Einstein equations. Let us now see what this leads to in the context of the tetrad formalism. In Einstein–Cartan formulation the Riemann curvature is encoded into the curvature $R^{IJ}(\omega)$ of the spin connection ω^{IJ}. We can take its SD part with respect to the 'internal' indices IJ using the SD projector. We define

$$R_+^{IJ}(\omega) := P_+^{IJ}{}_{KL} R^{KL}(\omega), \tag{5.20}$$

where

$$P_+^{IJ}{}_{KL} := \frac{1}{2}\left(\delta_K^{[I}\delta_K^{J]} + \frac{1}{2\sqrt{\sigma}}\epsilon^{IJ}{}_{KL}\right). \tag{5.21}$$

It is clear that when ω^{IJ} is the torsion-free metric spin connection the object $R_+^{IJ}(\omega)$ encodes exactly one of the two rows of the matrix (5.7), and thus we only need R_+^{IJ} to write down Einstein equations.

There exists a simple action principle that realises this idea. Consider the following action

$$S_{\text{chiral}}[e,\omega] = \frac{\sqrt{\sigma}}{8\pi G}\int e_I e_J P_+^{IJ}{}_{KL}\left(R^{KL}(\omega) - \frac{\Lambda}{6}e^K e^L\right). \tag{5.22}$$

Expanding the SD projector we see that the only difference between (5.22) and (3.60) is that we have added to the Einstein–Cartan action what is called the Holst term $e_I e_J F^{IJ}$ with an imaginary (in Lorentzian signature) coefficient. However, we can clearly do this without changing the dynamics of the theory. Indeed, when the connection has zero torsion, this term becomes a total derivative. This can be easily seen by considering the squared torsion term $d^\omega e^I d^\omega e_I$. Integrating by parts here one gets a multiple of the Holst term.

Thus, the action (5.22) gives an equivalent starting point to (3.60) for the purpose of obtaining the field equations. However, it clearly involves just half of the curvature of the spin connection. Further, we can use the fact that the Lorentz group Lie algebra is not simple, and rewrite the SD part of the curvature as the curvature of the SD part of the spin connection. This is possibly precisely because the Lie algebra can be written (5.1) as the sum of two commuting sub-algebras. Thus, if we write

$$\omega^{IJ} = (\omega^{IJ})_+ + (\omega^{IJ})_- \tag{5.23}$$

then

$$R_+^{IJ}(\omega) = R^{IJ}(\omega_+), \tag{5.24}$$

and we can write the action (5.22) as

$$S_{\text{chiral}}[e, \omega^+] = \frac{\sqrt{\sigma}}{8\pi G} \int (e_I e_J)_+ \left(R^{IJ}(\omega_+) - \frac{\Lambda}{6}(e^I e^J)_+ \right), \tag{5.25}$$

where the index 'plus' next to the 2-form $e^I e^J$ denotes the SD projection with respect to the internal indices $(e^I e^J)_+ = P_+^{IJ}{}_{KL} e^k e^L$. We thus obtain a first-order formulation of GR that is similar to (3.60), but in which only half of the spin-connection coefficients are present. This gives a significantly more economic formalism. Indeed, in the Einstein–Cartan case (3.60), the Lagrangian depends on 24 connection components per spacetime point. This is better than the case of Palatini theory (2.43), where in addition to the 10 metric components there are also 40 components of the affine connection. But this is nevertheless quite a few components to carry around in explicit calculations. What was achieved by passing to (5.25) is that now, in addition to the 16 components in the tetrad, the Lagrangian depends on just 12 connection components. One could object that the connection is now complex, and so its real and imaginary parts continue to comprise the same 24 components. But this is not the right interpretation. The Lagrangian depends on the 12 components of the SD connection ω^+ holomorphically, as no complex conjugate connection ever appears. Also, in Euclidean signature, no complexification has happened, and we indeed just halved the number of the connection components with the SD projection trick.

The 'chiral' formulation (5.25) thus keeps the main advantage of the Einstein–Cartan formulation of GR – it is polynomial in the fields, with at most quartic terms appearing in the action. It is also much more economical than the Einstein–Cartan formulation, because it depends only on $16 + 12$ field components per spacetime point, as compared to $16+24$ components in the Einstein–Cartan case. This makes (5.25) much better suited for explicit, e.g., perturbative calculations. One complication is that one needs to deal with the issue of reality conditions in the Lorentzian case. However, at least for perturbative calculations, these are not difficult to impose. One just imposes the condition that the tetrad is real. The correct reality conditions on the connection are then imposed automatically by the field equations. Further, loop calculations are customarily performed in Euclidean signature, and then one does not need to worry about reality conditions at all as all fields are real. We will come back to perturbative considerations in this formalism in Chapter 8.

The final remark is that, unlike in the full Einstein–Cartan formulation, in the chiral theory (5.25) the Hamiltonian analysis does not lead to any second-class constraints. This is directly linked to the halving of the number of 'momentum' variables introduced in this first-order theory. The Hamiltonian analysis of (5.22) directly leads to Ashtekar's new Hamiltonian formulation of GR; see Ashtekar (1987). For all these reasons, the chiral tetrad formulation (5.25) should be viewed as superior to the usual tetrad formalism.

There exists a good Yang–Mills analogy for the passage from (3.60) to (5.25). Let us describe it. The usual Yang–Mills Lagrangian (in Lorentzian signature) is

$$L_{YM} = -\frac{1}{4g^2}(F^a_{\mu\nu})^2, \tag{5.26}$$

where $F^a_{\mu\nu} = \partial_\mu A^a_\nu - \partial_\nu A^a_\mu + f^a{}_{bc} A^b_\mu A^c_\nu$ is the field strength and A^a_μ is a Lie algebra valued connection 1-form. However, one can always add to this Lagrangian the Pontryagin density for the connection A^a_μ, which is $\epsilon_{\mu\nu\rho\sigma} F^a_{\mu\nu} F^a_{\rho\sigma}$. This term is a total divergence that does not change the field equations one obtains by extremising the action. Further, we can always adjust the coefficient in front of this term so that the original term in the Lagrangian and the Pontryagin term combine into

$$L_{\text{chiral}} = -\frac{1}{2g^2} P_+^{\mu\nu\rho\sigma} F^a_{\mu\nu} F^a_{\rho\sigma} = -\frac{1}{2g^2}(F^a_{+\,\mu\nu})^2, \tag{5.27}$$

where P_+ is the SD projector (5.9).

The analogy with (3.60) and (5.25) arises when one writes down the first-order versions of the two Lagrangians (5.26) and (5.27). In both cases one 'integrates in' a 2-form field, but in the chiral case this field is SD. Thus, let us consider the following two actions

$$S[B, A] = \int B^a_{\mu\nu} F^{a\,\mu\nu} + g^2 (B^a_{\mu\nu})^2. \tag{5.28}$$

Integrating out the 2-form field $B^a_{\mu\nu}$ one gets back the Lagrangian (5.26). This gives us a first-order formalism for Yang–Mills. Interestingly, there is just a cubic interaction vertex in this formalism, at the expense of having both B and A fields propagating. The chiral first-order action, on the other hand, is given by

$$S_{\text{chiral}}[B^+, A] = \int B^{+\,a}_{\mu\nu} F^{a\,\mu\nu} + \frac{g^2}{2}(B^{+\,a}_{\mu\nu})^2. \tag{5.29}$$

Here $B^{+\,a}_{\mu\nu}$ is a SD Lie algebra valued 2-form field. The field equation for B^+ is $B^{+\,a}_{\mu\nu} = -(1/g^2) F^{+\,a}_{\mu\nu}$, i.e., B^+ is a multiple of the SD projection of the field strength. Substituting this back into the action one obtains (5.27).

The actions (5.28) and (5.29) are both good first-order actions, but the chiral action contains just half of the B variables of the first. Indeed, the non-chiral version contains six times the dimension of the Lie group components of $B^a_{\mu\nu}$. This is to be compared with the number of components in $B^{+\,a}_{\mu\nu}$, which is three times the dimension of the Lie group, because the dimension of the space of SD forms is three. The action (5.28) is not so useful as the starting point for perturbation theory because it contains six components of $B_{\mu\nu}$ for four components of A_μ. As we will see in Chapter 8 on perturbative descriptions, the 2-form field $B_{\mu\nu}$ of the formulation (5.28) has a nonvanishing propagator with itself, which complicates the perturbation theory. This is directly related to the fact that too many components have been 'integrated in' in passage from (5.26) to (5.28).

As it will become clear in Chapter 8, one would like the mismatch between the numbers of components in the connection and the auxiliary 2-form field to be one, which is the number of functions appearing in gauge transformations. This is not the case in the non-chiral version (5.28). However, in the chiral version of the first-order formalism we have three components of $B^+_{\mu\nu}$ for four components of A_μ, and the mismatch is indeed the desired one. It then turns out that the chiral version can be very elegantly gauge-fixed and provides a very nice and powerful perturbation theory. In particular, in the chiral version of the first-order perturbation theory the propagator of the 2-form field with itself vanishes, which simplifies calculations considerably.

The difference between the full Einstein–Cartan description (3.60) and its chiral version (5.25) is analogous to the difference between (5.28) and (5.29). As in the Yang–Mills case, the chiral action is a better starting point for perturbation theory, the same is true for the case of chiral formalism for GR; see Chapter 8.

The final remark we make about the chiral action (5.25) is that when $\Lambda \neq 0$ the frame field can in principle be integrated out (at least perturbatively), with the result being a pure connection action for the SD part of the spin connection only. We postpone discussing this until one of the following chapters.

5.4 Hodge Star and the Metric

To motivate the next step we recall that in the case of Einstein–Cartan formalism it was possible to pass to the BF-type formalism, in which the wedge product of two tetrads $e^I \wedge e^J$ was replaced by the 2-form field B^{IJ}, and a Lagrange multiplier term was added to the action to guarantee that the 2-form field comes from the tetrad. We shall repeat this trick for the chiral first-order action (5.25). This will change the nature of the object that solders the 'internal' and the tangent bundles in a profound way. To understand the geometry arising, we need to develop properties of the Hodge star in four dimensions in more detail. This will eventually lead to the so-called Plebański formalism for GR, which is a very powerful version of the chiral first-order description (5.25).

5.4.1 Hodge Star on Middle Degree Forms is Conformally Invariant

We start by stating a simple but profound

Lemma 5.4 *The Hodge operator on middle degree forms (and in particular the Hodge operator on 2-forms in four dimensions) is conformally invariant.*

In other words, if we change $g_{\mu\nu} \to \Omega^2 g_{\mu\nu}$ then $\epsilon_{\mu\nu}{}^{\rho\sigma}$ is unchanged. We only check this in four dimensions, the case of higher (even) dimensions is completely analogous. To verify this we note that the volume 4-form $\epsilon_{\mu\nu\rho\sigma}$ transforms as the square root of the determinant of the metric $\epsilon_{\mu\nu\rho\sigma} \to \Omega^4 \epsilon_{\mu\nu\rho\sigma}$. Then $\epsilon_{\mu\nu}{}^{\rho\sigma} = g^{\rho\alpha} g^{\sigma\beta} \epsilon_{\mu\nu\alpha\beta}$, and the inverse metric transforms as $g^{\mu\nu} \to \Omega^{-2} g^{\mu\nu}$, from which the assertion follows.

5.4.2 Hodge Star Determines the (Conformal) Metric

Lemma 5.4 states that a conformal metric, which is metric modulo conformal rescalings, determines the Hodge star operator. The following theorem shows that in four dimensions the converse is also true.

Theorem 5.5 *In four dimensions, the knowledge of the Hodge star operator, and thus the knowledge of the split* (5.4) *of the space of 2-forms into the eigenspaces of the Hodge star is equivalent to the knowledge of the conformal metric.*

This is the fact of the fundamental importance for our purposes as it will become clear from the following exposition. Because of its importance, we will present two proofs, one conceptual one constructive. The conceptual proof will explain why this can be true. In physics literature, a proof of this theorem (in its Euclidean signature version) has been given in Dray et al. (1989).

To do the conceptual proof, we need to build up a bit more knowledge about the split (5.4) of the space of 2-forms into its SD/ASD subspaces in different signatures. The first fact is that we can take the wedge product of a couple of 2-forms to get the top form. This means that the wedge product gives us the natural conformal metric in the space Λ^2

$$\langle B_1, B_2 \rangle_\wedge := B_1 B_2 / \epsilon, \tag{5.30}$$

where ϵ is an arbitrary top form on M. To divide by a top form we need to assume that a nowhere-vanishing top form exists, or, in other words, the manifold is orientable. We will always assume this in this book. Different choices of ϵ are related by multiplication by a nowhere-vanishing function, and so the wedge product metric is only a conformal metric, i.e., is defined modulo multiplication by a function. We note that the wedge product metric on the six-dimensional space of 2-forms is of split signature $(3, 3)$, which can easily be checked by taking a basis for this space. This is independent of any metric one may put on the manifold M itself.

Given the conformal metric (5.30) and a split (5.4) we can ask what the wedge product metric reduces to on $\Lambda^{\pm}$. This is easiest to see in the case of $\mathbb{R}^{p,q}$ with the usual flat metric. Then a convenient basis in Λ^+ in each signature case, in the orientation e^{1234}, is as follows

$$\begin{aligned} \Sigma^1_E &= e^{41} - e^{23}, & \Sigma^2_E &= e^{42} - e^{31}, & \Sigma^3_E &= e^{43} - e^{12}, \\ \Sigma^1_L &= \mathrm{i}e^{41} - e^{23}, & \Sigma^2_L &= \mathrm{i}e^{42} - e^{31}, & \Sigma^3_L &= \mathrm{i}e^{43} - e^{12}, \\ \Sigma^1_S &= e^{41} - e^{23}, & \Sigma^2_S &= e^{42} + e^{31}, & \Sigma^3_S &= e^{43} + e^{12}, \end{aligned} \tag{5.31}$$

where E, L, and S stand for Euclidean, Lorentzian, and Split, respectively. In formulas (5.31) the notation $e^{ijk\ldots}$ stands for $e^i e^j e^k$ and so on. It is now easy to see what the conformal metric on Λ^+ in each case is. Dividing by (twice) the volume form $\epsilon = 2e^{1234}$ we have

$$\langle \Sigma^i_E, \Sigma^j_E \rangle_\wedge = \delta^{ij}, \qquad \langle \Sigma^i_L, \Sigma^j_L \rangle_\wedge = \mathrm{i}\delta^{ij}, \qquad \langle \Sigma^i_S, \Sigma^j_S \rangle_\wedge = \eta^{ij}, \tag{5.32}$$

where $\eta^{ij} = \mathrm{diag}(+1, -1, -1)$. Thus, the space Λ^+ in the Euclidean case can be characterised by saying that the wedge product metric on it is definite (positive definite in the right orientation), in the split case the wedge product metric on Λ^+ is indefinite, and in the Lorentzian case the wedge product metric on Λ^+ is complex. We also note that in all the cases the space Λ^+ is wedge product metric orthogonal to Λ^-. Moreover, in the Lorentzian case $\Lambda^- = \overline{\Lambda^+}$, where the overline denotes the complex conjugation. In other words, the complex conjugates of elements in Λ^+ are in Λ^-.

We have checked these statements for the case of a flat metric on $\mathbb{R}^{p,q}$, but it is clear that these statements about the reduction of the wedge product metric to Λ^+ hold more generally. Indeed, one should just use an orthonormal set of 1-forms for a given metric in place of $e^{1,2,3,4}$ in (5.31) to see that (5.32) holds for an arbitrary metric. And since an arbitrary basis in Λ^+ in each signature case is given by an GL(3) transformation (GL(3, $\mathbb{C}$) transformation in the case of Lorentzian signature) of the orthonormal basis (5.31), we see that indeed the signature of the restriction of the wedge product metric to Λ^+ only depends on the metric signature.

With these results about $\Lambda^2(\mathbb{R}^{p,q}), p + q = 4$ in mind, we can characterise the split (5.4) further, depending on the signature. This will also make the formulation of Theorem 5.5 more precise.

Euclidean Signature. In this case, the Hodge operator is the same as the split of the space of 2-forms into a couple of three-dimensional orthogonal subspaces $\Lambda^{\pm}$, such that the wedge product metric is positive definite on Λ^+ and negative definite on Λ^-. Indeed, this is true for the Hodge operator coming from any Euclidean signature metric on M. It is also clear that the knowledge of such a split is equivalent to the knowledge of the Hodge operator. This is because, given such a split, one can decompose any form into its $\Lambda^{\pm}$ parts and then the Hodge simply acts by ± 1 on $\Lambda^{\pm}$.

The space of such splits is the Grassmanian $\mathrm{SO}(3,3)/\mathrm{SO}(3)\times\mathrm{SO}(3)$, where the stabiliser subgroup is the group mixing the forms in $\Lambda^{\pm}$, respectively, without changing the restrictions of the wedge product metric to these spaces. Now, to show that this Grassmanian is the same as the space of Euclidean signature (conformal) metrics in $\mathbb{R}^4$, we need the exceptional isomorphism

$$\mathrm{SO}(3,3) \sim \mathrm{SL}(4,\mathbb{R}). \tag{5.33}$$

These are groups belonging to two different series of classical groups, i.e., orthogonal D and special linear A, whose Lie algebras are represented by the same rank three simply laced Dynkin diagram $A_3 = D_3$. These groups coincide (modulo $\mathbb{Z}_2$). We will give a proof of this fact Section 5.5.4. We therefore have the isomorphism

$$\mathrm{SO}(3,3)/\mathrm{SO}(3) \times \mathrm{SO}(3) \sim \mathrm{SL}(4,\mathbb{R})/\mathrm{SO}(4), \tag{5.34}$$

where we have used $\mathrm{SO}(4) = \mathrm{SO}(3) \times \mathrm{SO}(3)/Z_2$, which will be proved Section 5.5.2. The space of Euclidean signature metrics modulo conformal rescalings is on the right-hand side of this relation. The left-hand side is the Grassmanian of three-planes in six-dimensional space of signature $(3,3)$ such that the restriction of the metric on the plane is definite. This is just the space of splits (5.4) in the case of this signature, which proves the Theorem 5.5 for this case.

Split Signature. The proof of the Theorem 5.5 in this case is analogous. The only difference is that now the split (5.4) is into a couple of three-dimensional subspaces with indefinite metric of opposite signature. Such a split characterises the Hodge operator for this signature completely. The space of such splits is the Grassmanian $\mathrm{SO}(3,3)/\mathrm{SO}(1,2) \times \mathrm{SO}(2,1)$, which, using the exceptional isomorphism (5.33) is the same as

$$\mathrm{SO}(3,3)/\mathrm{SO}(1,2) \times \mathrm{SO}(2,1) \sim \mathrm{SL}(4,\mathbb{R})/\mathrm{SO}(2,2), \tag{5.35}$$

the right-hand side being the space of conformal metrics of signature $(2,2)$ in four dimensions. Here we have used the isomorphisms $\mathrm{SO}(2,2) = \mathrm{SL}(2,\mathbb{R}) \times \mathrm{SL}(2,\mathbb{R})/\mathbb{Z}_2$ and $\mathrm{SO}(1,2) = \mathrm{SL}(2,\mathbb{R})/\mathbb{Z}_2$. Both these facts will be proven Section 5.5. This proves the theorem in this case.

Lorentzian Signature. The split (5.4) in this case is the split of the real six-dimensional space with metric of signature $(3,3)$ into two complex conjugate orthogonal subspaces. Such a split is the same as an almost complex structure on the space of 2-forms, compatible with the wedge product metric. The elements of $\mathrm{SO}(3,3)$ that commute with such an almost complex structure belong to $\mathrm{SO}(3,\mathbb{C})$, and these mix the 2-forms belonging to Λ^+ without changing the complex metric that the wedge product metric restricts to in this space. Thus, the space of such splits is the Grassmanian $\mathrm{SO}(3,3)/\mathrm{SO}(3,\mathbb{C})$, which in view of (5.33) is the same as

$$\mathrm{SO}(3,3)/\mathrm{SO}(3,\mathbb{C}) \sim \mathrm{SL}(4,\mathbb{R})/\mathrm{SO}(1,3), \tag{5.36}$$

because $\mathrm{SO}(3,\mathbb{C}) = \mathrm{SO}(1,3)$, which will be demonstrated Section 5.5.1. The right-hand side in this relation is the space of real Lorentzian signature metrics in four dimensions modulo conformal rescalings, which proves the Theorem 5.5 for this signature.

We note that the Theorem 5.5 is only true in four dimensions. Indeed, let us consider the case of six dimensions. Then the Hodge star on middle degree forms, which are 3-forms in this case, is the same as the decomposition of the $6 * 5 * 4/6 = 20$ dimensional space of 3-forms into two orthogonal subspaces. The space of such splits is the Grassmanian $\mathrm{SO}(10,10)/\mathrm{SO}(10) \times \mathrm{SO}(10)$. This space has dimension $20 * 19/2 = 190$ minus $2 * 10 * 9/2 = 90$, and so is 100-dimensional. On the other hand, the space of metrics in six dimensions is $6 * 7/2 = 21$-dimensional. Thus, the space of splits of Λ^3 into two orthogonal subspaces is much bigger than the space of metrics, and so there are splits that do not come from the Hodge operator. There is therefore no relation between the space of splits of the space of middle forms and metrics in higher dimensions.

5.4.3 Urbantke Metric

We now want to give a different proof of Theorem 5.5. Given a basis in Λ^+, this other construction presents the metric explicitly. Thus, let us assume that we are given a decomposition (5.4) of Λ^2 into two orthogonal (with respect to the wedge product metric) subspaces $\Lambda^\pm$. Let $\Sigma^i, i = 1, 2, 3$ be a basis in Λ^+. Then the (conformal) metric g_Σ such that its Hodge operator has $\mathrm{Span}(\Sigma^i)$ as its Λ^+ is explicitly given by the following formula

$$g_\Sigma(u,v)\epsilon_\Sigma \sim \epsilon^{ijk} i_u \Sigma^i i_v \Sigma^j \Sigma^k. \tag{5.37}$$

At this stage the proportionality coefficient in this formula is left unspecified, even though later we shall see that in each signature case there is a specific number that is most natural here. The object ϵ_Σ is the volume form for the metric g_Σ. In the physics literature the metric (5.37) is known as the Urbantke metric; see Urbantke (1984). Geometrically, the metric (5.37) arises as the unique conformal metric that makes the triple of 2-forms Σ^i self-dual. We will prove this statement in Sections 5.4.4 and 5.4.5.

It is clear that the formula (5.37) defines a symmetric tensor, but it is not clear what the signature of the metric arising this way is. The aim of the following discussion is to establish this for the three different ways that the split (5.4) may arise. We will start with the Lorentzian signature case, and then comment on changes one has to do to accommodate the other two signatures.

Given a three-dimensional subspace Λ^+ in the space of 2-forms, and a basis $\Sigma^i \in \Lambda^+$ we can always use a GL(3) transformation to make this basis orthonormal. We will make this statement more precise in each signature case.

Lorentzian Signature. Given a three-dimensional subspace $\Lambda^+ \subset \Lambda^2_{\mathbb{C}}$ that is wedge product metric orthogonal to the complex conjugate subspace $\langle \Lambda^+, \overline{\Lambda^+}\rangle_\wedge = 0$, we can always use a $\mathrm{GL}(3,\mathbb{C})$ rotation to choose a basis $\Sigma^i, i = 1,2,3$ of Λ^+ that satisfies

$$\text{Lorentzian case}: \Sigma^i\Sigma^j = 2\mathrm{i}\delta^{ij}\epsilon \quad \text{and} \quad \Sigma^i\overline{\Sigma^j} = 0. \tag{5.38}$$

Here ϵ is a real top form on the manifold. We shall refer to such a basis of Λ^+ as orthonormal. An example of an orthonormal basis for Λ^+ defined by a Lorentzian signature metric is given by (5.31).

Euclidean and Split signature. Given a three-dimensional subspace $\Lambda^+ \subset \Lambda^2$ on which the wedge product metric is definite in the Euclidean case, and indefinite in the split case, and using a $\mathrm{GL}(3,\mathbb{R})$ rotation, we can always choose a basis $\Sigma^i, i = 1,2,3$ of Λ^+ that satisfies

$$\text{Euclidean case}: \Sigma^i\Sigma^j = 2\delta^{ij}\epsilon, \qquad \text{Split case}: \Sigma^i\Sigma^j = 2\eta^{ij}\epsilon, \tag{5.39}$$

where ϵ is a top form. Again, we shall refer to such a basis of Λ^+ as orthonormal.

5.4.4 A Constructive Proof of Theorem 5.5: Lorentzian Case

The discussion that follows is for the Lorentzian case, we will discuss Euclidean and split versions in the following subsection. Given a triple of 2-forms satisfying (5.38), we shall prove that (i) there is a natural Lorentzian signature metric defined by Σ^i; (ii) the 2-forms Σ^i are SD in this metric; and (iii) this metric coincides with that given by (5.37). As they stand, these statements are a bit vague, they will be made precise in Theorem 5.6.

To proceed with the proof, as the first step, we split the 2-forms Σ^i into their real and imaginary parts

$$\Sigma^i = S^i + \mathrm{i}P^i. \tag{5.40}$$

In view of (5.38) the forms S^i, P^i satisfy

$$S^iS^j = 0, \qquad P^iP^j = 0, \qquad S^iP^j = \delta^{ij}\epsilon. \tag{5.41}$$

Now, since $P^1P^1 = 0$ this 2-form must be simple, and so $P^1 = e^4e^1$ for some 1-forms e^4, e^1. Similarly, S^1 is a simple 2-form, and we can write $S^1 = -e^2e^3$. Since $S^1P^1 = \epsilon$ we have $\epsilon = e^1e^2e^3e^4$ and the basis $e^{1,2,3,4}$ is nondegenerate. We note that the forms $e^{4,1}$ and $e^{2,3}$ are defined modulo unimodular transformations mixing e^4 and e^1, and similarly transformations mixing e^2 and e^3, as such transformations do not change e^4e^1 and e^2e^3.

Now, the form P^2 is also simple $P^2P^2 = 0$, and is orthogonal to both S^1 and P^1. This means that it can be written in the form

$$P^2 = (\alpha e^4 + \beta e^1)(\gamma e^2 + \delta e^3) \tag{5.42}$$

for some (real) coefficients α, β, γ, and δ. Similarly the simple 2-form S^2 can be written as

$$S^2 = -(\rho e^2 + \sigma e^3)(\mu e^4 + \nu e^1) \tag{5.43}$$

for some coefficients μ, ν, ρ, and σ. From $S^2 P^2 = \epsilon = e^{1234}$ we must have $(\beta\mu - \alpha\nu)(\gamma\sigma - \delta\rho) = -1$. We can now define the new basis 1-forms $\tilde{e}^4 = \alpha e^4 + \beta e^1$, $\tilde{e}^1 = \mu e^4 + \nu e^1, \tilde{e}^2 = \gamma e^2 + \delta e^3, \tilde{e}^3 = \rho e^2 + \sigma e^3$. These are unimodular transformations provided $\alpha\nu - \beta\mu = 1$ and $\gamma\sigma - \delta\rho = 1$. In this new basis we have $P^1 = \tilde{e}^4 \tilde{e}^1$, $S^1 = -\tilde{e}^2 \tilde{e}^3$ and $P^2 = \tilde{e}^4 \tilde{e}^2, S^2 = -\tilde{e}^3 \tilde{e}^1$. We have thus proved that P^1, S^1, P^2, and S^2 can always be mapped to this form. We will omit the tildas on the symbols of these 1-forms from now on.

We now come to the last pair P^3 and S^3. They are both simple, and both orthogonal to all the 2-forms P^1, S^1, P^2, and S^2. There are two different possibilities. Either $P^3 \sim e^4 e^3$ and $S^3 \sim e^1 e^2$ or $P^3 \sim e^1 e^2$ and $S^3 \sim e^4 e^3$. In the first of these possibilities, it is the 2-forms $P^i, i = 1, 2, 3$ that all share a common 1-form, while in the second case, these are the 2-forms S^i that share a common factor.

Let us now write the obtained solutions in a canonical form. There is still some freedom remaining after fixing the 2-forms P^1, S^1, P^2, and S^2. Indeed, we can rescale $e^{1,2} \to \lambda e^{1,2}$ and $e^{3,4} \to \lambda^{-1} e^{3,4}$ without changing P^1, S^1, P^2, and S^2. These rescalings allow to put the coefficient in front of P^3 and S^3 to plus minus unity. This allows us to present the four arising cases as

$$\text{Case A}\pm \quad \begin{aligned} P^1 &= e^4 e^1, & S^1 &= -e^2 e^3, \\ P^2 &= e^4 e^2, & S^2 &= -e^3 e^1, \\ P^3 &= \sigma e^4 e^3, & S^3 &= -\sigma e^1 e^2, \end{aligned} \tag{5.44}$$

where $\sigma = \pm 1$ and

$$\text{Case B}\pm \quad \begin{aligned} S^1 &= e^4 e^1, & P^1 &= -e^2 e^3, \\ S^2 &= e^4 e^2, & P^2 &= -e^3 e^1, \\ S^3 &= \sigma e^4 e^3, & P^3 &= -\sigma e^1 e^2, \end{aligned} \tag{5.45}$$

where we have relabelled $e^4 \to e^3, e^1 \to e^2$ and $e^3 \to -e^4, e^2 \to -e^1$ to write the second case forms. Note that this does not change the ϵ form. The case B is of course the same solution with S and P forms interchanged. As we have already mentioned, in the case A the 2-forms P^i all share a common factor, while in the case B the 2-forms S^i do.

Having obtained the four different possible canonical expressions for the 2-forms Σ^i satisfying (5.38) we can understand the metric that these forms define. It is clear that in all the four cases the 1-form e^4 is special, and so there is a natural metric

$$ds^2_\Sigma = -(e^4)^2 + (e^1)^2 + (e^2)^2 + (e^3)^2. \tag{5.46}$$

The four different cases that we have seen arising previously are then as follows. The case $A+$ corresponds to the 2-forms Σ^i being self-dual, and coinciding with the basis of forms listed in (5.31). The case $A-$ corresponds to Σ^i still begin SD with respect to the metric that they define, but with Σ^3 being minus what it is in the canonical basis (5.31). The case $B+$ forms are ASD, and $B-$ forms differ from $B+$ in the sign of Σ^3. We also note that the $B+$ case forms can be obtained as $-\mathrm{i}$ times the canonical basis of ASD 2-forms, given by (5.31) with the plus sign in front of the second terms.

These four different cases can be differentiated with the help of the Urbantke formula. Thus, let us see what the metric g_Σ given by

$$g_\Sigma(u,v)\epsilon_\Sigma = \frac{\mathrm{i}}{6}\epsilon^{ijk} i_u \Sigma^i i_v \Sigma^j \Sigma^k \tag{5.47}$$

is in each of the four cases. In this formula, the orientation ϵ_Σ is assumed to be that in $\Sigma^i\Sigma^j = 2\mathrm{i}\delta^{ij}\epsilon_\Sigma$. In the case $A+$ this formula reproduces the metric (5.46). In the case $A-$ the only difference is that Σ^3 is minus what it is in the case $A+$, and it is clear that this gives an additional minus sign on the right-hand side of (5.47), so that one obtains the metric of signature $(+,-,-,-)$. In the B cases, one multiplies the right-hand side in (5.47) by $(-\mathrm{i})^3 = \mathrm{i}$ as compared to the A cases, and so the metric obtained from (5.47) in both B cases is purely imaginary.

We can summarise the previous discussion as

Theorem 5.6 *Let Σ^i be a triple of 2-forms satisfying $\Sigma^i\Sigma^j = 2\mathrm{i}\delta^{ij}\epsilon_\Sigma$, where ϵ_Σ is a real top form, and $\Sigma^i\overline{\Sigma^j} = 0$. Then either the triple Σ^i gives a real metric g_Σ via (5.47), or $\mathrm{i}\Sigma^i$ does. So, we multiply the 2-forms Σ^i by the imaginary unit to get a real metric via (5.47) if necessary. This metric is of Lorentzian signature. The triple of 2-forms Σ^i that gives a real metric via (5.47) is self-dual in the orientation ϵ_Σ.*

We also see that there arises the notion of a sign of a triple Σ^i. Indeed, this arises as the sign that is necessary to put in front of the right-hand side of (5.47) to get the metric of signature $(-,+,+,+)$. Given a triple that satisfies all the conditions of the previous theorem, both signs are possible. In particular, if one has a triple Σ^i that gives the metric of signature $(-,+,+,+)$ via (5.47) with no additional sign necessary, then the triple $-\Sigma^i$ will require an extra sign to get the desired signature.

This notion of the sign of a triple Σ^i can be given a more invariant meaning. To this end, we can use the metric defined by Σ^i to raise one of the indices of Σ^i and convert these objects into endomorphisms of T^*M. Then, as it is easy to check, in the case that Σ^i are given by their standard expressions in (5.31) the resulting three endomorphisms satisfy the imaginary quaternion algebra

$$\Sigma^1\Sigma^2 = \Sigma^3, \qquad \Sigma^2\Sigma^3 = \Sigma^1, \qquad \Sigma^3\Sigma^1 = \Sigma^2. \tag{5.48}$$

On the other hand, in the case when Σ^3 is given by minus what it is in (5.31), the algebra of endomorphisms defined by Σ^i is with an extra minus sign on the right-hand side

$$\Sigma^1\Sigma^2 = -\Sigma^3, \qquad \Sigma^2\Sigma^3 = -\Sigma^1, \qquad \Sigma^3\Sigma^1 = -\Sigma^2. \tag{5.49}$$

Yet another way of stating the origin of this sign is as follows. The object Σ^i can be viewed as a map between two naturally oriented spaces. Either this map preserves the orientation, in which case it is given by its canonical expression as in (5.31), or it reverses it, in which case Σ^3 is minus what it is in (5.31). The reason why both spaces mapped one into another by $\Sigma : \mathbb{C}^3 \to \Lambda^+ \subset \Lambda^2_{\mathbb{C}}$ are oriented is as follows. The space $\mathbb{C}^3$ can be identified with the Lie algebra $\mathfrak{sl}(2)$, and this is naturally oriented. This is because one can take the triple $e^1, e^2, [e^1, e^2]$ as providing the positive orientation, for any $e^1, e^2 \in \mathfrak{su}(2)$. In our case this means that we take 123 as the positive orientation. On the other hand, the space Λ^+ also carries a natural orientation, because we can use the metric to convert SD 2-forms $\omega^{1,2} \in \Lambda^+$ into endomorphisms of T^*M, and then take the positive orientation of Λ^+ to be $\omega^{1}{}_{\mu}{}^{\nu}, \omega^{2}{}_{\mu}{}^{\nu}, \omega^{1}{}_{\mu}{}^{\rho}\omega^{2}{}_{\rho}{}^{\nu} - \omega^{2}{}_{\mu}{}^{\rho}\omega^{1}{}_{\rho}{}^{\nu}$. This explains why both the source and the target of the map Σ is oriented, and why Σ can be both orientation preserving (case $A+$) and orientation changing (case $A-$).

Note that the self-duality of Σ^i with respect to the metric they define versus anti–self-duality, is fixed by the sign on the right-hand side of $\Sigma^i\Sigma^j = 2\mathrm{i}\delta^{ij}\epsilon_\Sigma$. Indeed, the statement in Theorem 5.6 is that Σ^i are SD with respect to the metric they define in the orientation ϵ_Σ. One could then take complex conjugate objects instead, and they would satisfy $\overline{\Sigma}^i\overline{\Sigma}^j = -2\mathrm{i}\epsilon_\Sigma$, and be wedge product orthogonal to their complex conjugates. These objects satisfy all the conditions of the Theorem 5.6, and are thus SD with respect to the metric they define and in orientation $-\epsilon_\Sigma$. But being SD in orientation $-\epsilon_\Sigma$ is the same as being ASD in orientation ϵ_Σ. So, everything is consistent.

This also gives another way of looking at the cases $B\pm$ in the proof. Indeed, the 2-forms $\pm\mathrm{i}\overline{\Sigma^i}$, where Σ^i is the canonical SD forms as in (5.31) satisfy all the assumptions of Theorem 5.6. This is why these solutions to the equations $\Sigma^i\Sigma^j = 2\mathrm{i}\delta^{ij}\epsilon_\Sigma$ and $\Sigma^i\overline{\Sigma^j} = 0$ must appear together with the solutions $\pm\Sigma^i$. This explains the case $B\pm$ solutions that were found previously.

We have thus proved that the knowledge of the split of the space of 2-forms into two orthogonal subspaces of half the dimension, as appropriate for the Lorentzian signature, determines the conformal metric. In fact, we have proved a stronger statement in Theorem 5.6. Indeed, we have shown that given an orthonormal basis in the space Λ^+, which can always be chosen by an $\mathrm{GL}(3, \mathbb{C})$ transformation, there is a natural Lorentzian signature metric given by (5.47). Thus, this construction also fixes the conformal class of the metric, which was left undetermined by the conceptual proof.

There is an interesting twist to the previous story which is that the formula (5.47) also defines a metric even in the case that a triple Σ^i is not orthonormal, and just spans a three-dimensional subspace of Λ^2 that is wedge product orthogonal to its complex conjugate. We will come to this point in one of the following chapters, where it will be seen that this is related to the possibility of 'deforming' the Einstein condition in a nontrivial way.

5.4.5 A Constructive Proof of Theorem 5.5: Euclidean and Split Cases

A similar proof of the fact that the knowledge of the split of the space of 2-forms into two subspaces of half the dimension, together with a basis in one of the spaces, determines the metric, can be given in the Euclidean and split signature cases. Everything is real in this case, so the proof is somewhat simpler. So, let us start with a triple of 2-forms Σ^i such that $\Sigma^i\Sigma^j = 2\delta^{ij}\epsilon_\Sigma$ in the Euclidean case, or $\Sigma^i\Sigma^j = 2\eta^{ij}\epsilon_\Sigma$ in the split case. As in the Lorentzian case, we see that the triple Σ^i defines an orientation ϵ_Σ. Our task is now to prove that such a triple defines a metric of appropriate signature, and with respect to this metric and in orientation ϵ_Σ the triple Σ^i is SD.

Let us carry out the proof in the Euclidean case. It will then be clear what is necessary to change to get the split signature case. The most important first step is to form complex linear combinations

$$\Sigma^\pm = \Sigma^1 \pm \mathrm{i}\Sigma^2. \tag{5.50}$$

These complex 2-forms are simple

$$\Sigma^+\Sigma^+ = \Sigma^-\Sigma^- = 0, \tag{5.51}$$

and are thus decomposable. Thus, we can write

$$\Sigma^+ = sm, \qquad \Sigma^- = \bar{s}\bar{m}, \tag{5.52}$$

where $m, \bar{m}$ and $s, \bar{s}$ are some complex 1-forms spanning T^*M, with $\bar{s}$ being the complex conjugate of s and $\bar{m}$ being the complex conjugate of m. Of course, s, m are only defined modulo unimodular transformations that do not change the wedge product sm. The forms $m, \bar{m}$ and $s, \bar{s}$ span T^*M because we have $\Sigma^+\Sigma^- = 4\epsilon_\Sigma$. On the other hand $\Sigma^+\Sigma^- = -m\bar{m}s\bar{s}$.

For future reference, when the triple Σ^i is as given by (5.31) we have

$$\Sigma^+ = (e^4 - \mathrm{i}e^3)(e^1 + \mathrm{i}e^2), \qquad \Sigma^- = (e^4 + \mathrm{i}e^3)(e^1 - \mathrm{i}e^2), \tag{5.53}$$

and so $m = e^1 + \mathrm{i}e^2, s = e^4 - \mathrm{i}e^3$.

The next bit of information comes from the fact that Σ^3 is wedge product orthogonal to both $\Sigma^\pm$. It is also a real 2-form. This means that it is of the form

$$\Sigma^3 = \frac{a}{2\mathrm{i}}s\bar{s} + \frac{b}{2}s\bar{m} - \frac{\bar{b}}{2}m\bar{s} + \frac{d}{2\mathrm{i}}m\bar{m} \tag{5.54}$$

for some coefficients $a, d \in \mathbb{R}$ and $b \in \mathbb{C}$. We can then compute

$$\Sigma^3\Sigma^3 = -\frac{1}{2}(ad - |b|^2)m\bar{m}s\bar{s}. \tag{5.55}$$

However, we must have $\Sigma^3\Sigma^3 = 2\epsilon_\Sigma$ and so we have $ad - |b|^2 = 1$. This in particular means that a and d are always of the same sign.

The 2-form Σ^3 in (5.54) can be written in matrix form as

$$\Sigma^3 = \frac{1}{2\mathrm{i}}\begin{pmatrix} s & m \end{pmatrix}\begin{pmatrix} a & \mathrm{i}b \\ -\mathrm{i}\bar{b} & d \end{pmatrix}\begin{pmatrix} \bar{s} \\ \bar{m} \end{pmatrix}. \tag{5.56}$$

The matrix that appears here is Hermitian. We also know that it is unimodular. Such a Hermitian unimodular matrix can always be written as the product

$$\begin{pmatrix} a & \mathrm{i}b \\ -\mathrm{i}\bar{b} & d \end{pmatrix} = \pm g g^\dagger, \qquad g \in \mathrm{SL}(2,\mathbb{C}). \tag{5.57}$$

The presence of the sign on the right-hand side reflects the fact that a and d can be of both signs. The determinant condition on the left-hand side implies $|\det(g)|^2 = 1$. But we can always multiply g by a phase (without changing $gg^\dagger$) to achieve that $\det(g)$ is real and positive, and thus that g is unimodular.

So, we have shown that Σ^3 can always be written as

$$\Sigma^3 = \pm\frac{1}{2\mathrm{i}}\left((\alpha s + \beta m)(\bar{\alpha}\bar{s} + \bar{\beta}\bar{m}) + (\gamma s + \delta m)(\bar{\gamma}\bar{s} + \bar{\delta}\bar{m})\right) \tag{5.58}$$

for some coefficients $\alpha, \beta, \gamma, \delta \in \mathbb{C}$ satisfying $\alpha\delta - \beta\gamma = 1$. We can then define $\alpha s + \beta m$ in (5.58) to be the new s and $\gamma s + \delta m$ to be the new m, as this does not change the forms $\Sigma^\pm$. Thus, we learn that we can always represent the triple $\Sigma^\pm, \Sigma^3$ as

$$\Sigma^+ = sm, \qquad \Sigma^- = \bar{s}\bar{m}, \qquad \Sigma^3 = \pm\frac{1}{2\mathrm{i}}(s\bar{s} + m\bar{m}). \tag{5.59}$$

Having achieved this representation of the triple Σ^i the metric is

$$ds^2_\Sigma = s\bar{s} + m\bar{m}. \tag{5.60}$$

This is a metric of Riemannian all plus signature, and the 2-forms Σ^i are SD with respect to this metric, in the orientation ϵ_Σ. It is also not hard to check that this metric can also be obtained by the Urbantke formula, which in the case of this signature reads

$$g_\Sigma(u,v)\epsilon_\Sigma = \frac{1}{6}\epsilon^{ijk} i_u \Sigma^i i_v \Sigma^j \Sigma^k. \tag{5.61}$$

We can state the previous considerations as

Theorem 5.7 *Let Σ^i be a triple of real 2-forms satisfying $\Sigma^i\Sigma^j = 2\delta^{ij}\epsilon_\Sigma$. Then the metric defined by the Urbantke formula* (5.61) *is real, and of Riemannian* *signature. The triple Σ^i is self-dual with respect to this metric, and in the orientation ϵ_Σ.*

As in the Lorentzian case, we see that there is a subtlety that a triple Σ^i satisfying $\Sigma^i\Sigma^j = 2\delta^{ij}\epsilon_\Sigma$ carries the additional information of a sign, which is the sign that is necessary to put on the right-hand side of (5.61) to obtain the metric of signature all plus. If a triple Σ^i gives the all plus signature metric via (5.61), then the triple $-\Sigma^i$ will give the signature all minus, and an extra sign would be needed in (5.61) to flip this back to all plus. We similarly see that the case in which there is the minus sign in Σ^3 in (5.59) gives the all minus signature metric via (5.61), and so this triple carries the negative sign. As in the Lorentzian case, this sign has the geometric origin in the fact that the map Σ is a map between two naturally oriented spaces, and can therefore be orientation-changing as well as orientation-preserving.

In the split signature case, the proof is completely analogous, with some sign changes. The main difference is that in this case the relevant combinations of Σ^1, Σ^2 that are simple are real: $\Sigma^\pm = \Sigma^1 \pm \Sigma^2$. Indeed, when $\Sigma^1\Sigma^1 = 1, \Sigma^2\Sigma^2 = -1, \Sigma^3\Sigma^3 = -1$, and we have $\Sigma^+\Sigma^+ = \Sigma^-\Sigma^- = 0$. The Euclidean signature case proof works with a few changes, the main one being that it uses real coefficients everywhere. In this case we want to put $\Sigma^\pm$ into the form

$$\Sigma^\pm = (e^4 \pm e^3)(e^1 \pm e^2), \tag{5.62}$$

and Σ^3 into the form

$$\Sigma^3 = \mp\frac{1}{2}\left((e^4 + e^3)(e^4 - e^3) + (e^1 + e^2)(e^1 - e^2)\right), \tag{5.63}$$

where both signs are possible. Once this is achieved, the metric is

$$ds^2 = -(e^4 + e^3)(e^4 - e^3) - (e^1 + e^2)(e^1 - e^2), \tag{5.64}$$

which is the metric of split signature. The metric can also be obtained via the Urbantke formula (5.61). We state all this as a

Theorem 5.8 *Let Σ^i be a triple of real 2-forms satisfying $\Sigma^i\Sigma^j = 2\eta^{ij}\epsilon_\Sigma$. Then the metric defined by the Urbantke formula* (5.61) *is real, and of split signature. The triple Σ^i is self-dual with respect to this metric, in the orientation ϵ_Σ.*

5.5 The 'Lorentz' Groups in Four Dimensions

The purpose of this section is to remind the reader some facts about different signature 'Lorentz' groups in four dimensions, and also explain the origin of the 'accidental isomorphisms' (5.1). We have already covered this material to some extent in Chapter 1, but here we present a more complete treatment including the discussion of all three different possible signatures.

5.5.1 Lorentzian Signature

Let us start our discussion with the Lorentz group proper, i.e., the group $SO^+(1,3)$ of special (i.e., determinant one) pseudo-orthogonal transformations (i.e., living invariant the metric in $R^{1,3}$) that also preserve the time orientation. This group is doubly covered by the complex special linear group $SL(2,\mathbb{C})$, as the following construction explains.

Let us fix some Cartesian coordinate system x^4, x^1, x^2, x^3 in $\mathbb{R}^{1,3}$, so that the distance squared from a point with this coordinates to the origin, equals $-(x^4)^2 + (x^1)^2 + (x^2)^2 + (x^3)^2$. The reason why we use x^4 rather than x^0 notation is that we want to be uniform in our treatment of all the signatures. Let us form the following 2×2 matrix

$$\mathbf{x}_L = \mathrm{i}\begin{pmatrix} x^4 + x^3 & x^1 - \mathrm{i}x^2 \\ x^1 + \mathrm{i}x^2 & x^4 - x^3 \end{pmatrix}. \tag{5.65}$$

Note that

$$x_L = \mathrm{i}\left(x^4 \mathbb{I} + \sum_{i=1}^{3} \sigma^i x^i \right), \tag{5.66}$$

where $\sigma^i, i = 1, 2, 3$ are the usual Pauli matrices. It is our desire to write $\mathbf{x}_L$ as (5.66) that explains the sign choices in (5.65). We note that the matrix $\mathbf{x}_L$ given by (5.65) is i times a Hermitian matrix, and so is anti-Hermitian. Moreover, every 2×2 anti-Hermitian matrix can be written in the form (5.65) for some choice of x^4, x^i. This is clear from (5.66) and the fact that $\mathbb{I}, \sigma^i$ provide a basis in the space of 2×2 Hermitian matrices. We also note that the distance squared can be written as the determinant of (5.65)

$$\det(\mathbf{x}_L) = \eta_{IJ} x^I x^J. \tag{5.67}$$

Thus, we have constructed a map

$$\psi_L : \mathbb{R}^{1,3} \to \mathrm{AHerm}(2) \tag{5.68}$$

from Minkowski space to the space AHerm(2) of anti-Hermitian 2×2 matrices. This map is an isomorphism. The Minkowski metric is given by the pullback of the determinant with this map; see (5.67).

We now consider the following action of $SL(2,\mathbb{C})$ on the space of anti-Hermitian 2×2 matrices

$$\mathbf{x}_L \to g\mathbf{x}_L g^\dagger, \qquad g \in SL(2,\mathbb{C}). \tag{5.69}$$

This maps the space of anti-Hermitian matrices to itself. Also, this map preserves $\det(\mathbf{x}_L)$. Thus, pulling this map back with ψ_L (5.68), we get an action of $SL(2,\mathbb{C})$ on $\mathbb{R}^{1,3}$ that is distance preserving. This gives a map from the group $SL(2,\mathbb{C})$ to the group $O(1,3)$ of pseudo-orthogonal transformations of $\mathbb{R}^{1,3}$. It can be checked by an explicit computation that this map is actually into $\mathrm{SO}^+(1,3)$. We will denote this map by ψ_L as well, with a slight abuse of notation

$$\psi_L : \mathrm{SL}(2,\mathbb{C}) \to \mathrm{SO}^+(1,3). \tag{5.70}$$

It is also clear that this map is a group homomorphism.

It is clear that matrices $\pm\mathbb{I} \in \mathrm{SL}(2,\mathbb{C})$ get mapped to $\mathbb{I} \in \mathrm{SO}(1,3)$. Thus, the kernel of ψ_L is $\mathbb{Z}_2$, and we have constructed the double cover

$$\mathrm{SO}^+(1,3) = \mathrm{SL}(2,\mathbb{C})/\mathbb{Z}_2. \tag{5.71}$$

Example 5.9 As an example, let us consider the transformation generated by

$$g = \begin{pmatrix} e^{t/2} & 0 \\ 0 & e^{-t/2} \end{pmatrix} \in \mathrm{SL}(2,\mathbb{C}). \tag{5.72}$$

A simple computation shows that the pullback of this transformation on AHerm(2) with ψ_L gives the following pseudo-orthogonal transformation on $\mathbb{R}^{1,3}$

$$\begin{aligned} x^4 \to \tilde{x}^4 &= \cosh(t)x^4 + \sinh(t)x^3, \\ x^3 \to \tilde{x}^3 &= \sinh(t)x^4 + \cosh(t)x^3. \end{aligned} \tag{5.73}$$

The determinant of the corresponding $\mathrm{O}(1,3)$ matrix is clearly $+1$, and it preserves the time direction because the coefficient in front of x^4 is always positive.

It is clear from the previous construction that the concrete isomorphism (5.71) depends on the choice of coordinates x^4, x^i. However, all possible coordinate choices are related by Lorentz, and thus, $\mathrm{SL}(2,\mathbb{C})$ transformations. So, if we base the previous isomorphism (5.71) construction on a different choice of the coordinate system $\tilde{\mathbf{x}}_L = G\mathbf{x}_L G^\dagger, G \in \mathrm{SL}(2,\mathbb{C})$, then we get two different embeddings of $\mathrm{SL}(2,\mathbb{C})$ into $\mathrm{SO}^+(1,3)$ that are related by conjugation

$$\tilde{\mathbf{x}}_L \to \tilde{g}\tilde{\mathbf{x}}_L\tilde{g}^\dagger, \qquad \mathbf{x}_L \to g\mathbf{x}_L g^\dagger \quad \Rightarrow \quad \tilde{g} = GgG^{-1}. \tag{5.74}$$

So, while there is no canonical embedding $\mathrm{SL}(2,\mathbb{C}) \to \mathrm{SO}^+(1,3)$, different such embeddings are conjugate to each other inside $\mathrm{SL}(2,\mathbb{C})$.

The constructed isomorphism (5.71) also gives the isomorphism

$$\mathrm{SO}(3) = \mathrm{SU}(2)/\mathbb{Z}_2. \tag{5.75}$$

Indeed, one should just restrict the previous construction to matrices (5.65) with $x^4 = 0$. It is clear that these are tracefree (anti-) Hermitian matrices. Only the $\mathrm{SU}(2)$ subgroup of $\mathrm{SL}(2,\mathbb{C})$ acts on this space preserving it, and so we get (5.75).

5.5.2 Euclidean Signature

Let us spell out an analog of the previous construction for the Euclidean signature. We take the Cartesian coordinates on $\mathbb{R}^4$ to be x^4, x^i, and construct a 2×2 matrix

$$\mathbf{x}_E = \mathrm{i}\begin{pmatrix} \mathrm{i}x^4 + x^3 & x^1 - \mathrm{i}x^2 \\ x^1 + \mathrm{i}x^2 & \mathrm{i}x^4 - x^3 \end{pmatrix}. \tag{5.76}$$

This matrix can be viewed as an analytic continuation of $\mathbf{x}_L$ with $x^4 \to \mathrm{i}x^4$. It has the property that its determinant correctly reproduces the squared interval

$$\det(\mathbf{x}_E) = \delta_{IJ}x^I x^J. \tag{5.77}$$

We note that the matrix x_E is of the form

$$\mathbf{x}_E = \begin{pmatrix} \alpha & -\beta^* \\ \beta & \alpha^* \end{pmatrix}, \qquad \alpha, \beta \in \mathbb{C}, \tag{5.78}$$

and that every matrix of the form (5.78) can be represented as (5.76). Thus, we have effectively endowed $\mathbb{R}^4$ with a complex structure and represented it as $\mathbb{C}^2$, and also written the squared interval on it as $|\alpha|^2 + |\beta|^2$. There are of course many different ways of identifying $\mathbb{R}^4$ with $\mathbb{C}^2$, and this ambiguity is related to the ambiguity of choosing the coordinates in the construction (5.76). We will return to this ambiguity when discussing twistors in Chapter 9.

We now note that the Hermitian conjugate of (5.78) is given by

$$\mathbf{x}_E^\dagger = \begin{pmatrix} \alpha^* & \beta^* \\ -\beta & \alpha \end{pmatrix}, \tag{5.79}$$

while the inverse is given by

$$\mathbf{x}_E^{-1} = \frac{1}{|\alpha|^2 + |\beta|^2}\begin{pmatrix} \alpha^* & \beta^* \\ -\beta & \alpha \end{pmatrix}. \tag{5.80}$$

Thus, there is a relation between the Hermitian conjugation and inverse

$$\mathbf{x}_E^\dagger = \det(\mathbf{x}_E)\mathbf{x}_E^{-1}. \tag{5.81}$$

In particular, unit vectors $\det(\mathbf{x}_E) = 1$ correspond to unitary matrices $\mathbf{x}_E^{-1} = \mathbf{x}_E^\dagger$. It is also clear that every 2×2 matrix with property (5.81) can be written in the form (5.78). Thus, we have constructed an isomorphism

$$\psi_E : \mathbb{R}^4 \to \mathbb{H}, \qquad \mathbb{H} := \{\mathbf{x} \in \mathrm{Mat}(2, \mathbb{C}) : \mathbf{x}\mathbf{x}^\dagger = \det(\mathbf{x})\mathbb{I}\}. \tag{5.82}$$

At the end of this Section we shall explain that the space $\mathbb{H}$ so defined is actually the space of quaternions. This justifies the notation. As remarked previously, unit elements in $\mathbb{H}$ are (special) unitary 2×2 matrices.

We can now define an action of $\mathrm{SU}(2) \times \mathrm{SU}(2)$ on $\mathbb{H}$. This is the action of unit quaternions on $\mathbb{H}$ from left and right. It is given by

$$\mathbf{x}_E \to g_L \mathbf{x}_E g_R^{-1}, \qquad g_{L,R} \in \mathrm{SU}(2). \tag{5.83}$$

It is clear that this action preserves the space $\mathbb{H}$, and that it also preserves the determinant. Pulling this map back with ψ_E we get an action of $\mathrm{SU}(2) \times \mathrm{SU}(2)$ on $\mathbb{R}^4$ that preserves the distance squared, and thus is an orthogonal transformation.

It can be checked that only orthogonal transformations of determinant +1 arise this way, and so we actually get a map into SO(4). Again abusing the notation slightly, we denote this map by ψ_E

$$\psi_E : \mathrm{SU}(2) \times \mathrm{SU}(2) \to \mathrm{SO}(4). \tag{5.84}$$

As in the Lorentz group case, this map is a double cover, with the element $(-1,-1) \in \mathrm{SU}(2) \times \mathrm{SU}(2)$ being sent to the identity element in SO(4). So, we get the isomorphism

$$\mathrm{SO}(4) = \mathrm{SU}(2) \times \mathrm{SU}(2)/\mathbb{Z}_2. \tag{5.85}$$

As in the Lorentz group case, there is no canonical map of this sort, with different maps being based on different coordinate choices. Different possible maps of this sort are related by the conjugation inside $\mathrm{SU}(2) \times \mathrm{SU}(2)$.

Finally, let us explain why the set of complex 2×2 matrices satisfying $\mathbf{x}\mathbf{x}^\dagger = \det(\mathbf{x})\mathbb{I}$ is the same as the space $\mathbb{H}$ of quaternions. As is well-known and easy to check, the matrix representation of the quaternion $q = a + bi + cj + dk$ is given by

$$\begin{pmatrix} a + \mathrm{i}b & c + \mathrm{i}d \\ -c + \mathrm{i}d & a - \mathrm{i}b \end{pmatrix}. \tag{5.86}$$

These are precisely matrices of the form (5.78), which proves the claim.

5.5.3 Split Signature

In the split signature case we choose coordinates x^4, x^1, x^2, x^3 so that the squared interval is $-(x^4)^2 - (x^1)^2 + (x^2)^2 + (x^3)^2$, and associate to any point in $\mathbb{R}^{2,2}$ the matrix

$$\mathbf{x}_S = \begin{pmatrix} x^4 + x^3 & x^1 - x^2 \\ x^1 + x^2 & x^4 - x^3 \end{pmatrix}. \tag{5.87}$$

This can be viewed as an analytic continuation $\mathrm{i}x^2 \to x^2$ on (5.65), and removing the factor of i from in front of the matrix. This is a 2×2 real matrix. It is clear that every such matrix can be written in the form (5.87). The squared interval is (minus) the determinant. We could have kept the factor of i in front of the matrix in order it to be the case that the interval is the determinant. However, this would mean having to work with imaginary rather than real matrices. While this is possible, it does not seem natural given that everything can be chosen to be real in this case. We could have also changed the metric, but this is not natural either because it is the metric that naturally arises from the Urbantke formula with the basis as in (5.31). This justifies our choices.

The group $\mathrm{SL}(2,\mathbb{R}) \times \mathrm{SL}(2,\mathbb{R})$ acts on the space of real 2×2 matrices via

$$\mathbf{x}_S \to g_L \mathbf{x}_S g_R^{-1}, \qquad g_{L,R} \in \mathrm{SL}(2,\mathbb{R}). \tag{5.88}$$

This action preserves the squared interval, and gives a map from $\mathrm{SL}(2,\mathbb{R}) \times \mathrm{SL}(2,\mathbb{R})$ to $\mathrm{O}(2,2)$, which is actually into $\mathrm{SO}(2,2)$. This map has a nontrivial kernel Z_2, and so we get the isomorphism in the split signature case

$$\mathrm{SO}(2,2) = \mathrm{SL}(2,\mathbb{R}) \times \mathrm{SL}(2,\mathbb{R})/\mathbb{Z}_2. \tag{5.89}$$

It is possible to get yet another isomorphism for free from the previous construction. Thus, we can set $x^4 = 0$ and consider tracefree real matrices. The subgroup of $\mathrm{SL}(2,\mathbb{R})\times\mathrm{SL}(2,\mathbb{R})$ that preserves this space is the diagonal $\mathrm{SL}(2,\mathbb{R})$. On the other hand, the subgroup of $\mathrm{SO}(2,2)$ that acts on the plane $x^4 = 0$ is $\mathrm{SO}(1,2)$. This gives the isomorphism

$$\mathrm{SO}(1,2) = \mathrm{SL}(2,\mathbb{R})/\mathbb{Z}_2. \tag{5.90}$$

5.5.4 Isomorphism $\mathrm{SO}(3,3) \sim \mathrm{SL}(4,\mathbb{R})$

Let us also prove the isomorphism between the pseudo-orthogonal group of split signature in dimension six, and the real special linear group in dimension four. This isomorphism has played an important role in the conceptual proof of Theorem 5.5.

Let us consider the space $\Lambda^2\mathbb{R}^4$ of bivectors B^{IJ} in dimension four. The wedge product makes this into a metric space

$$\langle B_1, B_2\rangle_\wedge = \epsilon_{IJKL} B_1^{IJ} B_2^{KL}. \tag{5.91}$$

As we already know, this metric on $\Lambda^2\mathbb{R}^4$ is of split signature $(3,3)$. This creates a map

$$\psi : \Lambda^2\mathbb{R}^4 \to \mathbb{R}^{3,3}. \tag{5.92}$$

The group $\mathrm{SL}(4,\mathbb{R})$ naturally acts on $\mathbb{R}^4$, and thus there is also the natural action on $\Lambda^2\mathbb{R}^4$

$$\mathrm{SL}(4,\mathbb{R}) \ni G : \Lambda^2\mathbb{R}^4 \to \Lambda^2\mathbb{R}^4, \qquad B^{IJ} \to (GB)^{IJ} = G^I{}_K G^J{}_L B^{KL}.$$

Because $\det(G) = 1$, this action preserves the metric $\langle GB_1, GB_2\rangle_\wedge = \langle B_1, B_2\rangle_\wedge$. Thus, the push-forward of the action of $\mathrm{SL}(4,\mathbb{R})$ on $\Lambda^2\mathbb{R}^4$ to $\mathbb{R}^{3,3}$ is an isometry, and we get a map

$$\psi : \mathrm{SL}(4,\mathbb{R}) \to \mathrm{SO}(3,3). \tag{5.93}$$

It is clear that this map has a nontrivial kernel, because both $G = \mathbb{I}, -\mathbb{I}$ result in trivial action on $\mathbb{R}^{3,3}$. Thus, we obtain the double cover

$$\mathrm{SO}(3,3) = \mathrm{SL}(4,\mathbb{R})/\mathbb{Z}_2 \tag{5.94}$$

that was used in the proof of Theorem 5.5. The relation (5.94) is usually referred to in the literature as the *twistor* isomorphism, for its complexified version has a direct link to twistor theory. This will be explained in Chapter 9.

5.5.5 Lie Algebras: Lorentz Group

Let us now describe how the discussed previously isomorphisms become realised at the Lie algebra level. We will see that this is related to the SD/ASD decompositions.

We start our discussion with the Lorentz group $\mathrm{O}(1,3)$. This is the group of 4×4 matrices $m^I{}_J$ that preserve the Minkowski bilinear form $m^I{}_K m^J{}_L \eta^{KL} = \eta^{IJ}$. As before, our convention is that it is the direction x^4 that is timelike. In index-free notations the defining property of $\mathrm{O}(1,3)$ reads $m\eta m^T = \eta$, where m^T is the transpose. In infinitesimal form $m = \exp tX$ and we get $X\eta + \eta X^T = 0$ or $X^I{}_K\eta^{KJ} + X^J{}_K\eta^{IK} = 0$. Thus, the Lie algebra of Lorentz group can be parametrised by 4×4 antisymmetric matrices $X^{IJ} := X^I{}_K\eta^{KJ}$. A convenient basis in the space of such matrices is $(X^{MN})^{IJ} = \eta^{MI}\eta^{NJ} - \eta^{MJ}\eta^{NI}, M$, $N = 1,2,3,4$. Then the matrices $(X^{MN})^I{}_J$ are as follows. First, we have three antisymmetric matrices X^{12}, X^{23}, and X^{31}

$$X^{12} := -K^3 = \begin{pmatrix} 0 & 1 & 0 & 0 \\ -1 & 0 & 0 & 0 \\ 0 & 0 & 0 & 0 \\ 0 & 0 & 0 & 0 \end{pmatrix}, X^{23} := -K^1 = \begin{pmatrix} 0 & 0 & 0 & 0 \\ 0 & 0 & 1 & 0 \\ 0 & -1 & 0 & 0 \\ 0 & 0 & 0 & 0 \end{pmatrix},$$

$$X^{31} := -K^2 = \begin{pmatrix} 0 & 0 & -1 & 0 \\ 0 & 0 & 0 & 0 \\ 1 & 0 & 0 & 0 \\ 0 & 0 & 0 & 0 \end{pmatrix}.$$

Then, we have three symmetric matrices X^{41}, X^{42}, and X^{43}

$$X^{41} := -P^1 = \begin{pmatrix} 0 & 0 & 0 & -1 \\ 0 & 0 & 0 & 0 \\ 0 & 0 & 0 & 0 \\ -1 & 0 & 0 & 0 \end{pmatrix}, X^{42} := -P^2 = \begin{pmatrix} 0 & 0 & 0 & 0 \\ 0 & 0 & 0 & -1 \\ 0 & 0 & 0 & 0 \\ 0 & -1 & 0 & 0 \end{pmatrix},$$

$$X^{43} := -P^3 = \begin{pmatrix} 0 & 0 & 0 & 0 \\ 0 & 0 & 0 & 0 \\ 0 & 0 & 0 & -1 \\ 0 & 0 & -1 & 0 \end{pmatrix}.$$

The corresponding matrix Lie algebra is

$$[K^i, K^j] = \epsilon^{ij}{}_k K^k, \quad [K^i, P^j] = \epsilon^{ij}{}_k P^k, \quad [P^i, P^j] = -\epsilon^{ij}{}_k K^k. \tag{5.95}$$

The Lie algebra acts on vectors from $\mathbb{R}^{1,3}$ via matrix multiplication $x^I \to X^I{}_J x^J$. We also note that the described Lorentz group Lie algebra $\mathfrak{so}(1,3)$ can be obtained as the complexification of the Lie algebra of rotations $\mathfrak{so}(3)$. Indeed, (5.95) is just the Lie algebra over $\mathbb{R}$ with generators $K^i, P^i = \mathrm{i}K^i$, where K^i are the usual generators of $\mathfrak{so}(3)$.

Let us now see how the isomorphism (5.71) manifests itself at the Lie algebra level. Any element of SL(2, $\mathbb{C}$) can be written in the exponential form as

$$g = \exp\left(-\frac{\mathrm{i}}{2}\sigma^i\xi_i\right) \in \mathrm{SL}(2,\mathbb{C}), \qquad \xi_i \in \mathbb{C}^3. \tag{5.96}$$

Then, at the infinitesimal level, the Lie algebra of SL(2, $\mathbb{C}$) is composed of tracefree complex 2×2 matrices, of which a basis (over $\mathbb{C}$) is provided by $(-\mathrm{i}/2)\sigma^i$. A Lie algebra element $-(\mathrm{i}/2)\sigma^i\xi_i$ acts on anti-Hermitian matrices $\mathbf{x}_L$ via

$$\mathbf{x}_L \to -(\mathrm{i}/2)\sigma^i\xi_i\mathbf{x}_L + \mathbf{x}_L(\mathrm{i}/2)\sigma^i\xi_i^*. \tag{5.97}$$

Pulling back this action with ψ_L to $\mathbb{R}^{1,3}$ gives a concrete realisation of the corresponding Lie algebra isomorphism in (5.1).

It is easy to check that the generators K^i of rotations in this representation are given by anti-Hermitian matrices $K^i = -(\mathrm{i}/2)\sigma^i$, and generators of boosts are the Hermitian matrices $P^i = (1/2)\sigma^i$. It is easy to check that (5.95) is indeed satisfied. Thus, a general $\mathfrak{so}(1,3)$ Lie algebra element $X = K^i a_i + P^i b_i$ that acts on $\mathbb{R}^{1,3}$ via $x^I \to X^I{}_J x^J$ gets represented as a tracefree complex 2×2 matrix $-(\mathrm{i}/2)\sigma^i\xi_i \in \mathfrak{sl}(2,\mathbb{C})$ with

$$\xi_i = a_i + \mathrm{i}b_i. \tag{5.98}$$

The push-forward of the Lie algebra action on $\mathbb{R}^{1,3}$ via ψ_L is the action (5.97) on anti-Hermitian matrices $\mathbf{x}_L$, as is not hard to check by an explicit verification.

Another way in which the Lorentz Lie algebra isomorphism in (5.1) can be made concrete is as follows. Let us consider the complexification of the Lorentz Lie algebra $\mathfrak{so}(1,3)$ and introduce

$$L^i := \frac{1}{2}(K^i - \mathrm{i}P^i), \qquad R^i := \frac{1}{2}(K^i + \mathrm{i}P^i). \tag{5.99}$$

Using (5.95) it is easy to check that

$$[L^i, R^j] = 0, \qquad [L^i, L^j] = \epsilon^{ij}{}_k L^k, \qquad [R^i, R^j] = \epsilon^{ij}{}_k R^k. \tag{5.100}$$

So, indeed, the complexification of the $\mathfrak{so}(1,3)$ Lie algebra is given by two commuting $\mathfrak{sl}(2,\mathbb{C})$ Lie algebras. For real elements of $\mathfrak{so}(1,3)$ we can write

$$K^i a_i + P^i b_i = \frac{1}{2}(K^i - \mathrm{i}P^i)(a_i + \mathrm{i}b_i) + \frac{1}{2}(K^i + \mathrm{i}P^i)(a_i - \mathrm{i}b_i) = L^i\xi_i + R^i\xi_i^*, \tag{5.101}$$

where ξ_i is given by (5.98). We note that in the representation of the Lie algebra by 2×2 complex matrices the generator L^i is correctly reproduced when we take $K^i = -(\mathrm{i}/2)\sigma^i$ and $P^i = (1/2)\sigma^i$. The generator R^i is then its Hermitian conjugate and we have have $L^i = -(\mathrm{i}/2)\sigma^i, R^i = (\mathrm{i}/2)\sigma^i$. The formula (5.101) and the corresponding action this generates on $\mathbb{R}^{1,3}$ is the $\mathfrak{so}(1,3)$ counterpart of the $\mathfrak{sl}(2,\mathbb{C})$ formula (5.97).

Finally, we rewrite the generators L^i, R^i in terms of the matrix generators X^{MN} and observe that there is a relation to the SD/ASD decomposition. We have

$$L^1 = \frac{1}{2}(-X^{23} + \mathrm{i}X^{41}),\ L^2 = \frac{1}{2}(-X^{31} + \mathrm{i}X^{42}),\ L^3 = \frac{1}{2}(-X^{12} + \mathrm{i}X^{43}),$$

or, more compactly

$$L^i = \frac{\mathrm{i}}{2}\left(X^{4i} - \frac{1}{2\mathrm{i}}\epsilon^i{}_{jk}X^{jk}\right) = \mathrm{i}(P_+X)^{4i}, \tag{5.102}$$

where

$$(P_+X)^{MN} = \frac{1}{2}\left(X^{MN} + \frac{1}{2\mathrm{i}}\epsilon^{MN}{}_{RS}X^{RS}\right) \tag{5.103}$$

is the SD projection. Here we have used $\epsilon^{4ijk} = -\epsilon^{ijk}$. Similarly, we have $R^i = -\mathrm{i}(P_-X)^{4i}$. It is thus clear that the decomposition (5.100) is the decomposition of the algebra parametrised by 4×4 anti-symmetric matrices into its SD and ASD pieces. Up to a multiple of i and possibly a sign, the generator L^i, R^i are extracted as the $4i$ components of the SD, ASD projections of the real generators X^{MN}.

5.5.6 Lie Algebras: Euclidean Case

Let us also work out explicitly the Euclidean signature case. In this case, there is no need to complexify the Lie algebra $\mathfrak{so}(4)$ to exhibit the two commuting subalgebras, so things are bit easier.

The Lie algebra $\mathfrak{so}(4)$ is composed of antisymmetric matrices $X^I{}_J$. As generators we can similarly take $(X^{MN})^I{}_J = \delta^{MI}\delta^N{}_J - \delta^M{}_J\delta^{NI}$. We can also define $K^i = -(1/2)\epsilon^i{}_{jk}X^{jk}$ and $P^i = -X^{4i}$, with the commutational relations being

$$[K^i, K^j] = \epsilon^{ij}{}_kK^k, \quad [K^i, P^j] = \epsilon^{ij}{}_kP^k, \quad [P^i, P^j] = \epsilon^{ij}{}_kK^k. \tag{5.104}$$

Thus, the only difference with (5.95) is the absence on the sign on the right-hand side of the last commutator. We can now form two mutually commuting sets of generators

$$L^i = \frac{1}{2}(K^i - P^i), \qquad R^i = \frac{1}{2}(K^i + P^i), \tag{5.105}$$

with commutation relations being as in (5.100). A general Lie algebra element can be written as

$$K^ia_i + P^ib_i = \frac{1}{2}(K^i - P^i)(a_i - b_i) + \frac{1}{2}(K^i + P^i)(a_i + b_i) = L^i\xi_i^L + R^i\xi_i^R,$$

with

$$\xi_i^L = a_i - b_i, \qquad \xi_i^R = a_i + b_i. \tag{5.106}$$

This decomposition of the Lie algebra is clearly related to the SD–ASD decomposition of the space of 4×4 antisymmetric matrices because

$$L^i = \frac{1}{2}\left(X^{4i} - \frac{1}{2}\epsilon^i{}_{jk}X^{jk}\right) = (P_+X)^{4i}, \tag{5.107}$$

where $(P_+X)^{MN}$ is the SD projection of the antisymmetric matrix X^{MN}. Once again, we use the orientation convention $\epsilon^{4ijk} = -\epsilon^{ijk}$. Similarly, for the right generators we have $R^i = -(P_-X)^{4i}$. So, up to a sign, the generators of the two mutually commuting $\mathfrak{su}(2)$ Lie algebras are the $4i$ components of the SD and ASD projections of the generators X^{MN}.

It is also worth spelling out the 2×2 realisation of the previous Lie algebra. It is not hard to check that under the map (5.84), and the corresponding map of Lie algebras, the transformation

$$\mathbf{x}_E \to -\frac{\mathrm{i}}{2}\sigma^i\xi_i^L\mathbf{x}_E + \mathbf{x}_E\frac{\mathrm{i}}{2}\sigma^i\xi_i^R \tag{5.108}$$

corresponds to the transformation $x^I \to X^I{}_J x^J$.

The split signature case is analogous to that of Euclidean signature, with the exception of all relevant matrices being real. It is left to the reader as an exercise.

5.6 The Self-Dual Part of the Spin Connection

Recall from the previous chapter that we introduced the spin connection $\omega^I{}_J$ as a connection in a vector bundle E with fibres copies of $\mathbb{R}^{p,q}$. This bundle is required to be in the same topological class as the tangent bundle TM, and the frame, or soldering form e^I, is the object that provides this isomorphism.

The spin connection gives rise to a connection in $\Lambda^2 E$, which is the second antisymmetric power of the bundle E. Sections of $\Lambda^2 E$ are objects of the type $X^{IJ} = X^{[IJ]}$, and the covariant derivative acts on them as

$$d^\omega X^{IJ} = dX^{IJ} + \omega^I{}_K X^{KJ} + \omega^J{}_K X^{IK}. \tag{5.109}$$

In four dimensions the bundle $\Lambda^2 E$ splits into a direct sum of bundles

$$\Lambda^2 E = \Lambda^+ E \oplus \Lambda^- E, \tag{5.110}$$

and the spin connection induces connections on $\Lambda^\pm E$. We shall refer to these connections as the SD and ASD parts of the spin connection. Our task in this section is to develop a convenient way to work with these connections.

Because the spin connection preserves both the metric in E and the ϵ^{IJKL} tensor, the covariant derivative in E commutes with the decomposition in (5.110). In other words, we can first project a section X^{IJ} and then compute its covariant derivative, or first compute the derivative and then project. Also, we expect the connections in $\Lambda^\pm E$ to be related to the SD/ASD projections of the spin connection, i.e., objects $(P_\pm\omega)^{IJ}$. The specific relations are signature-dependent, as we shall now see.

5.6.1 Lorentzian Case

It is easiest to understand the connections arising in $\Lambda^\pm E$ by an explicit computation. Thus, let us compute the 41 component of the SD projection of $d^\omega X^{IJ}$. We have

$$\begin{aligned} 2\mathrm{i}(P_+ d^\omega X)^{41} &= \mathrm{i} d^\omega X^{41} - d^\omega X^{23} \\ &= +\mathrm{i}(dX^{41} + \omega^4{}_2 X^{21} + \omega^4{}_3 X^{31} + \omega^1{}_2 X^{42} + \omega^1{}_3 X^{43}) \\ &\quad - (dX^{23} + \omega^2{}_4 X^{43} + \omega^2{}_1 X^{13} + \omega^3{}_4 X^{24} + \omega^3{}_1 X^{21}). \end{aligned} \tag{5.111}$$

This can be rewritten as

$$d(\mathrm{i}X^{41} - X^{23}) + A_L^2(\mathrm{i}X^{43} - X^{12}) - A_L^3(\mathrm{i}X^{42} - X^{31}), \tag{5.112}$$

where

$$A_L^2 = \mathrm{i}\omega^{42} - \omega^{31}, \qquad A_L^3 = \mathrm{i}\omega^{43} - \omega^{12}, \tag{5.113}$$

where the indices of $\omega^I{}_J$ are raised with the Minkowski metric $\eta^{IJ} = \mathrm{diag}(1, 1, 1, -1)$. The other $4i$ components of $(P_+ DX)^{IJ}$ are computed similarly. Thus, if we introduce

$$A_L^i := 2\mathrm{i}(P_+\omega)^{4i} \tag{5.114}$$

and

$$X^i := 2\mathrm{i}(P_+X)^{4i}, \tag{5.115}$$

then we have

$$2\mathrm{i}(P_+ d^\omega X)^{4i} = dX^i + \epsilon^i{}_{jk} A^j X^k. \tag{5.116}$$

This explicitly shows that the connection on $\Lambda^+ E$ is a (complexified) SO(3) connection (5.114), with the covariant derivative acting on $4i$ components of the SD tensors $(P_+X)^{IJ}$ via (5.116). It is also clear that a SD tensor is completely characterised by its $4i$ components, because the ij components are related to the $4i$ components by self-duality. Indeed, for any SD tensor X_+^{IJ}

$$0 = 2\mathrm{i}(P_- X_+)^{4i} = \mathrm{i}X_+^{4i} + \frac{1}{2}\epsilon^i{}_{jk} X_+^{jk}, \tag{5.117}$$

which gives the desired relation. To summarise, we learn that the connection induced by the spin connection in $\Lambda^+ E$ is an $\mathrm{SO}(3, \mathbb{C})$ connection arising as the SD projection (5.114) of the spin connection. This is of course not surprising in view of the isomorphism (5.1).

Let us also note that we have based the identification (5.114) of the SD part of the spin connection with an $\mathrm{SO}(3, \mathbb{C})$ connection on a concrete choice of basis in the fibres of E. This is similar to our previous discussion of the isomorphism (5.71) and the corresponding isomorphism of Lie algebras. While the concrete isomorphism ψ_L that was constructed via anti-Hermitian matrices

(5.65) was basis-dependent, different choices of basis gave results that are conjugate in $\mathrm{SL}(2,\mathbb{C})$. Similarly here, while the identifications (5.114) and (5.115) are basis-dependent, different choices of basis give results that are conjugate in $\mathrm{SO}(3,\mathbb{C})$.

5.6.2 Euclidean Case

The Euclidean case reasoning is completely analogous. We have

$$\begin{gathered}2(P_+d^\omega X)^{41} = d^\omega X^{41} - d^\omega X^{23} = \\ +dX^{41} + \omega^4{}_2X^{21} + \omega^4{}_3X^{31} + \omega^1{}_2X^{42} + \omega^1{}_3X^{43} \\ -(dX^{23} + \omega^2{}_4X^{43} + \omega^2{}_1X^{13} + \omega^3{}_4X^{24} + \omega^3{}_1X^{21}).\end{gathered} \tag{5.118}$$

This can be rewritten as

$$d(X^{41} - X^{23}) + A_E^2(X^{43} - X^{12}) - A_E^3(X^{42} - X^{31}), \tag{5.119}$$

where

$$A_E^2 = \omega^{42} - \omega^{31}, \qquad A_E^3 = \omega^{43} - \omega^{12}. \tag{5.120}$$

More generally, if we introduce

$$A_E^i := 2(P_+\omega)^{4i} \tag{5.121}$$

and

$$X^i := 2(P_+X)^{4i}, \tag{5.122}$$

then we have

$$2(P_+d^\omega X)^{4i} = dX^i + \epsilon^i{}_{jk}A^jX^k. \tag{5.123}$$

Thus, again the connection on Λ^+E is an $\mathrm{SO}(3)$ connection given by the SD projection (5.121) of the spin connection. Again, the specific identification (5.121) is basis-dependent, but different choices of basis give conjugate results, and in this sense are immaterial.

The case of the split signature is analogous, except that it is an $\mathrm{SO}(1,2)$ connection that arises in this case. The formula for the covariant derivative in this case coincides with (5.123), with the only subtlety being that the indices on ϵ^{ijk} are lowered with the indefinite metric η_{ij}.

5.6.3 The Curvature

Because the covariant derivative d^ω commutes with the SD/ASD decomposition of E, the curvature of the connection arising on Λ^+E coincides with the SD projection of the curvature $R^{IJ}(\omega)$. Let us derive the relation between the curvature of $A^i_{L,E,S}$ introduced Sections 5.6.1 and 5.6.2, and this projection.

Let us do the Lorentzian signature computation. All other signatures are analogous. We have

$$2\mathrm{i}(P_+R)^{41} = \mathrm{i}R^{41} - R^{23} = \mathrm{i}(d\omega^{41} + \omega^4{}_2\omega^{21} + \omega^4{}_3\omega^{31}) \\ - (d\omega^{23} + \omega^2{}_4\omega^{43} + \omega^2{}_1\omega^{13}). \tag{5.124}$$

This can be rewritten as

$$d(\mathrm{i}\omega^{41} - \omega^{23}) + (\mathrm{i}\omega^{42} - \omega^{31})(\mathrm{i}\omega^{43} - \omega^{12}) = dA^1_L + A^2_L A^3_L. \tag{5.125}$$

Thus, we have

$$2\mathrm{i}(P_+R)^{4i} = F^i(A_L), \tag{5.126}$$

where the curvature of an SO(3) connection is given by

$$F^i(A) = dA^i + \frac{1}{2}\epsilon^i{}_{jk}A^jA^k. \tag{5.127}$$

In the Euclidean and split cases we have instead $2(P_+R)^{4i} = F^i(A_{E,S})$. Thus, we see that the curvature of the connection on Λ^+E can be identified with the curvature of the corresponding $A_{L,E,S}$ connections – $\mathrm{SO}(3,\mathbb{C})$ connection in the Lorentzian case, $\mathrm{SO}(3,\mathbb{R})$ connection in the Euclidean case and $\mathrm{SO}(1,2)$ connection in the split case. This is of course as expected in view of the isomorphisms (5.1).

5.7 The Chiral Soldering Form

The last bit of geometry that we need to understand the construction of the 'chiral' Plebański formulation of GR is the notion of what can be referred to as the chiral soldering form. This object is fundamental in this description of GR, and can be introduced in complete parallel to how the tetrad was introduced previously. Thus, let us remind the reader that the tetrad was viewed as an object that fixes the isomorphism between an abstract vector bundle with fibres copies of $\mathbb{R}^{p,q}$ and the tangent bundle TM. Similarly, we now introduce a vector bundle whose fibres are three-dimensional and that is required to be globally in the same topological class as the bundle of SD 2-forms on M. We then introduce an object Σ that gives this isomorphism. This object encodes all information about the metric. A connection is then introduced in this bundle, and there is an analog of the torsion-free condition that fixes the connection in terms of Σ. It then turns out that this Σ-determined connection is closely related to the SD part of the spin connection that we have studied in the previous section.

5.7.1 The Chiral Soldering Form

With the previous remarks in mind, we introduce a vector bundle $F \hookrightarrow S \to M$ with fibres F being copies of $\mathbb{R}^3, \mathbb{R}^{1,2}$ in the case of the Riemannian and split signatures, and copies of $\mathbb{C}^3$ in the case of the Lorentzian signature. The fibres are assumed to be equipped with a metric that is (positive) definite in the case of the Riemannian signature setup, indefinite in the case of split signature, and complex in the case of the Lorentzian signature. One requires the bundle S to be isomorphic to the bundle of SD 2-forms on M. We note that the topological type of the bundle Λ^+ of SD 2-forms on M with respect to some metric (of a given signature) is metric-independent.

The object to which we refer as the chiral soldering form is then defined as the map that provides the isomorphism between S and a three-dimensional subbundle $\Lambda^+ \subset \Lambda^2$. Thus, we define Σ to be a vector bundle map

$$\Sigma : S \to \Lambda^2. \tag{5.128}$$

In components, if X^i is a section of S then $\Sigma(X) = \Sigma^i X^j \delta_{ij} \in \Lambda^2$, where δ_{ij} is the metric in the fibres. In the case of the Lorentzian signature setup it is the space of complexified 2-forms that appears on the right-hand side of this map.

An additional and very important property that the map Σ is required to satisfy is the compatibility between the wedge product metric in Λ^2 and the metric in S. Indeed, we can pull back the wedge product conformal metric in Λ^2 to S. This leads to the following important definition

Definition 5.10 A chiral soldering form Σ is said to *satisfy the constraints* if the pullback of the wedge-product metric on Λ^2 to S coincides with the metric $\langle \cdot, \cdot \rangle$ that exists on the fibres of S

$$\Sigma(X)\Sigma(Y)/\epsilon \sim \langle X, Y \rangle, \qquad X, Y \in S. \tag{5.129}$$

In this formula, $\sim$ stands for proportional, and ϵ is a top form. It is clear that this condition is a geometric way of stating the orthonormality conditions on Σ^i as appear in (5.38) and (5.39).

In the case of the Lorentzian signature we also want to impose appropriate 'reality' conditions on Σ.

Definition 5.11 A complex-valued chiral soldering form Σ is said to *satisfy the reality conditions* if $\Sigma\overline{\Sigma} = 0$, where $\overline{\Sigma}$ denotes the map Σ followed by the complex conjugation, and $\mathrm{Tr}_\delta(\Sigma\Sigma) = 6i\epsilon_\Sigma$, where δ is the metric in E and ϵ_Σ is a real 4-form.

The discussion of the previous sections shows that a chiral soldering form Σ that satisfies the constraints, and in the case of the Lorentzian signature satisfies the reality conditions, encodes a metric of appropriate signature. The metric is explicitly given by (5.47), where one may have to multiply Σ by i to get a real rather than imaginary metric in the case of the Lorentzian signature, and by (5.61) in the other two cases.

The purpose of the condition (5.129) is to make sure that the object Σ describes just the metric, together with an orthonormal frame for S, and does not contain any additional geometric information. This can be verified by a count of variables present in Σ. This starts its life like a collection of three 2-forms. A 2-form in four dimensions needs 6 numbers (per point) to be specified. Thus, Σ carries 18 numbers per point of M. The condition (5.129) is then the statement that a certain symmetric 3×3 matrix constructed from Σ is proportional to the given matrix, the metric. This is five conditions, because the proportionality coefficient is left unspecified. We then have $18 - 5 = 13 = 10 + 3$, which is the number of components of a metric in four dimensions, plus three Euler angles describing a rotation that is needed to bring an orthonormal basis in the fibre to a given one.

The count in the previous paragraph is similar to that in the case of the tetrad. A tetrad is a collection of 16 components, and we have $16 = 10 + 6$, which is the number of components in the metric plus the dimension of the 'Lorentz' group that maps an orthonormal frame into a given one. So, we learn that the chiral soldering form Σ, subject to the conditions (5.129), carries less components than a tetrad.

The count is slightly more involved in the case of Lorentzian signature. In this case the 2-forms Σ^i are complex, and so carry 18 complex parameters per point. The conditions (5.129) are 5 complex conditions, which gives us 13 complex parameters. We then impose the reality conditions, which are 10 real conditions. This cuts the dimension of the parameter space in Σ down to 10 real describing a real metric plus 3 complex, this being the dimension of the Lorentz group. Thus, in this case, Σ, after all the constraints and reality conditions are imposed, carries the same number of parameters as the tetrad.

5.7.2 Relation to the Tetrad

The Theorems 5.6–5.8 show that there is a relation between a soldering form Σ that satisfies the constraints (and satisfies reality conditions in the case of the Lorentzian signature) and a tetrad for the metric defined by Σ. Indeed, the arguments in the proofs of Theorems 5.6–5.8 show that the object Σ can always be written in a canonical form in terms of a tetrad for the metric it defines. This is the form (5.31), with possibly the sign in front of Σ^3 changed, and possibly all Σ^i multiplied by the imaginary unit in the Lorentzian signature case and Σ^i replaced by ASD forms $\overline{\Sigma^i}$.

We then note that the basic 2-forms (5.31) can be obtained from the SD projections of the 2-forms $e^I e^J$, where e^I is the tetrad, with respect to the 'internal' indices IJ. Indeed, we have

$$(P_+ee)^{IJ} = \frac{1}{2}\left(e^I e^J + \frac{1}{2\sqrt{\sigma}}\epsilon^{IJ}{}_{KL}e^K e^L\right), \tag{5.130}$$

where, as before, σ is the sign with $\sqrt{\sigma} = \mathrm{i}$ for the Lorentzian signature and $\sqrt{\sigma} = 1$ for the other two cases. It is then easy to see that the $4i$ components of this SD projection gives use the basic 2-forms (5.31). Indeed, we have, in the Euclidean and split cases

$$2(P_+ee)^{4i} = e^4 e^i - \frac{1}{2}\epsilon^i{}_{jk}e^j e^k = \Sigma^i_{E,S}, \tag{5.131}$$

and in the Lorentzian case

$$2\mathrm{i}(P_+ee)^{4i} = \mathrm{i}e^4 e^i - \frac{1}{2}\epsilon^i{}_{jk}e^j e^k = \Sigma^i_L. \tag{5.132}$$

Here the indices on $\epsilon^i{}_{jk}$ are lowered with the metric δ_{ij} in the Euclidean and Lorentzian cases and metric $\eta_{ij} = \mathrm{diag}(1,-1,-1)$ in the split case. In all cases the assumed orientation of ϵ^{IJKL} is $\epsilon^{1234} = +1$. We have also used $\epsilon^{4ijk} = -\epsilon^{ijk}$. Thus, in all cases, the basic SD 2-forms (5.31) are just multiples of $4i$ components of the SD projections of 2-forms $e^I e^J$. This is, of course, not surprising, because the SD projections of 2-forms $e^I e^J$ are SD as 2-forms, and thus span Λ^+.

We note that this is very similar to the discussed decomposition of the Lie algebra of the 'Lorentz' groups into its SD/ASD parts. Indeed, we have seen that one way to understand the accidental isomorphisms (5.1) is by carrying out the SD/ASD decomposition of the space of 4×4 antisymmetric matrices that in all signature cases parametrises the Lie algebra. Similarly, an object $(ee)^{IJ}$ it is IJ antisymmetric, and its SD/ASD projection becomes possible, and related to Σ^i.

So, we have established that the canonical, i.e., as in (5.31), chiral soldering form Σ^i can be identified with the appropriate multiples of the $4i$ components of the SD projection of the form $e^I e^J$, where e^I is the tetrad for the metric that Σ^i defines. There remains a subtlety that in each case the form Σ^3, as comes from solving the constraints (and reality conditions in the Lorentzian case), may be minus of what it is in (5.31), and that in the Lorentzian case the forms Σ^i may be given by $\pm\mathrm{i}$ times the canonical ASD 2-forms.

5.7.3 The Torsion-Free Condition

Let us now introduce a metric connection in S. We will denote it by A^i, so that it defines a covariant derivative on sections X^i of S via

$$d^A X^i = dX^i + \epsilon^i{}_{jk}A^j X^k, \tag{5.133}$$

where again the indices on ϵ^{ijk} are lowered using the appropriate metrics, which is δ_{ij} in the case of the Euclidean and Lorentzian signatures, and η_{ij} in the split signature case.

We can then introduce the torsion

$$d^A\Sigma^i = d\Sigma^i + \epsilon^i{}_{jk}A^j\Sigma^k, \tag{5.134}$$

which is a 3-form with values in S. We can require the torsion to be zero, which gives a set of algebraic equations on A^i in terms of derivatives of Σ^i. One easily checks that the number of equations here is the same as the number of unknowns. It is not hard to prove that there is unique solution to this equation, which can be exhibited explicitly. We state this as

Theorem 5.12 *When the chiral soldering form satisfies the constraints (and satisfies the reality conditions in the Lorentzian case), there is a unique torsion-free connection satisfying* $d^A\Sigma = 0$.

In fact, in one of the following chapters we will see that a more general statement is possible, and there is a unique solution of the torsion-free equation for A^i, provided that the chiral soldering form satisfies some nondegeneracy condition. The statement of the Theorem 5.12 is in complete parallel with the statement of Lemma 3.2 in the case of tetrads.

Let us spell out the Lorentzian signature proof. We use the metric defined by Σ^i and take the Hodge dual of the torsion-free condition, which is a 3-form with values in S. Then, using the self-duality of Σ^i we get the following equation

$$\epsilon^i{}_{jk}\Sigma^{j}_{\mu}{}^{\nu}A^k_\nu = \frac{1}{2\mathrm{i}}\epsilon_\mu{}^{\nu\rho\sigma}d_\nu\Sigma^i_{\rho\sigma}. \tag{5.135}$$

Let us rewrite this equation as

$$J_\Sigma(A) = \frac{1}{2\mathrm{i}}{}^*(d\Sigma), \tag{5.136}$$

where we introduced an operator

$$J_\Sigma : \Lambda^1 \otimes S \to \Lambda^1 \otimes S, \qquad J_\Sigma : A^i_\mu \to \epsilon^i{}_{jk}\Sigma^{j}_{\mu}{}^{\nu}A^k_\nu. \tag{5.137}$$

To solve (5.136) we need to find the inverse of J_Σ. For this we can use the fact that when Σ^i satisfies the constraints and reality conditions, they are of the form determined by the Theorem 5.6. This means that the objects $\Sigma^{i}_{\mu}{}^{\nu}$ satisfy the algebra of quaternions

$$\Sigma^{i}_{\mu}{}^{\rho}\Sigma^{j}_{\rho}{}^{\nu} = -\delta^{ij}\delta_\mu{}^\nu + \epsilon^{ij}{}_k\Sigma^{k}_{\mu}{}^{\nu}. \tag{5.138}$$

This is true in the case $A+$. In the case $A-$, there is a minus sign in front of the second term on the right-hand side. In cases B, there is an extra factor of $\pm\mathrm{i}$ in the second term. For definiteness, we assume the relation (5.138). The other cases can be treated analogously.

Thus, assuming that (5.138) holds, a simple computation gives

$$J^2_\Sigma = 2\mathbb{I} + J_\Sigma. \tag{5.139}$$

This means that

$$J_\Sigma^{-1} = \frac{1}{2}(-\mathbb{I} + J_\Sigma). \tag{5.140}$$

This solves the torsion-free equation, explicitly determines A^i in terms of the derivatives of Σ^i, and thus proves the theorem.

A further characterisation of the torsion-free connection A^i is possible. Thus, when Σ^i are as in (5.131) and (5.132), this connection can be explicitly described in terms of the SD part of the spin connection studied in the previous section. Let us state this as

Theorem 5.13 *When Σ^i are given by the SD projections of the forms $e^I e^J$, the torsion-free connection A^i is given by the SD part of the torsion-free spin connection ω^{IJ}.*

The proof follows easily by combining facts established in the previous section. Thus, we know from (5.116) and (5.123) that for A^i given by the SD projection of the spin connection, the A-covariant derivative coincides with the SD projection of the spin connection covariant derivative. Thus, we have, in the Lorentzian case

$$d^A\Sigma^i = 2\mathrm{i}(P_+ d^\omega ee)^{4i} = 0, \tag{5.141}$$

where the last equality is the consequence of the torsion-free condition for the spin connection $d^\omega e^I = 0$. In the Euclidean and split cases the only difference is that there is no factor of i on the right-hand side of of the first equality here.

Let us also find an analog of this statement for the situation when Σ^3 is minus what it is in (5.31). It is not hard to check that this just changes some signs in the solution for A^i. Thus, we have the components $A^{1,2}$ both change signs as compared to the solution for the case when Σ^i is as in (5.31). This happens for all the signatures. It is clear also that the curvature components $F^{1,2}$ change signs as compared to what they are in the 'canonical' case (5.31). We do not discuss the case of Σ^i being related to ASD 2-forms because we will later show that this case does not arise in the full Plebański setup.

5.7.4 An Example of Curvature Computation Using the Chiral Formalism

The described chiral formalism is a powerful computational tool, as the following example aims to demonstrate. The idea of this formalism is to encode a given metric into a collection of orthonormal SD 2-forms as in (5.31). One computes the connection A^i from the torsion-free condition, and then its curvature gives access to the SD part of the curvature of the spin connection, i.e., effectively to the SD projection of the Riemann curvature with respect to one pair of indices. As we know, for all signature cases this is sufficient to impose the Einstein condition.

In the case of the Lorentzian signature one gets all of the Riemann curvature this way, because the ASD part is the complex conjugate of the SD one.

Let us see how this works on the example of the Lorentzian signature spherically symmetric metric (3.48) whose tetrad is given by (3.49). We label $t = 4, r = 1, \theta = 2$, and $\phi = 3$. The basis of the Lorentzian signature SD 2-forms (5.31) is then

$$\Sigma^1 = \mathrm{i}fgdtdr - r^2 \sin\theta d\theta d\phi, \qquad \Sigma^2 = \mathrm{i}frdtd\theta - gr\sin\theta d\phi dr, \qquad (5.142)$$
$$\Sigma^3 = \mathrm{i}fr\sin\theta dtd\phi - grdrd\theta.$$

The torsion-free connection A^i is computed quite straightforwardly. Thus, we have

$$d\Sigma^1 = -2r\sin\theta drd\theta d\phi. \qquad (5.143)$$

This must be equal to

$$d\Sigma^1 = -A^2\Sigma^3 + A^3\Sigma^2. \qquad (5.144)$$

Given that there is no dt on the right-hand side of (5.143) it is natural to expect that $A^2 \sim d\phi$ and $A^3 \sim d\theta$. The most natural guess is then

$$A^2 = -\frac{1}{g}\sin\theta d\phi, \qquad A^3 = \frac{1}{g}d\theta, \qquad (5.145)$$

which fulfils (5.144). Let us now consider $d\Sigma^2$. We have

$$d\Sigma^2 = \mathrm{i}(fr)'drdtd\theta - gr\cos\theta d\theta d\phi dr. \qquad (5.146)$$

This should be equal to

$$d\Sigma^2 = -A^3\Sigma^1 + A^1\Sigma^3. \qquad (5.147)$$

Assuming that A^3 is correctly given by (5.145), which will prove to be right, it suggests that A^1 has terms dt and $d\phi$. The relevant equation is satisfied for

$$A^1 = \frac{\mathrm{i}f'}{g}dt + \cos\theta d\phi. \qquad (5.148)$$

One can then check that the equation $D\Sigma^3 = 0$ is satisfied with the connection given by (5.145) and (5.148). We note that this connection is of course the same as the SD part of the spin connection determined in (3.51), (3.54) and (3.56). If one wishes, the full spin connection is recoverable in this case of the Lorentzian signature by extracting the real and imaginary parts of the connection A^i. We note that the spin connection is computed by the chiral method with less steps required, and also that the connection information is stored more compactly, into three complex 1-forms as compared to six real 1-forms in the tetrad method. These are the early signs of the superiority of the chiral method as compared to the tetrad method.

The main advantage of the chiral method becomes manifest when one computes the curvatures and writes the Einstein condition. The curvature components are given by

$$F^1 = -\mathrm{i}\left(\frac{f'}{g}\right)' dtdr - \left(1 - \frac{1}{g^2}\right)\sin\theta d\theta d\phi, \tag{5.149}$$
$$F^2 = -\frac{\mathrm{i}f'}{g^2}dtd\theta + \left(\frac{1}{g}\right)'\sin\theta d\phi dr,$$
$$F^3 = -\frac{\mathrm{i}f'}{g^2}\sin\theta dtd\phi + \left(\frac{1}{g}\right)' drd\theta.$$

At this step, the calculations are still very similar to those of the tetrad method; it's just that the result is more compactly stored as three complex 2-forms as compared to six real 2-forms. It is in the process of extracting the Einstein condition that the real power of the chiral method becomes manifest. In the tetrad method, one needed to form the Ricci tensor, which involved some laborious and prone to sign error manipulations. In the chiral method we simply require that the curvature is SD as a 2-form, i.e., that it is linear combination of the 2-forms Σ^i. Comparison with Σ^i then shows that each F^i can only be proportional to the corresponding Σ^i. One also easily extract the equations necessary for this to happen

$$-\frac{(f'/g)'}{1 - 1/g^2} = \frac{fg}{r^2}, \qquad \frac{f'/g^2}{(1/g)'} = \frac{fr}{gr}. \tag{5.150}$$

The second of these equations immediately gives the correct

$$\frac{f'}{f} + \frac{g'}{g} = 0, \tag{5.151}$$

which implies that $fg = conts$, which can be set to unity by rescaling the time coordinate. The first equation can then be written as

$$(1 - f^2)'' = \frac{2}{r^2}(1 - f^2), \tag{5.152}$$

whose solution is

$$1 - f^2 = \frac{r_+}{r} \pm \frac{r^2}{l^2}, \tag{5.153}$$

where r_+, l^2 are constants of integration. All in all, we see that the Schwarzschild–de Sitter solution presents itself more readily via the chiral method as compared to the tetrad method. The chiral method also gives directly the useful combinations of the Einstein equations obtained via the tetrad method.

The equations in (5.150) have been obtained from the requirement that the curvature F^i is SD as a 2-form. In the language of the curvature decomposition (5.7) this is the statement that the SD–ASD part of the Riemann curvature $B = 0$. This gives 9 out of 10 equations in general. There is one more equation, which is the statement that the scalar curvature is equal to a multiple of the cosmological constant. Let us see what this equation is via the chiral method. If the statement that the curvature F^i is SD is written as $F^i = M^{ij}\Sigma^j$, for some matrix M^{ij}, then the scalar curvature can be extracted by simply computing the trace of this matrix. In the case of the spherical symmetry under consideration, when the equation (5.150) is satisfied, the matrix M^{ij} is given by

$$M^{ij} = -\text{diag}\left(\frac{(f'/g)'}{fg}, \frac{f'}{fg^2r}, \frac{f'}{fg^2r}\right). \tag{5.154}$$

When $fg = 1$ this simplifies to

$$M^{ij} = -\frac{1}{2}\text{diag}\left((f^2)'', \frac{(f^2)'}{r}, \frac{(f^2)'}{r}\right). \tag{5.155}$$

For the solution (5.153) this becomes

$$M^{ij} = \text{diag}\left(\frac{r_+}{r^3} \pm \frac{1}{l^2}, -\frac{r_+}{2r^3} \pm \frac{1}{l^2}, -\frac{r_+}{2r^3} \pm \frac{1}{l^2}\right). \tag{5.156}$$

We see that

$$\text{Tr}(M) = \pm\frac{3}{l^2}. \tag{5.157}$$

This must be equal to the cosmological constant and so

$$\frac{\Lambda}{3} = \pm\frac{1}{l^2}, \tag{5.158}$$

which identifies the constant of integration l as the radius of curvature of the relevant constant curvature space.

All in all, we hope that our presentation demonstrates that the Schwarzschild solution of GR is obtained much more easily with the help of the chiral formalism.

5.8 Plebański Formulation of GR

Plebański's formulation (1977) of GR is based on the previously described geometric constructions. As we know from (5.25) it is possible to write a first-order Lagrangian for GR that contains only the SD part of the spin connection. This Lagrangian also only depends on the tetrad in the combination $(P_+ee)^{IJ}$, where P_+ is the SD projector. As our previous discussion shows, the SD part of the spin connection can be encoded by an $\text{SO}(3), \text{SO}(3,\mathbb{C})$, or $\text{SO}(1,2)$ connection, depending on the signature. The relevant formulas are (5.114) and (5.121). We have also seen that the SD part of the wedge product of two tetrads is naturally encoded in an $\mathfrak{so}(3), \mathfrak{so}(3,\mathbb{C})$, or $\mathfrak{so}(1,2)$ valued 2-form Σ^i. The relevant formulas are (5.131) and (5.132).

The idea now is to write a first-order action of BF-type with Σ^i and A^i as the basic fields. However, not every Lie algebra–valued 2-form Σ^i is the

SD projection of the wedge products of tetrads for some metric. However, we can add to the action a Lagrange multiplier term that will impose conditions that would guarantee that this is the case. This is completely analogous to the BF-type formalism that was developed for the full non-chiral theory earlier. The constraints that need to be imposed on Σ^i have also been previously discussed in this chapter. Thus, any Lie algebra–valued 2-form that satisfies the condition (5.129), as well as reality conditions in the Lorentzian case, is the SD projection of the wedge product of two copies of the tetrad, as discussed in Section 5.7.2. This gives all the required ingredients to state an action whose Euler–Lagrange equations are as desired.

The Plebański action reads

$$S_{\text{Pleb}}[\Sigma, A, \Psi] = \frac{1}{8\pi G\sqrt{\sigma}} \int \Sigma^i F_i - \frac{1}{2}\left(\Psi_{ij} + \frac{\Lambda}{3}\delta_{ij}\right)\Sigma^i\Sigma^j. \tag{5.159}$$

Here the index on F^i is lowered with the metric δ_{ij} in the Lorentzian and Euclidean cases, and η_{ij} in the split case. Also, in the case of the split signature, one must use the metric η_{ij} in place of δ_{ij} in the second term in (5.159). Note that in the case of the Lorentzian signature, all fields here need to be taken to be complex-valued, and so this action is not manifestly real. We will return to this point.

Varying this action with respect to the Lagrange multiplier field Ψ_{ij}, which is required to be tracefree, we get the constraint

$$\Sigma^i\Sigma^j \sim \delta^{ij} \tag{5.160}$$

in the cases of the Euclidean and Lorentzian signatures, and $\Sigma^i\Sigma^j \sim \eta^{ij}$ in the split signature case. These are the already discussed constraints (5.129). As the discussion of the previous sections shows, in Euclidean and split signatures, this constraint implies that Σ^i can be written in terms of the SD projection of the wedge product of two tetrads e^Ie^J as (5.131), with possibly Σ^3 being minus of what it is in the case of the SD projection of e^Ie^J.

In Lorentzian signature, all fields are complex-valued, and one must impose appropriate reality conditions. It is sufficient to impose the reality conditions on the metric-like field Σ^i, as the appropriate reality condition on the connection then gets imposed automatically by the field equations. The conditions on the 2-form field are discussed previously

$$\Sigma^i\overline{\Sigma^j} = 0, \qquad \text{Re}\left(\Sigma^i\Sigma^i\right) = 0. \tag{5.161}$$

The first of these equations gives nine conditions, which guarantee that conformal class of the metric (5.47) is real, while the last condition gives the reality of the volume form.

If one wishes, the previous Lorentzian signature reality conditions can be imposed with Lagrange multiplier terms added to the action. However, this makes the formalism less elegant, and so we will refrain from doing so in this book, always imposing the reality conditions (5.161) on complex solutions of the theory (5.159) to select those that admit a real interpretation.

In the Lorentzian case, as Theorem 5.6 shows, the conditions (5.160), together with the reality conditions (5.161), imply that Σ^i is related to the SD projection of the wedge product of two tetrads $e^I e^J$, for the metric defined by Σ; see (5.132). There is also a possibility that Σ^3 is minus what it is in the canonical case of the SD projection of $e^I e^J$. Another possibility that arises only in the Lorentzian case is that Σ^i can be $-\mathrm{i}$ times the canonical ASD 2-form. In all the cases, the chiral soldering form Σ^i defines a conformal metric of Lorentzian signature, via the Urbantke formula (5.47). The volume form ϵ_Σ is then extracted from $\Sigma^i\Sigma^j = 2\mathrm{i}\epsilon_\Sigma\delta^{ij}$.

The other field equations that follow from (5.159) are as follows

$$d^A\Sigma^i = 0, \qquad F_i = \left(\Psi_{ij} + \frac{\Lambda}{3}\delta_{ij}\right)\Sigma^j. \tag{5.162}$$

The first of these equations is the previously studied torsion-free condition. Together with the fact that Σ^i is the SD projection of the wedge product of two tetrads for the metric defined by Σ, this equation implies that the connection A^i is a multiple of the $4i$ component of the SD projection of the spin connection; see Theorem 5.13. This, in turn, implies that the curvature of A^i is a multiple of the $4i$ component of the SD projection of the curvature of the spin connection; see (5.126). Then, as we know from (5.7), the Einstein condition is equivalent to the condition that F^i is SD as a 2-form, which is precisely what the second equation in (5.162) says. The second equation in (5.162) also identifies the Lagrange multiplier field Ψ_{ij} with the SD part of the Weyl curvature tensor when all field equations are satisfied. It also correctly imposes the equation that the trace of the matrix that appears on the right-hand side of the second equation in (5.162) is equal to the cosmological constant. These considerations explain why Plebański formalism gives a correct description of GR.

We have verified that Plebański formalism gives the correct description in the case when the solution of the constraints (5.160) is taken to be the canonical solution (5.31). But it is also possible that Σ^3 is minus what it is in (5.31). In this case, as we previously discussed, the torsion-free SO(3) connection is still related to the SD part of the spin connection, and its curvature is still related to the SD part of the curvature of the spin connection. This means that even in this case the second equation in (5.162) gives the correct Einstein equations, modulo the subtlety that the sign in front of the cosmological constant should be changed.

Let us also discuss the subtlety that, in the Lorentzian signature case, the solution of the constraints (5.160) together with the reality conditions (5.161) may be $-\mathrm{i}$ times the canonical ASD 2-forms for the metric defined by Σ. In this case the torsion-free connection A^i will be a multiple of the $4i$ component of the ASD projection of the spin connection. Its curvature will then be a multiple of the $4i$ component of the ASD projection of the curvature of the spin connection. So, the second equation in (5.162) does impose the correct Einstein condition even in this case, by requiring the ASD projection of the curvature of the spin connection to be ASD as a 2-form. However, in this case the second equation in (5.162) cannot be satisfied with real Λ. Indeed, it would be satisfied with real Λ in the case in which one had the basic ASD 2-forms $\overline{\Sigma^i}$ in place of Σ^i. However, in the case under discussion one has $\mathrm{i}\overline{\Sigma^i}$ on the right-hand side of the second equation in (5.162). This is clearly impossible for real Λ. So, we learn that the case B as discussed in the proof of Theorem 5.6 cannot arise as the solution of Plebański field equations with real Λ.

We remark that the Plebański formulation, as well as the related formulation (3.93), is cubic in the fields, even with nonzero cosmological constant. This is the only known formulation of GR with $\Lambda \neq 0$ that is cubic. However, there is a drawback that it is not easy to couple fermions to gravity in this formulation. One can, of course, always couple matter to the metric defined by Σ^i, but this gives a very involved description. One would like to be able to couple matter directly to the fields present in (5.159), but in the case of fermions, this is not easy. The only known way of doing this is described in Capovilla et al. (1991) and uses further Lagrange multipliers.

5.9 Linearisation of the Plebański Action

The purpose of this section is to study the linearisation of the chiral Einstein–Cartan action (5.25), which we describe as the Plebański action (5.159) with the constraints $\Sigma^i\Sigma^j \sim \delta^{ij}$ assumed to be satisfied. As we have already said on several occasions, the chiral trick eliminates half of the spin connection components and thus leads to a more economic description. In this section we will start to appreciate this economy at the level of perturbation theory. We will also carry out the exercise of integrating out the perturbations of the 2-form field of Plebański formalism, and thus obtain the second-order chiral pure connection action. This leads to a useful result on an arbitrary Einstein background.

5.9.1 Action to Second Order in Perturbations

Let us start with a background configuration of fields Σ^i, A^i, and Ψ^{ij} satisfying the Plebanski field equations (5.162). We then add a small perturbation to the fields Σ^i and A^i. When Σ^i satisfies the constraints $\Sigma^i\Sigma^j \sim \delta^{ij}$, the Plebański Lagrangian reduces to

$$L = \Sigma^i F_i - \frac{\Lambda}{6}\Sigma^i\Sigma_i. \tag{5.163}$$

Its first variation is given by

$$\delta L = \delta\Sigma^i F_i + \Sigma^i \delta F_i - \frac{\Lambda}{3}\delta\Sigma^i\Sigma_i. \tag{5.164}$$

However, the perturbation $\delta\Sigma^i$ is not free here, because the linearisation of the constraints $\Sigma^i\Sigma^j \sim \delta^{ij}$ must hold. We will deal with this later.

Let us also give the expression for the second-order perturbation. We can write the result as

$$\delta^2 L = \delta^2\Sigma^i(F_i - \frac{\Lambda}{3}\Sigma_i) + 2\delta\Sigma^i\delta F_i + \Sigma^i\delta^2 F_i - \frac{\Lambda}{3}\delta\Sigma^i\delta\Sigma_i. \tag{5.165}$$

The reason why $\delta^2\Sigma^i$ appears is that the object Σ^i is supposed to satisfy the constraints and so is not varied freely. Another way to see this is to remember that the object Σ^i, when it satisfies the constraints, can be written as the SD part of the wedge product $e^I e^J$ for some frame, and we are really varying this frame rather than Σ^i itself. It is then clear that there is also the second-order part in its variation. It is, however, more efficient to work out this second-order part from the constraints $\Sigma^i\Sigma^j \sim \delta^{ij}$ rather than work with the SD projector; see Section 5.9.3.

Taking into account that the variations of the curvature are

$$\delta F^i = d^A\delta A^i, \qquad \delta^2 F^i = [\delta A, \delta A]^i \equiv \epsilon^{ijk}\delta A^j\delta A^k, \tag{5.166}$$

and that the background satisfies the Plebański field equations (5.162) we can write the second variation as

$$\delta^2 L = \Psi_{ij}\delta^2\Sigma^i\Sigma^j + 2\delta\Sigma_i d^A a^i + \Sigma_i[a,a]^i - \frac{\Lambda}{3}\delta\Sigma^i\delta\Sigma_i, \tag{5.167}$$

where Ψ^{ij} is the SD part of the Weyl curvature on the background (which can be zero), and we denoted $\delta A^i \equiv a^i$.

5.9.2 Alternative Second-Order Action

One can always add to the Plebański Lagrangian a constant multiple of the term $F^i F_i$. This is a total derivative term, whose integral is a multiple of the Pontryagin number for the corresponding SO(3) bundle. We can then adjust the coefficient in front of this term so that the Plebański action vanishes on the background $F^i = (\Lambda/3)\Sigma^i$. This is similar to what was done in the context of the full Einstein–Cartan theory when the action was rewritten in MacDowell–Mansouri form (3.77). The similar chiral Lagrangian reads

$$L' = -\frac{3}{2\Lambda}\left(F^i - \frac{\Lambda}{3}\Sigma^i\right)\left(F_i - \frac{\Lambda}{3}\Sigma_i\right). \tag{5.168}$$

Indeed, opening the brackets reproduces (5.163) plus a multiple of the Pontryagin term.

The benefit of using the Lagrangian (5.168) for the linearisation rather than (5.163) is that each of its two factors vanishes on the background $F^i = (\Lambda/3)\Sigma^i$, which means that on this background there is no need in δ^2 terms. Indeed, we have

$$\delta^2 L' = -\frac{3}{\Lambda}\left(\delta F^i - \frac{\Lambda}{3}\delta\Sigma^i\right)\left(\delta F_i - \frac{\Lambda}{3}\delta\Sigma_i\right) - \frac{3}{\Lambda}\Psi_{ij}\Sigma^j\left(\delta^2 F^i - \frac{\Lambda}{3}\delta^2\Sigma^i\right). \tag{5.169}$$

On a background with $\Psi^{ij} = 0$ the term in the second line is absent, which results in a particularly simple form of the second-order action, to be derived Sections 5.9.4. Such backgrounds are known as gravitational instantons. They can exist only in Euclidean and split signatures, and correspond to Einstein metrics for which the SD half of the Weyl curvature vanishes. We will consider these metrics in more details in the next chapter.

5.9.3 Implications of the Metricity Constraints

Let us decompose the perturbation $\delta\Sigma^i$ into the basis of SD and ASD 2-forms

$$\delta\Sigma^i = \phi^{ij}\Sigma_j + h^{ij}\bar{\Sigma}_j. \tag{5.170}$$

Here ϕ^{ij} and h^{ij} are general 3×3 matrices. The matrix ϕ^{ij} can be further decomposed into its trace, skew, and symmetric tracefree parts

$$\phi^{ij} = h\delta^{ij} - \epsilon^{ijk}\xi_k + \psi^{ij}, \tag{5.171}$$

where ψ^{ij} is symmetric tracefree and the sign is for future convenience. No such decomposition of h^{ij} is meaningful because the two indices in this matrix are really of different types, one refers to a basis in the space of SD 2-forms, while the other is the index that labels the ASD ones. So, the matrix h^{ij} is not further decomposed.

The first variation of the metricity constraints $\Sigma^i\Sigma^j \sim \delta^{ij}$ reads

$$\delta\Sigma^{(i}\Sigma^{j)} \sim \delta^{ij}, \tag{5.172}$$

which should be read as the equation that says that the tracefree part of the left-hand side must vanish.

In terms of the decomposition (5.170) the equation (5.172) simply says that

$$\psi^{ij} = 0. \tag{5.173}$$

All other irreducible components of $\delta\Sigma^i$ are not constrained to this order in perturbation.

The second variation of the metricity constraints produces

$$\delta^2\Sigma^{(i}\Sigma^{j)} + \delta\Sigma^i\delta\Sigma^j \sim \delta^{ij}. \tag{5.174}$$

The second-order perturbation of Σ^i can also be decomposed into the basis of SD and ASD 2-forms, similar to (5.170). The previous equation then implies that the ψ^{ij} part of $\delta^2\Sigma^i$ can be expressed in terms of the components of $\delta\Sigma^i$.

The are no other consequences of the metricity conditions to this order in the perturbation. The variations of the metricity constraints can be considered to third- and fourth-order where they imply more relations between $\delta^2\Sigma^i$ and $\delta\Sigma^i$. However, this will not be of interest to us here because we are just after the second-order terms.

The equation (5.174) implies that when projected on a symmetric tracefree part, the tensor $\delta^2\Sigma^{(i}\Sigma^{j)}$ can be replaced with minus the tensor $\delta\Sigma^i\delta\Sigma^j$. However, precisely such projections appear in both (5.167) and (5.169). This gives the following expressions for the second-order Lagrangians

$$\delta^2 L = -\left(\Psi_{ij} + \frac{\Lambda}{3}\delta_{ij}\right)\delta\Sigma^i\delta\Sigma^j + 2\delta\Sigma_i d^A a^i + \Sigma_i[a,a]^i, \tag{5.175}$$

and

$$\delta^2 L' = -\frac{3}{\Lambda}\left(d^A a^i - \frac{\Lambda}{3}\delta\Sigma^i\right)\left(d^A a_i - \frac{\Lambda}{3}\delta\Sigma_i\right) - \frac{3}{\Lambda}\Psi_{ij}\Sigma^i[a,a]^j - \frac{3}{\Lambda}\Psi_{ij}\delta\Sigma^i\delta\Sigma^j. \tag{5.176}$$

We will soon see that the form (5.176) is particularly useful on the background $\Psi^{ij} = 0$, while the form (5.175) is useful on an arbitrary Einstein background.

Let us also note that we can rewrite (5.176) as

$$\delta^2 L' = -\left(\Psi_{ij} + \frac{\Lambda}{3}\delta_{ij}\right)\delta\Sigma^i\delta\Sigma_j + 2\delta\Sigma_i d^A a^i - \frac{\Lambda}{3}\Psi_{ij}\Sigma^i[a,a]^j - \frac{3}{\Lambda}d^A a^i d^A a_i. \tag{5.177}$$

It is clear that integrating by parts in the last term in (5.177) and combining with the term before last gives the last term in (5.175). This explicitly shows that the previous second-order Lagrangians are equivalent modulo a surface term.

5.9.4 Pure Connection Lagrangian on an Instanton Background

The Lagrangians (5.176) and (5.177), together with the decomposition (5.170) of $\delta\Sigma^i$ into irreducible components and the fact that the Ψ^{ij} component in (5.171) vanishes, give a starting point for chiral perturbation theory. This can be developed on both $\Lambda = 0$ and $\Lambda \neq 0$ backgrounds. These two cases behave rather differently in terms of gauges that are available for gauge-fixing of the SO(3) and the diffeomorphism symmetry, and so need to be treated separately. We will consider the gravitational perturbation theory in more detail in one of the following chapters.

In the $\Lambda \neq 0$ case, the Lagrangians (5.175) and (5.176) can also be used as the starting point for integrating out the linearised 2-form field to produce a chiral pure connection description. The exercise of integrating out the 2-form field is possible at the fully nonlinear level, and the next chapter deals with the resulting formalism. However, we can see some of the arising simplifications already here.

Let us consider a gravitational instanton background on which $\Psi^{ij} = 0$. We then take the second-order Lagrangian in the form (5.176). It takes the simple form

$$\delta^2 L' = -\frac{3}{\Lambda}(d^A a^i - \frac{\Lambda}{3}\delta\Sigma^i)^2. \tag{5.178}$$

We also know that the first-order perturbation of the 2-form field is of the form

$$\delta\Sigma^i = (h\delta^{ij} - \epsilon^{ijk}\xi_k)\Sigma_j + h^{ij}\bar{\Sigma}_j. \tag{5.179}$$

We note that this is a decomposition of $\mathfrak{so}(3)$ valued 2-form field into its irreducible with respect to Lorentz group components. The matrix ϕ^{ij} in (5.170) is $\mathfrak{so}(3) \times \mathfrak{so}(3)$-valued, which is the tensor product of spin one representation of SO(3) with itself. This decomposes into spin two, spin one, and spin zero representations. The consequence of the metricity equation (5.172) is that the spin two component here vanishes.

A similar decomposition is available for the object $d^A a^i$, which is also an $\mathfrak{so}(3)$-valued 2-form. It also decomposes in $9+9$ components; in general, all of them nonvanishing. When we substitute these decompositions into (5.178), the irreducible components of $d^A a^i$ pair with similar irreducible components of $\delta\Sigma^i$ and then get squared. There are no mixed terms because different irreducible components cannot couple to each other. In this way, we get a sum over all irreducible components of $d^A a^i - (\Lambda/3)\delta\Sigma^i$ squared, where in each component apart from the spin two there is both parts coming from $d^A a^i$ and from $\delta\Sigma^i$. There is no spin two component in the $\delta\Sigma^i$, and so this component of the $d^A a^i$ is just squared in the previous action and does not couple to $\delta\Sigma^i$.

The procedure of integrating out $\delta\Sigma^i$ is then extremely simple. The equation for each irreducible component of $\delta\Sigma^i$ will just say that it is equal to $3/\Lambda$ times the corresponding component of $d^A a^i$. Substituting this into the Lagrangian (5.178) we see that most of the terms vanish. The only nonvanishing term is the spin two part of $d^A a^i$ squared. To write the resulting Lagrangian, we just need to understand the constants arising in the projection in the spin two part. The relevant formula is

$$d^A a^i\Big|_2 = \frac{1}{2}\Sigma^{(i\,\mu\nu} d^A_\mu a^{j)}_\nu\Big|_{tf}\Sigma^j, \tag{5.180}$$

where tf stands for the projection on the tracefree part. To check this, formula one uses (5.138). The square of this part (computed by taking the wedge product and contracting the SO(3) indices) is given by

$$\left(d^A a^i\Big|_2\right)^2 = \frac{1}{2}P_{ijkl}(\Sigma^{i\,\mu\nu} d^A_\mu a^j_\nu)(\Sigma^{k\,\rho\sigma} d^A_\rho a^l_\sigma)\epsilon, \tag{5.181}$$

where ϵ is the volume form and

$$P_{ijkl} = \delta_{i(k}\delta_{l)j} - \frac{1}{3}\delta_{ij}\delta_{kl} \tag{5.182}$$

is the projector on the symmetric tracefree part. Thus, overall, the second-order chiral pure connection action (defined as the second variation divided by two) is given by

$$S^{(2)}[a] = -\frac{3}{16\pi G\Lambda}\int \frac{1}{2}P_{ijkl}(\Sigma^{i\,\mu\nu}d^A_\mu a^j_\nu)(\Sigma^{k\,\rho\sigma}d^A_\rho a^l_\sigma). \tag{5.183}$$

We note that in the Euclidean signature, this has a definite sign, similar to what we observed in the non-chiral case; see (3.74). As in the non-chiral case, the linearised pure connection action is of the Weyl curvature–squared type, where the Weyl curvature is extracted from the curvature of the linearised connection. This action leads to a very nice gauge-fixing procedure and then gravitational perturbation theory, to which we will return in Chapter 8.

5.9.5 Pure Connection Action on an Arbitrary Einstein Background

We repeat the exercise of integrating out the perturbation of the 2-form field, but now on an arbitrary Einstein background. The most convenient starting point in this case is (5.175). We now need the full decomposition of the $d^A a^i$ into irreducible components, of which we only needed the spin two-part previously. The full decomposition reads

$$d^A a^i = (d^A a)\Sigma^i - \frac{1}{2}\epsilon^{ijk}(d^A a)_j\Sigma_k + d^A a^i\Big|_2 + (d^A a)^{ij}_-\bar{\Sigma}_j, \tag{5.184}$$

where we introduced the following notations

$$(d^A a) := \frac{1}{6}\Sigma^{i\,\mu\nu}d^A_\mu a_{i\,\nu}, \qquad (d^A a)^i := \frac{1}{2}\epsilon^i{}_{jk}\Sigma^{j\,\mu\nu}d^A_\mu a^k_\nu, \qquad (d^A a)^{ij}_- := \frac{1}{2}\bar{\Sigma}^{j\,\mu\nu}d^A_\mu a^i_\nu \tag{5.185}$$

for the different irreducible components. All these formulas are checked by projecting the left-hand side in (5.184) on the corresponding irreducible components and checking that the coefficients agree.

Using the decomposition (5.184), we get

$$\delta\Sigma^i d^A a^i/2\epsilon = 3h(d^A a) - \xi_i(d^A a)^i - h_{ij}(d^A a)^{ij}_-.$$

If we rewrite the first term in (5.175) as $M_{ij}\delta\Sigma^i\delta\Sigma^j$ it computes to

$$M_{ij}\delta\Sigma^i\delta\Sigma^j/2\epsilon = (h^2 + \xi^i\xi_i)\mathrm{Tr}(M) - M_{ij}\xi^i\xi^j - h^{ik}h^{jk}M_{ij}\delta_{kl}.$$

In both formulas, we, for simplicity, consider the case of the Euclidean signature. In the Lorentzian signature, there is an extra factor of the imaginary unit on the right-hand side of both formulas.

Overall, the second variation of the Lagrangian in the form (5.175) becomes

$$\delta^2 L/2\epsilon = -(h^2 + \xi^i\xi_i)\mathrm{Tr}(M) + M_{ij}\xi^i\xi^j + h^{ik}h^{jk}M_{ij}\delta_{kl} + 6h(d^A a) \\ - 2\xi_i(d^A a)^i - 2h_{ij}(d^A a)^{ij}_- + \frac{1}{2}\Sigma^{i\,\mu\nu}\epsilon_{ijk}a^j_\mu a^k_\nu. \tag{5.186}$$

It is now trivial to integrate out the components of the 2-form field perturbation. We get the following second-order pure connection action (defined as the second variation divided by two)

$$S^{(2)}[a] = \frac{1}{8\pi G}\int \frac{9}{\Lambda}(d^A a)^2 - (d^A a)^i(\Psi - (2\Lambda/3)\mathbb{I})^{-1}_{ij}(d^A a)^j \\ - (d^A a)^{ik}_-\delta_{kl}(\Psi + (\Lambda/3)\mathbb{I})^{-1}_{ij}(d^A a)^{jl}_- + \frac{1}{2}\Sigma^{i\,\mu\nu}\epsilon_{ijk}a^j_\mu a^k_\nu, \tag{5.187}$$

where we replaced the matrix M with its expression in terms of Ψ and Λ. When $\Psi = 0$ the factor of Σ^i in the last 'mass' term here can be replaced with the curvature. This term then becomes a multiple of $d^A a^i d^A a_i$ modulo a surface term. Decomposing this linearised curvature into its irreducible pieces and carrying out cancellations, one reproduces the action (5.183). This procedure is however not available on a nontrivial background, where it is $M^{-1}_{ij}F^j$ that would appear instead. The resulting term cannot be reduced to the commutator of two covariant derivatives.

The linearised action (5.187) is very interesting because there exists a gauge in which all of the first line equals zero. Also in this gauge the last 'mass' term for the connection is positive definite (in the Euclidean signature). We will establish all these facts when we treat the chiral pure connection perturbation theory in Chapter 8. This means that when the matrix of the curvatures $(\Psi + (\Lambda/3)\mathbb{I})$ is negative-definite, the linearised action is positive-definite. The fact that this holds in such generality and in particular is independent of the ASD as part of the Weyl curvature is rather remarkable. This can be shown to lead to a stronger than previously available result about the rigidity of Einstein metrics; see Fine et al. (2019).

5.10 Coupling to Matter

As we discussed, a drawback of Plebański formalism, shared by any BF-type description, is that the coupling to matter is not straightforward. Thus, it is not easy to write an action that would still use wedge products of differential forms and that would produce the desired coupling. It is possible to do this using additional Lagrange multiplier fields, as in Capovilla et al. (1991), but the procedure is not simple. However, there is a simple prescription for how Plebański field equations need to be modified in the presence of an arbitrary stress-energy tensor, as we now explain. Thus, while it is in general not easy to write down an action that describes matter coupling, it is easy to add the 'right-hand side' to Plebański–Einstein equations to incorporate effects of stress-energy tensor of matter.

Thus, let us assume that the matter distribution is described by a given stress-energy tensor $T_{\mu\nu}$. The stress-energy tensor can be decomposed into its trace and tracefree parts

$$T_{\mu\nu} = T^0_{\mu\nu} + \frac{T}{4} g_{\mu\nu}. \tag{5.188}$$

We then form the following 2-form with values in S

$$T^i_{\mu\nu} := \Sigma^i_{[\mu}{}^{\rho} T^0_{\rho\nu]}. \tag{5.189}$$

This 2-form is purely ASD. Indeed, this can be confirmed by projecting it onto the basis of SD 2-forms. We have

$$\Sigma^i_{[\mu}{}^{\rho} T^0_{\rho\nu]} \Sigma^{j\mu\nu} = T^0_{\rho\nu} \left(\delta^{ij} g^{\rho\nu} + \epsilon^{ij}{}_k \Sigma^{k\rho\nu} \right) = 0. \tag{5.190}$$

where we have used the algebra (5.138). Thus, if we add a multiple of $T^i_{\mu\nu}$ to the right-hand side of Plebański equations (5.162) we are adding an ASD part to F^i that encodes the tracefree part of $T_{\mu\nu}$. This is exactly what the tracefree part of the Einstein equations in the presence of matter is. Indeed, we know that the SD–ASD component of the Riemann curvature is essentially the tracefree part of Ricci. And Einstein equations say that the tracefree part of Ricci must be a multiple of the tracefree part of the stress-energy tensor of matter. This is precisely what happens when we add (5.189) on the right-hand side of Plebański equations. We will also need to change the trace part of the SD–SD projection of the Riemann. This can be done by adding to the right-hand side of (5.162) a multiple of $T\Sigma^i$.

To complete the story, we just need to work out the correct coefficients. This is done as follows. The standard form of Einstein equations is

$$R_{\mu\nu} - \frac{1}{2}(R - 2\Lambda) g_{\mu\nu} = 8\pi G T_{\mu\nu}. \tag{5.191}$$

In four dimensions, taking the trace gives

$$R = 4(\Lambda - 2\pi G T). \tag{5.192}$$

This means that we get the correct trace part by shifting $\Lambda \to \Lambda - 2\pi G T$. The correct coefficient in front of the ASD 2-form T^i is determined by working out some examples. We get

$$F_i = \left(\Psi_{ij} + \frac{\Lambda - 2\pi G T}{3} \delta_{ij} \right) \Sigma^j + 8\pi G T_i, \tag{5.193}$$

where T^i is given by (5.189) and the index is lowered with the metric δ_{ij} in the Lorentzian signature case (the only one relevant for physics). Because of the role it plays, it is natural to refer to the 2-form $T^i_{\mu\nu}$ as the ASD part of the *stress-energy 2-form*.

5.10.1 Example: Perfect Fluid

Let us carry out the exercise of determining the ASD part of the stress-energy 2-form for an ideal fluid with stress-energy tensor $T_{\mu\nu} = (\rho + P)u_\mu u_\nu + P g_{\mu\nu}$, where u_μ is the velocity four-vector $u_\mu u^\mu = -1$, and ρ, P are the energy and pressure densities respectively.

Let e^I be a frame for the metric. We can then decompose the velocity four-covector u_μ into the frame covectors e^I_μ

$$u_\mu = \frac{1}{\sqrt{1-|u|^2}}(e^0_\mu + u_i e^i_\mu). \tag{5.194}$$

This parametrises the unit timelike vector u_μ by its spatial projection $u_i/\sqrt{1-|u|^2}$. Here $|u|^2 = u_i u_j \delta^{ij}$. The tracefree part of the stress-energy tensor is given by $T^0_{\mu\nu} = (\rho + P)(u_\mu u_\nu + (1/4)g_{\mu\nu})$, and the trace $T = 3P - \rho$. We use $\Sigma^i_{\mu\nu} = 2\mathrm{i}e^0_{[\mu}e^i_{\nu]} - \epsilon^i{}_{jk}e^j_{[\mu}e^k_{\nu]}$. A long computation gives

$$T_i = \frac{(\rho + P)}{4(1-|u|^2)}\left(-(1+|u|^2)\delta_{ij} + 2u_i u_j - 2\mathrm{i}\epsilon_{ijk}u^k\right)\bar{\Sigma}^j, \tag{5.195}$$

where

$$\bar{\Sigma}^i = \mathrm{i}e^0 e^i + \frac{1}{2}\epsilon^i{}_{jk}e^j e^k \tag{5.196}$$

is a convenient basis of ASD 2-forms.

5.11 Historical Remarks

The basic objects of Plebański's formulation of GR are SD two-forms. These objects have appeared in the GR literature a lot before Plebánski (1977). In fact, Petrov's famous classification of 'spaces defining gravitational fields', see Petrov (2000) for a reprint of the original paper, already uses SD (and ASD) bivectors in a key way. Thus, the theorem proved by Petrov states that the gravitational field (a solution of vacuum Einstein equations) can be classified according to algebraic types of a complex symmetric 3×3 matrix obtained as a complex linear combination of the diagonal and off-diagonal blocks of the Riemann tensor viewed as a symmetric tensor in the linear space of bivectors. SD (and ASD) bivectors then naturally appear as principal bivectors of the Riemann tensor. A completely analogous but more modern treatment that forms the 3×3 matrix in question as the complex linear combination of the 'electric' and 'magnetic' parts of the Weyl tensor was given in Jordan et al. (2009), which was a reprint of the original paper Jordan published in the 1960s. Again, the SD and ASD bivectors are central in these considerations.

It was then remarked in Taubes (1966) that the 3×3 complex matrix encoding the Weyl curvature can be computed directly, i.e., avoiding computing the Riemann curvature first. This can be done by elementary operations of differentiation of SD complex linear combinations of the components of the torsion-free spin connection. This encodes 24 real components of the spin connection in 12 complex components of a SD connection. The 3×3 matrix of Weyl curvature components is then computed as the curvature of this SD connection.

The SD 2-forms first appeared in a pioneering paper in Cahen et al. (1967). This paper uses the null tetrad formalism and also provides a link to the spinor formalism of Penrose (1960). Thus, the spinor formalism combines the 24 real rotation coefficients into 12 complex Newmann–Penrose spin coefficients of Newman and Penrose (1962), which is similar to what happens in the SD formalism. The paper Cahen et al. (1967) for the first time writes equations for the SD connection 1-forms as those in terms of exterior derivatives of the SD 2-forms. It also clearly states that the isomorphism between the Lorentz group $\mathrm{SO}(1,3)$ and the complexified rotation group $\mathrm{SO}(3,\mathbb{C})$ is what is at the root of the SD formalism. Finally, the Einstein equations are very clearly stated in this paper, as the condition that the curvature of the SD connection is SD. Another exposition of the formalism for GR based on differential forms and self-duality is that of Israel (1970).

Yet another presentation of the SD formalism for GR appeared in Brans (1974). This reference is very close in spirit to our exposition. One important new point in this reference is the emphasis it places on the role played by the Hodge duality operator, which is interpreted as defining the complex structure in the space of 2-forms. Similar to Cahen et al. (1967), (vacuum) Einstein equations are stated here, as the condition that the curvature of the SD connection is SD.

The SD (chiral) formalism for general relativity was taken further by Plebánski (1977). The Plebański paper uses spinor notations, but it can be easily translated into more easily readable $\mathrm{SO}(3)$ notations used here and in, e.g., Brans (1974). The main novelty of Plebański's work is that for the first time the main object of the theory is taken to be not a metric from whose tetrads the SD 2-forms are constructed, but rather a triple of 2-forms satisfying certain additional equations. These equations guarantee that the 2-forms in question are obtained from tetrads, and thus provide a link to the usual metric formulation. Plebánski (1977) also gave a remarkably simple action principle realising these ideas. The basic dynamical field in this action is a triple of 2-forms, and no metric ever appears. Later, Ashtekar's new Hamiltonian formulation of general relativity (1987) was found by Jacobson and Smolin (1988) to be just the phase space version of Plebański's theory.

5.12 Alternative Descriptions Related to Plebański Formalism

The purpose of this section is to perform similar type of transformations to those we have done in the context of the BF-type description of the non-chiral theory. Thus, we will see that there is a very natural way of modifying GR in the Plebański description. We will also see that it is possible to perform field redefinitions and rewrite GR in a seemingly modified theory form. One can then integrate out the Lagrange multiplier fields completely, obtaining a Lagrangian that depends only on Σ^i and A^i. One can even integrate out the connection field and obtain a second-order formalism with the Σ^i as the only field.

5.12.1 Chiral Modifications of GR

We proceed in exact parallel to what was done in the non-chiral case. We first replace the matrix that appears in front of $\Sigma^i\Sigma^j$ term in (5.159) by a 3×3 symmetric matrix M^{ij}. We then add a Lagrange multiplier that imposes the condition that the trace of M^{ij} is Λ. We get

$$S[\Sigma, A, M, \mu] = \frac{1}{8\pi G\sqrt{\sigma}} \int \Sigma^i F_i - \frac{1}{2} M_{ij}\Sigma^i\Sigma^j + \frac{\mu}{2}\left(f(M) - \Lambda\right). \tag{5.197}$$

This action describes GR when the SO(3)-invariant function $f(M)$ is taken to be

$$f_{\rm GR} = \mathrm{Tr}(M). \tag{5.198}$$

However, one can consider other gauge-invariant functions $f(M)$ here. In particular, one can consider any function of the three independent invariants of M for which one can take $\mathrm{Tr}(M), \mathrm{Tr}(M^2)$, and $\mathrm{Tr}(M^3)$. What is very surprising about the theories one gets this way is that they continue to propagate exactly the same number of degrees of freedom as GR. Thus, no new propagating degrees of freedom is introduced by these modifications, which is a very strong statement because it seems to be in conflict with GR uniqueness theorems. We will come back to these chiral modifications of GR in the following chapters.

5.12.2 Field Redefinitions

The goal of this subsection is to repeat the field redefinitions trick that was already used in the non-chiral context, and thus rewrite the GR action in the form (5.197) with a nontrivial function $f(M)$. But, in spite of the new $f(M)$ being different from (5.198), the new action will still describe unmodified GR.

Consider the transformation

$$\Sigma^i = G^{ij}\tilde{\Sigma}_j + H^{ij}F_j, \tag{5.199}$$

where G^{ij}, H^{ij} are arbitrary at this stage 3×3 matrices, and $\tilde{\Sigma}$ is the new 2-form field. This transformation will map the first two terms in the Lagrangian (5.197) to

$$L \to \tilde{\Sigma}^t G^t F + F^t H^t F - \frac{1}{2}(\tilde{\Sigma}^t G^t + F^t H^t) M (G\tilde{\Sigma} + HF), \tag{5.200}$$

where we used the matrix notations with, e.g., $M_{ij}\Sigma^i\Sigma^j \equiv \Sigma^t M \Sigma$. Collecting the similar terms in the previous expression we rewrite it as

$$L \to F^t\left(H^t - \frac{1}{2}H^t M H\right)F + \tilde{\Sigma}^t\left(G^t - G^t M H\right)F - \frac{1}{2}\tilde{\Sigma}^t(G^t M G)\tilde{\Sigma}. \quad (5.201)$$

We now demand that after the transformation (5.199) the Lagrangian is still of BF-type, i.e., the matrix appearing in front of $\tilde{\Sigma}^i F^j$ is a multiple of the identity matrix. If we don't want to change the coefficient in front of the action, we should demand this multiple to be unity

$$G^t - G^t M H = \mathbb{I}. \quad (5.202)$$

We will also demand that the newly generated term quadratic in the curvature is a multiple of the Pontryagin number for the SO(3) bundle in question. Thus, we demand also that the matrix in front of $F^i F^j$ is a constant multiple of the identity

$$H^t - \frac{1}{2}H^t M H = t\mathbb{I}, \quad (5.203)$$

where t is an arbitrary parameter, and real, if we specialise the formalism to the cases of Euclidean or split signatures.

We are now going to solve the equations (5.203) and (5.202) for H and G in terms of M. First, the equation (5.203) tells us that H is a symmetric matrix, so we will drop the transpose symbol on H from now on. Assuming that G and H are invertible, we can rewrite the two equations (5.202) and (5.203) as

$$\mathbb{I} - MH = (G^t)^{-1}, \qquad \mathbb{I} - \frac{1}{2}MH = tH^{-1}. \quad (5.204)$$

We can then subtract twice the second equation from the first to get a relation between G and H

$$H = 2t(\mathbb{I} + (G^t)^{-1})^{-1}, \quad (5.205)$$

where we again assumed that $\mathbb{I} + (G^t)^{-1}$ is invertible. We then substitute this to, e.g., the first equation in (5.204) to obtain a simple equation involving just G

$$(G^t)^{-2} = \mathbb{I} - 2tM. \quad (5.206)$$

This tell us that G is also a symmetric matrix, and gives this matrix as one of the two branches of the square root

$$G = (\mathbb{I} - 2tM)^{-1/2}. \quad (5.207)$$

We can now concentrate on the last $\Sigma\Sigma$ term in (5.201). It is clear that the matrix in front of $\Sigma^i\Sigma^j$ transforms to

$$\tilde{M} = M\,(\mathbb{I} - 2tM)^{-1}. \quad (5.208)$$

We note for future use that the inverse of this is $M = \tilde{M}\left(\mathbb{I} + 2t\tilde{M}\right)^{-1}$.

All in all we learn that the field redefinition (5.199) with symmetric matrices G and H that depend on M according to (5.207) and (5.205) transform the Lagrangian in (5.197) into a Lagrangian of the same type

$$\Sigma^i F_i - \frac{1}{2}M_{ij}\Sigma^i\Sigma^j + \frac{\mu}{2}(f(M) - \Lambda) = \tilde{\Sigma}^i F_i - \frac{1}{2}\tilde{M}_{ij}\tilde{\Sigma}^i\tilde{\Sigma}^j + \frac{\mu}{2}\left(f\left(\tilde{M}(\mathbb{I} + 2t\tilde{M})^{-1}\right) - \Lambda\right) + tF^iF_i. \tag{5.209}$$

The only change in the new Lagrangian is that the function $f(M)$ became modified, and that a constant multiple of the topological term $\text{Tr}(F \wedge F)$ has been added.

Thus, we learn that there is a one-parameter group of transformations acting on the space of theories of the type (5.197), with all functions $f(M)$ belonging to the family

$$f_t(M) \equiv f(M_t), \qquad M_t = M(\mathbb{I} + 2tM)^{-1} \tag{5.210}$$

corresponding to (classically) physically equivalent theories. At the quantum level adding to the Lagrangian a topological term is not innocuous, as the example of the θ-term in QCD teaches us. So, we can only be sure about the classical equivalence of theories related by (5.210). Note that we can alternatively write $M_t^{-1} = M^{-1} + 2t\mathbb{I}$, from which the fact that the transformation $M \to M_t$ forms a one-parameter group $(M_{t_1})_{t_2} = M_{t_1+t_2}$ is obvious.

5.12.3 GR as BF Theory Plus Potential

We now use the result (5.210) to derive a new formulation of GR discovered in Herfray and Krasnov (2015). This is done by integrating out the auxiliary matrix M from the Lagrangian (5.197) with the defining function (5.210). We also note that the matrix M^{ij} cannot be integrated out from the Lagrangian with $f(M)$ given by (5.198), because this Lagrangian depends on M^{ij} linearly. In contrast, the effect of the previous field redefinitions is to produce a nonlinear dependence of the Lagrangian on M^{ij}, so that it can be integrate out by solving its field equation.

As before, let us introduce the notation

$$\Sigma^i \wedge \Sigma^j := X_\Sigma. \tag{5.211}$$

This is a 3×3 symmetric matrix valued in 4-forms. The equation for M is then

$$X_\Sigma = \mu(\mathbb{I} + 2tM)^{-2}, \tag{5.212}$$

which can be solved for M

$$2tM = \sqrt{\mu}(X_\Sigma)^{-1/2} - \mathbb{I}, \tag{5.213}$$

where we assumed that X_Σ is invertible and one of the two branches of the square root is taken. This gives

$$M(\mathbb{I}+2tM)^{-1}=\frac{1}{2t}\left(\mathbb{I}-\sqrt{\frac{X_\Sigma}{\mu}}\right). \tag{5.214}$$

We should now find μ from the constraint that the trace of the previous matrix is Λ. This gives

$$\sqrt{\mu}=\frac{\mathrm{Tr}\sqrt{X_\Sigma}}{3-2t\Lambda}, \tag{5.215}$$

and so

$$M=\frac{1}{2t}\left(\frac{\mathrm{Tr}\sqrt{X_\Sigma}}{3-2t\Lambda}(X_\Sigma)^{-1/2}-\mathbb{I}\right). \tag{5.216}$$

Thus, we can rewrite the action (5.197) with the defining function (5.210) and with the matrix M integrated out as

$$S[\Sigma,A]=\int\Sigma^iF_i-\frac{1}{4t(3-2t\Lambda)}\left(\mathrm{Tr}\sqrt{\Sigma^i\Sigma^j}\right)^2+\frac{1}{4t}\Sigma^i\Sigma_i. \tag{5.217}$$

The description of GR (5.217) was discovered in Herfray and Krasnov (2015). The presented here derivation via field redefinitions was spelled out in Krasnov (2018).

Let us discuss the effect of the field redefinition (5.199) on the metric. On-shell the curvature 2-forms become linear combinations of the 2-forms Σ^i. This is true in the case of GR, see (5.162), as well as for the modified theories. Because the conformal class of the metric is fixed by demanding that the 2-forms Σ^i span the space of SD 2-forms, the conformal class is unchanged by the field redefinition (5.199). However, this transformation does have the effect on the volume form that fixes a representative in the conformal class. In particular, the volume form that corresponds to an Einstein metric is constructed differently in the Plebański case and the formulation (5.217). This is explained in more details in Herfray and Krasnov (2015).

5.13 A Second-Order Formulation Based on the 2-Form Field

The action (5.217) can be used as the starting point for one more transformation. Thus, one can integrate out the connection field and produce a second-order formulation with the 2-form field Σ^i as the only field. We now describe this.

5.13.1 Parametrisation

For simplicity, we perform the following analysis in the case of the Euclidean signature. We now change the name of the 2-form field from Σ^i to B^i to signify the fact that the triple of 2-forms B^i no longer has to satisfy the simplicity (metricity) constraints, as these constraints no longer follow from the action (5.217).

A general 2-form field B^i then defines a conformal metric in which the triple of B^i's is SD. We can always introduce an orthonormal basis in the space Λ^+ of SD 2-forms of the corresponding metric. Let us denote these orthonormal 2-forms by Σ^a, where we introduced a new internal index $a = 1, 2, 3$. Because B^i are SD by construction, they can be expanded in the basis Σ^a. We have

$$B^i = b^i_a \Sigma^a, \tag{5.218}$$

where b^i_a are some collection of 3×3 coefficients, and Σ^a are assumed to be orthonormal $\Sigma^a \Sigma^b \sim \delta^{ab}$. It is convenient to give the two indices of $b^i_a \in \mathrm{GL}(3)$ transformation different names as it helps bookkeeping at later stages.

While the objects B^i are given, the objects b^i_a and Σ^a are defined only modulo certain ambiguities. Indeed, we can always conformally rescale both b^i_a and Σ^a so that their product remains unchanged. This is the reflection of the fact that in general, only the conformal class of the metric is defined by the triple of 2-forms B^i. Second, we can always perform an SO(3) rotation of the basis of Σ^a's $\Sigma^a \to \Lambda^a_b \Sigma^b$. The 2-forms B^i are unchanged if we simultaneously rotate the coefficients b^i_a. Thus, we have parametrised the 18 components of a triple of 2-forms B^i by nine components of b^i_a plus $18 - 5 = 13$ components of Σ^a satisfying the metricity condition $\Sigma^a \Sigma^b \sim \delta^{ab}$. But there is also a four-parameter redundancy in this parametrisation, one of conformal rescalings and three of SO(3) rotations. Thus, overall there is $13 + 9 - 4 = 18$ parameters in this parametrisation, as it should be.

5.13.2 A Torsion-Free Connection for an Arbitrary Triple of 2-Forms

We now pose and solve the problem of finding an SO(3) connection such that the torsion $d^A B^i$ vanishes. We know from previous considerations that when B^i satisfy $B^i B^j \sim \delta^{ij}$, this connection coincides with the SD part of the metric spin connection. However, we now make no assumption about B^i and want to solve for A^i in terms of the SD part of the spin connection, as well as derivatives of the objects b^i_a.

It is quite easy to find the connection with the required property. The torsion-free equation is, explicitly

$$d(b^i{}_a \Sigma^a) + \epsilon^i{}_{jk} A^j b^k{}_a \Sigma^a = 0. \tag{5.219}$$

Introducing the torsion-free metric connection γ^a, satisfying

$$d^\gamma \Sigma^a = d\Sigma^a + \epsilon^a{}_{bc} \gamma^b \Sigma^c = 0 \tag{5.220}$$

we can rewrite (5.219) as

$$d^\gamma b^i{}_a \Sigma^a + \epsilon^i{}_{jk} A^j b^k{}_a \Sigma^a = 0, \tag{5.221}$$

where we introduced

$$d^\gamma b^i{}_a = db^i_a - \epsilon^c{}_{ba} \gamma^b b^i{}_c. \tag{5.222}$$

To solve this equation we define

$$A^i = b^i{}_a A^a. \tag{5.223}$$

The torsion-free condition becomes

$$d^\gamma b^i{}_a \Sigma^a + \det(b)(b^{-1})^{ai} \epsilon_{abc} A^b \Sigma^c = 0. \tag{5.224}$$

We now multiply this equation with $b_i{}^a$, where the indices are raised-lowered with the Kronecker delta metrics, dualise on the spacetime indices, and use the self-duality of Σ^a. We get

$$\Sigma^b{}_\mu{}^\nu b_i{}^a d^\gamma_\nu b^i{}_b = \det(b) \epsilon^a{}_{bc} \Sigma^b{}_\mu{}^\nu A^c_\nu. \tag{5.225}$$

We can rewrite this in an index-free way as

$$J_\Sigma(A) = t, \tag{5.226}$$

where J_Σ is the operator (5.137) and

$$t^a_\mu := \frac{1}{\det(b)} \Sigma^b{}_\mu{}^\nu b_i{}^a d^\gamma_\nu b^i{}_b. \tag{5.227}$$

Thus, we have $A^a = J_\Sigma^{-1}(t)$, with the inverse J_Σ^{-1} given by (5.140). We will not need an explicit expression for this connection.

5.13.3 An Alternative Derivation

An alternative procedure for finding the torsion-free connection for B^i is possible. We follow Freidel (2008) in this section. The procedure in this section relates the sought torsion-free connection to the SD part of the metric spin connection and derivatives of the objects b^i_a and is more convenient in some computations.

The idea is to relate the sought connection A^i to a connection in a different SO(3) bundle. Thus, we introduce a new GL(3) connection $\omega^a{}_b$ according to the definition

$$d^A b^i_a X^a = b^i_a d^\omega X^a. \tag{5.228}$$

Explicitly

$$\omega^a{}_b = (b^{-1})^a_i A^i{}_j b^j{}_b + (b^{-1})^a_i db^i{}_b, \tag{5.229}$$

where $A^i{}_j = \epsilon^i{}_{kj} A^k$.

We can now find $\omega^a{}_b$ from the properties it must satisfy. First, we have

$$0 = d^A B^i = b^i{}_a d^\omega \Sigma^a, \tag{5.230}$$

and so the new connection still has zero torsion $d^\omega \Sigma^a = 0$. However, this connection does not coincide with the unique torsion-free connection $\gamma^a{}_b$ because this connection has zero torsion $d^\gamma \Sigma^a = 0$ and is also metric $d^\gamma \delta^{ab} = 0$. In contrast, the connection we are looking for has zero torsion but is metric for a metric that is different from δ^{ab}

$$0 = d^a \delta^{ij} = b^i{}_a b^j{}_b d^\omega m^{ab}, \tag{5.231}$$

where m^{ab} is the inverse metric to $m_{ab} = \delta_{ij} b^i{}_a b^j{}_b$. Thus, the metricity condition for $\omega^a{}_b$ is instead $d^\omega m^{ab} = 0$.

We look for the connection $\omega^a{}_b$ in the form of a sum of the torsion-free metric connection $\gamma^a{}_b$ and some GL(3)-valued 1-form $\rho^a{}_b$

$$\omega^a{}_b = \gamma^a{}_b + \rho^a{}_b. \tag{5.232}$$

We can find the object $\rho^a{}_b$ from the equations it must satisfy. First, the torsion-free condition gives

$$\rho^a{}_b \Sigma^b = 0. \tag{5.233}$$

Second, the metricity condition gives

$$d^\gamma m^{ab} + \rho^a{}_c m^{cb} + \rho^b{}_c m^{ac} = 0. \tag{5.234}$$

The last equation implies

$$\rho_{(ab)} = \frac{1}{2} d^\gamma m_{ab}, \tag{5.235}$$

where $\rho_{ab} := m_{ac}\rho^c{}_b$ and we have used $m_{ac} d^\gamma m^{cd} m_{db} = -d^\gamma m_{ab}$. To find the antisymmetric part of ρ_{ab} we use the torsion-free condition. We have

$$(\frac{1}{2} d^\gamma m_{ab} + \rho_{[ab]})\Sigma^b = 0. \tag{5.236}$$

Writing $\rho_{[ab]} = \epsilon_{acb}\rho^c$, dualising on the spacetime indices, and using the self-duality of Σ^a, we can rewrite this equation as

$$\frac{1}{2}\Sigma^b{}_\mu{}^\nu d^\gamma_\nu m_{ab} = \epsilon_{abc}\Sigma^b{}_\mu{}^\nu \rho^c_\nu. \tag{5.237}$$

Recalling the definition (5.137), we can rewrite this in an index-free way as

$$J_\Sigma \rho = \frac{1}{2} \Sigma \, d^\gamma m, \tag{5.238}$$

where $\Sigma \, d^\gamma m$ is an object in $\Lambda^1 \otimes \mathfrak{so}(3)$. The object ρ^a_μ is then given by

$$\rho = \frac{1}{2} J_\Sigma^{-1} \Sigma \, d^\gamma m, \tag{5.239}$$

where J_Σ^{-1} is explicitly given by (5.140). Using the algebra (5.138) of Σ's we get, explicitly

$$\rho_{\mu a} = \frac{1}{2}\Sigma^b{}_\mu{}^\nu d^\gamma_\nu (m_{ab} - \frac{m}{2}\delta_{ab}), \tag{5.240}$$

where $m = \text{Tr}(m) = \delta^{ab}m_{ab}$. This solves the problem of determining the connection $\omega^a{}_b$. Explicitly

$$\omega^a{}_b = \gamma^a{}_b + \frac{1}{2}m^{ac}d^\gamma m_{cb} + m^{ac}\epsilon_{cdb}\rho^d, \tag{5.241}$$

where the vector valued 1-form ρ^a is given by (5.240). These formulas can be used to explicitly compute the kinetic term B^iF_i in terms of the SD part of the spin connection for the metric defined by B^i as well as derivatives of the objects b^i_a. Details are worked out in Freidel (2008). Using this in (5.217) gives a second-order description of GR with the 2-form B^i as the only field. It is somewhat surprising that such a description is possible, because the 2-form field B^i is an object of a very different nature as compared to the metric. Nevertheless, we see that it can also be made dynamical and describe GR.

6
Chiral Pure Connection Formulation

The purpose of this chapter is to derive and develop the chiral pure connection formulation of general relativity (GR). This formalism is singled out from all the other ways of thinking about gravity because it leads to a remarkably simple linearised description of gravitons. Because of this it likely has a lot of yet unexplored potential, and so we will develop it in more detail than for other formulations. We will also present the related pure connection description of gravitational instantons.

6.1 Chiral Pure Connection Formalism for GR

The Plebańksi action (5.159) serves as a starting point for many chiral reformulations of four-dimensional GR. In particular, it is possible to obtain the chiral pure connection formulation starting from it, as we now demonstrate. The procedure for doing this is completely analogous to that already adopted in the non-chiral case. The chiral case is, however, much simpler in many respects, as we shall now see.

6.1.1 The Connection Formalism with Lagrange Multiplier Fields

We first integrate out the 2-form field Σ^i of the Plebański formulation. This results in the action

$$S[A,\Psi] = \frac{1}{16\pi G\sqrt{\sigma}} \int \left(\Psi^{ij} + \frac{\Lambda}{3}\delta^{ij}\right)^{-1} F^i F^j. \tag{6.1}$$

This action, which is an intermediate step towards the pure connection formulation Section 6.1.2, is itself a useful variational principle for GR. It depends on just $12+5$ variables. Even though it appears to be second-order in derivatives, this is an illusion. The most natural backgrounds on which this action can be expanded are maximally symmetric. On such backgrounds, $\Psi^{ij} = 0$ (zero

Weyl curvature), and the part of the linearised action that is quadratic in derivatives is just $d_A \delta A^i d_A \delta A^i$. Integrating by parts and replacing the commutator of covariant derivatives with a curvature, one reduces this to a term not containing derivatives.

The action (6.1) exists even with $\Lambda = 0$, but in this case it is not possible to expand it around a $\Psi^{ij} = 0$ background. Still, one can solve $\Lambda = 0$ Einstein equations in this formalism, as we will later demonstrate. This action is surprisingly similar to the MacDowell–Mansouri action (3.77) in that it is obtained as the wedge product of two copies of the curvature, contracted with some appropriate tensor. The similarity becomes even more pronounced if one compares to the action (3.78) that contains a dynamical field in front of the curvature squared term.

We can also rewrite the action (6.1) in a form containing an additional Lagrange multiplier

$$S[A, M, \mu] = \frac{1}{16\pi G\sqrt{\sigma}} \int \mathrm{Tr}(M^{-1}FF) + \mu\left(f(M) - \Lambda\right), \tag{6.2}$$

where for GR the function $f(M)$ is given by the trace (5.198), and $\mathrm{Tr}(M^{-1}FF) = M^{-1}_{ij}F^iF^j$. Note the perfect similarity between this action and (3.79). The action (6.2) is, of course, also the action (5.197) with the 2-form field integrated out. This action describes GR as well as the chiral modified theories obtained by changing $f(M)$. It is also a good starting point for developing the chiral connection perturbation theory, as we will see.

6.1.2 The Chiral Pure Connection Lagrangian

To go to the pure connection formulation we now integrate out M^{ij} from (6.2). Its Euler–Lagrange equation reads

$$M^{-1}XM^{-1} = \mu\mathbb{I}, \tag{6.3}$$

where we introduced the matrix of wedge products of the two copies of the curvature

$$X^{ij} \equiv X^{ij}_F := F^iF^j. \tag{6.4}$$

This is a symmetric 3×3 matrix with values in 4-forms. The equation for M is solved by

$$M = \sqrt{\frac{X}{\mu}}. \tag{6.5}$$

As usual, the Lagrange multiplier μ is found from the constraint it imposes

$$\sqrt{\mu} = \frac{\mathrm{Tr}\sqrt{X}}{\Lambda}, \tag{6.6}$$

so that

$$M^{-1} = \frac{\mathrm{Tr}\sqrt{X}}{\Lambda} X^{-1/2}. \tag{6.7}$$

The pure connection action becomes the integral of the Lagrange multiplier

$$S[A] = \frac{1}{16\pi G\Lambda\sqrt{\sigma}} \int \left(\mathrm{Tr}\sqrt{X}\right)^2. \tag{6.8}$$

This action was first obtained in Krasnov (2011). It is the most economic pure connection formulation of GR available. Indeed, it must be compared to the action (3.82) that depends on the 4×10 components of the connection, and to the action (3.104) that depends on the 24 components. In contrast, (6.8) depends on just 12 components of the $SO(3)$ connection. It is thus comparable to the usual metric formulation with its 10 components in economy. Moreover, it turns out that the perturbation theory in this chiral pure connection formalism can be set up in such a way that only 8 out of the 12 components propagate, 2 of them being the physical polarisations of the graviton, the remaining $3+3$ being unphysical gauge variables. This is more economical than GR in the metric formalism. But this perturbation theory only exists around $\Lambda \neq 0$ backgrounds, because of the presence of $1/\Lambda$ in front of the action.

6.1.3 The Split Signature Modification

In the case of the Split signature the previous discussion needs to be slightly modified because it is the metric η_{ij} rather than the identity metric δ_{ij} that must be used in all the formulas. In particular, in the split signature the function $f_{\mathrm{GR}} = \mathrm{Tr}(M)$ computes the trace of the matrix M with respect to η rather than with respect to the identity matrix. This means that the equation (6.3) will contain η rather than the identity matrix on the right-hand side. The procedure of solving this equation becomes more involved because η does not necessarily commute with the other matrices appearing in this equation. However, we understand how to solve this equation from the non-chiral case discussion. It is clear that some factors of $\sqrt{\eta}$ will be introduced in the process of the solution. In particular, the solution for M gets modified to

$$M^{-1} = \sqrt{\mu}\sqrt{\eta}(\sqrt{\eta}X\sqrt{\eta})^{-1/2}\sqrt{\eta}. \tag{6.9}$$

The Lagrange multiplier gets modified to

$$\sqrt{\mu} = \frac{\mathrm{Tr}\sqrt{\eta X}}{\Lambda}, \tag{6.10}$$

so that

$$M^{-1} = \frac{\mathrm{Tr}\sqrt{\eta X}}{\Lambda}\sqrt{\eta}(\sqrt{\eta}X\sqrt{\eta})^{-1/2}\sqrt{\eta}. \tag{6.11}$$

Thus, the chiral pure connection action relevant for the split signature case is

$$S_{\text{split}}[A] = \frac{1}{16\pi G\Lambda}\int \left(\text{Tr}\sqrt{\eta X}\right)^2. \tag{6.12}$$

We will later see that the necessity of introducing the factor of η under the square root in the split signature case makes perfect sense because it makes the argument of the square root a positive definite matrix so that the square root exists in real matrices. We also note that the matrix $\sqrt{\eta}$ does not exist as a real matrix, only as complex. This, however, does not create any difficulty because only the original real matrix η appears in the final action.

6.1.4 The Metric

In any pure connection formalism the metric is constructed algebraically from the curvature of the relevant connection. Let us see how it arises in the chiral case. The easiest way to see this is to recall that in the Plebański formalism the metric is constructed from the 2-form fields Σ^i via the Urbantke formula (5.47). In the process of integrating out the 2-form fields we have solved their field equations as

$$\Sigma^i = (M^{-1})^{ij} F_j. \tag{6.13}$$

If we substitute here the expression for M (6.7), as arises in the process of solving its field equations, we obtain the following explicit expression for the 2-form fields

$$\Sigma^i_F = \frac{\text{Tr}\sqrt{X}}{\Lambda}(X^{-1/2})^{ij} F_j. \tag{6.14}$$

It is easy to see that the 2-form fields Σ^i so constructed satisfy the constraints $\Sigma^i\Sigma^j \sim \delta^{ij}$. Thus, the metric of the chiral pure connection formalism is the Urbantke metric

$$g_\Sigma(u,v)\epsilon_\Sigma = \frac{\sqrt{\sigma}}{6}\epsilon^{ijk} i_u\Sigma^i i_v\Sigma^j\Sigma^k \tag{6.15}$$

for the 2-forms Σ^i given by (6.14).

We note, however, that this metric can be described more explicitly, and in particular directly in terms of the curvature 2-forms. Indeed, by construction, the Urbantke metric (5.47) is one in which the triple of 2-forms Σ^i becomes self-dual (SD) (in an appropriate orientation). However, since the Urbantke metric is to make Σ^i_F SD, and these 2-forms are given by a linear combination of the 2-forms F^i, the metric will also make the curvature 2-forms SD. Thus, the conformal class of the metric can be obtained as the unique conformal class that makes the triple of 2-forms F^i SD. In other words, the conformal class of the metric can be obtained directly from the curvature 2-forms by inserting them into the Urbantke formula. Thus, we have for the conformal metric

$$g_F(u,v) \sim \sqrt{\sigma}\epsilon^{ijk} i_u F^i i_v F^j F^k. \tag{6.16}$$

Let us now discuss the volume form. As we have seen in the previous chapter, the appropriate volume form can be obtained as $6\sqrt{\sigma}\epsilon_\Sigma = \Sigma^i\Sigma^i$. Applying this to the 2-form fields (6.14), we get the following volume form

$$\epsilon_F = \frac{1}{2\sqrt{\sigma}\Lambda^2}\left(\mathrm{Tr}\sqrt{X}\right)^2. \tag{6.17}$$

Again, we see that one does not need to construct Σ_F^i 2-forms and the volume form can be constructed directly from the curvature. We then note that, as for all previously discussed pure connection actions, the action is just a multiple of the total volume

$$S[A] = \frac{\Lambda}{8\pi G}\int \epsilon_F. \tag{6.18}$$

In the split signature case there are some modifications to the previous discussion. The factors of η that need to be introduced modify the previous formulas as follows

$$\Sigma_F^i = \frac{\mathrm{Tr}\sqrt{\eta X}}{\Lambda}\left(\sqrt{\eta}(\sqrt{\eta}X\sqrt{\eta})^{-1/2}\sqrt{\eta}\right)^{ij}F_j. \tag{6.19}$$

This satisfies $\Sigma_F^i\Sigma_F^j \sim \eta^{ij}$. Again, we can avoid the need for computing Σ_F^i by noticing that the metric is such that it makes Σ_F^i SD. But this means that the metric also makes F^i SD, and so it can be computed directly from the curvature 2-forms, by using the Urbantke formula. The split signature volume form is obtained as

$$\epsilon_F = \frac{1}{2\Lambda^2}\left(\mathrm{Tr}\sqrt{\eta X}\right)^2, \tag{6.20}$$

and the action is still given by a multiple of the total volume.

6.1.5 Field Equations

The first variation of the chiral pure connection action (in Euclidean and Lorentzian signatures) is given by

$$\delta S[A] = \frac{1}{16\pi G\Lambda\sqrt{\sigma}}\int \mathrm{Tr}\sqrt{X}\,\mathrm{Tr}(X^{-1/2}\delta X), \tag{6.21}$$

where

$$\delta X^{ij} = 2F^{(i}d^A\delta A^{j)}. \tag{6.22}$$

This means that the Euler–Lagrange equations following from the chiral pure connection action are

$$d^A\left(\mathrm{Tr}\sqrt{X}(X^{-1/2})^{ij}F_j\right) = 0. \tag{6.23}$$

However, noting the relation (6.14) we can rewrite the arising field equations as

$$d^A\Sigma_F^i = 0. \tag{6.24}$$

Thus, the field equations one obtains by extremising the action just state that the connection coincides with the unique torsion-free connection for the 2-form fields Σ_F^i constructed from the connection. This makes it clear how Einstein equations arise in this formalism. The field equations of the pure connection description are second-order partial differential equations for the connection components. Then, given a solution, the metric constructed from Σ_F^i is guaranteed to be Einstein because on solutions of (6.24) the curvature F^i coincides with the SD part of the curvature of the spin connection for the metric constructed from Σ_F^i. On the other hand, by the very construction of the metric the curvature 2-forms F^i are SD as the 2-forms. The self-duality of the curvature 2-forms is the Einstein condition.

6.1.6 Lorentzian Signature Reality Conditions

The previously described formalism works for the SO(3) and SO(1, 2) connections, but is needs additional discussion for the case of $SO(3, \mathbb{C})$ connections appropriate for the Lorentzian signature. Indeed, in this case we want to impose the reality conditions on the Σ^i 2-forms. These conditions are of two types. First, one imposes nine conditions $\Sigma^i \overline{\Sigma^j} = 0$ that guarantee that the conformal class of the metric is that of a real Lorentzian one. But because Σ_F^i and F^i are linear combinations of each other, these reality conditions can be stated directly as conditions on the curvature

$$F^i \overline{F^j} = 0. \tag{6.25}$$

The last condition is that the volume form as constructed from the Σ^i is real. This condition translates to

$$\mathrm{Re}\left(\mathrm{Tr}\sqrt{X}\right)^2 = 0. \tag{6.26}$$

Again, this is a condition directly on the curvature of the connection, as is appropriate for a pure connection formalism.

6.1.7 (Gauge) Invariances of the Pure Connection Action

The action (6.8) is gauge and diffeomorphism invariant. It is a useful exercise to verify this explicitly, as some convenient for future use identities will result from this exercise. Let us first discuss the diffeomorphisms. Whenever we are discussing the action of diffeomorphisms on a connection, we can modify the usual Lie derivative Cartan formula $\mathcal{L}_\xi = i_\xi d + d i_\xi$ by adding to it a gauge transformation with the parameter $i_\xi A$. This gives a very convenient for practical applications formula for the transformation of the connection under diffeomorphisms

$$\delta_\xi A^i = i_\xi F^i. \tag{6.27}$$

The SO(3) gauge transformations, on the other hand, are given by the usual

$$\delta_\phi A^i = d^A \phi^i. \tag{6.28}$$

6.1.8 Action in Terms of a Homogeneity Degree One Function

To discuss the invariance of the action, it is very convenient to write the pure connection action (6.8) in a more general form

$$S[A] = \frac{1}{16\pi G\Lambda\sqrt{\sigma}} \int g(X), \tag{6.29}$$

where $g(X)$ is some SO(3)-invariant function of the 3×3 symmetric matrix $X^{ij} = F^i F^j$. In addition, in order for the action to make sense (note that X^{ij} is the matrix with values in 4-forms), the function $g(X)$ must be homogeneous of degree one in its argument. The chiral pure connection GR action (6.8) is clearly of this form with

$$g_{\rm GR}(X) = \left(\mathrm{Tr}\sqrt{X}\right)^2, \tag{6.30}$$

but also any of the chiral modifications of GR is of this form. So, our discussion is going to be more general than is necessary for the purposes of GR, but it is easier to follow this more general discussion.

6.1.9 Diffeomorphism Invariance

For purposes of this and the next subsection we set the coefficient in front of the action to unity. The first variation of the action (6.29) is

$$\delta S[A] = 2 \int \frac{\partial g}{\partial X^{ij}} F^i d^A \delta A^j. \tag{6.31}$$

Let us first discuss its diffeomorphism invariance. To do this we need a much more concrete way of working with the 4-form-valued matrices than we required before. So, we introduce the densitiesed ϵ-symbol via

$$dx^\mu dx^\nu dx^\rho dx^\sigma = \tilde{\epsilon}^{\mu\nu\rho\sigma} d^4x, \tag{6.32}$$

where d^4x is the coordinate volume form. This symbol takes values ± 1 in any coordinate system. Using this symbol we can rewrite the first variation of the action as

$$\delta S[A] = \int \frac{\partial g}{\partial \tilde{X}^{ij}} \tilde{\epsilon}^{\mu\nu\rho\sigma} F^i_{\mu\nu} d^A_\rho \delta A^j_\sigma, \tag{6.33}$$

where we introduced a densitiesed 3×3 matrix

$$\tilde{X}^{ij} := \frac{1}{4} \tilde{\epsilon}^{\mu\nu\rho\sigma} F^i_{\mu\nu} F^j_{\rho\sigma}. \tag{6.34}$$

We now substitute (6.27) for the variation. This means that we need to compute $d^A_{[\mu}\xi^\alpha F_{\alpha\nu]}$. This can be transformed using the Bianchi identity $d^A_{[\mu}F^i_{\rho\sigma]} = 0$. We have

$$\xi^\alpha d^A_{[\mu}F^i_{\nu]\alpha} = -\frac{1}{2}\xi^\alpha d^A_\alpha F^i_{\mu\nu}. \tag{6.35}$$

Another identity that we need is

$$\tilde{\epsilon}^{\mu\nu\rho\sigma}F^{(i}_{\mu\nu}F^{j)}_{\alpha\sigma} = \delta^\rho_\alpha \tilde{X}^{ij}. \tag{6.36}$$

This identity follows from the fact that the quantity $F^{(i}_{\mu\nu}F^{j)}_{\alpha\sigma}$ antisymmetrised in three indices $\mu\nu\sigma$ is in fact completely antisymmetric, and thus proportional to the ϵ-tensor. Thus, the left-hand side must be a multiple of $\tilde{X}^{ij}$, and so the formula (6.36) results.

The previous two identities mean that

$$\begin{aligned}\tilde{\epsilon}^{\mu\nu\rho\sigma}F^{(i}_{\mu\nu}d^A_\rho\delta_\xi A^{j)}_\sigma &= \tilde{X}^{ij}d_\alpha\xi^\alpha + F^{(i}_{\mu\nu}\frac{1}{2}\xi^\alpha d^A_\alpha F^{j)}_{\rho\sigma} \\ &= \tilde{X}^{ij}d_\alpha\xi^\alpha + \frac{1}{4}\xi^\alpha d^A_\alpha(\tilde{\epsilon}^{\mu\nu\rho\sigma}F^{(i}_{\mu\nu}F^{j)}_{\rho\sigma}) = d^A_\alpha(\xi^\alpha\tilde{X}^{ij}).\end{aligned} \tag{6.37}$$

Thus, integrating by parts we have

$$\delta_\xi S[A] = -\int \xi^\alpha\tilde{X}^{ij}d^A_\alpha\frac{\partial g}{\partial\tilde{X}^{ij}}. \tag{6.38}$$

The right-hand side here is zero by the homogeneity of $g(X)$. Indeed, the Euler relation for $g(X)$ reads

$$\frac{\partial g}{\partial\tilde{X}^{ij}}\tilde{X}^{ij} = g. \tag{6.39}$$

Differentiating this identity one time we get

$$\tilde{X}^{ij}\partial_\alpha\frac{\partial g}{\partial\tilde{X}^{ij}} = 0. \tag{6.40}$$

The partial derivative here can be replaced with the covariant derivative because the function $g(X)$ is SO(3) invariant. Indeed, this gauge invariance means that

$$\frac{\partial g}{\partial\tilde{X}^{ij}}\epsilon^{ikl}\phi^k\tilde{X}^{lj} = 0 \tag{6.41}$$

for any gauge parameter ϕ^i. These are precisely the terms that need to be added to the usual derivative in (6.40) to convert it to the covariant derivative. This establishes the diffeomorphism invariance.

The fact that the action is diffeomoprhism-invariant implies that not all field equations arising from the variational principle are independent. Let us derive the corresponding relations. The field equations one obtains from (6.29) are

$$d^A\left(\frac{\partial g}{\partial X^{ij}}F^j\right) = 0. \tag{6.42}$$

Using Bianchi identity this can be rewritten as

$$d^A\left(\frac{\partial g}{\partial X^{ij}}\right)F^j = 0. \tag{6.43}$$

Let us multiply this 3-form with the 1-form $i_\xi F^i$. Using the identity

$$i_\xi F^{(i}F^{j)} = \frac{1}{2}i_\xi(F^iF^j) = \frac{1}{2}i_\xi X^{ij} \tag{6.44}$$

we have

$$i_\xi F^i d^A\left(\frac{\partial g}{\partial X^{ij}}\right)F^j = \frac{1}{2}i_\xi X^{ij} d^A\left(\frac{\partial g}{\partial X^{ij}}\right). \tag{6.45}$$

But the right-hand side here is zero by the covariant derivative version of the homogeneity consequence (6.40). This shows that there are four relations between the field equations, as a consequence of the diffeomorphism invariance of the theory

$$i_\xi F^i \mathcal{E}^i = 0, \tag{6.46}$$

where $\mathcal{E}^i = 0$ are the field equations and $\mathcal{E}^i$ is a Lie algebra-valued 3-form.

6.1.10 SO(3) *Invariance*

Let us also demonstrate the invariance under the SO(3) gauge rotations. We have

$$\begin{aligned}\delta_\phi S[A] &= 2\int \frac{\partial g}{\partial X^{ij}}F^i d^A d^A \phi^j = 2\int \frac{\partial g}{\partial X^{ij}}F^i \epsilon^{jkl}F^k\phi^l \\ &= 2\int \frac{\partial g}{\partial X^{ij}}\epsilon^{jkl}X^{ik}\phi^l = 0.\end{aligned} \tag{6.47}$$

The last equality is the direct consequence of the gauge invariance of the function g.

Let us also discuss the relations between the field equations that arise as the consequence of gauge invariance. Let us consider the exterior covariant derivative of the 3-forms $\mathcal{E}^i$ whose vanishing gives the field equations. We have

$$\begin{aligned}d^A\mathcal{E}^i = d^A d^A\left(\frac{\partial g}{\partial X^{ij}}F^j\right) &= \epsilon^{ikl}F^k\left(\frac{\partial g}{\partial X^{lm}}F^m\right) \\ &= \epsilon^{ikl}X^{km}\frac{\partial g}{\partial X^{lm}} = 0,\end{aligned} \tag{6.48}$$

again as consequence of the gauge invariance of g.

6.1.11 Definite, Semi-Definite Connections and the Sign of a Connection

The procedure of deriving the pure connection action (6.8) was formal in the sense that we did not concern ourselves with the question of which branch of the square root of the matrix X to take, and even whether the square root exists in real matrices (for the Euclidean and split signature) at all. Our goal is now to discuss this, and thus make sense of the so-far formal manipulations. To this end, we need to introduce some notions that help to simplify the discussion that follows.

Let us start by considering an SO(3) connection, as is appropriate for describing the Euclidean signature. We compute its curvature 2-forms, and then the corresponding matrix X. This matrix is definite if all 3 of its eigenvalues are nonvanishing and of the same sign. Following a related discussion in Fine et al. (2014) we will call an SO(3) connection **definite** if the corresponding matrix X is definite at all points of M. We note that a definite connection gives M an orientation, which is the orientation in which all eigenvalues of X are positive.

Let us now recall that in the chiral pure connection formalism, the metric is defined by requiring that the curvature 2-forms F^i become SD, and that the volume form is given by (6.17). We also know that the signature of the arising metric is controlled by the restriction of the wedge product metric to the three-dimensional subspace in Λ^2 spanned by the 2-forms F^i. Namely, if the matrix X is definite, then the arising conformal metric is of the Euclidean signature. If the matrix X is indefinite, then the metric is of the split signature. This shows that a definite connection on M defines a Euclidean signature metric on M.

Another important notion that we need is that of a sign of a connection. Let us assume that an SO(3) connection is definite, so that it defines an orientation of M and a Euclidean signature conformal metric on M via

$$g(u,v)_F \sim \epsilon^{ijk} i_u F^i i_v F^j F^k/\mu, \tag{6.49}$$

where μ is an arbitrary 4-form in the orientation defined by the connection. The connection is said to be **positive** if no additional sign is needed in this formula to render a metric of the all plus signature. The connection is said to be **negative** if one needs an extra minus sign in the Urbantke formula to result in an all plus metric. We will later see that the sign of the connection correlates with the sign of the cosmological constant in that positive connections are those relevant for describing the $\Lambda > 0$ geometries and negative connections are relevant for $\Lambda < 0$.

Definite connections are easiest to work with, but we shall soon see that the set of definite connections is too small and does not cover most of the examples of interest. For this reason we introduce a weaker notion of semi-definite connections. An SO(3) connection is called **semi-definite** if M is split into open regions in each of which the matrix X is definite. Moreover, since each region of definiteness receives an orientation in which X is positive-definite, we will also require that for a semi-definite connection these orientations agree. Thus, a semi-definite connection gives M an orientation. We will later see that this is the most interesting situation, as connections coming from 'general' Einstein metrics are semi-definite. A semi-definite connection defines a Euclidean signature metric in each open region of definiteness.

Finally, we can extend the notion of a sign of a connection to the semi-definite case as well. There is a sign in each region of definiteness, which is what is required for the Urbantke metric computed from the curvature to have the signature all plus. In principle, each region could carry a different sign, and we will later see that there are actually examples of semi-definite connections with different signs in different regions of definiteness.

Semi-definite connections are 'almost' definite, and so are easiest to work with. However, we shall later see that not all SO(3) connections are semi-definite. There are phenomena of two types that do occur to prevent an SO(3) connection from being semi-definite. First, it can be that the matrix X is degenerate at all points of M. This actually occurs even for some Einstein manifolds; see next section. Second, the matrix X can be definite in some regions of M and indefinite in some others. Third, it may be that X is definite in all open regions of definiteness, but then the orientations that these regions receive by requiring X to be positive-definite do not match. As we shall demonstrate in the next section, all of these in general do occur for a general SO(3) connection, but not for an Einstein connection, i.e., a connection that comes from an Einstein metric. So, we will eventually need to discuss all these possibilities in order to define the action.

Let us now discuss analogs of the previously introduced notions for the split and Lorentzian signatures. In the case of the split signature, we call an SO(1, 2) connection definite if the corresponding matrix X is of indefinite signature everywhere on M. Such a connection gives M an orientation in which X is of the same signature as η. However, there is no sign of a connection in this case, as both signs that could be used in the Urbantke formula would result in the same split signature. Similarly, a semi-definite SO(1, 2) connection splits M into regions in which X is of indefinite signature. Each region defines an orientation and these are required to match.

In the Lorentzian signature case we are dealing with SO(3, $\mathbb{C}$) connections. Such a connection is said to satisfy the reality conditions if the conformal metric

$$g(u,v)_F \sim \mathrm{i}\,\epsilon^{ijk} i_u F^i i_v F^j F^k, \tag{6.50}$$

is of Lorentzian signature. There is no longer a notion of definiteness because the matrix X in this case is generally complex. But we can still have connections that are semi-definite in the sense that M splits into regions where X is nondegenerate. But there is no longer an orientation that such connections can define. As a result, there is also no notion of a sign of a connection. But one can still demand that there exists a global choice of orientation of M so that the Urbantke formula with the right-hand side divided by the corresponding 4-form produces a fixed Lorentzian signature metric over M.

6.1.12 Towards a Non-Perturbative Definition of the Action

We can now come to a discussion of how the action (6.8) obtained by formal manipulations can be made sense of. Our first remark is that the action is certainly defined perturbatively, as means of producing a perturbative expansion around some given background. As we shall see in the details in Chapter 8, the most natural background for this chiral pure connection theory is the maximally symmetric one.[1] On this background, the matrix X is a multiple of the identity, and so its square root is clearly defined. One can expand the square root of X in powers of deviations of X from the identity. With this in mind, all formal manipulations performed previously make sense. In fact, the action (6.8) can be expanded around an arbitrary Einstein background and this expansion is not ambiguous. So, the pure connection action (6.8) is definitely well-defined perturbatively.

Let us now consider the question whether the action (6.8) makes sense beyond perturbation theory. In the following sections we will see examples of solving the field equations (6.24), and these examples indicate that the theory makes sense beyond a perturbative expansion. However, these examples also reveal that it is in general inconsistent to limit oneself to just one of the branches of the square root in (6.8). We now turn to a discussion of this. In particular, our goal is to establish that in the case of the Euclidean setup, the action (6.8) is bounded from below.

We only discuss the case of the Euclidean signatures. The Lorentzian case needs to be treated separately because of the issues with the reality conditions. But it is the Euclidean signature action that usually participates in the gravitational path integral, and so establishing its boundedness is particularly relevant.

The discussion in the previous section shows that the chiral pure connection functional (6.8) is well-defined on definite SO(3) connections. In this case, the prescription is to take the positive branch of the square root of the positive-definite matrix X, and to take the integral in the orientation defined by the connection, so as to get a positive result (volume). However, explicit examples show that in most of the situations of interest the connection is not definite everywhere, and is only semi-definite, i.e., M is covered by open regions of definiteness. These examples also show that it is in general not correct to take the positive branch of the square root everywhere on M, and once one crosses from one region of definiteness to another, there must be a change in the branch of the square root.

That this is expected can be easily seen by considering SO(3) connections that come from Einstein metrics. In this case, we know from the Plebański formalism that

$$F^i = \left(\Psi^{ij} + \frac{\Lambda}{3}\delta^{ij}\right)\Sigma_j. \tag{6.51}$$

Thus, the corresponding matrix X is

$$X \sim \left(\Psi + \frac{\Lambda}{3}\mathbb{I}\right)^2. \tag{6.52}$$

[1] In fact, in the chiral case there is no difference between the maximally symmetric and a more general instanton background, see more on this in Section 6.3.

We then see that this is a definite matrix everywhere apart from the places where one of the eigenvalues of the SD part of Weyl matrix Ψ exactly balances the Λ term. When this happens, the corresponding eigenvalue of X vanishes, and the matrix is degenerate. In a 'general' Einstein metric, however, this happens across hypersurfaces in M that split the manifold into open regions of definiteness of X, or possibly also at isolated singularities. We also see that for such a 'general' Einstein SO(3) connection an orientation can be chosen (consistently over all of M) so that almost everywhere the eigenvalues $\lambda_{1,2,3}$ of X are positive, and $\sqrt{X}$ exists as a real matrix. Thus, we see that 'general' Einstein connections are semi-definite.

We should now discuss the meaning of 'general' in the previous paragraph. There exist Einstein metrics for which the matrix $\Psi + (\Lambda/3)\mathbb{I}$ is everywhere degenerate. The easiest example is, of course, flat space $\mathbb{R}^4$, where the connection and thus the previous matrix are zero everywhere. The simplest example with nonzero scalar curvature is $S^2 \times S^2$. In such cases, the metric cannot be reconstructed from the curvature of the SD part of the spin connection. It is clear that these Einstein manifolds cannot be treated via the chiral pure connection formalism. So, 'generic' Einstein manifolds for us will be those for which the curvature of at least one chiral half of the spin connection (i.e., either SD or anti-self-dual) at a general point of M spans a three-dimensional subspace in Λ^2 and thus, allows the metric to be reconstructed from it. So, we restrict our attention to the Einstein metrics with a semi-definite chiral spin connection that can be described by the present formalism.

Let us now return to the question of which branch of the square root to take in $\sqrt{X}$. The previous discussion shows that one should not restrict one's attention to only the positive branch of the square root because the branch of $\sqrt{X}$ that one wants to reproduce is given by

$$\sqrt{X} \sim \Psi + \frac{\Lambda}{3}\mathbb{I}, \tag{6.53}$$

and the matrix on the right-hand side does not have to be definite. Thus, one certainly wants to keep not just the positive branch of $\sqrt{X}$ in defining the action. In finding explicit solutions of the theory (6.8) it is usually easy to decide on this from the requirement that the fields are continuous. So, when one crosses from one region of definiteness to another, the branch of the square root must change appropriately. However, if one is given a definite connection in only an open region of M, it is a priori not known which branch of the square root to take to compute the action. The most we can say at the moment is that given a connection that is definite in some open region of M, whatever branch is taken, the matrix $\sqrt{X}$ is real.

The previous discussion is important because it allows us to conclude that on semi-definite connections, the action (6.8) is bounded. Thus, assume that we are given a semi-definite SO(3) connection, i.e., a connection for which the the manifold M is split into open regions where the matrix X is definite. We do not yet know which branch of the square root in $\sqrt{X}$ to take in which region, but we know that in any branch this matrix is real, and so is its trace. This means that on such connections the action (6.8) is of a definite sign, positive for positive Λ, and negative for negative Λ. This is in contrast with the Euclidean Einstein–Hilbert action that is never definite.

We can now repeat this discussion for an SO(1, 2) connection. In this case, we want the matrix X to be indefinite almost everywhere. The easiest case is when the matrix X is of the same indefinite signature everywhere on M. This is the analog of the definite SO(3) connections in the SO(1, 2) setting. In this case, one knows that the metric it defines is of the split signature. Moreover, such a connection determines an orientation of M by requiring the signature of X to be the same as that of η. The action is also well-defined because one has an additional factor of η multiplying X under the square root in (6.20). The matrix ηX is positive-definite, and the square root can be taken in real matrices. The action then has a definite sign.

If the connection is not indefinite of the same signature everywhere, but only splits M into open regions where X is of a given indefinite signature, one still knows that it defines a split signature metric in every of these regions. One also knows that in all regions for any choice of the branch of the square root of the positive definite matrix ηX, the matrix $\sqrt{\eta X}$ is real, and so the action again has a definite sign.

This establishes that on semi-definite connections that split M into regions of definiteness, the action is well-defined and is positive or negative depending on the sign of Λ. This is the case for both SO(3) or SO(1, 2) connections. In these cases, there also exists a Euclidean or split signature metric almost everywhere on M.

Let us now discuss the cases when the connection is not semi-definite. Let us first consider the situations when the matrix X is nondegenerate almost everywhere on M, apart from possibly-hypersurfaces splitting M into open regions (smaller dimension singularities where X becomes degenerate are, of course, also possible). The first possibility is then that the matrix X is definite in some regions and indefinite in some others. If this is the case for an SO(3) connection, we would conclude that this connection defines a split signature metric in regions where X is indefinite. We would then be able to define the action in these regions by taking the square root of ηX rather than X, as is appropriate for the split signature. There is similarly a possibility that for an SO(1, 2) connection there are some regions where X is definite rather than indefinite. In this case, one just treats these regions as appropriate for the Euclidean signature, and uses $\sqrt{X}$ rather than $\sqrt{\eta X}$ in the action. So, this situation is easy to deal with.

Another possibility is that for, e.g., an SO(3) connection, the manifold M splits into open regions where the matrix X is definite, but the orientations that are required to make X positive-definite do not match. The most natural option in this case is to fix an orientation of M and to compute the action (6.8) as the integral of the top form ϵ_F in the given orientation. The orientation of the form ϵ_F, however, will not agree with the fixed orientation, and so the integral becomes a sum of positive and negative contributions. In this case, we can no longer conclude that the action has a fixed sign. An analogous situation is possible in the SO(1, 2) setting. In this case, the orientations required to make X to be of the same signature as η may not agree. Again, the action would be given by a sum of positive and negative contributions, and thus not have a fixed sign. Thus, we can only conclude that the action has a sign on semi-definite connections.

6.1.13 Not All Connections Are Semi-Definite

The following argument shows that the situations we worried about at the end of the previous subsection actually do arise. Thus, let us consider an SO(3) connection that is the SD part of the spin connection for some Euclidean metric, but not necessarily an Einstein one. In this case, we can decompose the curvature 2-forms F^i into the basis of SD and ASD 2-forms

$$F^i = M^{ij}\Sigma_j + N^{ij}\bar{\Sigma}_j, \tag{6.54}$$

where $\bar{\Sigma}^i$ are the ASD 2-forms. Let us now compute the corresponding matrix X. We have

$$X \sim M^2 - N^2, \tag{6.55}$$

where we used the fact that the wedge product metric on ASD 2-forms is negative of that on the SD 2-forms. We thus see that the matrix X is given by the difference of two positive-definite contributions, one that is the square of the SD plus scalar part of the Riemann curvature, and the other that is the square of the tracefree part of the Ricci. Since the metric we start from is completely arbitrary, nothing prevents the Ricci part from winning over the other part. Moreover, this can happen in some regions of the manifold, while in some other regions it can be M^2 that wins. So, it seems that nothing prevents the matrix X from being positive-definite in some regions and negative-definite in others. This is the case when we said the action becomes a sum of positive and negative contributions and thus not of a fixed sign.

Another possibility is that in a certain region some of the eigenvalues of X are of one sign and some of the other. Thus, the matrix X can be definite in some regions and indefinite in some others. In this case the metric that such a connection would define would be of split signature in regions where X is indefinite. This possibility seems also to be allowed by the relation (6.55).

So, we conclude that connections constructed from the 'general' Euclidean signature metrics do not have to be semi-definite. One can define the action (6.8) even on such connections, but then the action can have any sign.

6.1.14 The Action on Semi-Definite Connections is Bounded from Above

Let us now consider a semi-definite SO(3) connection. This means that M is split into open regions where X is definite and we can choose an orientation of M such that X is everywhere positive-definite. As we know, in this case the action (6.8) is nonnegative (for positive Λ). We shall now see that in this case, the action is also bounded from above.

We divide the 4-form valued matrix X by an arbitrary 4-form ϵ in orientation that makes X positive-definite. This converts X into an ordinary positive-definite 3×3 matrix. It then has positive eigenvalues $\lambda_{1,2,3}$ and there exist real square roots $\sqrt{\lambda_1}, \sqrt{\lambda_2}$, and $\sqrt{\lambda_3}$. These, however, can be of both signs. The Lagrangian density is given by

$$\mathcal{L} = (\sqrt{\lambda_1} + \sqrt{\lambda_2} + \sqrt{\lambda_3})^2 = \lambda_1 + \lambda_2 + \lambda_3 + 2\sqrt{\lambda_1\lambda_2} + 2\sqrt{\lambda_2\lambda_3} + 2\sqrt{\lambda_3\lambda_1}. \tag{6.56}$$

The quantity $\mathcal{L}\epsilon$ is to be integrated over all regions of definiteness. Note again that the quantities $\sqrt{\lambda_i}$ can be of either sign. We now use the inequalities

$$\lambda_i + \lambda_j \geq 2\sqrt{\lambda_i\lambda_j} \tag{6.57}$$

that follow from $(\sqrt{\lambda_i} - \sqrt{\lambda_j})^2 \geq 0$ to conclude

$$\mathcal{L} \leq 3(\lambda_1 + \lambda_2 + \lambda_3), \tag{6.58}$$

with the equality holding if and only if all eigenvalues are equal. This means that on semi-definite connections

$$0 < \int_M \left(\text{Tr}\sqrt{X}\right)^2 \leq 3\int_M \text{Tr}(FF), \tag{6.59}$$

with the equality holding if and only if X is a multiple of the identity matrix. The right-hand side in this inequality is a topological number that depends only on the SO(3) bundle over M that is taken. In order to be able to reproduce Einstein metrics via this formalism, this bundle must be in the same topological class as the bundle of SD 2-forms on M. In this case, the quantity on the right-hand side of (6.59) can be expressed as a specific linear combination of the signature of M and its Euler characteristic.

There is a similar bound for semi-definite connections of the SO(1, 2) setup, but there is no longer such a bound when the connection is not semi-definite. So, we conclude that on general connections, the action (6.8) is not bounded from either above or below. It is thus not usable for doing, e.g., lattice simulations of Euclidean quantum gravity in the SO(3) connection formalism. The only option for doing such simulations would be to change the definition of the action so that all regions contribute to it positively. This is the case on semi-definite connections, and one could then argue that the total action must always be the total volume as defined by the connection, and thus of a fixed sign. This would make the action bounded from below, and allow Monte-Carlo–type studies. It is, however, far from clear that this would give a sensible definition of the theory. But this option can be tried.

6.1.15 Choice of the Branch of the Square Root

We have discussed the general properties of the action (6.8) that are independent of a choice of the square root that one has to make to compute it. We now need to discuss this choice.

As we already discussed and as is clearly seen from examples, it is inconsistent to restrict one's attention to only the positive branch of the square root. As we shall now discuss, for each sign of the cosmological constant, there are in fact just two possible branches that can arise and that should be decided between. However, as we shall also see, it is unfortunately not possible to decide which branch of the square root to take just by looking at the matrix X. On the other hand, such a decision is easy to make when considering the problem of finding a solution of GR using this formalism. The rule is that when one crosses from one region of definiteness of X to another, the branch should change so that all fields remain continuous. But if one is just given a connection that is definite in some open region, there is no way to tell which branch of the square root must be taken, as we shall see from examples. Thus, the only way to define, e.g., a state sum model in this formalism would be to take both branches into account, hoping that the 'correct' branch will win in the state sum. Whether anything like this happens, however, is far from clear.

In order to proceed with our discussion, we first need to understand better the possible configurations that the matrix on the right-hand side of (6.51) can take. It is convenient to factor out Λ and consider the matrices of unit trace of the form

$$\frac{\Psi}{\Lambda} + \frac{1}{3}\mathbb{I}. \tag{6.60}$$

This is a symmetric matrix that can be diagonalised by an SO(3) transformation. Let us denote the eigenvalues by $x, y, 1 - x - y$.

In all the examples we are aware of the behaviour of the sign of the determinant of the matrix $\Psi/\Lambda + (1/3)\mathbb{I}$ correlates with the sign of the cosmological constant. Thus, in the case of positive Λ this matrix always has a positive determinant, which corresponds to the shaded region in the previous figure (i.e., regions I, II). In the case of negative Λ, one can certainly have examples with this matrix taking values in region I. For example, we have all negative scalar curvature instantons. In all other $\Lambda < 0$ examples we are aware of, it is the region III in the previous figure that plays role. Thus, there is an example of asymptotically hyperbolic space in which one of the eigenvalues of $\Psi/\Lambda + (1/3)\mathbb{I}$ changes sign

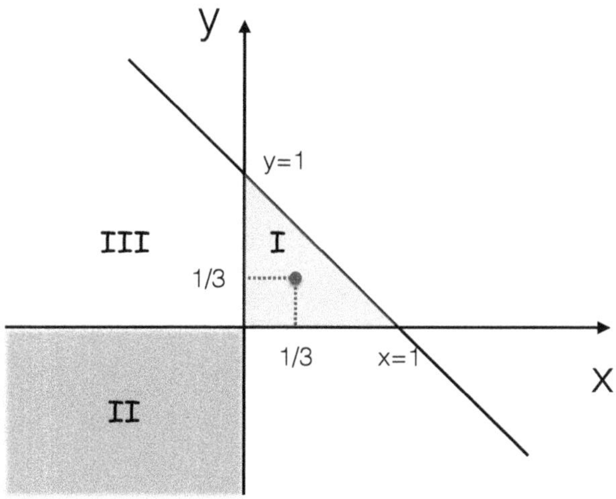

Figure 6.1 The space of diagonalisable 3×3 symmetric matrices of unit trace (modulo conjugation) can be visualised on the plane. The x and y are two of the eigenvalues, while the third one is given by $1 - x - y$. The central triangular-shaded region (labelled region I) is one where all eigenvalues are less than unity. The other shaded region $x < 0, y < 0$ (region II) is one where the largest eigenvalue takes values bigger than unity and the matrix continues to have a positive determinant. Region III is where only one of the eigenvalues is negative and the matrix has negative determinant.

and one crosses from region I to region III. There is also an example where one always remains in region III, with two eigenvalues of $\Psi/\Lambda + (1/3)\mathbb{I}$ changing sign simultaneously to remain in region III.

For concreteness, let us restrict the following discussion to the case of matrices of positive determinant, i.e., regions I and II in Figure 6.1. Region III can be analysed analogously. The subspace of matrices with all eigenvalues positive is represented by the triangular-shaded region in the figure. Indeed, both x and y and $1 - x - y$ must be greater than zero, which gives this triangular region. Its boundaries are places where one of the eigenvalues goes to zero. The point $x = y = 1/3$ is the 'central' point where all eigenvalues are the same and the matrix is a multiple of the identity.

When two of the eigenvalues are negative, without loss of generality, we can parametrise the matrices by the two negative eigenvalues, so that $1 - x - y$ is positive. We note that this is also automatically the eigenvalue largest in modulus because

$$(1 - x - y)^2 - x^2 = (1 - y)(1 - y - 2x) > 1 \qquad \forall x, y < 0, \tag{6.61}$$

and similarly for the difference $(1 - x - y)^2 - y^2$. We also note that in this region of the parameter space

$$(1 - x - y)^2 - x^2 - y^2 = 1 - 2(x + y) + 2xy > 1 \qquad \forall x, y < 0. \tag{6.62}$$

Let us now reconsider the procedure of integrating out the Lagrange multiplier fields from (6.1) or (6.2). For our purposes, it will be more convenient to think about integrating out Ψ from (6.1). Let us assume that the matrix X has been diagonalised and has three positive eigenvalues $\lambda_{1,2,3}$. We then know that the solution for $\Psi + (\Lambda/3)\mathbb{I}$ in terms of X will be such that the former is diagonal when the latter is diagonal. Thus, the procedure of finding the pure connection action is that of integrating out the parameters x and y from

$$Q(\{\lambda_i\}; x, y) := \frac{\lambda_1}{x} + \frac{\lambda_2}{y} + \frac{\lambda_3}{1-x-y}. \tag{6.63}$$

Differentiating Q with respect to x and y and setting the results to zero we get two equations

$$\frac{\lambda_1}{x^2} = \frac{\lambda_3}{(1-x-y)^2}, \quad \frac{\lambda_2}{y^2} = \frac{\lambda_3}{(1-x-y)^2}, \tag{6.64}$$

from which we read

$$\frac{\lambda_1}{x^2} = \frac{\lambda_2}{y^2}. \tag{6.65}$$

As we have already discussed, for $\Lambda > 0$ two of the eigenvalues of $(\Psi/\Lambda) + (1/3)\mathbb{I}$ can be demanded to be of the same sign (so that its determinant is positive), and so we have

$$y = \sqrt{\frac{\lambda_2}{\lambda_1}}x, \tag{6.66}$$

where the positive branch of the square root is taken. We now substitute this into any of the two equations and get a quadratic equation for x, with the solutions being

$$x = \frac{\sqrt{\lambda_1}}{\sqrt{\lambda_1} + \sqrt{\lambda_2} \pm \sqrt{\lambda_3}}, \tag{6.67}$$

where the two different solutions correspond to the two possible signs in the denominator. Here the square root always stands for the positive branch thereof. This is, of course, the already familiar solution for M/Λ with M given by (6.7). We now clearly see that there are only two possible solutions that give $(\Psi/\Lambda) + (1/3)\mathbb{I}$ in the desired region of the parameter space, i.e., a matrix of positive determinant.

Let us now discuss which of the two solutions in (6.67) to take. First of all, if we take the positive branch, then for any values of $\lambda_{1,2,3}$, we are in the triangular-shaded region of the parameter space where all three eigenvalues are positive (and necessarily less than one). Thus, the positive branch of the square root always lands us on the solution where eigenvalues are not too big as compared to the cosmological constant. To get a solution that corresponds to the region *II* of the parameter space we need to have

$$\sqrt{\lambda_3} - \sqrt{\lambda_1} - \sqrt{\lambda_2} > 0. \tag{6.68}$$

If this inequality is satisfied, then there is also the second branch of the square root of X that becomes available, corresponding to the negative sign in (6.67). So, one could hope that the right prescription is to take the positive branch of the square root when (6.68) is not satisfied, and the negative branch when (6.68) holds. However, this prescription does not pass the test of explicit examples. Thus, in Section 6.2 we consider an example of the so-called Page metric where the manifold is split into two regions of definiteness, and so one must change the branch of the square root of X as one goes from one region to the other. And at the same time (6.68) is always satisfied. This shows that there exists no criterion that would allow us to select a branch of the square root just by looking at the matrix X. The positive branch is always available and must be considered. But when (6.68) is satisfied (with λ_3 being the largest eigenvalue), then also the second branch becomes available and must be considered. The best one can do in forming a state sum is then to take both possible branches into account.

In the case $\Lambda < 0$, we have a similar situation, but it is now that regions I and III of the parameter space play role. One must decide which of the two possible branches of $\sqrt{X}$ to take just having access to the positive definite matrix X. Again, it is not possible to decide on this just by knowing X. In a concrete solution, this decision is taken by demanding that all fields are continuous across a surface on which one or two eigenvalues of X vanish. But without such continuity considerations, the best one can do is to allow both possible branches of the square root (i.e., those giving $\sqrt{X}$ in regions I and III) to contribute.

6.2 Example: Page Metric

The purpose of this section is to treat an example of a nontrivial positive scalar curvature Einstein metric known as Page metric. We will present the metric in the frame formalism and then show how to obtain it via the pure connection route. This metric gives a very good illustration of issues arising when selecting an appropriate branch of the square root as discussed in the previous section.

6.2.1 Page Metric

Let us consider a metric of the form

$$ds^2 = g^2(Q^{-2}dr^2 + \sigma_2^2 + \sigma_3^2) + \frac{4Q^2}{g^2}\sigma_1^2 \tag{6.69}$$

where σ_i are the usual 1-forms on S^3

$$d\sigma_i = -\frac{1}{2}\epsilon_i{}^{jk}\sigma_j \wedge \sigma_k \tag{6.70}$$

given by

$$\begin{aligned} \sigma_1 &= d\psi + \cos\theta d\phi, \\ \sigma_2 &= \cos\psi d\theta + \sin\psi \sin\theta d\phi, \\ \sigma_3 &= -\sin\psi d\theta + \cos\psi \sin\theta d\phi, \end{aligned} \tag{6.71}$$

where we used a somewhat unconventional numbering of the 1-forms and

$$g^2(r) = 1 - r^2, \qquad Q^2 = \frac{\lambda}{3} r^4 + r^2(1-2\lambda) + 1 - \lambda, \tag{6.72}$$

with λ being the (dimensionless) cosmological constant.

For $\lambda \leq 1/4$ and $\lambda \geq 3/4$ the equation $Q^2(r^2) = 0$ has real roots $r_-^2 \leq r_+^2$. The function $Q^2(r^2)$ is then nonnegative in the regions $r^2 \leq r_-^2$ and $r^2 \geq r_+^2$. We would like the metric (6.69) to describe a compact manifold, and so we would like this metric to be cut off by the 'horizon' at some value of r. It is clear that this can only be the smaller root r_-^2, so that the range of r^2 is then $r^2 \in [0, r_-^2]$. In order for the function g^2 to be nonnegative in this region, we must require $r_-^2 \leq 1$. It is then not hard to check that the allowed region of λ for which the smaller root r_-^2 is less or equal to unity is $\lambda \in [3/4, 1]$.

It is instructive to see what happens to (6.69) at both ends of this interval. When $\lambda = 3/4$ we have

$$Q^2\Big|_{\lambda=3/4} = \frac{1}{4}(1-r^2)^2, \tag{6.73}$$

and the metric takes the following form

$$ds^2\Big|_{\lambda=3/4} = \frac{4dr^2}{1-r^2} + (1-r^2)(\sigma_1^2 + \sigma_2^2 + \sigma_3^2). \tag{6.74}$$

Introducing a new coordinate $r = \cos(\theta)$, the metric takes the form

$$ds^2\Big|_{\lambda=3/4} = 4d\theta^2 + \sin^2(\theta)(\sigma_1^2 + \sigma_2^2 + \sigma_3^2). \tag{6.75}$$

The metric

$$\sigma_1^2 + \sigma_2^2 + \sigma_3^2 = (d\psi + \cos\theta d\phi)^2 + d\theta^2 + \sin^2\theta d\phi^2 \tag{6.76}$$

should be compared with the metric (1.72) on three-sphere in Hopf coordinates. Rewritten in terms of ψ, θ, and ϕ coordinates the Hopf metric becomes

$$ds_{S^3}^2 = \left(d\psi + \frac{1}{2}(1+\cos\theta)d\phi\right)^2 + \frac{1}{4}(d\theta^2 + \sin^2\theta d\phi^2). \tag{6.77}$$

Thus, we see that

$$\sigma_1^2 + \sigma_2^2 + \sigma_3^2 = 4ds_{S^3}^2 \tag{6.78}$$

if the ψ coordinate is taken with period 4π rather than 2π. Indeed, writing $d\psi/2$ in brackets in (6.77) in place of $d\psi$ changes the period of ψ to 4π.

This ψ coordinate can then be combined with ϕ into a new coordinate and the Hopf metric on unit S^3 takes the form of the quarter of (6.76). The metric (6.75) is then clearly four times the metric on the four-sphere of unit radius.

On the other end of the interval, when $\lambda = 1$, the root r_-^2 becomes zero, and the metric completely degenerates. Let us now consider intermediate values of λ. Then, the metric is cut off at the horizons located at $\pm r_-$. Near these horizons, e.g., near the one located at $r = r_-$, introducing the coordinate $\epsilon = r_- - r$ we have

$$Q^2 = (-2r_-\epsilon)\left(\frac{2}{3}r_-^2 + (1-2\lambda)\right). \tag{6.79}$$

Then, near this horizon the metric takes the following form

$$ds^2 = \frac{d\epsilon^2}{\xi^2\epsilon} + 4\xi^2\epsilon\sigma_1^2 + (1-r_-^2)(\sigma_2^2+\sigma_3^2), \tag{6.80}$$

where

$$\frac{1}{\xi^2} = \frac{1-r_-^2}{(-2r_-)(2r_-^2/3+1-2\lambda)}. \tag{6.81}$$

Or, introducing a new coordinate $R = 2\sqrt{\epsilon}/\xi$ we get

$$ds^2 = dR^2 + \xi^4 R^2\sigma_1^2 + (1-r_-^2)(\sigma_2^2+\sigma_3^2). \tag{6.82}$$

With the period of the ψ being 4π, there is no conical singularity at the horizon if $\xi^2 = 1/2$. This fixes the value of λ to be, numerically

$$\lambda \approx 0.933, \tag{6.83}$$

with the corresponding value of r_- being

$$r_- \approx 0.281. \tag{6.84}$$

Thus, the radial coordinate of a complete metric in the family (6.69) ranges in $r \in [-0.281, 0.281]$, with the value of λ given by (6.83).

Having fixed the metric, we can proceed with the determination of the SD connection. The SD 2-forms are

$$\begin{aligned}\Sigma^1 &= 2dr\sigma_1 - g^2\sigma_2\sigma_3,\\ \Sigma^2 &= g^2Q^{-1}dr\sigma_2 - 2Q\sigma_3\sigma_1,\\ \Sigma^3 &= g^2Q^{-1}dr\sigma_3 - 2Q\sigma_1\sigma_2.\end{aligned} \tag{6.85}$$

The SD part of the Levi–Civita connection is given by

$$A^1 = \alpha\sigma_1, \quad A^2 = \beta\sigma_2, \quad A^3 = \beta\sigma_3, \tag{6.86}$$

where

$$\alpha = 1 - \left(\frac{Q^2}{g^2}\right)' - \frac{2Q^2}{g^4}, \quad \beta = \frac{Q}{g^2}\left(1 - \frac{(g^2)'}{2}\right). \tag{6.87}$$

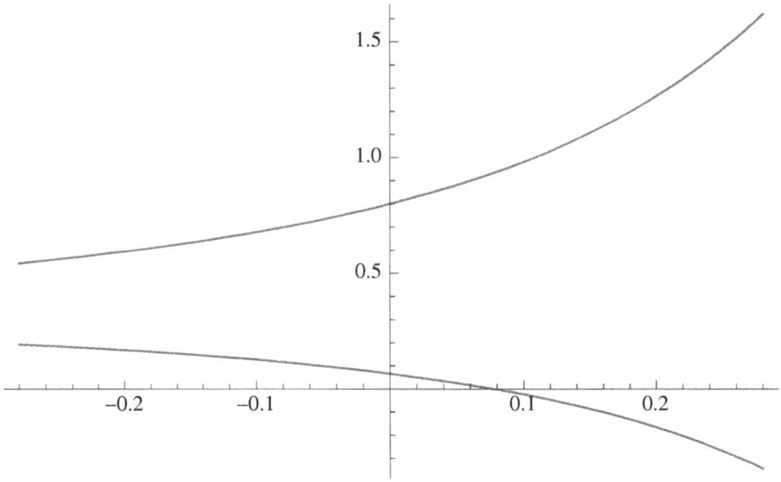

Figure 6.2 Plot of the different eigenvalues of the curvature endomorphism matrix for the Page metric. Two of the eigenvalues cross zero. Thus, the corresponding connection is only semi-definite.

The curvature is given by

$$F^i = M^{ij}\Sigma^j, \qquad M^{ij} = \text{diag}(M_1, M_2, M_3), \tag{6.88}$$

where

$$M_1 = \frac{1}{2}\alpha' = \frac{\lambda}{3} + \frac{2(4\lambda - 3)}{3(1-r)^3}, \quad M_2 = M_3 = \frac{Q}{g^2}\beta' = \frac{\lambda}{3} - \frac{1(4\lambda - 3)}{3(1-r)^3}. \tag{6.89}$$

It is clear that the first of the curvature matrix eigenvalues remains positive, while the two other eigenvalues pass through zero. Thus, it is clear that the SD connection for the Page metric is definite in a large region near the horizon $r = -r_-$, where all three eigenvalues of the curvature matrix $\Psi + (\Lambda/3)\mathbb{I}$ are positive. There is then a surface in the manifold, topologically an S^3, occurring at a fixed value of r, where two of the eigenvalues become zero. They change signs on the opposite side of this surface, with the connection again being definite in some region near the other horizon $r = r_-$. This behaviour of the curvature matrix is illustrated in Figure 6.2.

We can now relate this example to our previous discussion as to which branch of the square root of the matrix X to take. First, the matrix $X \sim M^2$ is the square of the matrix M. We have seen that the branch of $\sqrt{X}$ in which the smaller in modulus eigenvalues $\sqrt{\lambda_1}$ and $\sqrt{\lambda_2}$ need to be taken negative becomes possible when the inequality (6.68) is satisfied. In our case $\sqrt{\lambda_i} \sim M_i$, the largest in modulus eigenvalue is M_1, and so

$$M_1 - M_2 - M_3 = -\frac{\lambda}{3} + \frac{4(4\lambda - 3)}{3(1-r)^3}. \tag{6.90}$$

One can check that this is positive for all $r \in [-r_-, r_-]$, and equal to unity when $M_2 = M_3 = 0$. Thus, in the case of the Page metric, the inequality (6.68) is satisfied throughout M, both in the region when $\Psi + (\Lambda/3)\mathbb{I}$ is positive-definite and in the region when two of its eigenvalues changed signs. This gives an explicit example of a situation when it is impossible to decide which branch of the square root of X should be taken just by looking at the matrix X itself. Both branches are possible throughout M, and both are actually realised.

6.2.2 Page Metric via the Pure Connection Route

In this section we would like to obtain the Page metric by solving equations for the connection. Thus, we start with a connection of the form (6.86), with some functions α and β that only depend on the radial coordinate. The curvatures read

$$\begin{aligned} F^1 &= \alpha' dx \sigma_1 - (\alpha - \beta^2)\sigma_2\sigma_3, \\ F^2 &= \beta' dx \sigma_2 - \beta(1-\alpha)\sigma_3\sigma_1, \\ F^3 &= \beta' dx \sigma_3 - \beta(1-\alpha)\sigma_1\sigma_2, \end{aligned} \tag{6.91}$$

where x is the coordinate with respect to which the derivatives of α and β are taken. The matrix X^{ij}, which is defined only modulo multiplication by an arbitrary function, can be taken to be

$$X^{ij} = \text{diag}(1, c, c), \tag{6.92}$$

where

$$c := \frac{\beta\beta'(1-\alpha)}{\alpha'(\alpha - \beta^2)}. \tag{6.93}$$

The equations we need to solve are

$$d^A \left(\text{Tr}\sqrt{X}\, (X^{-1/2})^{ij} F_j \right) = 0. \tag{6.94}$$

As one can check, there is only a single independent field equation that we get in this case, which can be taken to be the equation for $i = 2$. Dividing this by β, which we thus assume to be nonzero (this can be checked to be true for the Page metric everywhere except at the ends) we can write

$$\left(\frac{1+2\sqrt{c}}{\sqrt{c}} \right)' (1-\alpha) + \alpha'(1+2\sqrt{c}) \left(1 - \frac{1}{\sqrt{c}} \right) = 0. \tag{6.95}$$

Here, $\sqrt{c}$ is some choice of the square root, and our aim is in particular to clarify how things depend on this choice. This equation can be solved for α as a function of $\sqrt{c}$. Let us therefore introduce a new coordinate

$$x := \sqrt{c}. \tag{6.96}$$

We have

$$\frac{\alpha'}{1-\alpha} = \frac{1}{x(x-1)(1+2x)}, \tag{6.97}$$

and thus,

$$1-\alpha = K_1 x \left(\frac{1}{(1-x)(1+2x)^2} \right)^{1/3}, \tag{6.98}$$

where K_1 is the integration constant. Knowing $\alpha(\sqrt{c})$ we can find $\beta(\sqrt{c})$ from (6.93). The equation to be solved reads

$$\frac{1}{2}(\beta^2)' + \frac{x}{(x-1)(1+2x)}\beta^2 = \frac{\alpha x}{(x-1)(1+2x)}, \tag{6.99}$$

with the solution being

$$\beta^2 = \alpha - \frac{K_1}{(1-x)^{1/3}(1+2x)^{2/3}} + \frac{K_2}{(1-x)^{2/3}(1+2x)^{1/3}}, \tag{6.100}$$

where K_2 is another integration constant. This solves the problem of finding the connection, albeit in terms of a not very geometric coordinate $\sqrt{c}$.

To be able to recover the Page metric in its form (6.69) we need to write down the metric defined by the connection (6.86). We look for the metric in the form

$$ds^2 = N^2 dx^2 + a_1^2\sigma_1^2 + a_2^2\sigma_2^2 + a_3^2\sigma_3^2, \tag{6.101}$$

whose SD 2-forms are

$$\Sigma^1 = N a_1 dx \sigma_1 - a_2 a_3 \sigma_2\sigma_3, \qquad \text{etc.} \tag{6.102}$$

We want the curvatures (6.91) to be multiples of these 2-forms. This gives the following set of equations

$$\frac{\alpha'}{Na_1} = \frac{\alpha-\beta^2}{a_2a_3}, \quad \frac{\beta'}{Na_2} = \frac{\beta(1-\alpha)}{a_3a_1}, \quad \frac{\beta'}{Na_3} = \frac{\beta(1-\alpha)}{a_1a_2}. \tag{6.103}$$

The last two equations imply that $a_2 = a_3$ and

$$a_1 = N\frac{\beta(1-\alpha)}{\beta'}. \tag{6.104}$$

The first equation then gives

$$a_2^2 = a_3^2 = N^2\frac{(\alpha-\beta^2)\beta(1-\alpha)}{\alpha'\beta'}. \tag{6.105}$$

This shows that the sought metric is in the conformal class

$$ds^2 \sim \alpha'(\beta')^2 dx^2 + \alpha'\beta^2(1-\alpha)^2\sigma_1^2 + \beta\beta'(\alpha-\beta^2)(1-\alpha)(\sigma_2^2+\sigma_3^2),$$

where at this stage x is an arbitrary coordinate with respect to which the derivatives of α and β are taken. The conformal factor is fixed by requiring that its volume form is given by (6.17), which in our case is

$$\frac{1}{\lambda^2}\left(\sqrt{\alpha'(\alpha-\beta^2)} + 2\sqrt{\beta\beta'(1-\alpha)}\right)^2 \sigma_1\sigma_2\sigma_3 dx, \tag{6.106}$$

where λ is the (dimensionless) cosmological constant. This fixes the metric to be

$$\lambda ds^2 = \left(\sqrt{\alpha'(\alpha-\beta^2)} + 2\sqrt{\beta\beta'(1-\alpha)}\right)\left(\sqrt{\frac{\alpha'}{\alpha-\beta^2}}\frac{\beta'}{\beta(1-\alpha)}dx^2 + \sqrt{\frac{\alpha'}{\alpha-\beta^2}}\frac{\beta(1-\alpha)}{\beta'}\sigma_1^2 + \sqrt{\frac{\alpha-\beta^2}{\alpha'}}(\sigma_2^2+\sigma_3^2)\right). \tag{6.107}$$

Now using $\beta\beta'(1-\alpha) = c\alpha'(\alpha-\beta^2)$ together with $c = x^2$ and (6.97) we can rewrite everything in terms of x, α, and β. The metric takes the following form

$$\lambda ds^2 = \frac{\alpha-\beta^2}{\beta^2(1-x)^2(1+2x)}dx^2 + \frac{\beta^2(1-\alpha)^2(1+2x)}{x^2(\alpha-\beta^2)}\sigma_1^2 + (\alpha-\beta^2)(1+2x)(\sigma_2^2+\sigma_3^2). \tag{6.108}$$

Taking into account (6.98) and (6.100) gives the metric explicitly, as a function of the radial coordinate x. However, one still needs to fix the constants of integration to obtain a complete metric.

Let us now see how the radial coordinate of the Page metric (6.69) arises. In the Page metric, the product of the coefficients in front of dr^2 term, and in front of σ_1^2 is equal to four. Thus, we can introduce a new radial coordinate from the condition

$$\frac{(1-\alpha)dx}{x(1-x)} = \pm 2dr, \tag{6.109}$$

where any choice of the sign can be taken. In the case of Page metric, the quantity x is the ratio of two eigenvalues of $\Psi + (\Lambda/3)\mathbb{I}$, and decra closed manifold the decreasing x corresponds to increasing r, and motivated by this, we take the negative sign. Taking into account (6.98) gives

$$\frac{K_1}{\lambda}\left(\frac{1+2x}{1-x}\right)^{1/3} = 2(\text{const} - r), \tag{6.110}$$

where an arbitrary integration constant appears. In the case of the Page metric, this constant is unity, and $K_1 = 2\lambda^{2/3}(4\lambda-3)^{1/3}$, and we make the same choice. We are free to do this, as at this stage, this is just a definition of a new radial coordinate, which we can choose any way we like. This gives $x = M_2/M_1$, with M_1 and M_2 given by (6.89). To fix the metric completely we choose K_2 to be equal what it is for the Page metric, i.e., $K_2 = \lambda^{1/3}(4\lambda-3)^{2/3}$. With these choices we get

$$(\alpha-\beta^2)(1+2x) = \lambda(1-r^2), \quad \beta^2 = \frac{Q^2}{(1-r)^2}, \qquad \frac{(1-\alpha)^2(1+2x)^2}{x^2} = 4\lambda^2(1-r)^2, \tag{6.111}$$

and the Page metric is reproduced.

6.3 Pure Connection Description of Gravitational Instantons

Gravitational instantons are Euclidean signature Einstein metrics whose Weyl curvature is chiral, i.e., only one of the two halves of the Weyl curvature is nonvanishing. The chiral formalism developed in this and the previous chapter encodes the metric in a triple of 2-forms that reconstruct the metric from the requirement that they are to become SD with respect to it. In particular, in the chiral pure connection description, the metric is encoded into the curvature of an SO(3) connection. We will now see that this formalism allows for a very simple description of the ASD Einstein metrics, i.e., the metrics for which the SD part of the Weyl curvature is vanishing.

6.3.1 Perfect Connections

We restrict our attention to the Euclidean signature. Thus, we consider an SO(3) connection on a vector bundle that is in the same topological class as the bundle of SD 2-forms for some metric on M. The topological class of this bundle is independent of the metric, and is the property of M itself.

Definition 6.1 An SO(3) connection is called **perfect** if its curvature satisfies

$$F^i F^j \sim \delta^{ij}.$$

Given a perfect connection, let us define a set of Lie algebra valued 2-forms

$$\Sigma^i_F := \frac{3}{\Lambda} F^i. \tag{6.112}$$

These 2-forms satisfy $\Sigma^i_F \Sigma^j_F \sim \delta^{ij}$ by the perfectness of the connection. They also satisfy $d^A \Sigma^i_F = 0$ by the Bianchi identity for the curvature. Thus, we can apply to such 2-forms the general statements of the previous chapter and conclude that Σ^i_F define a metric and, moreover, the connection A^i is the SD part of the spin connection for that metric. But then the definition of Σ^i_F becomes the statement that the curvature of the SD part of the spin connection is SD as the 2-form, which is the Einstein condition, and moreover, the SD part of the Weyl curvature tensor vanished. Thus, perfect connections describe ASD Einstein metrics of nonzero scalar curvature.

Thus, in the previous description of gravitational instantons, the only equations that need to be solved are $F^i F^j \sim \delta^{ij}$, which are first-order partial differential equations (PDEs) on an SO(3) connection. A solution to this first-order PDEs is then automatically a solution to second-order chiral pure connection PDEs (6.23) that give Einstein connections. This is all similar to what happens in the instanton sector of the Yang–Mills theory.

6.3.2 Analogy with Self-Dual Yang–Mills Theory

SD Yang–Mills theory is a modification of the full Yang–Mills theory that keeps only the so-called SD solutions of the Yang–Mills theory field equations. The field equations of the SD Yang–Mills theory read

$$F_+ = 0, \tag{6.113}$$

where the plus denotes the SD projection of the Yang–Mills field strength. This equation says that at most, the ASD part of the field strength is nonvanishing. A Yang–Mills theory connection that satisfies this first order in derivatives field equation also satisfies the full Yang–Mills theory second-order field equation $d_A^\mu F^a_{\mu\nu} = 0$, by Bianchi identity. Indeed, $F_+ = 0$ means that the curvature 2-form is ASD and thus, $F^a_{\mu\nu} = -(1/2)\epsilon_{\mu\nu}{}^{\rho\sigma} F^a_{\rho\sigma}$. But this means that the Yang–Mills theory field equation is satisfied by the Bianchi identity for $F^a_{\mu\nu}$.

Next, one can show that on a closed manifold the Yang–Mills action is bounded from above by the Pontryagin number for the corresponding gauge bundle. Indeed, the action is

$$S_{\text{YM}}[A] = -\frac{1}{4}\int (F^a_{\mu\nu})^2 = -\frac{1}{4}\int (F^a_{+\,\mu\nu})^2 + (F^a_{-\,\mu\nu})^2. \tag{6.114}$$

It is then clear that

$$S_{\text{YM}}[A] \le \frac{1}{4}\int (F^a_{+\,\mu\nu})^2 - (F^a_{-\,\mu\nu})^2 = \frac{1}{2}\int F^a F^a, \tag{6.115}$$

with the equality if and only if the connection is a Yang–Mills instanton $F_+ = 0$. Thus, Yang–Mills instantons are global maxima of the Yang–Mills action.

The analogy with gravitational instantons is clear. Perfect SO(3) connections are those satisfying the first-order perfectness PDEs. They are automatically solutions of the second-order PDEs (6.23) and thus, are, in particular, Einstein connections. Moreover, as we have seen in (6.59), these connections are also global maxima of the chiral pure connection action. All these statements exactly mimic what happens in the case of Yang–Mills instantons.

6.3.3 Action Principle for Perfect Connections

There is a simple action principle that gives the desired pure connection description of instantons field equations. The action is given by

$$S_{\text{inst}}[A, \Psi] = \int \Psi^{ij} F_i F_j, \tag{6.116}$$

where Ψ^{ij} is a symmetric tracefree matrix. Varying with respect to Ψ^{ij} we get the desired perfectness condition for the connection. Varying with respect to the connection we get the field equations for the Lagrange multiplier fields

$$d^A(\Psi^{ij} F_j) = 0. \tag{6.117}$$

It is interesting to note that this gives a polynomial in the fields action principle with at most quintic interaction. Also, because the action is linear in one of the fields one knows from general principles that the corresponding quantum theory is one-loop exact. Indeed, it is easy to convince oneself that it is impossible to construct any diagrams with more than one loop in such a theory. As was discussed in Krasnov (2017b), this quantum theory is, in fact, quantum finite in that the divergences that are possible at one loop are removed by field redefinitions. This gives an interesting and rare example of a gravitational theory that is quantum finite.

Another remark is that we can rewrite the action (6.116) in the general form (6.2) with

$$f_{\text{inst}}(M) = \text{Tr}(M^{-1}). \tag{6.118}$$

Thus, given that a general $f(M)$ corresponds to a modified theory of gravity, with see that there is a specific choice in which the effect of modification is to allow only the ASD Einstein solutions to the GR field equations.

Yet another remark is that it is possible to think about the full GR in formulation (6.1) as the instanton theory (6.116) plus additional interactions. Indeed, let us expand the action (6.1) in powers of Ψ. We have

$$S[A,\Psi] = \frac{3}{16\pi G\Lambda\sqrt{\sigma}}\int\left(\delta_{ij} - \frac{3\Psi_{ij}}{\Lambda} + \sum_{n=2}^{\infty}\left(\frac{-3}{\Lambda}\right)^n(\Psi^n)_{ij}\right)F^iF^j.$$

The first term here is topological; the second term is the instanton action (6.116), while all other terms introduce additional interactions among the fields already present in the instanton theory.

6.3.4 Example: Fubini-Study Metric

Let us see how the pure connection formalism can be used to obtain the Fubini–Study metric on $\mathbb{C}P^2$. We first describe the metric in the usual formalism, and then show how it arises by solving the equations $F^iF^j \sim \delta^{ij}$.

Fubini–Study metric is Kähler, which, in particular, means that it can be written in the form

$$ds^2 = \frac{\partial^2 K}{\partial\zeta^A\partial\bar{\zeta}^{\bar{A}}}d\zeta^A d\bar{\zeta}^{\bar{A}}, \tag{6.119}$$

where $\zeta^A, A = 1,2$ are the complex coordinates and

$$K = \frac{6}{\Lambda}\log\left(1 + \frac{\Lambda}{6}(|\zeta^1|^2 + |\zeta^2|^2)\right) \tag{6.120}$$

is the Kähler potential. Introducing the radial coordinate r and the Euler angles ψ, θ, ϕ via

$$\zeta^1 = r\cos(\theta/2)e^{\mathrm{i}(\psi+\phi)/2}, \qquad \zeta^2 = r\sin(\theta/2)e^{\mathrm{i}(\psi-\phi)/2} \tag{6.121}$$

with the range $\psi \in [0, 4\pi], r \in [0, \infty]$ and the usual range for the spherical angles θ, ϕ the metric can be written in the following form

$$ds^2 = \frac{dr^2}{Q^2} + \frac{r^2}{4Q^2}\sigma_1^2 + \frac{r^2}{4Q}\left(\sigma_2^2 + \sigma_3^2\right), \tag{6.122}$$

where $\sigma_{1,2,3}$ are the standard 1-forms on S^3 given by (6.71) and

$$Q = 1 + \frac{\Lambda r^2}{6}. \tag{6.123}$$

6.3.4.1 Chiral Half of the Spin Connection: SD

We would now like to confirm that the above metric is an instanton in the sense of one of its chiral halves of the spin connection being perfect. We will see that this is only true for one of the two chiral halves. So, we compute both the SD and the ASD parts of the spin connection. Alternatively, this can be phrased by saying that we compute the SD connection first with the standard choice of the orientation, and then by reversing the orientation. Reversing the orientation makes the ASD forms SD.

Let us start with the orientation choice we have been using previously. The basis of SD 2-forms is

$$\Sigma^1 = \frac{r}{2Q^2} dr\sigma_1 - \frac{r^2}{4Q}\sigma_2\sigma_3, \quad \Sigma^2 = \frac{r}{2Q^{3/2}} dr\sigma_2 - \frac{r^2}{4Q^{3/2}}\sigma_3\sigma_1, \tag{6.124}$$

and similarly for Σ^3. Starting with an ansatz

$$A^1 = \alpha\sigma^1, \qquad A^2 = \beta\sigma_2, \qquad A^3 = \beta\sigma_3 \tag{6.125}$$

one finds

$$\alpha = \frac{r^2\Lambda}{4Q}, \qquad \beta = 0. \tag{6.126}$$

This means that the only nonvanishing curvature component is

$$F^1 = \alpha' dr\sigma_1 - \alpha\sigma_2\sigma_3. \tag{6.127}$$

This is SD when $\alpha'/\alpha = 2/Qr$, which is satisfied. We have

$$F^1 = \Lambda\Sigma^1, \qquad F^2 = F^3 = 0. \tag{6.128}$$

We see that the metric (6.122) is Einstein, and we also see that the Weyl curvature in $\Psi^{22} + (\Lambda/3)$ and $\Psi^{33} + (\Lambda/3)$ exactly cancels the contribution from the scalar curvature $\Psi^{22} = \Psi^{33} = -\Lambda/3$. Because two of the curvature 2-forms vanish, one cannot recover the metric from the curvature and so we cannot describe the Fubini–Study metric using the chiral pure connection formalism with this orientation.

6.3.4.2 Chiral Spin Connection: ASD

We now compute the ASD part of the spin connection, or the SD connection but with the 'wrong' choice of the orientation. So, we instead take the following basis of 2-form

$$\Sigma^1 = \frac{r}{2Q^2} dr\sigma_1 + \frac{r^2}{4Q}\sigma_2\sigma_3, \quad \Sigma^2 = \frac{r}{2Q^{3/2}} dr\sigma_2 + \frac{r^2}{4Q^{3/2}}\sigma_3\sigma_1, \tag{6.129}$$

and similarly for Σ^3. In this case, the connection functions are found to be

$$\alpha = 1 - \frac{\Lambda r^2}{12Q}, \qquad \beta = \frac{1}{\sqrt{Q}}. \tag{6.130}$$

The first component of the curvature is then

$$F^1 = \alpha' dr\sigma_1 + (\beta^2 - \alpha)\sigma_2\sigma_3, \tag{6.131}$$

which is SD if

$$\frac{\alpha'}{\beta^2 - \alpha} = \frac{2}{rQ}, \tag{6.132}$$

which can be checked to be satisfied. One then has $F^1 = -(\Lambda/3)\Sigma^1$, which is the correct Plebański equation with $\Psi^{11} = 0$ in this orientation. Similarly, the second curvature component is

$$F^2 = \beta' dr\sigma_2 + \beta(\alpha - 1)\sigma_2\sigma_3. \tag{6.133}$$

This is SD if

$$\frac{\beta'}{\beta(\alpha - 1)} = \frac{2}{r}, \tag{6.134}$$

which can again be checked to be satisfied. Overall, we have in this orientation

$$F^i = -\frac{\Lambda}{3}\Sigma^i. \tag{6.135}$$

Thus, the Fubini–Study metric is an instanton in the sense of one of the two halves of its spin connection being perfect.

6.3.5 Pure Connection Description of Fubini–Study

We now show how to recover the Fubini–Study metric using the connection formalism. We start with the ansatz (6.125). The curvature components are given in (6.131) and (6.133) and the equations $F^iF^j \sim \delta^{ij}$ take the form

$$\alpha'(\beta^2 - \alpha) = \beta\beta'(\alpha - 1). \tag{6.136}$$

This integrates to

$$\beta^2 = \kappa(1 - \alpha)^2 + 2\alpha - 1, \tag{6.137}$$

where κ is the integration constant. We have already computed the metric described by connection (6.125) in our discussion of the Page metric; see (6.107). Because we are now matching the curvature components to the basis of SD 2-forms in the 'wrong' orientation, the signs in terms $\alpha - \beta^2$ and $1 - \alpha$ in (6.107) must be reversed. If we take α as the radial coordinate, we get the following metric

$$-(\Lambda/3)ds^2 = \frac{\beta^2 - \alpha}{\beta^2(\alpha - 1)^2}d\alpha^2 + \frac{\beta^2(\alpha - 1)^2}{\beta^2 - \alpha}\sigma_1^2 + (\beta^2 - \alpha)(\sigma_2^2 + \sigma_3^2).$$

For a general value of κ this is what is known as the Taub-Newman-Unti-Tamburino (Taub–NUT) metric, and can be put in the standard Taub–NUT form by an appropriate choice of the radial coordinate. For $\kappa = 1$ this metric gives us the metric on the four-sphere. Indeed, in this case $\beta^2 = \alpha^2$ and the metric reduces to

$$(\Lambda/3)ds^2 = \frac{d\alpha^2}{\alpha(1-\alpha)} + \alpha(1-\alpha)(\sigma_1^2 + \sigma_2^2 + \sigma_3^2).$$

Introducing a new coordinate

$$\alpha = \frac{1}{2}(1 + \cos\theta) \tag{6.138}$$

the metric becomes

$$(\Lambda/3)ds^2 = d\theta^2 + \frac{1}{4}\sin^2\theta(\sigma_1^2 + \sigma_2^2 + \sigma_3^2), \tag{6.139}$$

which is the metric on the unit four-sphere.

For $\kappa = 0$ we obtain the Fubini–Study metric. Indeed, in this case the metric is

$$(\Lambda/3)ds^2 = \frac{d\alpha^2}{(2\alpha-1)(1-\alpha)} + (2\alpha-1)(1-\alpha)\sigma_1^2 + (1-\alpha)(\sigma_2^2 + \sigma_3^2).$$

Introducing a new coordinate

$$\alpha = 1 - \frac{\Lambda r^2}{12Q}, \tag{6.140}$$

where Q is given by (6.123) the metric becomes (6.122).

6.4 First-Order Chiral Connection Formalism

The chiral pure connection action (6.8) leads to a second order in derivatives field equations. It also contains the difficulty of defining the square root in the action. As we have discussed, this action is certainly defined perturbatively around any given background. One can also use this formulation to explicitly solve Einstein equations, as we have demonstrated on the example of the Page metric. However, the necessity to worry about how to choose a branch of the square root of the matrix X makes this formalism not ideal. Also, the pure connection formalism is only available when the cosmological constant is nonzero.

In this section we will advocate a first-order version of the chiral pure connection formalism. The field equations that it leads to are first-order in derivatives and as the result there are extra fields as compared to the pure connection case. But there are no awkward matrix square roots to deal with, and in this sense, the new formalism is much easier to deal with. We will also see that the procedure of solving Einstein equations in this first-order formalism is in some cases much simpler than that in the second-order one. Finally, the new formalism is available even when $\Lambda = 0$, and is still more economic that the Plebański description because the 2-form fields of the latter have been integrated out.

6.4.1 The Action and Field Equations

The new formalism is the middle point between the Plebański action with all fields Σ^i, A^i, Ψ^{ij} present, and the pure connection action that contains only the A^i field. It is obtained by integrating out the 2-form field Σ^i from Plebański action. The most convenient form of the resulting action has already been stated in (6.2), which we repeat

$$S[A, M, \mu] = \frac{1}{16\pi G\sqrt{\sigma}} \int \mathrm{Tr}(M^{-1}FF) + \mu\,(f(M) - \Lambda)\,. \tag{6.141}$$

This is a functional of a connection and two Lagrange multiplier fields, a symmetric 3×3 matrix M^{ij}, and another 4-form valued μ, which imposes the constraint that some gauge-invariant function of M^{ij} (e.g., the trace in the case of GR) is not dynamical.

The field equations that one obtains by varying the action are as follows

$$d^A(M^{-1}F) = 0, \qquad M^{-1}XM^{-1} = \mu\frac{\partial f}{\partial M}, \tag{6.142}$$

where $X = FF$. The first of these equations should be viewed as a first-order differential equation on M, while the second determines X in terms of M and can then, in principle, be integrated to obtain the connection. We will later see that this interpretation is actually a good strategy for solving the system of equations (6.142) system of equations in many examples. In the case of GR, the matrix of partial derivatives of $f(M)$ appearing on the right-hand side of the second equation is equal to the identity matrix. We note that (6.142) is a very compact and elegant way of writing Einstein equations. We also note that nothing prevents us from setting $\Lambda = 0$ in the action (6.141).

When $\Lambda \neq 0$, the matrix M can be solved for from the second equation, and the solution substituted to the first, resulting in the pure connection description. But the idea is not to do this too soon, and instead solve the first equation in (6.142) treating the components of M as independent fields. If this is possible then X can be solved for in terms of M from the second equation, from where the connection can be determined. So, in cases that this programme can be realised, we get an efficient strategy for solving Einstein equations. Of course, in general, things are not so simple and what one gets is a coupled system of first-order differential equations for both M and the connection. But there are examples in which this strategy works, as we shall now describe.

6.5 Example: Bianchi I Connections

In this section, we consider the example of spatially homogeneous anisotropic universes with flat spatial slices, the so-called Bianchi I setup.

6.5.1 Connections and Curvature

We consider the Lorentzian signature GR and start with the following ansatz for the connection

$$A^1 = \mathrm{i}h_1(\tau)dx^1, \qquad \text{etc.} \tag{6.143}$$

Here, $h_i(\tau)$ are three functions of an arbitrary time coordinate τ, while x^i are the Cartesian coordinates on the spatial slices (surfaces of homogeneity). The corresponding curvature 2-form is

$$F^1 = \mathrm{i}\dot{h}_1 d\tau dx^1 - h_2 h_3 dx^2 dx^3, \qquad \text{etc.} \tag{6.144}$$

where an overdot denotes derivative with respect to τ. Calculating the wedge product, we obtain

$$F^i F^j = 2\mathrm{i}\delta^{ij} X_i h \epsilon_c, \tag{6.145}$$

where $\epsilon_c = dx^1 dx^2 dx^3 d\tau$ is the coordinate volume form, no summation is implied in this formula, $h = h_1 h_2 h_3$, and

$$X_i = \frac{\dot{h}_i}{h_i}. \tag{6.146}$$

If we now define $X^{ij} = F^i F^j/\epsilon$ with

$$\epsilon = 2\mathrm{i}h\epsilon_c, \tag{6.147}$$

then $X^{ij} = \mathrm{diag}\,(X_1, X_2, X_3)$.

6.5.2 Evolution Equations in the Pure-Connection Parametrisation

There is no difficulty in considering the most general class of theories at least in the first steps. The reason for the choice (6.147) of the volume form defining the matrix X^{ij} is that the pure-connection formulation equation (6.42) reduces to the system

$$\left(\frac{\partial g}{\partial X_i}\right)^{\cdot} = g(X) - \frac{\partial g}{\partial X_i}\sum_j X_j, \tag{6.148}$$

which is a system of first-order differential equations for X_i. Specialising to the case of the function $g(X)$ given by (6.30), it is not hard to obtain the familiar GR solution. We will, however, obtain the solution in a simpler way working in the formulation with auxiliary fields.

An alternative form of equations (6.148) is obtained by multiplying these equations by $h = h_1 h_2 h_3$ and using definition (6.146). Equations (6.148) then reduce to

$$\left(\frac{\partial g}{\partial X_i} h\right)^{\cdot} = g(X)h. \tag{6.149}$$

This form is convenient for analysing the case of arbitrary $g(X)$.

We note that the field equations (6.149) can be obtained both directly by substituting the ansatz for the connection into (6.42), or from an action principle. Indeed, on our ansatz the action (6.29) reduces to

$$S \sim \int d\tau g(X_1, X_2, X_3)h. \tag{6.150}$$

The variation of this action with respect to, e.g., h_1 is

$$\delta S \sim \int d\tau \frac{\partial g}{\partial X_1}\left(\frac{\delta \dot{h}_1}{h_1} - \frac{X_1}{h_1}\delta h_1\right) h + g(X)\frac{\delta h_1}{h_1}h. \tag{6.151}$$

This gives the following equation of motion

$$\left(\frac{\partial g}{\partial X_1}h_2h_3\right)^{\cdot} h_1 = g(X)h - \frac{\partial g}{\partial X_1}X_1 h, \tag{6.152}$$

which can be rewritten as (6.149).

6.5.3 The Metric

Before we begin our analysis of the evolution equations, it is useful to compute the metric determined by the connection. The easiest way is to directly look for a metric that makes the curvature forms (6.144) SD.

We are looking for the metric in the Bianchi I form

$$ds^2 = -N^2(\tau)d\tau^2 + \sum_i a_i^2(\tau)\left(dx^i\right)^2. \tag{6.153}$$

This means that the basis of SD 2-forms is

$$\Sigma^1 = \mathrm{i}Na_1 d\tau dx^1 - a_2a_3dx^2dx^3, \qquad \text{etc.} \tag{6.154}$$

We now require that the curvature 2-forms are proportional to the corresponding Σ^i's. This gives

$$\frac{\dot{h}_1}{Na_1} = \frac{h_2h_3}{a_2a_3}, \qquad \text{etc.,} \tag{6.155}$$

from which we get

$$\frac{a_1^2}{N^2} = \frac{h_1^2}{X_2X_3}, \quad \text{etc.} \tag{6.156}$$

Another equation for determining the metric is obtained by fixing the metric volume form

$$\epsilon_m = Na_1a_2a_3dx^1dx^2dx^3d\tau = Na_1a_2a_3\epsilon_c. \tag{6.157}$$

By our prescription, this should be equal to a multiple of $g(X)$:

$$2\mathrm{i}\Lambda^2\epsilon_m = g(FF), \tag{6.158}$$

where Λ is the cosmological constant and for the case of GR the function $g(X)$ is given by (6.30). Using (6.145), we get

$$\Lambda^2 N a_1 a_2 a_3 = g(X) h. \tag{6.159}$$

Combining this equation with (6.156), we have

$$N^2 = \left(\frac{g(X) X_1 X_2 X_3}{\Lambda^2} \right)^{1/2}. \tag{6.160}$$

If we require that the metric be real, and that the signature of the τ coordinate be negative, then the final expression for the metric is

$$\Lambda ds^2 = \sqrt{|g(X) X_1 X_2 X_3|} \left[-d\tau^2 + \prod_j X_j^{-1} \sum_k h_k^2 X_k \left(dx^k\right)^2 \right]. \tag{6.161}$$

Note that this metric is time-reparametrisation invariant, as it should be.

6.5.4 Solution in the General Case

One of the miracles of the pure-connection formulation of gravity under consideration is that it allows one to write the *general* solution to the problem at hand for an arbitrary theory, i.e., for an arbitrary choice of the function $g(X)$. This becomes possible by using a clever choice of the time variable.

We begin with solution for the case of general $g(X)$ and then specialise to GR. Solution of GR in which one works in the physical time from the beginning is also possible, but is more involved and will not be considered. Details can be found in Herfray et al. (2016b). Let us consider the evolution equations in the form (6.149). By using time-reparametrisation freedom, it is always possible to choose the time variable τ in such a way that

$$g(X) h = \text{const}. \tag{6.162}$$

The geometric significance of this choice is that this is the time coordinate in which the metric volume form is proportional to the coordinate volume form, i.e., $\sqrt{\det g} = N a_1 a_2 a_3 = \text{const}$. This is clear from (6.159).

With this choice, equation (6.149) can be integrated to give an implicit solution for $X(\tau)$:

$$\frac{\partial g(X)}{\partial X_i} = g(X) \left(\tau - \tau_i\right), \tag{6.163}$$

where τ_i are arbitrary integration constants. The homogeneity of the function $g(X)$ implies another relation

$$\sum_i X_i \left(\tau - \tau_i\right) = 1. \tag{6.164}$$

Equations (6.163) and (6.146) give a complete solution to the problem for an arbitrary theory from our class. We now give some general analysis of the solution obtained, and then specialise to GR.

6.5.5 De Sitter Solution

Consider $\tau \to \infty$, and assume that $g(X)\tau$ remains constant as $\tau \to \infty$. Then equation (6.163) implies that all derivatives $\partial g(X)/\partial X_i$ become mutually equal. The symmetry of the function $g(X)$, in turn, implies that all X_i become equal to each other in this limit. Relation (6.164) then gives the solution

$$X_i \approx \frac{1}{3\tau} \quad \text{as} \quad \tau \to \infty. \tag{6.165}$$

The homogeneity of $g(X)$ then justifies the assumption $f(X)\tau \to$ const that we made in deriving this solution.

The corresponding metric describes the de Sitter spacetime. Indeed, we have $g(X) = g_0/\tau$, where g_0 is a constant. Then, by rescaling the spatial coordinates, we can always choose the solution in the form $h_i = \tau^{1/3}$. Then metric (6.161) becomes

$$ds^2 = \sqrt{\frac{3f_0}{\Lambda^2}} \left(-\frac{d\tau^2}{9\tau^2} + \tau^{2/3} dr^2 \right) = \sqrt{\frac{3f_0}{\Lambda^2}} \left(-dt^2 + e^{2t} dr^2 \right), \tag{6.166}$$

where $\tau = e^{3t}$ is the time coordinate change, and $dr^2 = \sum_i (dx^i)^2$. This is nothing but the de Sitter metric, which is thus the solution of theory with any $g(X)$.

6.5.6 Integration Constants

Without loss of generality, one can shift the time variable so that

$$\sum_i \tau_i = 0. \tag{6.167}$$

Second, apart from the trivial case $\tau_i = 0$ for all i, which gives the de Sitter solution, by the remaining freedom of time rescaling, which does not violate (6.162), we can achieve the condition

$$\sum_i \tau_i^2 = 2. \tag{6.168}$$

This normalization is convenient because squaring (6.167) we can rewrite (6.168) as

$$\tau_1\tau_2 + \tau_2\tau_3 + \tau_3\tau_1 = -1. \tag{6.169}$$

Without loss of generality, we can arrange the integration constants so that

$$\tau_3 \le \tau_2 \le \tau_1. \tag{6.170}$$

Because of condition (6.167), we have $\tau_3 < 0 < \tau_1$. When $\tau_2 = \tau_1$, we have $\tau_2 = \tau_1 = 1/\sqrt{3}$ and $\tau_3 = -2/\sqrt{3}$. This is the largest absolute value that τ_3 can reach. In the opposite extreme $\tau_2 = \tau_3$ we have $\tau_2 = \tau_3 = -1/\sqrt{3}$ and $\tau_1 = 2/\sqrt{3}$, which is the largest value τ_1 can reach. All in all, we have

$$\tau_c \leq \tau_1 \leq 2\tau_c, \qquad -\tau_c \leq \tau_2 \leq \tau_c, \qquad -2\tau_c \leq \tau_3 \leq -\tau_c, \tag{6.171}$$

where $\tau_c = 1/\sqrt{3}$.

6.5.7 The Case of GR

We have, in general,

$$g(X) = \Lambda \mathrm{Tr}(M^{-1}X), \tag{6.172}$$

and therefore, $\partial g/\partial X = \Lambda M^{-1}$. Thus, the solution (6.163) becomes

$$M_1 = \frac{\Lambda}{g(X)(\tau - \tau_1)}, \qquad \text{etc.} \tag{6.173}$$

Here $g(X)$ needs to be determined from the constraint $f(M) = \Lambda$. In the case of GR this gives

$$g(X) = \sum_i \frac{1}{\tau - \tau_i} = \frac{3\tau^2 - 1}{\prod_i (\tau - \tau_i)} \equiv s_1, \tag{6.174}$$

which gives

$$M_1 = \frac{\Lambda}{s_1(\tau - \tau_1)}, \qquad \text{etc.} \tag{6.175}$$

This determines the auxiliary matrix M completely. We now determine the components of the matrix X in terms of those of M using the equations $M^{-1}XM^{-1} = \mu\mathbb{I}$. This gives $X_i = \mu M_i^2$. The Lagrange multiplier μ is determined from (6.172), which gives $\mu = g(X)/\Lambda^2$. Overall, we have

$$X_1 = \frac{1}{s_1(\tau - \tau_1)^2} = \frac{(\tau - \tau_2)(\tau - \tau_3)}{(3\tau^2 - 1)(\tau - \tau_1)}, \qquad \text{etc.} \tag{6.176}$$

The quantities X_i^{GR} have simple poles at $\tau = \tau_i$, and all blow up as $\tau \to \pm 1/\sqrt{3}$, which corresponds to the Kasner singularity. This behaviour is illustrated in Figure 6.3.

6.5.8 Solution for the Metric

Let us also write the corresponding metric components; see (6.161). We have $g(X) = s_1$ and so

$$g(X)X_1X_2X_3 = \frac{1}{(3\tau^2 - 1)^2}, \qquad \frac{g(X)X_1}{X_2X_3} = \frac{(3\tau^2 - 1)^2}{(\tau - \tau_1)^4}, \tag{6.177}$$

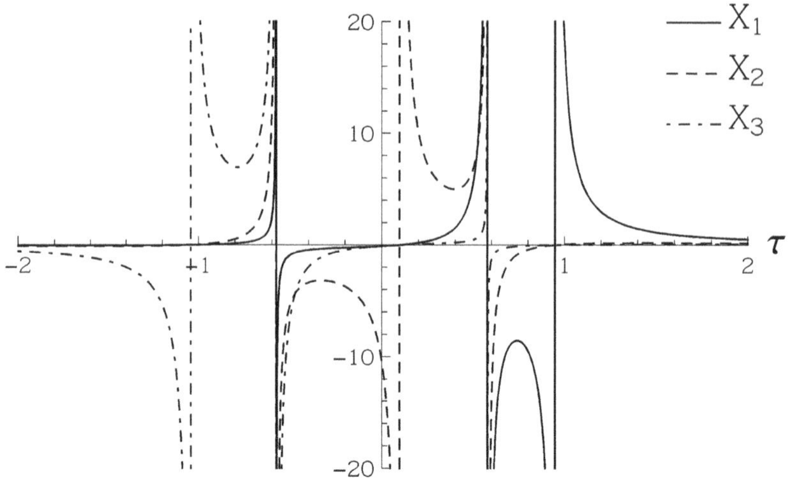

Figure 6.3 Plots of the components of the matrix X.

and similarly for the other components. All expressions are manifestly positive, so taking the square root, we have

$$N^2 = \frac{1}{\Lambda(3\tau^2 - 1)}, \qquad a_1^2 = h_1^2 \frac{3\tau^2 - 1}{\Lambda(\tau - \tau_1)^2}. \tag{6.178}$$

In the time interval $\tau \in (-\tau_c, \tau_c)$, $\tau_c = 1/\sqrt{3}$, instead of taking the modulus of expressions to get nonnegative metric components, we reverse the sign of the cosmological constant Λ. This is the correct interpretation, as this time interval corresponds to a solution of GR with negative cosmological constant.

We now study this solution in more detail, and, in particular, integrate the equations for h_i near the singularity.

6.5.9 Behaviour Near the Poles

When all three integration constants are different, the function $s_1(\tau)$ has three simple poles at $\tau = \tau_i$, and two simple zeros at $\tau = \pm\tau_c$. Let us analyse the behaviour near the poles.

Consider, for example, the limit $\tau \to \tau_1$. In this case, we have $s_1 \sim 1/(\tau - \tau_1) \to \infty$. Solution (6.176) behaves as

$$X_1 \approx \frac{1}{\tau - \tau_1}, \qquad X_2 \sim X_3 \sim \tau - \tau_1. \tag{6.179}$$

This is an integrable behaviour, with $h_1 \to 0$ and h_2 and h_3 finite as $\tau \to \tau_1$. We thus see that all X_i change sign at $\tau = \tau_1$.

Let us determine the behaviour of the components (6.160) of the canonical metric (6.161) at this point. Integrating the first equation in (6.179), we obtain

$$h_1 \sim \tau - \tau_1, \tag{6.180}$$

while h_2 and h_3 tend to constants. So, the significance of the point $\tau = \tau_1$ is in the fact that one of the connection components h_1 passes through zero there.

Now, using this behaviour, we see that the metric lapse function as well as the scale factors (6.160) are finite and regular as $\tau \to \tau_1$. So, the $\tau = \tau_1$ is just a special point where one of the components of the connection goes to zero.

It is also interesting to analyse what happens with the components of the matrix $F^i F^j$ of curvature wedge products at this point. We know that, e.g., $F^1 F^1 = 2\mathrm{i}X_1 h\epsilon_c$, where ϵ_c is the coordinate volume element. Thus, we see that $F^1 F^1$ remains finite at $\tau = \tau_1$ because the pole in X_1 is cancelled by the zero in h_1 (recall $h = h_1 h_2 h_3$). We also see that the other two components, $F^2 F^2$ and $F^3 F^3$, vanish at $\tau = \tau_1$. We have an order two zero at this point, so that the matrix M that is proportional to the square root of X has order one zero in its components $M_2, M_3 \sim \tau - \tau_1$. This is the already familiar pattern from our consideration of the Page metric, in which two of the eigenvalues of $\sqrt{X}$ change sign so that the sign of the determinant of this matrix is unchanged. In terms of our previous discussion illustrated by Figure 6.1, the matrix M/Λ starts in region I for large positive τ and crosses to region II for $\tau_c < \tau < \tau_1$.

It is also interesting to consider what happens near $\tau = \tau_2$. Let us consider the components of the matrix M given by (6.175), which we write as

$$\frac{M_1}{\Lambda} = \frac{(\tau - \tau_2)(\tau - \tau_3)}{3\tau^2 - 1}, \quad \frac{M_2}{\Lambda} = \frac{(\tau - \tau_1)(\tau - \tau_3)}{3\tau^2 - 1}, \quad \frac{M_3}{\Lambda} = \frac{(\tau - \tau_1)(\tau - \tau_2)}{3\tau^2 - 1}.$$

We see that M_1/Λ changes sign at the singularity $\tau = \tau_c$ and is negative 'on the other side' of the singularity. The quantities $M_2/\Lambda, M_3/\Lambda$ both change signs at $\tau = \tau_1$ and then change signs again at the singularity, so they are both positive as $\tau \to \tau_c$ from the left. Thus, we have one of the eigenvalues of M/Λ negative and two positive, which means that the determinant of M/Λ is negative in the interval $\tau \in (-\tau_c, \tau_c)$. At $\tau = \tau_2$ both M_1/Λ and M_3/Λ change sign, but the determinant of M/Λ remains negative. This means that the matrix M/Λ remains in region III of Figure 6.1 for all $\tau \in (-\tau_c, \tau_c)$.

6.5.10 Behaviour Near the Singularity

At the singularity $\tau = \tau_c = 1/\sqrt{3}$, the function s_1 has a simple zero. Thus, we have

$$X_1 \approx -\frac{(\tau_c - \tau_2)(\tau_c - \tau_3)}{2\sqrt{3}(\tau_1 - \tau_c)(\tau - \tau_c)}, \quad \text{etc}, \qquad g(X) \sim \tau - \tau_c. \tag{6.181}$$

Integrating (6.146), we get

$$h_1 \sim (\tau - \tau_c)^{-\frac{(\tau_c - \tau_2)(\tau_c - \tau_3)}{2\sqrt{3}(\tau_1 - \tau_c)}}, \quad \text{etc.} \tag{6.182}$$

We thus see that the lapse function (6.178) diverges, while the scale factors behave as

$$a_i^2 \sim (\tau - \tau_c)^{p_i}, \tag{6.183}$$

where

$$p_1 = 1 - \frac{(\tau_c - \tau_2)(\tau_c - \tau_3)}{\sqrt{3}(\tau_1 - \tau_c)}, \quad \text{etc.} \tag{6.184}$$

These exponents satisfy

$$p_1 + p_2 + p_3 = 1, \qquad p_1 p_2 + p_2 p_3 + p_3 p_1 = 0. \tag{6.185}$$

From (6.178) we see that the physical time near the singularity is $t \sim \sqrt{\tau - \tau_c}$, and thus the behaviour (6.183) is the usual Kasner one,

$$a_i^2 \sim t^{2p_i}, \tag{6.186}$$

with the correct exponents (6.185).

Note that the components of the gauge field (6.182) all diverge at the singularity, so this is a true singularity not only of the canonical metric (6.161) but also of the fundamental gauge field.

6.6 Spherically Symmetric Case

The purpose of this section is to solve the Lorentzian signature spherically symmetric problem with negative cosmological constant. Again, we use the mixed first-order version of the chiral pure connection formulation.

6.6.1 Equations

We take the following spherically symmetric ansatz for the connection

$$A^1 = \mathrm{i}a(R)dt + \cos(\theta)d\phi,\ A^2 = -b(R)\sin(\theta)d\phi,\ A^3 = b(R)d\theta. \tag{6.187}$$

The curvatures are

$$\begin{aligned} F^1 &= -\mathrm{i}a' dt dR + (b^2 - 1)\sin(\theta) d\theta d\phi, \\ F^2 &= -\mathrm{i}ab\, dt d\theta + b' \sin(\theta) d\phi dR, \\ F^3 &= -\mathrm{i}ab \sin(\theta) dt d\phi + b' dR d\theta, \end{aligned} \tag{6.188}$$

where the primes denote the derivative with respect to the radial coordinate, at this stage arbitrary. The diagonal X matrix can then be taken to be

$$X_1 = \frac{a'}{a}, \qquad X_2 = X_3 = \frac{bb'}{b^2 - 1}. \tag{6.189}$$

The field equations $d^A(M^{-1}F) = 0$ can be written in the following form

$$(M_1^{-1}a(b^2-1))' = \frac{g(X)}{\Lambda}a(b^2-1),$$
$$(M_2^{-1}a(b^2-1))' = \frac{g(X)}{\Lambda}a(b^2-1), \tag{6.190}$$

where we manipulated the equations to map them into a form in which the right-hand side is the same. Here

$$\frac{g(X)}{\Lambda} = \mathrm{Tr}(M^{-1}X) = M_1^{-1}\frac{a'}{a} + M_2^{-1}\frac{(b^2-1)'}{b^2-1}. \tag{6.191}$$

6.6.2 Solution

We now choose the radial coordinate so that

$$g(X)a(b^2-1) = \Lambda. \tag{6.192}$$

The solution for the matrix M is then immediately written down

$$M_1^{-1} = \frac{g(X)}{\Lambda}(R+R_1), \qquad M_2^{-1} = \frac{g(X)}{\Lambda}(R+R_2), \tag{6.193}$$

where $R_{1,2}$ are integration constants. We remark that we want the solution with $\Lambda < 0$, so that there is an asymptotic region at spatial infinity.

6.6.3 Metric

The metric is computed from the requirement that the curvature 2-forms (6.188) are SD, and that the metric volume form is a constant multiple of the coordinate volume form. This last condition follows from our choice of the radial coordinate (6.192). Let us give some details of this calculation. We look for the metric in the standard spherically symmetric form

$$ds^2 = -f^2dt^2 + g^2dR^2 + r^2d\omega^2, \tag{6.194}$$

where $r = r(R)$ and $d\Omega^2$ is the standard area element on the unit S^2. The basis of SD 2-forms for this metric is

$$\Sigma^1 = ifgdtdR - r^2\sin\theta d\theta d\phi,$$
$$\Sigma^2 = ifrdtd\theta - gr\sin\theta d\phi dR, \tag{6.195}$$
$$\Sigma^3 = ifr\sin\theta dtd\phi - grdRd\theta.$$

We now demand that our curvature 2-forms are proportional to the corresponding Σ^i's. This gives the following equations

$$\frac{a'}{fg} = \frac{b^2-1}{r^2}, \qquad \frac{ab}{fr} = \frac{b'}{gr}. \tag{6.196}$$

These can be rewritten as

$$fg = \frac{X_1 a r^2}{b^2 - 1}, \qquad \frac{f}{g} = \frac{ab^2}{X_2(b^2-1)}, \tag{6.197}$$

and give

$$f^2 = \frac{X_1 a^2 b^2 r^2}{X_2(b^2-1)^2}, \qquad g^2 = \frac{X_1 X_2 r^2}{b^2}. \tag{6.198}$$

Another equation we need is that for the volume form. This we get by demanding that $g(FF) = 2\mathrm{i}\Lambda^2\epsilon$, where ϵ is the metric volume form $\epsilon = fgr^2\epsilon_c$, and where $\epsilon_c = \sin\theta dRd\theta d\phi dt$ is the coordinate volume element. The quantity $g(FF)$ evaluates to $g(FF) = 2\mathrm{i}g(X)a(b^2-1)\epsilon_c$. Thus, we get

$$\Lambda^2 fgr^2 = g(X)a(b^2-1). \tag{6.199}$$

Taking the square of this and substituting the expressions for f^2 and g^2 we get

$$r^8 = \frac{g(X)^2(b^2-1)^4}{X_1^2\Lambda^4} \quad \Rightarrow \quad r^2 = \sqrt{\frac{|g(X)|(b^2-1)^2}{|X_1|\Lambda^2}}. \tag{6.200}$$

We thus get the following metric

$$|\Lambda|ds^2 = \sqrt{\frac{|g(X)|(b^2-1)^2}{|X_1|}}\left(-\frac{X_1 a^2 b^2}{X_2(b^2-1)^2}dt^2 + \frac{X_1 X_2}{b^2}dR^2 + d\Omega^2\right),$$

where as usual $d\Omega^2$ is the metric on the unit sphere.

6.6.4 Solving for the Connection Components

We now specialise to the case of GR. We have the condition $\mathrm{Tr}(M) = \Lambda$. And so we take the trace of M as obtained in (6.193) to get

$$1 = \frac{1}{g(X)}\left(\frac{1}{R+R_1} + \frac{2}{R+R_2}\right) \equiv \frac{3R + 2R_1 + R_2}{g(X)(R+R_1)(R+R_2)}. \tag{6.201}$$

It is now convenient to shift the radial coordinate so as to impose

$$2R_1 + R_2 = 0. \tag{6.202}$$

We now take $R_1 \equiv \bar{R}$ to be the single parameter of the solution. This gives

$$g(X) = \frac{3R}{(R+\bar{R})(R-2\bar{R})}, \tag{6.203}$$

and thus a complete solution to the problem of determining the matrix M

$$M_1 = \frac{\Lambda(R-2\bar{R})}{3R}, \qquad M_2 = \frac{\Lambda(R+\bar{R})}{3R}. \tag{6.204}$$

We can now find the components of the matrix X. We have $M^{-1}XM^{-1} = \mu\mathbb{I}$, from which $X = \mu M^2$. But we can also write the original equation as

$M^{-1}X = \mu M$ and then take the trace. Using $\Lambda \text{Tr}(M^{-1}X) = g(X)$ we get as before $\mu = g(X)/\Lambda^2$. So, the final expressions for the quantities X_1 and X_2 are

$$X_1 = \frac{(R-2\bar{R})}{3R(R+\bar{R})}, \qquad X_2 = \frac{(R+\bar{R})}{3R(R-2\bar{R})}. \tag{6.205}$$

We now integrate the relations (6.189) to obtain the components of the connection

$$a = K_1 \frac{R+\bar{R}}{(3R)^{2/3}}, \qquad b^2 - 1 = K_2 \frac{R-2\bar{R}}{(3R)^{1/3}}, \tag{6.206}$$

where $K_{1,2}$ are integration constants. This gives a complete solution to the problem, modulo the issue of fixing (or interpreting) the integration constants $K_{1,2}$ and $\bar{R}$.

6.6.5 A Relation Between K_1 and K_2

There is a relation between the integration constants $K_{1,2}$ that follows from our gauge-fixing condition $g(X)a(b^2-1) = \Lambda$. Indeed, substituting the $g(X)$ that we have found in (6.203), as well as the connection components, we get

$$K_1 K_2 = \Lambda. \tag{6.207}$$

6.6.6 Fixing $K_{1,2}$ from the Asymptotics

Let us fix $K_{1,2}$ from the large R asymptotics of the metric as found previously. We have $X_1, X_2 = 1/3R$ asymptotically, and $g(X) = 3/R$. So, for large R the metric reads

$$\left(\frac{|\Lambda|}{3}\right) ds^2 = -K_1^2 \frac{R^{2/3}}{3^{4/3}} dt^2 + \frac{dR^2}{(3R)^2} + |K_2| \frac{R^{2/3}}{3^{1/3}} d\Omega^2. \tag{6.208}$$

We would like this to be the usual (asymptotically) hyperbolic metric

$$ds^2 = l^2 \left(-r^2 dt^2 + \frac{dr^2}{r^2} + r^2 d\Omega^2\right), \tag{6.209}$$

with

$$\frac{\Lambda}{3} = -\frac{1}{l^2}. \tag{6.210}$$

This means that asymptotically at least

$$K_1^2 \frac{R^{2/3}}{3^{4/3}} = r^2, \qquad |K_2| \frac{R^{2/3}}{3^{1/3}} = r^2, \tag{6.211}$$

and therefore,

$$K_1^2 = 3|K_2|. \tag{6.212}$$

Using (6.207) this gives

$$|K_1| = \left(\frac{3}{l}\right)^{2/3}, \qquad |K_2| = \left(\frac{3}{l^4}\right)^{1/3}, \tag{6.213}$$

and thus

$$R^2 = r^6 l^4. \tag{6.214}$$

There are two possible branches here $R = \pm r^3 l^2$, with only one of these branches giving the desired Anti-de Sitter-Schwarzschild metric, see the next section.

6.6.7 The Final Metric

The conformal factor in the metric computes to $\sqrt{(3R)^{4/3}K_2^2} = 3r^2$, which immediately gives the correct angular part of the metric. For the other terms, after numerous cancellations, we get the following metric

$$ds^2 = l^2\left(-b^2 dt^2 + \frac{dr^2}{b^2} + r^2 d\Omega^2\right). \tag{6.215}$$

In particular, it is seen that (6.214) is the relation that is valid everywhere.

We can now fix the components of the connection completely. In terms of the radial coordinate r the functions a and b^2 take the form

$$a = \mathrm{sign}(K_1)\left(\pm r + \frac{\bar{R}}{r^2 l^2}\right), \quad b^2 - 1 = \mathrm{sign}(K_2)\left(r^2 \mp \frac{2\bar{R}}{l^2 r}\right).$$

Here, the two possible signs are those of the two branches in $R = \pm r^3 l^2$. We see that the sign of K_2 must be chosen to be plus in order for the function b^2 to behave as r^2 for large r. This means that we must take K_1 to be negative (because $K_1 K_2 = \Lambda$ is negative). But then if we want the function a to behave asymptotically as r we need to take the $R = -r^3 l^2$ branch. This finally gives

$$a = r - \frac{\bar{R}}{r^2 l^2}, \qquad b^2 = 1 + r^2 + \frac{2\bar{R}}{l^2 r}. \tag{6.216}$$

This becomes the usual Euclidean Schwarzschild–AdS metric if we further replace $r \to r/l$ and identify

$$\bar{R} = -Ml, \tag{6.217}$$

where M is the black hole mass. The fact that we need both $R, \bar{R}$ negative to get the familiar solution correlates with the fact that we are considering the negative Λ case.

6.6.8 Behaviour of the Matrix M

It is interesting to consider the behaviour of the matrix M in the solution we have obtained. With the previous identifications we have

$$\frac{M_1}{\Lambda} = \frac{1}{3}\left(1 - \frac{2M}{r^3 l}\right), \qquad \frac{M_2}{\Lambda} = \frac{1}{3}\left(1 + \frac{M}{r^3 l}\right). \tag{6.218}$$

This means that asymptotically for large r the matrix M/Λ is in the region I of its parameter space; see Figure 6.1. But at $r = (2M/l)^{1/3}$ one of its eigenvalues changes sign, and the matrix crosses to region III of the parameter space. It is also interesting that for negative M, the behaviour would be different, and we would have two of the eigenvalues of M/Λ changing sign instead. But in this case there is no horizon and one has a 'naked' singularity at $r = 0$. It is interesting that the expected behaviour of matrix M/Λ (in the sense of taking values in regions I and III of the parameter space for $\Lambda < 0$) correlates with positive mass and no naked singularity condition.

6.7 Bianchi IX and Reality Conditions

The purpose of this section is to analyse the so-called Bianchi IX model using the first-order connection formalism (6.142). The Bianchi IX setup gives a very good illustration to the problem of imposing Lorentzian signature case reality conditions. In all the setups studied so far the problem of imposing the Lorentzian signature reality conditions was trivial, as it was always relatively straightforward to select the desired conditions. In particular, the same reality conditions worked for any modified theory of the type (6.2). We shall now analyse an example where this problem becomes nontrivial, and where it is no longer obvious which reality conditions to impose in the connection formalism. The correct reality conditions in the case of GR are the ones that require the metric constructed from the connection to be real Lorentzian. But we shall explicitly see that the compatibility of these conditions with the dynamics of the model is nontrivial. If one changes the theory changing $f(M)$, one changes the dynamics, and the modified dynamics is in general no longer compatible with the conditions that the metric constructed from the connection is real. This means that there is, in general, no sensible Lorentzian signature interpretation of the modified theories of the type (6.2), and these modified theories exist only in the Euclidean or split signatures. The analysis in Section 6.7.4 also illustrates how subtle the chiral connection formalism Lorentzian signature reality conditions are even in the case when they can be imposed, which is in the case of GR.

The novelty of the present setup as compared to all the cases considered previously is that the corresponding GR solution has the SD half of the Weyl curvature complex, while in all the situations considered before this chiral half of the Weyl curvature was real. It is in such setups that the problem of the connection formalism Lorentzian sector reality conditions becomes highly nontrivial.

6.7.1 Ansatz for the Bianchi IX Model

By using the gauge symmetry, we can present the connection 1-forms for Bianchi IX model in the following form:

$$A^1 = h_1\sigma_1, \quad A^2 = h_2\sigma_2, \quad A^3 = h_3\sigma_3, \tag{6.219}$$

where $\sigma_{1,2,3}$ are again the canonical 1-forms on S^3 given in (6.71) and satisfying $d\sigma_1 = -\sigma_2\sigma_3$, etc. The functions $q_{1,2,3}(t)$ are complex-valued, and part of the motivation for the exercise in this section is to understand which reality conditions need to be imposed on them. The curvature 2-forms of this connection are

$$F^1 = \dot{h}_1\, dt\sigma_1 - H_1\, \sigma_2\sigma_3, \tag{6.220}$$

as well as cyclic permutations. Here we have introduced the notation

$$H_1 := h_1 - h_2h_3 \tag{6.221}$$

and similarly with cyclic permutation of indices. The nonzero exterior products are

$$F^1F^1 = 2\dot{h}_1H_1\, dt\sigma_1\sigma_2\sigma_3, \tag{6.222}$$

and similarly for the other diagonal components of the matrix F^iF^j.

6.7.2 The Metric

We know that the reality conditions that we want to impose on the components of the connection must guarantee that the metric constructed from the curvature of this connection is real Lorentzian. So, let us compute the metric. We look for the metric in the following Lorentzian signature form:

$$g = -N^2(t)dt^2 + a_1^2(t)(\sigma_1)^2 + a_2^2(t)(\sigma_2)^2 + a_3^2(t)(\sigma_3)^2. \tag{6.223}$$

The basis of SD 2-form for the previous metric is given by

$$\Sigma^1 = \mathrm{i}Na_1dt\sigma_1 - a_2a_3\sigma_2\sigma_3, \tag{6.224}$$

and similarly for the other components. We want the curvatures (6.220) to be proportional to the basic 2-forms Σ^i, which gives the equations

$$\frac{Na_1}{a_2a_3} = \frac{\dot{h}_1}{\mathrm{i}H_1}, \tag{6.225}$$

together with cyclic permutations. Because the left-hand side of these equations is real for a real Lorentzian metric, we see that the quantities, $\dot{h}_i/H_i$, $i = 1, 2, 3$, should be purely imaginary.

6.7.3 Dynamics of the Model

We can obtain the dynamical equations for this model by substituting the ansatz (6.219) for the connection into the first-order action (6.141). If we define $X_i : \dot{q}_i H_i$ and take the matrix M^{ij} to be diagonal with diagonal entries M_i we get the following action

$$S[h, M, \mu] = \frac{1}{\mathrm{i}} \int \left[\sum_i \frac{\dot{h}_i H_i}{M_i} + \mu \left(\sum_i M_i - \Lambda \right) \right] dt. \tag{6.226}$$

We have set $16\pi G = 1$ here. Let us now define the generalised momenta

$$p_i = \frac{\partial L}{\partial \dot{h}_i} = \frac{H_i}{M_i}. \tag{6.227}$$

It is interesting to note that in this model, we see quite explicitly that the interpretation of the auxiliary field entries of the matrix M^{ij} is that of certain functions on the phase space of the system. They are essentially the inverses of the canonical momenta of the system. The action takes the following Hamiltonian form:

$$S[h, p, \mu] = \frac{1}{\mathrm{i}} \int \left[\sum_i p_i \dot{h}_i + \mu \left(\sum_i \frac{H_i}{p_i} - \Lambda \right) \right] dt. \tag{6.228}$$

This is an action for a system with three configurational variables h_i, with the Hamiltonian that is a constraint. We also note that for this setup the action one obtains would be the same for any of the modified theories of the type (6.2). The only thing that changes is the phase space constraint that is imposed. We emphasise that at this stage all fields are complex-valued, with reality conditions still to be imposed.

Action (6.228) generates the Hamiltonian equations of motion for h_i and p_i:

$$\dot{h}_1 = \mu \frac{H_1}{p_1^2}, \qquad \dot{p}_1 = \mu \left(\frac{1}{p_1} - \frac{h_2}{p_3} - \frac{h_3}{p_2} \right), \tag{6.229}$$

and similarly with cyclic permutations of indices 1, 2, and 3, as well as the constraint

$$\sum_i \frac{H_i}{p_i} = \Lambda. \tag{6.230}$$

In a modified theory, both the evolution equations as well as the constraint would be modified.

6.7.4 Analysis of the Reality Conditions

We require the metric to be real and using (6.225), this implies that the quantities

$$\frac{N^2 a_i^2}{N a_1 a_2 a_3} = \frac{\dot{h}_i}{\mathrm{i}H_i} = \frac{\mu}{\mathrm{i}p_i^2} \tag{6.231}$$

should be real and positive. The simplest possible realisation of this is to take p_i all real and μ imaginary, on the positive imaginary semiaxis.

After the variation of (6.228) with respect to μ, without loss of generality, one can set $\mu = \mathrm{i}$ (this is achieved by redefinition of time t). After this, the Hamiltonian equations of motion (6.229) take the form

$$\dot{h}_1 = \frac{\mathrm{i}H_1}{p_1^2}, \qquad \dot{p}_1 = \mathrm{i}\left(\frac{1}{p_1} - \frac{h_2}{p_3} - \frac{h_3}{p_2}\right), \tag{6.232}$$

and similarly with cyclic permutation of indices 1, 2, and 3.

We now show that the reality of p_i is compatible with the equations of motion (6.232), i.e., that these reality conditions are preserved in time. While this is as expected because we know we are dealing with GR for which the metric can be chosen to be real, this is far from obvious in the chiral connection formulation under consideration. To check that the reality conditions are compatible with the dynamics we set

$$h_i = x_i + \mathrm{i}y_i, \qquad i = 1, 2, 3. \tag{6.233}$$

Then the reality of p_i, as well as the requirement that this reality is preserved by the time evolution, by virtue of the second equation in (6.232) results in the additional constraints

$$\frac{1}{p_1} - \frac{x_2}{p_3} - \frac{x_3}{p_2} = 0, \tag{6.234}$$

and similarly with cyclic permutation of indices 1, 2, and 3. The resulting system of equations can be solved and x_i determined in terms of p_i

$$x_1 = \frac{p_2 p_3}{2}\left(-\frac{1}{p_1^2} + \frac{1}{p_2^2} + \frac{1}{p_3^2}\right) \quad \text{etc.} \tag{6.235}$$

After this, the imaginary parts of the first set of equation in (6.232) as well as the second set of equations give evolution equations for y_i and p_i :

$$\dot{y}_1 = \frac{x_1 - x_2 x_3 + y_2 y_3}{p_1^2}, \tag{6.236}$$

$$\dot{p}_1 = \frac{y_2}{p_3} + \frac{y_3}{p_2}, \tag{6.237}$$

and similarly with cyclic permutation of indices. In addition, the real part of the first equation in (6.232) gives

$$\dot{x}_1 = \frac{-y_1 + x_2 y_3 + x_3 y_2}{p_1^2}. \tag{6.238}$$

This set of equations can be seen to be satisfied identically in view of the equations (6.236) and the relations (6.235). Indeed, differentiating, e.g., the first of the relations (6.235) and then using the second set of equations in (6.236) one obtains precisely the right-hand side of the equation (6.238). Thus, these equations are not independent and can be dropped. We should note, however, that the fact that the arising overdetermined system of equations is consistent is highly nontrivial. This would not be the case for modified theories.

Finally, the constraint (6.230) can also be split into its real and imaginary parts. The real part gives

$$\frac{x_1 - x_2 x_3 + y_2 y_3}{p_1} + (\text{cyclic permutations}) = \Lambda, \tag{6.239}$$

where it is understood that the quantities x_i should be replaced with their expressions (6.235) in terms of p_i. The imaginary part of the constraint gives

$$\frac{y_1 - x_2 y_3 - x_3 y_2}{p_1} + (\text{cyclic permutations}) = 0. \tag{6.240}$$

This can be rewritten as

$$y_1(\frac{1}{p_1} - \frac{x_2}{p_3} - \frac{x_2}{p_3}) + y_2(\frac{1}{p_2} - \frac{x_1}{p_3} - \frac{x_3}{p_3}) + y_3(\frac{1}{p_3} - \frac{x_1}{p_2} - \frac{x_2}{p_1}) = 0,$$

which is satisfied by virtue of equations (6.234).

All in all, with the reality conditions imposed the model reduces to the set of six first-order equations for the real quantities, y_i and p_i, as well as the constraint (6.239). The arising system of equations is highly nontrivial and cannot be solved in an explicit form, so we refrain from analysing it any further.

Attempting the previous analysis for a modified theory from the same class shows that the dynamics of the model is no longer consistent with the reality conditions (real p_i) one would like to impose. To verify this explicitly one can, e.g., take $f(M) = \det(M)$. It is not impossible that there exists a modified set of reality conditions that is compatible with the dynamics, but it is clear that these must be a model, and thus function $f(M)$-dependent. But even if this is the case, these reality conditions, while dynamics compatible, will no longer be the conditions that the metric is real. The physical interpretation, if any, is then unclear. This discussion makes it clear that the modified gravity theories obtained by changing $f(M)$ only exist for Euclidean and split signatures, but do not admit a Lorentzian interpretation.

6.8 Connection Description of Ricci Flat Metrics

The purpose of this section is to show that the first-order connection formalism described previously extends also to the $\Lambda = 0$ case. In this case, one cannot integrate out the matrix of auxiliary fields M^{ij}, and has to use the mixed first-order formalism with both M and the connection fields present. One, however, still gets a useful and powerful description.

6.8.1 Action, Field Equations, Metric

Since the matrix M^{ij} has to be tracefree in this case, it makes sense to write in its place the traceless matrix Ψ^{ij} from the start. The action reads

$$S[A,\Psi] = \frac{1}{16\pi G\sqrt{\sigma}} \int \mathrm{Tr}(\Psi^{-1}FF). \tag{6.241}$$

The corresponding field equations are

$$d^A(\Psi^{-1}F) = 0, \qquad \Psi^{-1}X\Psi^{-1} \sim \mathbb{I}, \tag{6.242}$$

where as before the 3×3 matrix X is given by $X^{ij} = F^iF^j$.

In this formalism, the metric is determined from the curvature of the connection as follows. First, the conformal class is the one that makes the triple of curvatures F^i SD. When $\Lambda \neq 0$ we can also determine the volume form explicitly in terms of the curvatures. This is not possible in the case $\Lambda = 0$. The best we can do is to have a formula for the volume form, which involves both Ψ and the curvature. This formula is easy to derive. Indeed, we want the volume form to be ϵ such that $\Sigma^i\Sigma_i = 6\sqrt{\sigma}\epsilon$. We also have $\Sigma_i = \Psi^{-1}_{ij}F^j$. This means that we can obtain the volume form as

$$\epsilon = \frac{1}{6\sqrt{\sigma}}\mathrm{Tr}(\Psi^{-1}X\Psi^{-1}). \tag{6.243}$$

Let us present an example that illustrates the usage of this formalism.

6.8.2 Schwarzschild Solution

Let us see how the familiar Schwarzschild solution is obtained using this formalism. The ansatz for the connections remains unchanged as compared to that in Section 6.6.1, and so is the analysis of the curvature. However, in the analysis of the metric we can only determine its conformal class now and get relations (6.198). We can no longer match the metric volume form to a multiple of $\mathrm{Tr}(M^{-1}X)$ because this quantity is now zero. Indeed, it is still true that $X = \mu M^2$. But now we should set $\mathrm{Tr}(M) = 0$. So, we have $\mathrm{Tr}(M^{-1}X) = \mu\mathrm{Tr}(M) = 0$. But this means that the right-hand side of the equations (6.191) for the matrix M becomes zero. This means that the equations $d^A(M^{-1}F) = 0$ take the form

$$(M_1^{-1}a(b^2-1))' = 0, \qquad (M_2^{-1}a(b^2-1))' = 0, \tag{6.244}$$

with the solution being

$$M_1 = R_1 a(b^2-1), \qquad M_2 = R_2 a(b^2-1), \tag{6.245}$$

where R_1 and R_2 are integration constants. The condition of zero trace gives

$$R_1 + 2R_2 = 0. \tag{6.246}$$

Denoting $R_2 = \bar{R}$ we thus have $R_1 = -2\bar{R}$. We also have $X = \mu M^2$, which gives

$$X_1 = \mu(R) 4\bar{R}^2 a^2 (b^2 - 1)^2, \qquad X_2 = \mu(R)\bar{R}^2 a^2 (b^2 - 1)^2, \tag{6.247}$$

where $\mu(R)$ is some function of the radial coordinate. Recalling the definitions (6.189) we have

$$\frac{a'}{a} = \frac{4bb'}{b^2 - 1}, \tag{6.248}$$

which integrates to

$$a = \kappa(b^2 - 1)^2, \tag{6.249}$$

where κ is another constant of integration. In principle, this gives a complete solution of the problem.

Let us now determine the metric. On the solutions we should have $\Sigma = M^{-1}F$. But we should also have $\Sigma^i \Sigma^j = 2\mathrm{i}\delta^{ij}\epsilon$, where ϵ is the metric volume form. Evaluating $\Sigma^1\Sigma^1 = 2\mathrm{i}M_1^{-2} X_1 a(b^2 - 1)\epsilon_c$, where ϵ_c is the coordinate volume form, we see that

$$\mu(R) a(b^2 - 1) = fgr^2, \tag{6.250}$$

which is the desired relation that fixes the metric completely. Indeed, substituting $fg = X_1 ar^2/(b^2 - 1)$ from (6.198) the previous relation becomes

$$\mu(R) a(b^2 - 1) = \frac{X_1 ar^4}{b^2 - 1}. \tag{6.251}$$

Substituting X_1 from (6.247) we get

$$a = \frac{1}{2\bar{R}r^2}. \tag{6.252}$$

This implies

$$X_1 = \frac{a'}{a} = -\frac{2}{r}\frac{dr}{dR}. \tag{6.253}$$

We can now fix all of the metric components. Using (6.198) and (6.247) we have

$$f^2 = \frac{4a^2b^2r^2}{(b^2 - 1)^2}, \qquad g^2 = \frac{X_1^2 r^2}{4b^2} = \frac{1}{b^2}\left(\frac{dr}{dR}\right)^2. \tag{6.254}$$

We can also substitute $r^2 = 1/2\bar{R}a$ and then (6.249) into the first of these relations. We get

$$f^2 = \frac{2\kappa}{\bar{R}} b^2. \tag{6.255}$$

The metric is thus

$$ds^2 = -\frac{2\kappa}{\bar{R}} b^2 dt^2 + \frac{dr^2}{b^2} + r^2 d\Omega^2. \tag{6.256}$$

This means that if we want to have the usual $fg = 1$, asymptotically we have to set $\kappa = \bar{R}/2$. With this choice

$$b^2 - 1 = \pm \frac{1}{\bar{R}r}, \tag{6.257}$$

where both signs are possible. In the usual Schwarzschild solution $a = (b^2-1)'/2$. This is always possible to achieve by changing $t \to -t$ if necessary. We obtain this relation if we take the negative sign in (6.257). We then identify

$$\frac{1}{\bar{R}} = 2M \tag{6.258}$$

to get

$$a = \frac{M}{r^2}, \qquad b^2 = 1 - \frac{2M}{r}. \tag{6.259}$$

This illustrates that the $\Lambda = 0$ case can also be treated via the pure connection formalism, with appropriate modifications to account for the fact that it is no longer possible to solve for X in terms of M completely, as there remains the freedom in choosing $\mu(R)$. This does not affect the final metric that is fixed completely by an appropriate choice of the radial coordinate.

6.8.3 Eguchi-Hanson Metric

As another example of the $\Lambda = 0$ connection formalism, let us obtain the Eguchi–Hanson instanton. We work in the Euclidean signature. We start with a bi-axial ansatz

$$A^1 = \alpha\sigma_1, \qquad A^2 = \beta\sigma_2, \qquad A^3 = \beta\sigma_3, \tag{6.260}$$

where σ_i are the already familiar 1-forms (6.71) on the three-sphere, and α and β are functions of some radial coordinate R. The curvatures are given by

$$F^1 = \alpha' dR\sigma_1 + (\beta^2 - \alpha)\sigma_2\sigma_3, \quad F^2 = \beta' dR\sigma_2 + \beta(\alpha - 1)\sigma_3\sigma_1, \tag{6.261}$$

and similarly for F^3. It is easiest to obtain the field equations by substituting this ansatz into the action (6.241), and then vary with respect to the independent functions. In general, this procedure is not guaranteed to produce the correct field equations, and so its result has to be compared to the correct equations. But in this case the procedure works. The traceless matrix Ψ can be taken to be of the form $\Psi = \text{diag}(-2\psi, \psi, \psi)$, where $\psi = \psi(R)$. This produces the following Lagrangian

$$L \sim \psi^{-1}\left(-\frac{1}{2}\alpha'(\beta^2 - \alpha) + (\beta^2)'(\alpha - 1)\right). \tag{6.262}$$

Varying this with respect to ψ produces

$$-\frac{1}{2}\alpha'(\beta^2-\alpha)+(\beta^2)'(\alpha-1)=0. \tag{6.263}$$

Varying with respect to α, β produces two more equations

$$\frac{(\psi^{-1})'}{\psi^{-1}}=-\frac{3}{2}\frac{\alpha'}{\alpha-1}, \qquad \frac{(\psi^{-1})'}{\psi^{-1}}=-\frac{3(\beta^2)'}{\beta^2-\alpha}. \tag{6.264}$$

We see that these two equations imply (6.263), and so it is sufficient to consider just the last two equations. The first of this is solved to produce

$$\psi=\kappa(\alpha-1)^{3/2}, \tag{6.265}$$

where κ is an integration constant, while the first and second combined imply

$$\frac{1}{2}\frac{\alpha'}{\alpha-1}=\frac{(\beta^2)'}{\beta^2-\alpha}. \tag{6.266}$$

This can be solved for β as a function of α

$$\beta^2=2-\alpha+K\sqrt{\alpha-1}, \tag{6.267}$$

where K is another constant of integration.

We now determine the metric. We look for it in the form

$$ds^2=f^2dR^2+g^2\sigma_1^2+\frac{r^2}{4}(\sigma_2^2+\sigma_3^2), \tag{6.268}$$

where r is the new radial coordinate. The basis of ASD 2-forms for this metric is

$$\bar{\Sigma}^1=fgdR\sigma_1+\frac{r^2}{4}\sigma_2\sigma_3, \quad \bar{\Sigma}^2=\frac{fr}{2}dR\sigma_2+\frac{gr}{2}\sigma_3\sigma_1, \tag{6.269}$$

and similarly for $\bar{\Sigma}^3$. We find the metric by requiring the curvatures (6.261) to be proportional to the above ASD 2-forms. We use the ASD rather than the SD 2-forms because the solution we want to exhibit is hyper-Kähler, which, in particular, means that the SD part of its spin connection is identically zero. This is why we work with the ASD part of the spin connection rather than the SD. Matching the ASD basic 2-forms to the curvatures gives

$$\frac{\alpha'}{\beta^2-\alpha}=\frac{4fg}{r^2}, \qquad \frac{\beta'}{\beta(\alpha-1)}=\frac{f}{g}. \tag{6.270}$$

We obtain one more equation by matching the volume forms. The metric volume form $\epsilon=fgr^2/4\epsilon_c$, where $\epsilon_c=dR\sigma_1\sigma_2\sigma_3$ must match the form in (6.243). This gives

$$\frac{3}{4}fgr^2=\frac{1}{4\psi^2}\alpha'(\beta^2-\alpha)+\frac{1}{\psi^2}(\beta^2)'(\alpha-1). \tag{6.271}$$

Using the first equation in (6.270) as well as (6.266) we get

$$\frac{r^2}{2} = \left| \frac{\beta^2 - \alpha}{\psi} \right|. \tag{6.272}$$

Substituting here the solution (6.267) we get

$$\frac{r^2}{2} = \left| \frac{K - 2\sqrt{\alpha - 1}}{\kappa(\alpha - 1)} \right|. \tag{6.273}$$

The other two metric functions are then

$$f^2 = \frac{r^2}{16} \frac{(\alpha')^2}{\beta^2(\alpha - 1)^2}, \qquad g^2 = \frac{r^2\beta^2(\alpha - 1)^2}{(\beta^2 - \alpha)^2}, \tag{6.274}$$

which determines the metric completely.

Eguchi–Hanson instanton corresponds to the case $K = 0$. This gives

$$\alpha = 1 + \frac{a}{r^4}, \qquad \beta^2 = 1 - \frac{a}{r^4}, \qquad \psi = \frac{4a}{r^6}, \tag{6.275}$$

where we redefined the constant of integration $\sqrt{a} := \kappa/4$ to match the usual form of the Eguchi–Hanson metric. The metric is then given by

$$ds^2 = \frac{1}{\beta^2} dr^2 + \frac{r^2\beta^2}{4}\sigma_1^2 + \frac{r^2}{4}(\sigma_2^2 + \sigma_3^2). \tag{6.276}$$

For $a = 0$ this is the usual metric on $\mathbb{R} \times S^3 = \mathbb{R}^4$. However, when $a \neq 0$ the absence of conical singularity at $r_+ = a^{1/4}$ can be seen to require that the period of ψ is 2π rather than 4π. This means that for $a \neq 0$ the metric is asymptotically that of $\mathbb{R}^4/\mathbb{Z}_2$.

6.8.4 Linearisation Around an Einstein Background

The action (6.241) is also interesting because it produces a useful perturbative expansion around an arbitrary Einstein background. Indeed, taking the Lagrangian $L = \Psi^{-1}FF$, where the trace is implied, we have, using index-free notations

$$\delta L = -\Psi^{-1}\delta\Psi\Psi^{-1}FF + 2\Psi^{-1}F\delta F, \tag{6.277}$$

and

$$\delta^2 L = 2\Psi^{-1}\delta\Psi\Psi^{-1}\delta\Psi\Psi^{-1}FF - 4\Psi^{-1}\delta\Psi\Psi^{-1}F\delta F + 2\Psi^{-1}\delta F\delta F + 2\Psi^{-1}F\delta^2 F. \tag{6.278}$$

We now specialise to an Einstein background $F = \Psi\Sigma$. Defining the second-order Lagrangian $L^{(2)}$ as half the second variation we get

$$L^{(2)} = 2\mathrm{Tr}(\delta\Psi\Psi^{-1}\delta\Psi)\epsilon - 2(\Psi^{-1}\delta\Psi)_{(ij)}\Sigma^i d^A a^j + \Psi^{-1}_{ij} d^A a^i d^A a^j + \epsilon_{ijk}\Sigma^i a^j a^k, \tag{6.279}$$

where $a^i := \delta A^i$ and ϵ is the metric volume form appearing via $\Sigma^i\Sigma^j = 2\epsilon\delta^{ij}$. For simplicity we work in the Euclidean signature here.

We now want to integrate out the perturbation of the Lagrange multiplier field Ψ. The required manipulations become trivial in a special very useful gauge

$$\Sigma^{i\mu\nu}a_{\nu\, i} = 0, \qquad d^{A\mu}a^i_\mu = 0. \tag{6.280}$$

The first of these conditions fixes the diffeomorphism freedom, while the second is the usual Lorentz gauge for the SO(3) gauge rotations. We will motivate this gauge in the chapter on gravitational perturbation theory. In this gauge, the decomposition (5.184) of the covariant derivative of the connection perturbation into irreducible pieces simplifies greatly. There is no spin zero and no spin one part. We then simply have

$$d^A a^i = (d^A a)_2^{ij}\Sigma^j + (d^A a)_-^{ij}\bar{\Sigma}_j. \tag{6.281}$$

Here $(d^A a)_2^{ij}$ is the spin two part of the covariant derivative, which is symmetric traceless, and $(d^A a)_-$ is the ASD part. In particular, we have $\Sigma^i d^A a^j = 2(d^A a)_2^{ij}\epsilon$ in this gauge. We can then rewrite the linearised Lagrangian as

$$\begin{aligned} L^{(2)} &= 2\mathrm{Tr}((\delta\Psi - (d^A a)_2)\Psi^{-1}(\delta\Psi - (d^A a)_2))\epsilon \\ &\quad - 2\mathrm{Tr}((d^A a)_2\Psi^{-1}(d^A a)_2)\epsilon + \Psi^{-1}_{ij}d^A a^i d^A a^j + \epsilon_{ijk}\Sigma^i a^j a^k. \end{aligned}$$

It is now trivial to integrate out $\delta\Psi$. The remaining Lagrangian is further simplified by writing

$$\Psi^{-1}_{ij}d^A a^i d^A a^j = 2\mathrm{Tr}((d^A a)_2\Psi^{-1}(d^A a)_2)\epsilon - 2\mathrm{Tr}((d^A a)_-\Psi^{-1}(d^A a)_-)\epsilon.$$

This means that the spin two part $(d^A a)_2$ cancels out and we get

$$L^{(2)} = -2(d^A a)_-^{ik}\Psi^{-1}_{ij}(d^A a)_-^{jk}\epsilon + \epsilon^{ijk}\Sigma^i a^j a^k. \tag{6.282}$$

This result is essentially the same as the previously derived linearised action (5.187) in the $\Lambda \neq 0$ case, after the gauge condition (6.280) is imposed. It is remarkable that such a simple linearised Lagrangian is also available in the $\Lambda = 0$ case.

6.9 Chiral Pure Connection Perturbation Theory

We now return to $\Lambda \neq 0$ setup. The purpose of this section is to perform the linearisation of the chiral pure connection action around an instanton background. Our goal is in particular to reproduce (5.183) as the relevant linearised action. The difference with the previous treatment is that we start directly with the action in terms of connections and linearise, rather than with the linearised Plebański action from which later the perturbations of the 2-form field is integrated out. The following manipulations give a good illustration of how the chiral connection description of gravity works.

We work with the action in the form (6.29). We will not specialise to the case of GR until the very end of the calculation, thus showing that any member of the class of theories obtained by 'deforming' $g(X)$ has the same linearisation around instantons.

We work on an instanton background for which the connection is perfect $F^iF^j \sim \delta^{ij}$. The metric is obtained by defining $\Sigma^i = (3/\Lambda)F^i$, where Λ is the cosmological constant. The metric is then the one that makes Σ^i SD and with the volume form ϵ such that $\Sigma^i\Sigma^j = 2\delta^{ij}\epsilon$. The first variation of the action is given by (6.31). The second variation is given by

$$\delta^2 S[A] = \int 4\frac{\partial g}{\partial X^{ij}\partial X^{kl}}(F^i d^A a^j)(F^k d^A a^l) + 2\frac{\partial g}{\partial X^{ij}}(d^A a^i d^A a^j + F^i \epsilon^{jkl} a^k a^l), \tag{6.283}$$

where $a^i := \delta A^i$. On an instanton background the matrix of curvature wedge products $X^{ij} \sim \delta^{ij}$. On such a background, there are certain statements that can be made about the matrices of partial derivatives of the function g, for any function $g(X)$. First, the matrix of first derivatives must be proportional to the identity matrix

$$X^{ij} \sim \delta^{ij} \qquad \Rightarrow \qquad \frac{\partial g}{\partial X^{ij}} \sim \delta^{ij}. \tag{6.284}$$

This follows from SO(3) invariance of the function $g(X)$. This means that the second line in the linearised action (6.283) is a multiple of

$$\begin{aligned} \int d^A a^i d^A a^i + F^i \epsilon^{ijk} a^j a^k &= \int a^i d^A d^A a^i + F^i \epsilon^{ijk} a^j a^k \\ &= \int a^i \epsilon^{ijk} F^j a^k + F^i \epsilon^{ijk} a^j a^k = 0, \end{aligned} \tag{6.285}$$

where to obtain the first equality we integrated by parts. This means that we only need to consider the first line in (6.283).

We can also constrain the form of the matrix of second derivatives of $g(X)$. This follows from the fact that $g(X)$ is a homogeneity degree one function. The corresponding Euler relation gives

$$\frac{\partial g}{\partial X^{ij}} X^{ij} = g(X). \tag{6.286}$$

Differentiating this relation with respect to X^{kl} we obtain

$$\frac{\partial g}{\partial X^{ij}\partial X^{kl}} X^{ij} = 0. \tag{6.287}$$

This means that on the instanton background the matrix of second derivatives of $g(X)$ is a symmetric endomorphism from the space of symmetric 3×3 matrices to itself. Moreover, it is constructed from the identity matrix only, and it maps tracefree matrices again to such matrices. Thus, it can only be a multiple of the projector on symmetric tracefree matrices

$$\frac{\partial g}{\partial X^{ij} \partial X^{kl}} \sim P_{ijkl}, \tag{6.288}$$

where the projector P_{ijkl} is given by (5.182). This implies that for any $g(X)$ the linearisation of the pure connection action is given by

$$\delta^2 S[A] \sim \int P_{ijkl}(\Sigma^i d^A a^j)(\Sigma^k d^A a^l), \tag{6.289}$$

which matches the previous GR result (5.183). The overall coefficient in front of this linearised action is, of course, theory-specific, and in the case of GR is as in (5.183). This argument, in particular, shows that for any $g(X)$ the theory (6.29) has exactly the same dynamical content as GR.

7

Deformations of General Relativity

In this chapter we look in more details at 'deformations' of general relativity (GR), a class of theories obtained by changing the function $f(M)$ in (5.197) or (6.2). We already have some understanding of these theories from the analysis in the previous chapter. Thus, we have seen that many results are completely general and do not depend on which function $f(M)$ is chosen. In particular, we know that all theories from this class have the same linearisation (6.289) around instanton backgrounds, and thus have the same dynamical content. The main goal of this chapter is to describe what can be called the 'geometrically natural' modified theory. We also compute a solution of the Bianchi I setup for this particular modified theory to see that its behaviour is even simpler than in the case of GR. It would be very interesting to find a geometric interpretation of this particular chiral modification of Euclidean signature GR, which is, at the moment, an open problem.

7.1 A Natural Modified Theory

We have defined chiral modified gravity theories as those described by the action (5.197) with function $f(M)$ being an arbitrary gauge invariant function of a 3×3 matrix M^{ij}. The field equations of such a theory are

$$d_A(M^{-1}F) = 0, \qquad M^{-1}FFM^{-1} = \mu \frac{\partial f}{\partial M}. \tag{7.1}$$

The second equation can in general be solved for M in terms of the matrix FF, and the solution substituted into the first equation, which then gives a second-order partial differential equation (PDE) on the connection. We have seen that GR can be described in this fashion, as well as the self-dual theory (6.118) whose only solutions are the gravitational instantons. In the latter case, though, the second equation just implies $FF \sim \mathbb{I}$ and cannot be solved for the matrix M, which in this case is an inverse of a tracefree matrix.

We have also seen that in general such a modified theory admits metric interpretation. The conformal class of the metric is determined from the Urbantke formula (6.16). While there is in general an ambiguity in the choice of the conformal factor, there is always a natural choice, which is to require that the action (5.197) is a multiple of the total volume. Thus, one can in general choose $\epsilon_F = \mathrm{Tr}(M^{-1}FF)$. The formula (6.17) shows that this is indeed the correct procedure in the case of GR.

7.1.1 Geometrically Natural Conformal Factor

As we now explain, there is another procedure for choosing a geometrically natural conformal factor in the Urbantke formula. Indeed, we can require that the left-hand side in (6.16) is the metric times its volume form, and this is equal to the right-hand side. Thus, we set

$$g_{\mathrm{U}}(\xi,\eta)\epsilon_{\mathrm{U}} = \frac{\sigma}{6}\epsilon^{ijk} i_\xi F^i \wedge i_\eta F^j \wedge F^k, \tag{7.2}$$

where ϵ_{U} is the metric volume form, and where $\sigma = \pm 1$ is the sign of the connection that is required to obtain a metric of some desired signature, e.g., all plus. We will only discuss the case of the Euclidean signature as we already know that there are problems with a physical interpretation of the Lorentzian modified theories.

7.1.2 A Computation

As we discussed, any modified theory comes with its natural choice of the volume form $\epsilon_F = \mathrm{Tr}(M^{-1}FF)$ so that the action is the total volume. Let us now see which choice of $f(M)$ gives the same volume form as the geometrically natural choice (7.2).

Any metric in the conformal class of (7.2) makes the triple of curvature 2-forms anti-self dual, in the orientation in which the matrix $X \sim FF$ is positive-definite. Let us choose *some* metric g in this conformal class, and introduce a canonical orthonormal basis Σ^i in the space of SD 2-forms for the metric g. Explicitly, given a frame basis, Σ^i's are the forms that are given by (5.31). They in particular satisfy $\Sigma^i\Sigma^j = 2\epsilon_g\delta^{ij}$, where ϵ_g is the metric volume form, positively oriented in the orientation provided by the connection.

Then the curvature 2-forms can be expanded in the basis of Σ^i as

$$F^i = \sigma\left(\sqrt{X}\right)^{ij}\Sigma^j, \tag{7.3}$$

where $\sigma = \pm 1$ is the sign of the definite connection, and $\sqrt{X}$ is the positive-definite matrix square root of the positive-definite matrix X. We stress that the relation (7.3) can be written for an arbitrary choice of metric g in the conformal class of the Urbantke metric. This relation can also be used as an alternative definition of the sign of the definite connection. The decomposition (7.3) follows using the algebra of Σ's. Indeed, we have

$$F^i F^j = \sigma^2 \sqrt{X}^{ik} \sqrt{X}^{jl} (2) \delta^{kl} \epsilon_g = 2X^{ij} \epsilon_g. \tag{7.4}$$

We now use (7.3) with Σ^i's being those for the Urbantke metric (7.2). Thus, we now take $X = X_{\rm U}$ with respect to the volume form of the metric $g_{\rm U}$. Substituting (7.3) into (7.2) and using the algebra of Σ's we get the relation $\epsilon_U = (detX_U)^{1/2}\epsilon_U$, from which we conclude that

$$\det X_{\rm U} = 1. \tag{7.5}$$

As we already remarked, for any function $f(M)$ in (5.197), or $g(X)$ in (6.29), we can use the volume form $\epsilon^* = g(X)\epsilon$ to define X via $FF = 2X\epsilon$. One then has $\epsilon^* = g(X)\epsilon^*$ and hence $g(X) = 1$. This immediately allows us to translate the condition (7.5) into a choice of the function $g(X)$. Thus, the condition (7.5) derived previously corresponds to a homogeneous degree one function

$$g_U(X) = (\det X)^{1/3}. \tag{7.6}$$

We then note that for this function

$$\frac{\partial f}{\partial X} = \frac{1}{3} (\det X)^{1/3} X^{-1}, \tag{7.7}$$

and so the field equations of this theory become

$$d_A \left((\det X)^{1/3} X^{-1} F \right) = 0. \tag{7.8}$$

It is not hard to see that these are the field equations of the theory (5.197) with

$$f_U(M) = \det(M). \tag{7.9}$$

Indeed, in this case the relation between M and FF becomes

$$M^{-1} FF M^{-1} = \mu \det(M) M^{-1}, \tag{7.10}$$

from which we get

$$FF = \mu \det(M) M. \tag{7.11}$$

We can then fix μ and thus M completely from the requirement that $\det(M) = \Lambda$. Taking the determinant of both sides gives $\det(FF) = \mu^3 (\det(M))^4$, and so

$$M = \left(\frac{\Lambda}{\det(FF)} \right)^{1/3} FF, \tag{7.12}$$

which is homogeneity degree zero in FF and so it does not matter which volume form is chosen to extract a matrix from the 4-form valued matrix FF. Substituting this into the action gives

$$S[A] = 3\int \left(\frac{\det(FF)}{\Lambda}\right)^{1/3}, \tag{7.13}$$

which is in the form (6.29) with a multiple of the function (7.6) as the defining function.

All in all, we see that the geometrically natural choice of the conformal factor for the Urbantke metric (7.2) corresponds to the modified theory of the form (5.197) with $f_U(M) = \det(M)$.

7.1.3 Modified Bianchi I

To illustrate the behaviour of the modified theory, we now study the behaviour of the spatially homogeneous anisotropic Bianchi I solution in the deformed case. Our main conclusion here is that this particular modified theory is in fact simpler than GR!

All considerations are those of Section 6.5, with appropriate modifications to make the metric signature Euclidean. This just requires removing the factor of imaginary unit from (6.143). All other formulas are unchanged. In particular, we have the general solution (6.173)

$$M_1 = \frac{1}{g(X)(\tau - \tau_1)}, \qquad \text{etc.} \tag{7.14}$$

where now $g(X) = \text{Tr}(M^{-1}X)$. The function $g(X)$ needs to be determined from the constraint $f(M) = 1$. We specialise to the case $f(M) = \det(M)$, which we will see is completely solvable. The constraint gives

$$g^{-3} = (\tau - \tau_1)(\tau - \tau_2)(\tau - \tau_3), \tag{7.15}$$

and thus

$$M_1 = \frac{(\tau - \tau_2)^{1/3}(\tau - \tau_3)^{1/3}}{(\tau - \tau_1)^{2/3}}, \qquad \text{etc.} \tag{7.16}$$

We can also find the components of the matrix X. The relation between M and X for $f(M) = \det(M)$ becomes

$$M^{-1}XM^{-1} = \mu M^{-1} \qquad \Rightarrow \qquad X = \mu M. \tag{7.17}$$

From this we also get $g(X) = 3\mu$, and thus find μ. This finally gives

$$X_1 = \frac{1}{3(\tau - \tau_1)}, \qquad \text{etc.} \tag{7.18}$$

Notably, this is a much simpler solution that (6.176) in the case of GR. In particular, it is trivial to integrate and find the connection components. Indeed, recalling that $X_1 = h_1'/h_1$ we have

$$h_1 = K_1(\tau - \tau_1)^{1/3}, \qquad \text{etc.,} \tag{7.19}$$

where K_i are integration constants.

It is now straightforward to compute the metric. Its general expression is given by (6.161). We have

$$g(X)X_1X_2X_3 = \left(\frac{1}{3}\right)^3 \left((\tau-\tau_1)(\tau-\tau_2)(\tau-\tau_3)\right)^{-4/3}, \tag{7.20}$$

and

$$g(X)\frac{X_1}{X_2X_3} = 3(\tau-\tau_2)^{2/3}(\tau-\tau_3)^{2/3}(\tau-\tau_1)^{-4/3}. \tag{7.21}$$

This gives the following metric

$$\begin{aligned} 3^{-1/2}ds^2 &= \frac{1}{9}\left((\tau-\tau_1)(\tau-\tau_2)(\tau-\tau_3)\right)^{-2/3} d\tau^2 \\ &\quad + (\tau-\tau_1)^{1/3}(\tau-\tau_2)^{1/3}(\tau-\tau_3)^{1/3}\sum_{i=1}^{3}\frac{K_i^2(dx^i)^2}{(\tau-\tau_i)^{1/3}}. \end{aligned} \tag{7.22}$$

This should be contrasted with the similar solution (6.178) in the case of GR, which is significantly more complicated because function h_i obtained by integrating (6.176) are considerably more involved than (7.19). The metric arising as the solution of the modified theory is not Einstein, and has singularities at $\tau = \tau_i$. Nevertheless, the connection components (7.19) just have zeroes at these points, and so it is only the metric interpretation that becomes problematic at these points. There is a real solution for all the connection components for all τ.

8

Perturbative Descriptions of Gravity

Spin one particles are described by gauge fields, which are 1-forms with values in some internal space, or rank one tensors. Gravitons are spin two particles, and so it seems very natural to describe them by rank two tensors. This seems to suggest that the metric formulation of gravity is the most natural one, and any other formalism only introduces unnecessary complications. The purpose of this chapter is to challenge this conclusion. In particular, we will see that formalisms that are based on collections of differential forms, in fact, lead to simpler perturbative descriptions of gravitons than is possible using the metric. But in order to see this most clearly, we will need to introduce the language of spinors.

We start with some motivating remarks. First, we note that when one deals with objects with two different type of indices, such as, e.g., the tetrad, choosing a background to expand the theory about gives an object that serves to identify the indices of different types. For example, the background tetrad $e^I_{0\mu}$ can be thought of as an object that identifies the internal index I with the spacetime index μ. Indeed, with the help of $e^I_{0\mu}$ all objects with mixed indices can be converted into objects with just indices of one of the two types. The same happens in any of the formalisms that are based on objects with mixed types of indices. This argument shows that, at least at the level of perturbative description, the formalism that are based on objects with two different types of indices are in no way inferior to the metric formalism with its sole type of indices.

To motivate the necessity of introduction of spinors, we recall that the latter are fundamental representations of the Lorentz group. And in order to understand what happens in a certain formalism, it is often necessary to decompose the objects arising into irreducible representations. In some cases the use of tensors is sufficient to describe such representations, e.g., in the metric formalism, the two irreducible Lorentz representations that are present in the metric perturbation $h_{\mu\nu}$ are the traceless and the trace part of this rank two tensor. However, for more complicated objects, such as, e.g., a connection, such tensor description is insufficient, and the most adequate language for decomposing the field into its irreducible Lorentz representations is that of spinors.

Spinors are also essential for the purpose of understanding what types of differential operators are present in a formalism. This is because every operator is composed of differential operators of the first order, and then the language of spinors allows for a very efficient classifications of the arising possibilities.

For all these reasons the language of spinors is indispensable for understanding more conceptual aspects of any formalism. At the same time, this language is not always the one most efficient for explicit computations. Tensor symbolic manipulation is usually a more powerful option. So, it is usually best to have both the spinorial and the usual tensor descriptions of the same objects available, as well as an efficient dictionary that allows us to switch between the two descriptions. We will start this chapter by developing some aspects of this dictionary.

The organisation of the remainder of this chapter is as follows: We define the irreducible representations of Lorentz group and show that there are two types of first-order differential operators that arise naturally in this context. We then briefly discuss the standard metric perturbation theory, and also present the chiral version of the Yang–Mills perturbation theory, in order to be able to later contrast this story with what happens in the case of gravity.

The main goal of this chapter is to explore alternatives to the standard metric perturbation theory. We first develop the chiral version of the first-order perturbation theory, which uses a very different representation of the kinetic term for the spin two field. In this formalism the spin two particle is still described by a rank two tensor field, but the diffeomorphism invariant kinetic term is built from very different first-order operators. We then develop an even more drastic departure from the usual formalism where the chiral half of the spin connection is used instead. This gives by far the most economic description, but one that works only around non-flat backgrounds, e.g., constant curvature ones. We will see that the use of the chiral connection as the main variable for gravity leads to dramatic simplifications also around arbitrary Einstein backgrounds.

The main outcome of our analysis in this chapter is the conclusion that in its chiral versions, the gravitational perturbation theory behaves in complete parallel with chiral Yang–Mills perturbation theory. The latter, as we shall see, is the simplest and most conceptually clear way of doing Yang–Mills perturbative calculations. It is this parallel with Yang–Mills, invisible in the usual metric version of the gravity perturbation theory, which serves as a strong justification to take the chiral formulations of gravity seriously.

8.1 Spinor Formalism

The standard source on the spinor formalism is the book, *Spinors and Space Time*, by Penrose and Rindler (1986). The main difference between this source and our presentation is that we use the metric of signature mostly plus, which necessitates the usage of anti-Hermitian rather than Hermitian matrices to represent vectors in $\mathbb{R}^{1,3}$. At the same time we would like our tetrad to be a hermitian object. This leads to a certain minus sign in the formalism that requires some practice to get used to. There are also some deviations in normalisations as compared to Penrose and Rindler.

8.1.1 Spinors in $R^{1,3}$

The description of 4D spinors is analogous for all three different signature cases. We only describe the Lorentzian signature case. As we already know from (5.65), the four coordinates of $\mathbb{R}^{1,3}$ can be collected into an anti-hermitian matrix

$$\mathbf{x} = \mathrm{i} \begin{pmatrix} x_4 + x_3 & x_1 - \mathrm{i}x_2 \\ x_1 + \mathrm{i}x_2 & x_4 - x_3 \end{pmatrix} \tag{8.1}$$

so that

$$\det(\mathbf{x}) = -x_4^2 + x_1^2 + x_2^2 + x_3^2. \tag{8.2}$$

is the usual norm of a vector in $\mathbb{R}^{1,3}$. The double cover $\mathrm{SL}(2,\mathbb{C})$ of the Lorentz group acts in the space of such matrices via

$$\mathbf{x} \to g\mathbf{x}g^\dagger, \qquad g \in \mathrm{SL}(2,\mathbb{C}). \tag{8.3}$$

The spinors are introduced as the two different types of irreducible representations of the double cover $\mathrm{SL}(2,\mathbb{C})$ of the Lorentz group. We define the **unprimed** λ_A spinors as two-component columns (with complex entries) on which $\mathrm{SL}(2,\mathbb{C})$ acts by multiplication from the left

$$\lambda_A \to g_A{}^B \lambda_B. \tag{8.4}$$

This is an irreducible representation of $\mathrm{SL}(2,\mathbb{C})$. We shall refer both to the representation itself and to the space in which such spinors take value as S_+.

The other irreducible representation is referred to as that of **primed** spinors $\lambda_{A'}$. These are again two-component columns on which $\mathrm{SL}(2,\mathbb{C})$ acts by multiplication from the left, but this time with the complex conjugate group element

$$\lambda_{A'} \to (g^*)_{A'}{}^{B'} \lambda_{B'}. \tag{8.5}$$

We shall refer to primed spinors as taking values in S_-. From the definition of the representations S_+, S_- it is clear that the complex conjugation of a spinor in S_+ is an S_- spinor $(S_+)^* = S_-$.

8.1.2 Raising and Lowering Spinor Indices

The next notion we need is that of a bilinear form in the space of spinors. For S_+ spinors, this is defined as

$$\langle \lambda \eta \rangle := (\epsilon \lambda)^T \eta = -\lambda^T \epsilon \eta, \tag{8.6}$$

where the $2 \otimes 2$ matrix η is given by

$$\epsilon = \begin{pmatrix} 0 & 1 \\ -1 & 0 \end{pmatrix}. \tag{8.7}$$

The row $(\epsilon\lambda)^T$ can be referred to as the spinor λ^A with its index raised, so that the spinor contraction takes the form $\langle \lambda \eta \rangle = \lambda^A \eta_A$.

It is clear from the definition of the spinor pairing (8.6) that it is antisymmetric $\langle \lambda \eta \rangle = -\langle \eta \lambda \rangle$. In index notations, this can be written as

$$\lambda^A \eta_A = -\lambda_A \eta^A, \tag{8.8}$$

and so we have the famous rule that raising one spinor index in an expression and simultaneously lowering the index it is contracted with gives the minus sign.

The reason for the choice of signs as in (8.6) is that using the index notation it can be written as

$$\lambda^A \eta_A = -\lambda^T \epsilon \eta = -\lambda_A \epsilon^{AB} \eta_B. \tag{8.9}$$

On the other hand, by (8.8) this is equal to $-\lambda_A \eta^A$ and so we have the rule

$$\epsilon^{AB} \eta_B = \eta^A, \tag{8.10}$$

which is the standard spinor index raising rule in the literature.

We also introduce the operator of lowering a spinor index. This is required to be such that first raising an index of a spinor and then lowering it back produces the original spinor. It is clear from (8.6) that the required operation is obtained by multiplying the row λ^A by the matrix ϵ from the right and than taking the transpose. This means that the lowering operation in index notations is written as

$$\lambda^A \epsilon_{AB} = \lambda_B. \tag{8.11}$$

To check that all the definitions are consistent, we write

$$\epsilon^{AB} \lambda_B \epsilon_{AC} = \lambda_C, \tag{8.12}$$

and so we must have

$$\epsilon^{AB} \epsilon_{AC} = \mathbb{I}_C{}^B. \tag{8.13}$$

Using the definition of the lowering operation, we can also write this as

$$\epsilon^{AB} \epsilon_{AC} = \epsilon_C{}^B, \tag{8.14}$$

from which we learn that the epsilon tensor with first lower and second upper index is the identity matrix. This is indeed true in the matrix representation. Indeed, the left-hand side equals to $(\epsilon)^T \epsilon = \mathbb{I}$.

We can also check that the matrix ϵ_{AB} with both indices lowered is represented as the same matrix ϵ. Indeed, according to the previous definitions

$$\epsilon_{CD} = \epsilon^{AB}\epsilon_{AC}\epsilon_{BD}. \tag{8.15}$$

In matrix notations the right-hand side is $(\epsilon)^T\epsilon\epsilon = \epsilon$, which shows that everything is consistent.

The same definitions and properties hold for the primed spinors. In this case, the spinor metric is the object $\epsilon^{A'B'}$ and the same raising lowering conventions hold.

8.1.3 Transformation Properties of the Spinors with Raised Indices

We can work out how the spinors with their index raised transform. Because for any SL(2, $\mathbb{C}$) matrix g we have $g^T\epsilon = \epsilon g^{-1}$, we see that

$$\lambda^T\epsilon \to \lambda^T g^T\epsilon = \lambda^T\epsilon g^{-1}, \tag{8.16}$$

and so the spinor with its index raised transforms by multiplication by g^{-1} from the right. For the primed spinors, the definitions are analogous, and a primed spinor with its index raised $\lambda^{A'}$ (which is a row) transforms by multiplying it with $(g^*)^{-1}$ from the right.

8.1.4 Matrix x *as a Bi-Spinor*

Let us now interpret the matrix $\mathbf{x}$ as a bi-spinor. In view of its transformation property (8.3), it is clear that this matrix should be interpreted as a bi-spinor $\mathbf{x}_{AA'}$ with two lower spinor indices of opposite type. Let us also compute the matrix with both spinor indices raised $\mathbf{x}^{AA'}$. In matrix notations it is the matrix $\epsilon\mathbf{x}\epsilon^T$. We have

$$\epsilon\mathbf{x}\epsilon^T = \mathrm{i}\begin{pmatrix} x_4 - x_3 & -x_1 + \mathrm{i}x_2 \\ -x_1 - \mathrm{i}x_2 & x^4 + x_3 \end{pmatrix} = \det(\mathbf{x})\mathbf{x}^{-1}. \tag{8.17}$$

This implies that we can write the Minkowski space interval as

$$\frac{1}{2}\mathbf{x}_{AA'}\mathbf{x}^{AA'} = -x_4^2 + x_1^2 + x_2^2 + x_3^2. \tag{8.18}$$

8.1.5 Spinor Soldering Form

We now introduce objects $e_\mu^{AA'}$ via

$$\mathbf{x}^{AA'} = \mathrm{i}\sqrt{2}e_\mu^{AA'}x^\mu, \tag{8.19}$$

so that the objects $e_\mu^{AA'}$ are Hermitian and we have

$$-e_{AA'\mu}e_\nu^{AA'} = \eta_{\mu\nu}. \tag{8.20}$$

The minus sign here is unavoidable if one insists on using Hermitian objects (which is convenient) and insists on the mostly plus signature, which is convenient for quantum field theory purposes. One way to accept this seemingly unnatural sign is to introduce the notion of natural spinor contraction. The unprimed indices are contracted naturally, as in $\lambda^A \eta_A$, while for the primed indices the natural contraction is the opposite $\lambda_{A'}\eta^{A'}$. The metric is then reproduced as the natural contraction of the two copies of the soldering form

$$e^A_{\mu\, A'} e_{\nu A}{}^{A'} = \eta_{\mu\nu}. \tag{8.21}$$

The explicit matrix representation of the objects $e^{AA'}_\mu$ with both upper indices follows from (8.19) and the expression (8.17) for the matrix $\mathbf{x}^{AA'}$. We have

$$\begin{aligned} e_4 &= \frac{1}{\sqrt{2}}\begin{pmatrix} 1 & 0 \\ 0 & 1 \end{pmatrix}, \quad e_1 = \frac{1}{\sqrt{2}}\begin{pmatrix} 0 & -1 \\ -1 & 0 \end{pmatrix}, \\ e_2 &= \frac{1}{\sqrt{2}}\begin{pmatrix} 0 & \mathrm{i} \\ -\mathrm{i} & 0 \end{pmatrix}, \quad e_3 = \frac{1}{\sqrt{2}}\begin{pmatrix} -1 & 0 \\ 0 & 1 \end{pmatrix}, \end{aligned} \tag{8.22}$$

We will also introduce the inverse of the soldering form $e^\mu_{AA'}$ via the relation

$$e^{AA'}_\mu e^\mu_{BB'} = -\epsilon_B{}^A \epsilon_{B'}{}^{A'}. \tag{8.23}$$

The sign here has the same origin as in (8.20) and is dictated by the desire to have $e^\mu_{AA'}$ to be equal to the object $e^{AA'}_\mu$ with its indices raised and lowered with the available metrics.

8.1.6 The Basis of SD and ASD Forms

Let us now introduce the following 2-forms constructed from $e^{AA'}_\mu$

$$\Sigma^{AB}_{\mu\nu} := e^A_{[\mu\, A'} e^{B\, A'}_{\nu]}. \tag{8.24}$$

Alternatively, in the form notation, this definition becomes

$$\Sigma^{AB} = \frac{1}{2} e^A{}_{A'} e^{BA'}. \tag{8.25}$$

The choice of sign and the coefficient in this definition is dictated by the desire to have a very simple expression for this object in the fully spinor notation when the indices $\mu\nu$ are converted into spinor indices via the inverse soldering form. Indeed, a simple computation gives

$$\Sigma^{AB}_{MM'NN'} := \Sigma^{AB}_{\mu\nu} e^\mu_{MM'} e^\nu_{NN'} = \epsilon_M{}^{(A} \epsilon_N{}^{B)} \epsilon_{M'N'}. \tag{8.26}$$

This is a natural and easy to remember expression, which motivates the definition (8.24).

It is instructive to compute explicitly the matrix representation of the 2-forms $\Sigma^{AB}_{\mu\nu}$ for various $\mu\nu$. For the matrices Σ^{AB}_{4i} we have

$$\Sigma_{41} = \frac{1}{2}(e_4 \epsilon e_1^T - e_1 \epsilon e_4^T) = \frac{1}{2}\begin{pmatrix} -1 & 0 \\ 0 & 1 \end{pmatrix}, \tag{8.27}$$
$$\Sigma_{42} = \frac{1}{2}(e_4 \epsilon e_2^T - e_2 \epsilon e_4^T) = \frac{1}{2}\begin{pmatrix} \mathrm{i} & 0 \\ 0 & \mathrm{i} \end{pmatrix},$$
$$\Sigma_{43} = \frac{1}{2}(e_4 \epsilon e_3^T - e_3 \epsilon e_4^T) = \frac{1}{2}\begin{pmatrix} 0 & 1 \\ 1 & 0 \end{pmatrix},$$

which are all symmetric matrices as could have been expected from (8.26). The other components of $\Sigma_{\mu\nu}^{AB}$ follow from self-duality, e.g., $\Sigma_{23}^{AB} = -\mathrm{i}\Sigma_{41}^{i}$. That this relation holds can of course be also checked explicitly by computing Σ_{23}^{AB} in the matrix form.

We will introduce the basis in the space of anti–self-dual (ASD) 2-forms as *minus* the complex conjugation of the self-dual (SD) ones

$$\bar{\Sigma}_{\mu\nu}^{A'B'} := e_{[\mu}^{AA'} e_{\nu]A}{}^{B'}. \tag{8.28}$$

Note that in this definition the unprimed spinors are naturally contracted, which gives some motivation for this choice of sign. In the form notations the definition (8.28) becomes

$$\bar{\Sigma}^{AB} = \frac{1}{2} e^{AA'} e_A{}^{B'}. \tag{8.29}$$

8.1.7 Useful Identities

Given the previous definitions, a useful set of identities relating the soldering forms and the objects $\Sigma_{\mu\nu}^{AB}, \bar{\Sigma}_{\mu\nu}^{A'B'}$ can be derived. We first note an identity for the wedge product of two soldering forms

$$e^{AA'} e^{BB'} = \epsilon^{A'B'} \Sigma^{AB} - \epsilon^{AB} \bar{\Sigma}^{A'B'}. \tag{8.30}$$

To prove this identity we note that the 2-form on the left must be decomposable into its SD and ASD parts. The coefficients are then checked by contracting both sides with $\epsilon_{AB}, \epsilon_{A'B'}$.

Another useful identity is that for a product of two soldering forms contracted in one of the spinor indices. We have

$$e_{\mu}^{A}{}_{A'} e_{\nu}^{BA'} = -\frac{1}{2} \eta_{\mu\nu} \epsilon^{AB} + \Sigma_{\mu\nu}^{AB}. \tag{8.31}$$

The second term follows from the definition (8.24), while the coefficient in the first term is checked by contracting both sides with ϵ_{AB} and taking note of (8.20). The complex conjugate of the identity (8.31) gives

$$e_{\mu}^{AA'} e_{\nu A}{}^{B'} = \frac{1}{2} \eta_{\mu\nu} \epsilon^{A'B'} + \bar{\Sigma}_{\mu\nu}^{A'B'}. \tag{8.32}$$

Using this last relation we can also derive a useful for the following identity

$$\bar{\Sigma}^{A'B'}{}_{\mu}{}^{\rho}\bar{\Sigma}^{C'D'}{}_{\rho}{}^{\nu} = \frac{1}{2}\eta_{\mu}{}^{\nu}\epsilon^{A'(C'}\epsilon^{D')B'} + \frac{1}{2}\left(\bar{\Sigma}^{A'(C'}{}_{\mu}{}^{\nu}\epsilon^{D')B'} + \bar{\Sigma}^{B'(C'}{}_{\mu}{}^{\nu}\epsilon^{D')A'}\right). \tag{8.33}$$

This is the spinor analog of (5.138), but for the ASD 2-forms. We will need this identity when we discuss Euclidean twistors and their relation to almost complex structures.

8.2 Spinors and Differential Operators

Using the spinor soldering form $e_{\mu}^{AA'}$ and its inverse, any tensor object can be converted into a purely spinorial one. And for any formalism where there are also objects of mixed type with spacetime and internal indices, a nonvanishing such object provided by the background around which everything is expanded can be used to identify the internal and spacetime indices. Thus, at the end indices of all types can be converted into the spinor ones. This, together with simple representation theory of the Lorentz group, provides a very efficient way of describing what happens in each formalism. In preparation for such a discussion, for the specific formalisms we have previously described, we need to introduce some important first-order differential operators that arise naturally in the spinor context.

8.2.1 Irreducible Representations of the Lorentz Group

Each irreducible representation of the Lorentz group can be obtained by tensoring copies of its fundamental spinor representations. The spinor indices of each type are then symmetrised to remove all the traces and thus produce an irreducible representation. Thus, an irreducible representation of Lorentz group is characterised by two integers, the numbers of copies of unprimed and primed spinor representations used to build the representation under consideration. We will refer to irreducible representations of Lorentz as $S_+^k \otimes S_-^n$. This is an irreducible representation of dimension $(k+1)(n+1)$.

We then introduce a convenient notion of the total *spin* of an object taking value in an irreducible representation as

$$\text{spin}(S_+^k \otimes S_-^n) := (k+n)/2. \tag{8.34}$$

This is the total number of fundamental spinor representations used to build $S_+^k \otimes S_-^n$, divided by two. Note that since $k+n$ is an integer, the spin is either an integer of half-integer.

8.2.2 Spin-Increasing Differential Operator

There is a very simple and natural first-order differential operator that maps an object of some spin to an object whose spin is larger by one. Indeed, acting on a spinor object with ∂_μ and converting the spacetime index into a pair MM' adds one unprimed and one primed spinor index. The resulting collections of unprimed and primed spinor indices can then be symmetrised (to extract the irreducible part). Thus, we get maps

$$\begin{array}{c} S_+^k \otimes S_-^n \\ \Big\downarrow{\scriptstyle d_{(k,n)}} \\ S_+^{k+1} \otimes S_-^{n+1} \end{array}$$

where the representation that the operator acts on is indicated as its subscript. Explicitly

$$\psi_{A_1\ldots A_k A'_1\ldots A'_n} \to \partial_{(A'(A}\psi_{A_1\ldots A_k)A'_1\ldots A'_n)}, \tag{8.35}$$

where there is a double symmetrisation on the right-hand side. There can also be a numerical coefficient in the definition of such operators. We will discuss various possible choices in the following sections. We denote the operator that acts on functions (spin zero) as $d_{(0,0)} \equiv d$. Note that the operator d thus introduced is not nilpotent

$$d_{(k+1,n+1)}d_{(k,n)} \neq 0, \tag{8.36}$$

and that an infinite collection of integer spin spaces $C^\infty(M), S_+ \otimes S_-, \ldots, S_+^k \otimes S_-^k, \ldots$ gets created by the action of d's on the functions. Each of the operators $g_{(k,n)}$ has an adjoint, which is a first-order differential operator lowering the spin by one.

8.2.3 Dirac Operators

Apart from operators that change the spin, we can introduce a set of operators that just change an unprimed index to a primed one or the reverse. These are the Dirac operators

$$S_+^k \otimes S_-^n \xrightarrow{\quad \delta_{(k,n)} \quad} S_+^{k+1} \otimes S_-^{n-1}$$

where our convention is that the operators δ increase the number of unprimed indices while lowering the number of primed. Explicitly, an operator of this type acts as

$$\psi_{A_1\ldots A_k A'_1\ldots A'_n} \to \partial_{(A}{}^{A'}\psi_{A_1\ldots A_k)A'_1\ldots A'_{n-1}A'}, \tag{8.37}$$

where there can also be a numerical coefficient different from identity on the right-hand side. Similarly to the operators d already introduced, the δ's are not nilpotent

$$\delta_{(k+1,n-1)}\delta_{(k,n)} \neq 0. \tag{8.38}$$

$$
\begin{array}{ccccccc}
C^\infty(M) & & & & & & \text{spin } 0 \\
\Big\downarrow d & & & & & & \\
S_+ \otimes S_- & \xrightarrow{\ \delta\ } & S_+^2 & & & & \text{spin } 1 \\
\Big\downarrow d_{(1,1)} & & \Big\downarrow d_{(2,0)} & & & & \\
S_+^2 \otimes S_-^2 & \xrightarrow{\ \delta_{(2,2)}\ } & S_+^3 \otimes S_- & \xrightarrow{\ \delta_{(3,1)}\ } & S_+^4 & & \text{spin } 2
\end{array}
$$

Figure 8.1 The diagram of irreducible representations of Lorentz group and related differential operators.

We denote $\delta \equiv \delta_{(1,1)}$. Each of the operators $\delta_{(k,n)}$ has an adjoint that is a first-order differential operator acting by increasing the number of primed indices by one and respectively lowering the number of unprimed ones. We will work out the corresponding expressions when we need them.

8.2.4 The Diagram of Spaces

The whole set of spaces generated from the spin zero by the action of d's and δ's can be drawn as an infinite triangular diagram, (see Figure 8.1). Here we show just the first three rows of it, i.e., drawing the spaces of spin not higher than 2. We also show only the chiral half of the diagram, i.e., the spaces $S_+^k \otimes S_-^n$ with $k \geq n$. As we shall see, these are the spaces relevant for our chiral descriptions.

We would now like to study the operators appearing in the diagram in Figure 8.1 and understand their relations with various Laplacians that arise on the nodes.

8.2.5 The Operators d and δ: Spinor Computation

Given an object $v_{AA'}$ in $S_+ \otimes S_-$, the operator δ is given by

$$(\delta v)_{AB} = \partial_{(A}{}^{A'} v_{B)A'}. \tag{8.39}$$

This operator has an adjoint with respect to the inner product on both spaces. This is computed from

$$\int (\delta v)_{AB} \phi^{AB} = \int v^A{}_{A'} (\delta^* \phi)_A{}^{A'}, \tag{8.40}$$

where on the right-hand side the indices are in different positions as is required in the metric contraction of two vectors (8.21). The adjoint reads

$$(\delta^* \phi)_{AA'} = -\partial^B{}_{A'} \phi_{AB}. \tag{8.41}$$

We then have

$$(\delta^* \delta v)_{AA'} = -\partial^B{}_{A'} \frac{1}{2} (\partial_A{}^{B'} v_{BB'} + \partial_B{}^{B'} v_{AB'}). \tag{8.42}$$

This can be simplified by using the following identity

$$\partial_A{}^{B'} v_{BB'} - \partial_B{}^{B'} v_{AB'} = \epsilon_{AB} \partial_E{}^{B'} v^E{}_{B'}. \tag{8.43}$$

This identity arises by noting that the left-hand side is antisymmetric in AB and thus must be a multiple of ϵ_{AB}. The coefficients are then checked by contracting both sides with ϵ^{AB}. Using the identity (8.43) we have

$$(\delta^* \delta v)_{AA'} = -\partial^B{}_{A'} \partial_B{}^{B'} v_{AB'} + \frac{1}{2} \partial_{AA'} \partial_B{}^{B'} v^B{}_{B'}. \tag{8.44}$$

On the other hand, because partial derivatives commute, the two copies of the operator $\partial_{AA'}$ contracted in any pair of indices produce the box operator. For the case at hand the relevant identity is

$$\partial^B{}_{A'} \partial_B{}^{B'} = \frac{1}{2} \epsilon_{A'}{}^{B'} \Box, \tag{8.45}$$

where $\Box := \partial^A{}_{A'} \partial_A{}^{A'}$. Finally, we get

$$(\delta^* \delta v)_{AA'} = -\frac{1}{2} \Box v_{AA'} + \frac{1}{2} \partial_{AA'} \partial_B{}^{B'} v^B{}_{B'}. \tag{8.46}$$

This can be rewritten as

$$2\delta^* \delta = -\Box_{(1,1)} - dd^*, \tag{8.47}$$

where the d^* is the operator that maps vectors into scalars $d^* v = -\partial_B{}^{B'} v^B{}_{B'}$.

8.2.6 The Operators d and δ: Tensor Computation

The computation of the previous section can be carried out in much simpler way by using the tensor notations. To this end, we will need to recall that the basis of SD 2-forms Σ^i, where we now use SO(3) notations, is the map from the space of SD 2-forms Λ^2 to the the space $\mathbb{R}^3$, and the later can be identified with the space of symmetric rank two unprimed spinors. Thus $\Lambda^+ \sim S^2_+$. Using this fact we can alternatively define the operator δ as follows. Given a 1-form v_μ we define

$$(\delta v)^i = \Sigma^{i\mu\nu} \partial_\mu v_\nu, \tag{8.48}$$

where $\Sigma^i_{\mu\nu}$ are the already familiar SD 2-forms defined in, e.g., (5.31). We note that in order for this operator to agree with its spinor version we would need to put a coefficient of $1/\sqrt{2}$ in front. However, the previous normalisation is simpler, and we will stick to it from now on. So, (8.39) and (8.48) differ in normalisation.

If we use the metric δ^{ij} in S_+^2 and the spacetime metric in $S_+ \otimes S_-$, we can obtain the adjoint of this operator from

$$(\phi, \delta v) \equiv \int \phi^i \Sigma^{i\mu\nu} \partial_\mu v_\nu = \int v_\mu \Sigma^{i\mu\nu} \partial_\nu \phi^i \equiv (\delta^* \phi, v), \tag{8.49}$$

where we integrated by parts. Thus,

$$(\delta^* \phi)_\mu = \Sigma^{i\ \nu}_{\mu} \partial_\nu \phi^i. \tag{8.50}$$

We now consider the operator $\delta^* \delta$ on $S_+ \otimes S_-$. We have

$$(\delta^* \delta v)_\mu = \Sigma^{i\ \nu}_{\mu} \partial_\nu \Sigma^{i\rho\sigma} \partial_\rho v_\sigma. \tag{8.51}$$

We now use the identity

$$\Sigma^i_{\mu\nu} \Sigma^i_{\rho\sigma} = \eta_{\mu\rho}\eta_{\nu\sigma} - \eta_{\mu\sigma}\eta_{\nu\rho} - \mathrm{i}\epsilon_{\mu\nu\rho\sigma} \tag{8.52}$$

that follows from the definition of Σ^i's. This identity gives

$$(\delta^* \delta v)_\mu = \partial_\mu \partial^\nu v_\nu - \Box v_\mu, \tag{8.53}$$

which agrees with (8.47) up to normalisation. In spite of having to carry around the objects $\Sigma^i_{\mu\nu}$, the tensor computation is completely algorithmic and is easier to follow. For this reason we will mostly work with the tensor versions of the operators d, δ introduced previously. But it is important to have in mind their spinor interpretation, otherwise, it is hard to understand the specific ways that the objects $\Sigma^i_{\mu\nu}, \partial_\mu$ and the fields are contracted.

Another illuminating computation is that of the operator $\delta\delta^*$ on S_+^2. We have

$$(\delta\delta^* \phi)^i = \Sigma^{i\mu\nu} \partial_\mu \Sigma^{j\ \rho}_{\nu} \partial_\rho \phi^j. \tag{8.54}$$

This is simplified using the identity (5.138) and we get

$$(\delta\delta^* \phi)^i = -\Box \phi^i. \tag{8.55}$$

An obvious, but very important, property of the two operators d, δ is that their composition gives zero

$$\delta d = 0. \tag{8.56}$$

This is an obvious consequence of the previous definitions and the fact that the partial derivatives commute. Thus, the image of d is contained in the kernel of δ, and we have the usual cohomology setup. This means that the space $S_+ \otimes S_-$ admits an orthogonal decomposition into elements of the form df, of the form $\delta^* \phi$, and harmonic elements, i.e., those that satisfy $(\delta^* \delta + dd^*)v = 0$. This is an analog of the usual Hodge decomposition of Λ^1, but applied to a slightly different complex.

8.2.7 The Operator $d_{(1,1)}$

To motivate the discussion that follows let us also carry out a similar exercise for the operator $d_{(1,1)}$. The space $S^2_+ \otimes S^2_-$ can be identified with the space of rank two symmetric tracefree tensors. Thus, we can define $d_{(1,1)}$ as

$$(d_{(1,1)}v)_{\mu\nu} := \partial_{(\mu}v_{\nu)} - \frac{1}{4}\eta_{\mu\nu}\partial^\alpha v_\alpha, \tag{8.57}$$

where we again work in tensor notations to simplify the computations that need to be done. The adjoint operator is then given simply by

$$(d^*_{(1,1)}h)_\mu = -\partial^\nu h_{\mu\nu}, \qquad h_{\mu\nu} \in S^2_+ \otimes S^2_-. \tag{8.58}$$

We now compute $d^*_{(1,1)}d_{(1,1)}$. The result is

$$2d^*_{(1,1)}d_{(1,1)} = -\Box_{(1,1)} + \frac{1}{2}dd^*, \tag{8.59}$$

where $\Box_{(1,1)}$ is the $\Box$ operator on vectors. This representation of $\Box_{(1,1)}$ is reminiscent of the representation in terms of d, δ in (8.47). However, unlike the latter, the representation does not come with a cohomological orthogonal decomposition of vectors into those of form $d_\mu f$ and $d^\nu h_{\mu\nu}$ because $d_{(1,1)}d \neq 0$.

8.2.8 Box Operator and Arrows of the Operator Diagram

The results of the previous sections can be summarised qualitatively as follows. Consider a node of the diagram in Figure 8.1, and the corresponding $\Box$ operator. In terms of the first-order operators the $\Box$ operator can be represented in many different ways. Thus, we saw that the operator $\Box_{(1,1)}$ on vectors can be represented in terms of the operator δ and its adjoint (8.47), or in terms of the operator $d_{(1,1)}$ and its adjoint (8.59), as well as the operators d, d^*. This means that if we want to describe propagating particles of spin one, we could base this description either on the first-order operator δ that does not change the spin of the field, or on $d_{(1,1)}$, which increases the spin. One can see that the $\Box$ operator on every node of the diagram 8.1 can be obtained by going to the neighbouring node lying to the right, left, or below with either δ, δ^*, or d and coming back, plus terms in which one goes to the neighbouring node lying above with d^* and coming back. Thus, for a node of the diagram that does not lie on the edge there are three different representations of the $\Box$. In particular, there are two qualitatively different ways to represent the $\Box$ operator on every node, one is in terms of operators δ, δ^* that do not increase the spin, and another in terms of the operators d, d^* that increase the spin by one. In both cases there are always terms involving the operators d^*, d that decrease the spin by one.

From the different available representations of the $\Box$ operator for each node, the one that uses operators δ, δ^* is based on the operator of the exterior differentiation of forms. This is clearly the case for the operator δ, see (8.48).

Indeed, this definition can be rephrased by saying that one takes the exterior derivative of a 1-form and then takes the SD part of the arising 2-form, thus producing an object in S_+^2. In contrast, there is no exterior derivative interpretation of the operators d, d^* (apart from the trivial case when these operators act on functions). This conclusion is true in general, i.e., the operators δ, δ^* acting on any node can be reformulated in terms of the exterior derivative, while there is no such interpretation of the operators d, d^*. In other words, the Dirac operator is always a version of the exterior derivative operator. This general statement is supported with more examples of explicit construction of operators δ, δ^* in the following sections.

8.2.9 Choices to Make When Describing a Particle of Given Spin

There are several choices to make if we want to describe particles of a given spin. First, we want to use some field, which is in an irreducible representation of Lorentz group, or can be decomposed into a collection of such representations. Then, when a choice of representations is made, there is another choice as to what representation of the $\Box$ operator should appear in the linearised description. As we discussed previously, there are essentially two inequivalent choices. In one such choice the $\Box$ operator is built from δ, δ^* and this is related to the exterior derivative operator, while there is no such relation to the exterior derivative for the representation of $\Box$ that uses d, d^*.

For example, in our usual description of gauge theory we make a choice to describe spin one particles using fields taking values in $S_+ \otimes S_-$ representation (i.e., vectors, or 1-forms). The operator that is used in the usual description is the one that is related to the exterior derivative operator, and is the gauge invariant operator $\delta^*\delta$. The gauge invariance of this is the consequence of the fact that $\delta d = 0$. To be precise, here we are referring to the chiral version of the usual description of gauge theory, which is one in terms of the action (5.29). Integrating out the SD 2-form field from this action, one produces the Lagrangian $(F_+)^2$, the kinetic operator appearing in which is precisely $\delta^*\delta$. We could have decided to use instead the representation (8.59) of $\Box$ in terms of the operator $d_{(1,1)}$ and its adjoint. However, the gauge invariance would be harder to realise in this description, and this is the reason why it is not what we usually use. In other words, for gauge theory, the relevant kinetic operator is the one arising by going to the right of the diagram 8.1 and then back, and not to the bottom and then back.

We thus see why it is unnatural to use the $d_{(1,1)}$ operator for describing spin one particles. But what about the choice of which irreducible representation to use? The irreducible representation S_+^2 is also of spin one, and so it seems that it can be possible to use a field taking values in that representation instead. The answer as to viability of such a description is again provided by the formulation (5.29), where the auxiliary 2-form field in fact takes values in the space S_+^2. One then sees that only one of the two possible helicities of the spin one particle can be described by the field taking values in S_+^2. The other helicity gets projected away once the operator δ is applied to the gauge field. This becomes very pronounced in the so-called spinor helicity formalism. So, the representation space that sits at the edge of the diagram 8.1 can only describe one of the two helicities one wants to describe. For this reason it cannot be used for the description of both helicities, even though we do have a simple representation of the $\Box$ operator in terms of δ, δ^* on these edge nodes.

Let us now descend one row lower in the diagram and discuss possible descriptions of spin two particles. In the usual metric description everything is based on the sum of two irreducible representations, namely $S_+^2 \otimes S_-^2$, where symmetric tracefree tensors of rank two live, as well as the trivial representation for the trace part. Then in the usual description the kinetic operator on gravitons is built from the operator $d_{(2,2)}$ going down the diagram, and its adjoint. This in very clear from the fact that the gravitational Lagrangian can be written in the $\Gamma\Gamma$ form, see (2.19), and the linearised Christoffel symbol is constructed from the objects of the type $\partial_\rho h_{\mu\nu}$, where the spin of the field $h_{\mu\nu}$ is clearly increased by one by taking the derivative. One can then understand the reason why the kinetic operator for gravitons is not a square of some first-order differential operator. Indeed, we cannot write the kinetic operator on gravitons, which we saw in (2.38) to be a version of Lichnerowitz Laplacian, as some first-order differential operator times its adjoint. This is in contrast to the kinetic operator in the spin one case, where such representation as $\delta^*\delta$ is possible, see (8.53). The reason for this impossibility to write the desired kinetic operator as a square is the fact that the combination of two operators $d_{(1,1)}$ and $d_{(2,2)}$ is not zero. Indeed, the operator $d_{(1,1)}$ is used in metric gravity to represent the effect of diffeomorphisms, in that the change of the metric perturbation taking value in $S_+^2 \otimes S_-^2$ under an infinitesimal diffeomorphism by a vector v field taking values in $S_+ \otimes S_-$ is $d_{(1,1)}v$. The fact that $d_{(2,2)}d_{(1,1)}v \neq 0$ means that we cannot construct the kinetic term with desired gauge invariance as the simple $d^*_{(2,2)}d_{(2,2)}$. A more complicated construction is necessary, and this is what is achieved by the operator (2.38) that appears by linearising the Ricci scalar.

Thus, in the usual metric description of gravity we have made two choices. One was to describe spin two particles using the representation $S_+^2 \otimes S_-^2$ (together with the trivial representation of Lorentz). The other was to base the kinetic term on the operator $d_{(2,2)}$ and its adjoint. Already at this point there is an alternative, which is to construct the kinetic term from the operator $\delta_{(2,2)}$ instead.[1] Thus, for the space $S_+^2 \otimes S_-^2$, there are only two inequivalent representations of the $\Box$ operator that we could use. As we have argued previously, the operators that do not increase the spin of the field are versions of the Dirac operator that are based

[1] We could also use the Dirac operator that goes to the left of the diagram into the representation $S_+ \otimes S_-^3$, but this is just the complex conjugate of the operator $\delta_{(2,2)}$ and so this possibility does not produce anything new.

on the exterior derivative operator. We will see this explicitly for the operator $\delta_{(2,2)}$ in Section 8.5. Thus, we anticipate that there is a different description of gravitons, based on $\delta_{(2,2)}$ rather than the usual description with $d_{(2,2)}$. In Section 8.5 we will see that this alternative is just the linearised description following from the chiral Plebanski formalism.

However, for the description of the spin two particles there arises another alternative. Thus, there are now several different irreducible representations of Lorentz group having spin two. These are the spaces $S_+^2 \otimes S_-^2$ on which the usual metric description is based, as well as the spaces $S_+^3 \otimes S_-$ and S_+^4. The last of these spaces lies at the edge of the diagram, and we argued that these spaces are unsuitable for describing both helicities. This leaves us with the possibility that spin two particles can also be described by a field taking values in $S_+^3 \otimes S_-$, plus possibly some other irreducible representation of lower spin. This is indeed true. This description arises as the linearisation of the chiral pure connection formalism.

All the claims made will be substantiated by the discussion in the following sections. However, already at this stage we see that there are many different alternatives arising in the description of the spin two. The standard metric formalism makes a choice, but this choice is far from being unique. Other choices can be made, and are in no way less natural than the choice of the metric description. In fact, we will see that the choice to base everything on the operator $\delta_{(2,2)}$ in the metric description provides great simplifications. Even more simplicity can be achieved if one instead chooses to describe gravitons using the representation $S_+^3 \otimes S_-$.

We now build some intuition about operators appearing in the diagram 8.1. These are then used in alternative descriptions of spin two particles that we argued should be contemplated.

8.2.10 Operators $d_{(2,0)}$ and $\delta_{(3,1)}$

Before we start the discussion of the operator $\delta_{(2,2)}$, we consider the simpler case of operator $\delta_{(3,1)}$ and related to it operators. We have seen that the upper two rows of the diagram 8.1 give us two operators whose composition δd is zero. This gives us a complex of operators that is important in the standard description of particles of spin one. We will now see that essentially the same story repeats itself for the operators connecting the spaces $S_+^2, S_+^3 \otimes S_-$ and S_+^4, i.e., another triple of spaces lying near the edge of the diagram 8.1. This triple of spaces plays role in the chiral connection description of gravity, as we shall see later.

First, we need to fix the forms and normalisations of the operators $d_{(2,0)}$ and $\delta_{(3,1)}$. We build the operator $d_{(2,0)}$ that increases the spin as the operator that takes a derivative of an object $\phi^i \in S_+^2$ followed by the projection on the representation $S_+^3 \otimes S_-$. The relevant projector from the space of S_+^2 valued 1-forms to $S_+^3 \otimes S_-$ is constructed from the operator J_Σ that was already introduced in (5.137). Thus, we have

$$P_{(3,1)} := \frac{2}{3}\left(\mathbb{I} - \frac{1}{2}J_\Sigma\right). \tag{8.60}$$

Using (5.139) one easily checks that this is a projector $P^2_{(3,1)} = P_{(3,1)}$. Using (5.138) one can also check explicitly that this projector annihilates the S^2_+ valued 1-forms of the form $\Sigma^{i}_{\mu}{}^{\nu}\xi_\nu$, which is the way that the irreducible representation $S_+ \otimes S_-$ sits inside the space $S^2_+ \otimes \Lambda^1 \sim S^2_+ \otimes S_+ \otimes S_-$.

Using the projector (8.60) we define

$$(d_{(2,0)}\phi)^i_\mu := \sqrt{\frac{3}{2}}(\partial_\mu \phi^i)_{(3,1)} = \sqrt{\frac{2}{3}}\left(g_{\mu\nu}\delta^{ij} + \frac{1}{2}\epsilon^{ijk}\Sigma^k_{\mu\nu}\right)\partial^\nu \phi^j. \tag{8.61}$$

In the second equality we have spelled out the projector explicitly. We have introduced a prefactor in front for future convenience. The adjoint operator is given by

$$(d^*_{(2,0)}a)^i = -\sqrt{\frac{3}{2}}\partial^\mu a^i_\mu, \tag{8.62}$$

where $a^i_\mu \in S^3_+ \otimes S_-$. Now, an easy calculation using the algebra (5.138) gives

$$d^*_{(2,0)}d_{(2,0)} = -\Box_{(2,0)}. \tag{8.63}$$

It is our desire to have no extra numerical factors in this relation that has led to the inclusion of $\sqrt{3/2}$ in the definition of $d_{(2,0)}$.

We now bring in the second operator, namely $\delta_{(3,1)}$. This operator can be defined as

$$(\delta_{(3,1)}a)^{ij} := (\Sigma^{i\mu\nu}\partial_\mu a^j_\nu)_{(4,0)} = P^{ijkl}(\Sigma^{k\mu\nu}\partial_\mu a^l_\nu), \tag{8.64}$$

where P_{ijkl} is the projector on symmetric tracefree tensors given by (5.182). The adjoint is given by

$$(\delta^*_{(3,1)}\psi)^i_\mu = \Sigma^{j}_{\mu}{}^{\nu}\partial_\nu \psi^{ij}. \tag{8.65}$$

It is easy to check that the $S_+ \otimes S_-$ part of the right-hand side here vanishes for ψ^{ij} symmetric and tracefree, and so the right-hand side is automatically in $S^3_+ \otimes S_-$ and no explicit projector is necessary.

To check that the normalisation of $\delta_{(3,1)}$ is well chosen, we compute $\delta_{(3,1)}\delta^*_{(3,1)}$ as it acts on $\psi^{ij} \in S^4_+$. A straightforward computation gives

$$\delta_{(3,1)}\delta^*_{(3,1)} = -\Box_{(4,0)}, \tag{8.66}$$

which confirms our choice of the pre factor in (8.64).

The most important fact about $\delta_{(3,1)}$ is that its kernel contains the image of $d_{(2,0)}$

$$\delta_{(3,1)}d_{(2,0)} = 0. \tag{8.67}$$

Thus, we are once again in the cohomology setup, in exact analogy to the complex formed by the operators d, δ. This means that elements of $S_+^3 \otimes S_-$ admit an orthogonal decomposition into elements of the form $d_{(2,0)}\phi$, those of the form $\delta^*_{(3,1)}\psi$, as well as harmonic elements that are in the kernel of the Box-type operator $\delta^*_{(3,1)}\delta_{(3,1)} + d_{(2,0)}d^*_{(2,0)}$. An explicit calculation of this operator gives

$$\delta^*_{(3,1)}\delta_{(3,1)} + d_{(2,0)}d^*_{(2,0)} = -\Box_{(3,1)}. \tag{8.68}$$

To do this computation, we have used the fact that for S_+^2 valued 1-forms that are in $S_+^3 \otimes S_-$, we have

$$a^i_\mu + \epsilon^{ijk}\Sigma^{j\ \nu}_{\mu} a^k_\nu = 0, \tag{8.69}$$

which is just the expression of the fact that the projection onto the space $S_+ \otimes S_-$ vanishes. The relevant projector is $P_{(1,1)} = (1/3)(\mathbb{I} + J_\Sigma)$, which immediately gives the relation (8.69). This allows us to conclude that on $S_+^3 \otimes S_-$

$$\Sigma^{i\ \nu}_{\mu} a^j_\nu = \Sigma^{j\ \nu}_{\mu} a^i_\nu - \epsilon^{ijk} a^k_\mu, \tag{8.70}$$

which is a relation that one has to use in the computation leading to (8.68).

8.2.11 The Space $S_+^2 \otimes S_-^2$ and Related Operators

The space $S_+^2 \otimes S_-^2$ is one used to describe gravitons in the usual metric formalism. The operator $d_{(1,1)}$ mapping vectors to symmetric tracefree tensors is a familiar one. In the standard metric story, one is not considering 'chiral' objects that can be built using $\Sigma^i_{\mu\nu}$, and, therefore, the only other operator that is usually part of this story is $d_{(2,2)}$, which maps the spin two field into a spin three object. If one allows the description to become chiral one gets access to another operator, namely $\delta_{(2,2)}$. As we have already said, this operator appears naturally in the linearisation of the Plebanski formalism. Let us define and understand this operator in some details. As in the previous sections we do all the computations in tensor notations, as this is much more algorithmic and better suited for computations than the spinor notation.

The space $S_+^2 \otimes S_-^2$ behaves in a much more complicated way than it was in the case for $S_- \otimes S_+$ and $S_+^3 \otimes S_-$. The primary reason for this is the absence of the cohomological decomposition for $S_+^2 \otimes S_-^2$. Indeed, the two natural operators that lead to and then from $S_+^2 \otimes S_-^2$, namely $d_{(1,1)}$ and $\delta_{(2,2)}$, have the property $\delta_{(2,2)}d_{(1,1)} \neq 0$. Thus, no cohomological decomposition of the type 'exact, co-exact, and harmonic' is possible in the case of $S_+^2 \otimes S_-^2$, unlike $S_- \otimes S_+$ and $S_+^3 \otimes S_-$. It is this fact that is behind the increased complexity of the usual perturbation theory based on $S_+^2 \otimes S_-^2$, as compared to the perturbation theory based on $S_+ \otimes S_-$ and $S_+^3 \otimes S_-$.

To define $\delta_{(2,2)}$ we note that from the object $h_{\mu\nu} \in S_+^2 \otimes S_-^2$ we can construct an $\mathbb{R}^3$ valued 2-form $\Sigma^{i\ \alpha}_{[\mu} h_{\nu]\alpha}$. It can be checked that this 2-form is purely ASD. This will be explained in details in Section 8.5, when we consider the linearisation of

Plebanski theory. We can thus obtain a first-order differential operator by taking the exterior derivative of this 2-form, and then dualising the arising 3-form into a 1-form. We obtain an element of $S_+^2 \otimes \Lambda^1$. This can then be further projected onto $S_+^3 \otimes S_-$. This results in the following definition

$$(\delta_{(2,2)}h)^i_\mu := \sqrt{6}\,(\Sigma^{i\alpha\beta}\partial_\alpha h_{\beta\mu})_{(3,1)} = 2\sqrt{\frac{2}{3}}\left(g_{\mu\nu}\delta^{ij} + \frac{1}{2}\epsilon^{ijk}\Sigma^k_{\mu\nu}\right)(\Sigma^{j\alpha\beta}\partial_\alpha h_{\beta\nu}), \tag{8.71}$$

where we spelled out the projector. This can be further simplified by making use of the following identity

$$\epsilon^{ijk}\Sigma^j_{\mu\nu}\Sigma^k_{\rho\sigma} = -2\Sigma^i_{\mu[\rho}\eta_{\sigma]\nu} + 2\Sigma^i_{\nu[\rho}\eta_{\sigma]\mu}. \tag{8.72}$$

This identity can be proven, e.g., by replacing both Σ's with $2\mathrm{i}\Sigma^i_{\mu\nu} = \epsilon_{\mu\nu}{}^{\alpha\beta}\Sigma^i_{\alpha\beta}$ and then using the decomposition of the product of two ϵ's into products of the metric tensor. At the same time, if one knows that such an identity must be true, the coefficient on the right-hand side can be checked by contracting $\nu\rho$ and using the algebra of Σ's. Using (8.72) in the second term of (8.71) gives

$$(\delta_{(2,2)}h)^{i\mu} = \sqrt{6}\left(\Sigma^{i\alpha\beta}\partial_\alpha h_{\beta\mu} - \frac{1}{3}\Sigma^i_\mu{}^\alpha\partial^\nu h_{\nu\alpha}\right). \tag{8.73}$$

The adjoint operator is easier to write, and it is given by

$$(\delta^*_{(2,2)}a)_{\mu\nu} = \sqrt{6}\,\Sigma^i_{(\mu}{}^\alpha\partial_\alpha a^i_{\nu)}. \tag{8.74}$$

To compute the adjoint we have used (8.69). Here we only needed to symmetrise the result on the right-hand side, as it is automatically tracefree as follows from $\Sigma^i_\mu{}^\nu a^i_\nu = 0$.

The choice of the numerical coefficient in (8.71) is justified by the following property that can be checked by an explicit computation

$$d_{(2,0)}\delta = \delta_{(2,2)}d_{(1,1)}. \tag{8.75}$$

In other words, the box connecting the spaces $S_+\otimes S_-$, $S_+^2\otimes S_-^2$, S_+^2, and $S_+^3\otimes S_-$ in the diagram 8.1 commutes.

Having defined the relevant operators, we can compute the second-order operator $\delta^*_{(2,2)}\delta_{(2,2)}$. A simple computation, using the previous definitions, gives

$$\delta^*_{(2,2)}\delta_{(2,2)} = -6\Box_{(2,2)} - 8d_{(1,1)}d^*_{(1,1)}. \tag{8.76}$$

Another useful result is a computation of the operator $\delta^*_{(2,2)}\delta_{(2,2)}$, which is a $\Box$-type operator on $S_+^3\otimes S_-$. We get

$$\delta^*_{(2,2)}\delta_{(2,2)} = -6\Box_{(3,1)} - 4d_{(2,0)}d^*_{(2,0)}. \tag{8.77}$$

These results will be useful in our discussion of the chiral version of the metric perturbation theory in Section 8.5.

8.3 Minkowski Space Metric Perturbation Theory

We now develop different versions of the spin two perturbation theory, starting from the usual metric description, which we present very briefly.

We have derived the linearisation of the Einstein–Hilbert action in (2.24). Working in units $32\pi G = 1$ the linearised Lagrangian reads

$$L^{(2)} = \frac{1}{2} h^{\mu\nu} \Box h_{\mu\nu} + \partial_\mu h^{\mu\rho} \partial^\nu h_{\nu\rho} - \partial_\mu h^{\mu\nu} \partial_\nu h - \frac{1}{2} h \Box h. \tag{8.78}$$

This can be further simplified by a choice of gauge. We choose the gauge-fixing term to be

$$L_{\text{g.f.}} = -(\partial^\nu h_{\mu\nu} - \frac{1}{2}\partial_\mu h)^2, \tag{8.79}$$

where $h := \eta^{\mu\nu} h_{\mu\nu}$ is the trace part of the perturbation. Adding this to the linearised Lagrangian, we get

$$L^{(2)} + L_{\text{g.f.}} = \frac{1}{2} h^{\mu\nu} \Box h_{\mu\nu} - \frac{1}{4} h \Box h, \tag{8.80}$$

which is diagonal in the metric perturbation and can be easily inverted to find the propagator.

To invert the kinetic term, we couple the gauge-fixed Lagrangian to a current by adding $J^{\mu\nu} h_{\mu\nu}$. We then solve for $h_{\mu\nu}$ in terms of $J_{\mu\nu}$ in the momentum space, with the result being

$$h_{\mu\nu} = \frac{1}{k^2} \left(J_{\mu\nu} - \frac{1}{2} \eta_{\mu\nu} \eta^{\rho\sigma} J_{\rho\sigma} \right). \tag{8.81}$$

Substituting this to the Lagrangian and remembering that there is a factor of i in front of the Lorentzian signature action gives us the following propagator

$$\langle h_{\mu\nu} h_{\rho\sigma} \rangle = -\frac{\mathrm{i}}{2k^2} (\eta_{\mu\rho}\eta_{\nu\sigma} + \eta_{\mu\sigma}\eta_{\nu\rho} - \eta_{\mu\nu}\eta_{\rho\sigma}). \tag{8.82}$$

This is the so-called de Donder gauge propagator.

The interaction vertices obtained by naively expanding the Einstein–Hilbert action in powers of the metric perturbation quickly become very complicated, see, e.g., the appendix of Goroff and Sagnotti (1986). Some simplifications can be achieved by using a nonlinear parametrisation, e.g., treating the inverse densitiesed metric as the basic field; see Cheung and Remmen (2017). However, whatever the choice of the parametrisation of the metric field, there are vertices of arbitrarily high order, which makes calculations difficult. A way out is to introduce an additional auxiliary field by going to the first-order formalism. One version of the first-order formalism is that of Palatini in which the auxiliary field is an independent affine conneciton. In this case, using the inverse densitiesed metric as the basic field, one can render theory to be cubic; see Cheung and Remmen (2017). Also, by a shift of the $\Gamma^\rho{}_{\mu\nu}$ field one can achieve a situation where only the metric–metric and connection–connection propagators are different from zero; see Cheung and Remmen (2017). The structure of the arising cubic vertices, however, is quite complicated. Our goal in the remaining sections of this chapter is to develop simpler alternatives.

8.4 Chiral Yang–Mills Perturbation Theory

In preparation for the discussion in the following sections, we first develop the chiral version of the Yang–Mills perturbation theory. This will serve as a point of comparison for the developments in the gravity case. We shall see that the Yang–Mills case is very simple and beautiful. Our goal in the sections to follow will be to come as close as possible to this description.

8.4.1 Lagrangian and Perturbative Expansion

Our starting point will be the first-order chiral formalism for Yang–Mills (5.29). We will parametrise the SD 2-form field as $B^{+a} = \Sigma^i \phi^{ia}$, where Σ^i is the usual basis of SD 2-forms and ϕ^{ia} is the new set of fields. The Lagrangian reads

$$L = \Sigma^{i\mu\nu}\phi^{ia}F^a_{\mu\nu} + 2g^2(\phi^{ia})^2 = 2\Sigma^{i\mu\nu}\phi^{ia}\partial_\mu A^a_\nu + 2g^2(\phi^{ia})^2 + \Sigma^{i\mu\nu}\phi^{ia}f^{abc}A^b_\mu A^c_\nu. \tag{8.83}$$

The first line is the kinetic term, while the second line gives the only present cubic interaction.

We can immediately note that the kinetic term of the Lagrangian (8.83) is composed of the operator δ that we introduced in (8.48). This operator is gauge-invariant, and upon integrating the auxiliary field ϕ^{ia} out, one gets $\delta^*\delta$ as the gauge-invariant kinetic term for the spin one particles.

8.4.2 Gauge-Fixing: Tensor Calculation

To determine the gauge-fixing that is required in order to invert the kinetic term and obtain propagators let us integrate out the auxiliary field ϕ^{ia} from the quadratic part of the Lagrangian. We have

$$\Sigma^{i\mu\nu}\partial_\mu A^a_\nu + 2g^2\phi^{ia} = 0, \tag{8.84}$$

and thus

$$L^{(2)} = -\frac{1}{2g^2}(\Sigma^{i\mu\nu}\partial_\mu A^a_\nu)^2 = -\frac{1}{2g^2}((\partial_\mu A^a_\nu)^2 - (\partial^\mu A^a_\mu)^2), \tag{8.85}$$

which is the standard result for the spin one kinetic term. It is then clear that we must add to this Lagrangian the following gauge-fixing term

$$L_{\text{g.f.}} = -\frac{1}{2g^2}(\partial^\mu A^a_\mu)^2. \tag{8.86}$$

We can represent it in the first-order form by introducing a new auxiliary field ϕ^a

$$L_{\text{g.f.}} = 2\phi^a \partial^\mu A^a_\mu + 2g^2(\phi^a)^2. \tag{8.87}$$

We can then note that the gauge-fixed Lagrangian can be described very efficiently by 'enlarging' the auxiliary field $\Sigma^i \phi^{ia}$ in (8.83). Indeed, we can write the gauge-fixed Lagrangian as

$$L^{(2)} + L_{\text{g.f.}} = 2\Phi^{a\mu\nu}\partial_\mu A^a_\nu + \frac{g^2}{2}(\Phi^{a\mu\nu})^2, \tag{8.88}$$

where we introduced a new combination

$$\Phi^{a\mu\nu} := \Sigma^{i\mu\nu}\phi^{ia} + \eta^{\mu\nu}\phi^a. \tag{8.89}$$

We note that we can even write the interaction using the new field $\Phi^{a\mu\nu}$. Indeed, the object $f^{abc}A^b_\mu A^c_\nu$ is $\mu\nu$ symmetric, and so one can enlarge $\Sigma^{i\mu\nu}\phi^{ia}$ in it into $\Phi^{a\mu\nu}$ for free. Thus, we can write

$$L^{(3)} = \Phi^{a\mu\nu} f^{abc} A^b_\mu A^c_\nu. \tag{8.90}$$

In Section 8.4.5 we will see that the object $\Phi^{a\mu\nu}$ has a very simple and natural spinor interpretation.

8.4.3 Propagators

Having gauge-fixed the kinetic term, let us carry out the exercise of computing the propagators. Our goal is in particular to see that the propagator of the auxiliary field ϕ^{ia} with itself is zero.

To simplify the calculation we will only add the current for the auxiliary field ϕ^{ia}, and not for ϕ^a introduced previously. Thus, we consider the following Lagrangian

$$L = 2\Sigma^{i\mu\nu}\phi^{ia}\partial_\mu A^a_\nu + 2g^2(\phi^{ia})^2 + 2\phi^a\partial^\mu A^a_\mu + 2g^2(\phi^a)^2 + J^{a\mu}A^a_\mu + J^{ia}\phi^{ia}. \tag{8.91}$$

We now integrate out the auxiliary fields. The equation for ϕ^{ia} gets modified to

$$\Sigma^{i\mu\nu}\partial_\mu A^a_\nu + 2g^2\phi^{ia} + \frac{1}{2}J^{ia} = 0. \tag{8.92}$$

This gives the Lagrangian with auxiliary fields integrated out

$$\begin{aligned} L &= -\frac{1}{2g^2}\left(\Sigma^{i\mu\nu}\partial_\mu A^a_\nu + \frac{1}{2}J^{ia}\right)^2 - \frac{1}{2g^2}(\partial^\mu A^a_\mu)^2 + J^{a\mu}A^a_\mu \\ &= -\frac{1}{2g^2}(\partial_\mu A^a_\nu)^2 - \frac{1}{2g^2}J^{ia}\Sigma^{i\mu\nu}\partial_\mu A^a_\nu - \frac{1}{8g^2}(J^{ia})^2 + J^{a\mu}A^a_\mu. \end{aligned} \tag{8.93}$$

We now integrate out the connection by going to the momentum space. The momentum space Lagrangian becomes

$$L = -\frac{1}{2g^2} A^a_\mu(-k) k^2 A^a_\mu(k) - \frac{\mathrm{i}}{2g^2} J^{ia}(-k) \Sigma^{i\mu\nu} k_\mu A^a_\nu(k) - \frac{1}{8g^2} J^{ia}(-k) J^{ia}(k) + J^{a\mu}(-k) A^a_\mu(k). \tag{8.94}$$

Extremising with respect to A^a_μ gives

$$A^a_\mu = \frac{1}{k^2}\left(-\frac{\mathrm{i}}{2} J^{ia} \Sigma^{i\ \nu}_{\mu} k_\nu + g^2 J^a_\mu\right). \tag{8.95}$$

This gives the following Lagrangian in terms of currents only

$$L = \frac{1}{2g^2 k^2}\left(\frac{\mathrm{i}}{2} J^{ia}(-k) \Sigma^{i\ \alpha}_{\mu} k_\alpha + g^2 J^a_\mu(-k)\right)\left(-\frac{\mathrm{i}}{2} J^{ia}(k) \Sigma^{i\mu\beta} k_\beta + g^2 J^{a\mu}(k)\right) - \frac{1}{8g^2} J^{ia}(-k) J^{ia}(k).$$

The connection to connection and connection to auxiliary field terms here are

$$\frac{g^2}{2k^2} J^{a\mu}(-k) J^a_\mu(k) - \frac{\mathrm{i}}{2k^2} J^{ia}(-k) \Sigma^{i\mu\nu} k_\mu J^a_\nu(k). \tag{8.96}$$

The remaining terms are

$$\frac{1}{8g^2 k^2} J^{ia}(-k) \Sigma^{i\mu\nu} k_\nu J^{ja}(k) \Sigma^{j\ \rho}_{\mu} k_\rho - \frac{1}{8g^2} J^{ia}(-k) J^{ia}(k) = 0, \tag{8.97}$$

where we have used (5.138). Thus, there are only the connection–connection propagator

$$\langle A^a_\mu(-k) A^b_\nu(k)\rangle = \frac{g^2}{\mathrm{i}k^2} \delta^{ab} \eta_{\mu\nu}, \tag{8.98}$$

as well as the connection to auxiliary field propagator

$$\langle \phi^{ia}(-k) A^b_\mu(k)\rangle = \frac{1}{2k^2} \delta^{ab} \Sigma^{i\ \nu}_{\mu} k_\nu. \tag{8.99}$$

The cubic vertex is also extremely simple in this version of perturbation theory and is given by

$$\langle \phi^{ia} A^b_\mu A^c_\nu\rangle = 2\mathrm{i} f^{abc} \Sigma^i_{\mu\nu}. \tag{8.100}$$

8.4.4 Non-Chiral Version

The fact that there is no propagator of the auxiliary field with itself is directly related to the fact that the chiral version of the first-order formalism is used rather than the non-chiral. Let us see this explicitly. Thus, we work out also the propagators for the version of the theory where the auxiliary 2-form field is arbitrary antisymmetric rank two tensor. The corresponding Lagrangian is

$$L = B^{a\mu\nu} F^a_{\mu\nu} + g^2 (B^a_{\mu\nu})^2. \tag{8.101}$$

The kinetic term is

$$L^{(2)} = 2B^{a\mu\nu}\partial_\mu A^a_\nu + g^2(B^a_{\mu\nu})^2. \tag{8.102}$$

We now do the gauge-fixing as before, see (8.86) and add currents for both fields

$$L^{(2)} + L_{\text{g.f.}} = 2B^{a\mu\nu}\partial_\mu A^a_\nu + g^2(B^a_{\mu\nu})^2 - \frac{1}{2g^2}(\partial^\mu A^a_\mu)^2 + J^{a\mu\nu}B^a_{\mu\nu} + J^{a\mu}A^a_\mu.$$

Extremising with respect to the auxiliary field gives

$$B^a_{\mu\nu} = -\frac{1}{g^2}\left(\partial_{[\mu}A^a_{\nu]} + \frac{1}{2}J^a_{\mu\nu}\right). \tag{8.103}$$

The Lagrangian with the auxiliary field integrated out is

$$\begin{aligned} L &= -\frac{1}{g^2}\left(\partial_{[\mu}A^a_{\nu]} + \frac{1}{2}J^a_{\mu\nu}\right)^2 - \frac{1}{2g^2}(\partial^\mu A^a_\mu)^2 + J^{a\mu}A^a_\mu \\ &= -\frac{1}{2g^2}(\partial_\mu A^a_\nu)^2 - \frac{1}{g^2}J^{a\mu\nu}\partial_\mu A^a_\nu - \frac{1}{4g^2}(J^{a\mu\nu})^2 + J^{a\mu}A^a_\mu. \end{aligned} \tag{8.104}$$

We now integrate out the connection field by going to the momentum space. We get

$$A^a_\mu = \frac{1}{k^2}\left(-\mathrm{i}J^{a\nu}_\mu k_\nu + g^2 J^a_\mu\right). \tag{8.105}$$

This gives the following Lagrangian of the currents

$$\begin{aligned} L = &\frac{1}{2g^2k^2}\left(\mathrm{i}J^{a\mu\nu}(-k)k_\nu + g^2J^{a\mu}(-k)\right)\left(-\mathrm{i}J^{a\nu}_\mu(k)k_\nu + g^2J^a_\mu(k)\right) \\ &- \frac{1}{4g^2}J^{a\mu\nu}(-k)J^a_{\mu\nu}(k). \end{aligned} \tag{8.106}$$

The connection to connection and connection to the auxiliary field terms here are

$$\frac{g^2}{2k^2}J^{a\mu}(-k)J^a_\mu(k) - \frac{\mathrm{i}}{k^2}J^{a\mu}(-k)J^{a\nu}_\mu k_\nu, \tag{8.107}$$

while the auxiliary field terms are

$$\frac{1}{2g^2k^2}J^{a\mu\nu}(-k)k_\nu J^{a\rho}_\mu k_\rho - \frac{1}{4g^2}J^{a\mu\nu}(-k)J^a_{\mu\nu}(k). \tag{8.108}$$

There is no cancellation here and the propagator of the auxiliary field with itself is different from zero. This makes the non-chiral perturbation theory much more complicated, which is a very compelling reason to prefer the chiral version.

8.4.5 Spinor Interpretation

The previous chiral story has a very simple and natural spinor interpretation. To see it, we translate the kinetic term in (8.83) into spinor notations. This is simple given the spinor expression (8.26) for the 2-forms Σ^i. We have

$$L^{(2)} = 2\epsilon_M{}^A \epsilon_N{}^B \phi^a_{AB} \epsilon_{M'N'} \partial^{MM'} A^{aNN'} + 2g^2(\phi^a_{AB})^2. \tag{8.109}$$

The kinetic term here can be simplified with the result being

$$L^{(2)} = 2\phi^{aAB} \partial_{AM'} A^a_B{}^{M'} + 2g^2(\phi^a_{AB})^2. \tag{8.110}$$

Integrating out the auxiliary field we get an extremely simple form of the gauge field kinetic term

$$L^{(2)} = -\frac{1}{2}(\partial_{(AM'} A^a_{B)}{}^{M'})^2. \tag{8.111}$$

This form of the Lagrangian makes the gauge invariance obvious. Indeed, we have the identity

$$\partial_{AM'} \partial_B{}^{M'} = -\frac{1}{2}\epsilon_{AB} \Box, \tag{8.112}$$

and then in computing the effect of the gauge transformation the antisymmetric ϵ_{AB} is killed by the symmetrisation present in the Lagrangian.

As we know from the previous discussion, this Lagrangian is gauge-fixed by replacing the combination $\Sigma^{a\mu\nu}\phi_{ia}$ with (8.89). The spinor version of this replacement is

$$\begin{aligned} \epsilon_M{}^A \epsilon_N{}^B \phi^a_{AB} \epsilon_{M'N'} &\to \epsilon_M{}^A \epsilon_N{}^B \phi^a_{AB} \epsilon_{M'N'} - \phi^a \epsilon_{MN} \epsilon_{M'N'} \\ &= \epsilon_M{}^A \epsilon_N{}^B \phi^a_{AB} \epsilon_{M'N'} - \frac{1}{2}\phi^a \epsilon_{AB} \epsilon^{AB} \epsilon_{MN} \epsilon_{M'N'}, \end{aligned} \tag{8.113}$$

where we rewrote the last term in a suggestive way. This expression can also be rewritten as

$$(\phi^a_{AB} + \phi^a \epsilon_{AB}) \left(\epsilon_M{}^{(A} \epsilon_N{}^{B)} - \frac{1}{2}\epsilon_{MN} \epsilon^{AB} \right) \epsilon_{M'N'} = (\phi^a_{AB} + \phi^a \epsilon_{AB}) \epsilon_M{}^B \epsilon_N{}^A \epsilon_{M'N'}, \tag{8.114}$$

where we have used the Schouten identity

$$\epsilon_{MN} \epsilon^{AB} = \epsilon_M{}^A \epsilon_N{}^B - \epsilon_M{}^B \epsilon_N{}^A. \tag{8.115}$$

Thus, if we introduce a new spinor field

$$\Phi^a_{AB} := \phi^a_{AB} + \phi^a \epsilon_{AB} \tag{8.116}$$

that does takes values in a reducible representation $S_+ \otimes S_+ = S^2_+ \oplus C^\infty$, we can write the gauge-fixed Lagrangian as

$$L^{(2)} + L_{\text{g.f.}} = 2\Phi^{aAB} \partial_{BM'} A^a_A{}^{M'} + 2g^2(\Phi^a_{AB})^2. \tag{8.117}$$

Integrating out the new auxiliary field produces

$$L^{(2)} + L_{\text{g.f.}} = -\frac{1}{2}\left(\partial_{BM'} A^a_A{}^{M'}\right)^2, \tag{8.118}$$

where the difference with (8.111) is that there is no symmetrisation anymore.

This can be interpreted as follows. What appears in the gauge-fixed kinetic term is the Dirac operator

$$\delta : \mathfrak{g} \otimes S_+ \otimes S_- \to \mathfrak{g} \otimes S_+ \otimes S_+, \tag{8.119}$$

where the primed index of $A^a_{MM'}$ is changed into an unprimed index with the Dirac operator $\partial_N{}^{M'}$. Upon integrating out the auxiliary field Φ^a_{AB} we then get the Dirac operator squared, which is a multiple of the $\Box$ operator. This is the conceptual explanation of the mechanism that is at play in the gauge-fixed of the chiral version of the Yang–Mills perturbation theory. In other words, gauge-fixed is achieved by simply adding a new, antisymmetric in the AB component to the auxiliary field ϕ^a_{AB}; see (8.116). This removes the symmetry of this auxiliary field and produces a gauge-fixed operator that is just the usual Dirac operator acting on the space of spinors valued in $\mathfrak{g} \otimes S_+$. In the next section we will see that a very similar mechanism is at play in the chiral versions of the gravitational perturbation theory.

8.5 Minkowski Space Chiral First-Order Perturbation Theory

The goal of this section is to develop the chiral version of the first-order perturbation theory for gravity. Thus, we use the tetrad formalism, in which the action is polynomial in the fields. However, the non-chiral version of this formalism carries too many connection field components, similar to the non-chiral version of the Yang–Mills first-order perturbation theory we considered in Section 8.4.4. This has the effect that the auxiliary field (2-form field in the case of non-chiral Yang–Mills theory, spin connection in the case of non-chiral gravity) will have a nonvanishing propagator with itself. This produces complications and the resulting perturbation theory is far from being the most efficient one. The usage of chiral variables solves this problem in both Yang–Mills theory and gravity, as we shall see. This makes the chiral version of the gravitational perturbation theory preferable, and we only develop this chiral version.

We perform calculations mostly using tensor notations, and only at the end translate everything into the language of spinors for interpretation. Our starting point for the chiral first-order flat space perturbation theory is the Plebanski action (5.159), which we write as

$$S_{\text{chiral}}[h, A] = \frac{4}{\mathrm{i}} \int \Sigma^i F^i, \tag{8.120}$$

where we have put $\Lambda = 0, 32\pi G = 1$ and it is understood that the constraint $\Sigma^i \Sigma^j \sim \delta^{ij}$ is imposed to imply that Σ^i is constructed from the metric denoted schematically by h as one of the arguments of the action functional. We should keep in mind that Σ^i is not an independent variable, and only the first and second perturbations of Σ^i are nonzero. The SO(3) indices are raised and lowered with the Kronecker delta, metric, and so we can always keep them in the upper position.

We now expand the action (8.120) working around the background described by Σ^i corresponding to the flat metric and the zero connection. Expanding the action by replacing Σ^i with $\Sigma^i + \delta\Sigma^i + (1/2)\delta^2\Sigma^i$ and keeping the quadratic, cubic, and quartic terms, we get

$$S_{\text{chiral}}[h, a] = \frac{4}{\mathrm{i}} \int \delta\Sigma^i da^i + \frac{1}{2}\epsilon^{ijk}\Sigma^i a^j a^k + \frac{1}{2}\epsilon^{ijk}\delta\Sigma^i a^j a^k + \frac{1}{2}\delta^2\Sigma^i da^i \\ + \frac{1}{4}\epsilon^{ijk}\delta^2\Sigma^i a^j a^k. \tag{8.121}$$

Thus, we have a quartic formulation of GR as is appropriate for a version of the tetrad formalism. We now need to understand how to gauge-fix and invert the kinetic term. To do this, it is necessary to understand the gauge invariances that we want to gauge-fix.

8.5.1 Linearised Gauge Invariance

The linearised (i.e., second-order) Lagrangian is invariant under the following symmetries. The diffeomorphisms act only on the metric perturbation

$$\delta_\xi \delta\Sigma^i = d i_\xi \Sigma^i. \tag{8.122}$$

The linearised action is clearly invariant under this transformation by integration by parts in the first term. The gauge transformations act on both fields

$$\delta_\phi \delta\Sigma^i = \epsilon^{ijk}\Sigma^j \phi^k, \qquad \delta_\phi a^i = d\phi^i. \tag{8.123}$$

Substituting this into the linearised action produces

$$\int \epsilon^{ijk}\Sigma^j \phi^k da^i + \epsilon^{ijk}\Sigma^i d\phi^j a^k \tag{8.124}$$

which vanishes by integration by parts in one of the two terms.

8.5.2 Parametrisation of the Perturbation of the 2-Form Field

Let us now discuss implications of the metricity constraints and devise a convenient parametrisation of the object $\delta\Sigma^i$. As we have already discussed in (5.172), the metricity constraints imply that the spin two part of the perturbation of the 2-form field vanishes. This means that this perturbation is of the form

$$\delta\Sigma^i = \frac{1}{4}(h\delta^{ij} + \epsilon^{ijk}\xi_k)\Sigma^j + \frac{1}{4}h^{ij}\bar{\Sigma}^j, \tag{8.125}$$

where the reason for the factors of 1/4 will become clear in Section 8.5.3. We now show that a general 2-form of this type can be written as

$$\delta\Sigma^i_{\mu\nu} = \Sigma^{i\ \alpha}_{[\mu} h_{\nu]\alpha}, \tag{8.126}$$

where $h_{\mu\nu}$ is an arbitrary, i.e., not necessarily symmetric, tensor. Here, again, there is some freedom in the choice of the numerical coefficient. The choice we made makes $h_{\mu\nu}$ transform under linearised diffeomorphisms in the standard way; see Section 8.5.3.

To establish (8.126) as the correct parametrisation we can compute both the SD and ASD parts of this 2-form by projecting it onto $\Sigma^j, \bar{\Sigma}^j$. We get

$$\begin{aligned}\Sigma^{j\mu\nu}\delta\Sigma^i_{\mu\nu} = \Sigma^{j\mu\nu}\Sigma^{i\ \alpha}_{\mu} h_{\nu\alpha} &= (\delta^{ij}\eta^{\nu\alpha} - \epsilon^{jik}\Sigma^{k\nu\alpha})h_{\nu\alpha} \\ &= \delta^{ij}h + \epsilon^{ijk}\Sigma^{k\mu\nu}h_{\mu\nu}.\end{aligned} \tag{8.127}$$

This shows that the SD part of $\delta\Sigma^i$ is of the form (8.125) with h in both expressions matching and $\xi^i = \Sigma^{i\mu\nu}h_{\mu\nu}$. Note that h is the same object in both expressions if there is a factor of 1/4 in (8.125). For the ASD part we have

$$\bar{\Sigma}^{j\mu\nu}\delta\Sigma^i_{\mu\nu} = \bar{\Sigma}^{j\mu\nu}\Sigma^{i\ \alpha}_{\mu} h_{\nu\alpha}. \tag{8.128}$$

This expression cannot be simplified any further. The only thing we know about the contractions $\bar{\Sigma}^{j\mu\nu}\Sigma^{i\ \alpha}_{\mu}$ is that for every i and j this is a symmetric in $\nu\alpha$ tensor. Thus, in the ASD projection only the symmetric part of $h_{\mu\nu}$ contributes. Again, we see that the ASD part of $\delta\Sigma^i$ is as in (8.125) with

$$h^{ij} = \Sigma^{i\ \mu}_{\alpha}\bar{\Sigma}^{j\alpha\nu}h_{\nu\mu}. \tag{8.129}$$

This establishes (8.126) as a convenient parametrisation of the first-order perturbation of the 2-form field.

8.5.3 The Action of Gauge on the Tensor $h_{\mu\nu}$

Let us also spell out the action of the gauge transformations on the field $h_{\mu\nu}$. For the diffeomorphisms (8.122) we have

$$\delta_\xi\delta\Sigma^i_{\mu\nu} = 2d_{[\mu}\xi^\alpha\Sigma^i_{\alpha\nu]} = 2\Sigma^{i\ \alpha}_{[\mu}\partial_{\nu]}\xi_\alpha. \tag{8.130}$$

Comparing with (8.126) we see that the corresponding transformation of the tensor $h_{\mu\nu}$ is

$$\delta_\xi h_{\mu\nu} = 2\partial_\mu\xi_\nu, \tag{8.131}$$

where no symmetrisation is taken. This shows that the symmetric part of our tensor $h_{\mu\nu}$ is just the usual metric perturbation. This explains our choice of having no extra numerical factor in (8.126).

Let us also compute the effect of SO(3) rotations (8.123). We have

$$\delta_\phi h_{\mu\nu} = \Sigma^k_{\mu\nu}\phi^k, \tag{8.132}$$

which is just the shift of the SD part of the antisymmetric part of $h_{\mu\nu}$. We also note that the expression (8.126) is independent of the ASD part of the antisymmetric part of $h_{\mu\nu}$.

8.5.4 The Linearised Action in Terms of the Tensor $h_{\mu\nu}$

We now compute the linearised part of the action, which is the first two terms in (8.121) in terms of the parametrisation (8.126). We have

$$\delta\Sigma^i da^i = \frac{1}{2}\Sigma^{i\ \alpha}_{\mu} h_{\nu\alpha}\partial_\rho a^i_\sigma dx^\mu dx^\nu dx^\rho dx^\sigma = \frac{1}{2}\Sigma^{i\ \alpha}_{\mu} h_{\nu\alpha}\partial_\rho a^i_\sigma \epsilon^{\mu\nu\rho\sigma} d^4x.$$

This expression can be further simplified as follows. Using self-duality, we have

$$\Sigma^{i\ \alpha}_{\mu} = \frac{1}{2\mathrm{i}}\epsilon_\mu{}^{\alpha\beta\gamma}\Sigma^i_{\beta\gamma}. \tag{8.133}$$

This implies that

$$\frac{1}{2}\epsilon^{\mu\nu\rho\sigma}\Sigma^{i\ \alpha}_{\mu} h_{\nu\alpha} = \frac{\mathrm{i}}{2}\left(\Sigma^{i\rho\sigma}h - \Sigma^{i\rho\alpha}h_\alpha{}^\sigma + \Sigma^{i\sigma\alpha}h_\alpha{}^\rho\right), \tag{8.134}$$

where we have expanded the product of two ϵ tensors into the sum of products of copies of the metric tensor.

Similarly, for the second term in the linearised action we have

$$\frac{1}{2}\epsilon^{ijk}\Sigma^i a^j a^k = \frac{1}{4}\epsilon^{\mu\nu\rho\sigma}\epsilon^{ijk}\Sigma^i_{\mu\nu}a^j_\rho a^k_\sigma d^4x = \frac{\mathrm{i}}{2}\epsilon^{ijk}\Sigma^{i\mu\nu}a^j_\mu a^k_\nu d^4x, \tag{8.135}$$

where we again used self-duality of the $\Sigma^i_{\mu\nu}$ forms.

Overall, the linearised Lagrangian becomes

$$L^{(2)} = 2\left(\Sigma^{i\mu\nu}h - \Sigma^{i\mu\alpha}h_\alpha{}^\nu + \Sigma^{i\nu\alpha}h_\alpha{}^\mu\right)\partial_\mu a^i_\nu + 2\epsilon^{ijk}\Sigma^{i\mu\nu}a^j_\mu a^k_\nu. \tag{8.136}$$

8.5.5 Form Notations

The expression (8.136) for the arising Lagrangian is combersome and hides what is happening. The meaning of all transformations can be clarified by rewriting everything in the differential form notation. Thus, we give the perturbation 2-form $\delta\Sigma^i$ a new name

$$(\Sigma h)^i := \frac{1}{2}\Sigma^{i\ \alpha}_{\mu} h_{\nu\alpha}dx^\mu dx^\nu. \tag{8.137}$$

We can then rewrite the kinetic term as

$$(\Sigma h)^i da^i = -d(\Sigma h)^i a^i = -\langle\star d(\Sigma h), a\rangle d^4x, \tag{8.138}$$

where we integrated by parts, $\star$ is the Hodge star on forms and the angle brackets denote the metric pairing with respect to the spacetime as well as the internal index. This way of rewriting already exhibits clearly the transformation properties. Thus, under the diffeomorphisms the 2-form $(\Sigma h)^i$ transforms as

$$\delta_\xi(\Sigma h)^i = di_\xi \Sigma^i. \tag{8.139}$$

This makes it obvious that the kinetic term is diffeomorphism invariant. Under SO(3) rotations the perturbation 2-form transforms as

$$\delta_\phi(\Sigma h)^i = \epsilon^{ijk}\Sigma^j \phi^k. \tag{8.140}$$

This means that

$$\delta_\phi \star d(\Sigma h)^i = \star\epsilon^{ijk}\Sigma^j d\phi^k = -\mathrm{i}(J_\Sigma(d\phi))^i, \tag{8.141}$$

where J_Σ is the operator on $S_+^2 \otimes \Lambda^1$ introduced in (5.137). The convenience of this notation becomes clear if we similarly rewrite the potential term of the Lagrangian. Indeed, we have

$$\frac{1}{2}\epsilon^{ijk}\Sigma^i a^j a^k = -\frac{\mathrm{i}}{2}\langle J_\Sigma(a), a\rangle dx^4. \tag{8.142}$$

Overall, the Lagrangian becomes

$$L^{(2)} = 4\mathrm{i}\langle \star d(\Sigma h), a\rangle - 2\langle J_\Sigma(a), a\rangle. \tag{8.143}$$

The invariance under gauge rotations now becomes clear. Indeed, in the first term, we only need to vary the 2-form part, because the variation of the connection produces $d\phi^i$, which vanishes against the d present in the 2-form part by integration by parts. In the second term, we only need to vary one of the two occurrences of a^i and multiply the result by a factor of two, because this term is symmetric in both copies of the connection perturbation. So, we get

$$\begin{aligned}\delta_\phi L^{(2)} &= 4\mathrm{i}\langle \delta_\phi \star d(\Sigma h), a\rangle - 4\langle J_\Sigma(\delta_\phi a), a\rangle \\ &= 4\langle J_\Sigma(d\phi), a\rangle - 4\langle J_\Sigma(d\phi), a\rangle = 0.\end{aligned} \tag{8.144}$$

8.5.6 Solving for the Connection

Even though we consider the first-order formalism, and the eventual perturbation theory will have two independent fields in it, it is a good exercise to integrate out the connection perturbation from the linearised Lagrangian. We will later see that essentially the same calculation is needed in the process of finding the propagators.

We can rewrite the Lagrangian (8.143) as

$$L^{(2)} = 2\langle \tilde{a}, a\rangle - 2\langle J_\Sigma(a), a\rangle, \tag{8.145}$$

where we introduced

$$\tilde{a} := 2\mathrm{i} \star d(\Sigma h). \tag{8.146}$$

The equation obtained by varying the Lagrangian with respect to the connection reads

$$\tilde{a} = 2J_\Sigma(a), \tag{8.147}$$

which is easily solved by using the fact that the inverse of J_Σ is given by

$$J_\Sigma^{-1} = \frac{1}{2}(-\mathbb{I} + J_\Sigma). \tag{8.148}$$

It is then clear that the Lagrangian expressed solely in terms of $h_{\mu\nu}$ is

$$L^{(2)} = \frac{1}{2}\langle \tilde{a}, J_\Sigma^{-1}(\tilde{a})\rangle = -2\langle \star d(\Sigma h), J_\Sigma^{-1}(\star d(\Sigma h))\rangle. \tag{8.149}$$

This form of writing the Lagrangian makes all the invariances manifest. Indeed, it is explicitly invariant under diffeomorphisms because $d(\Sigma h)$ is invariant. And under gauge rotations $\delta \star d(\Sigma h) = -\mathrm{i}J_\Sigma(d\phi)$ and then varying only the second slot J_Σ^{-1} is cancelled by J_Σ and the result is zero by integration by parts. We will soon see that the spin two kinetic operator in (8.149) is constructed from the already familiar to us operator $\delta_{(2,2)}$ that is of Dirac type.

8.5.7 Index Notation Calculation

Let us now compute the second-order Lagrangian (8.149) explicitly, using the index notation. However, even prior to the computation, we can anticipate the result. We know that the resulting Lagrangian is going to be diffeomorphism invariant. We also know that it is invariant under shifts of the antisymmetric part of $h_{\mu\nu}$ by an arbitrary SD 2-form, as this is how the SO(3) rotations act. It is also invariant under shifts by an arbitrary ASD 2-form, because the object $(\Sigma h)^i$ only depends on the SD part of $h_{[\mu\nu]}$. So, the Lagrangian (8.149) is invariant under shifts of $h_{\mu\nu}$ by an arbitrary antisymmetric tensor, and is thus $h_{[\mu\nu]}$ independent. The only Lagrangian that is constructed from $h_{(\mu\nu)}$ and is diffeomorphism invariant is the usual linearised metric Lagrangian (8.78).

To verify this expectation by an explicit calculation we use index notations. We then have

$$\tilde{a}^i_\mu := \Sigma^{i\ \nu}_{\mu}\partial_\nu h + \Sigma^{i\alpha\beta}\partial_\alpha h_{\beta\mu} - \Sigma^{i\ \alpha}_{\mu}\partial^\beta h_{\alpha\beta}. \tag{8.150}$$

To find the connection a^i_μ we need to compute the action of J_Σ on (8.150). This is given by

$$\begin{aligned} J_\Sigma(\tilde{a})^i_\mu &= \epsilon^{ijk}\Sigma^{j\ \alpha}_{\mu}\left(\Sigma^{k\ \beta}_{\alpha}\partial_\beta h + \Sigma^{k\rho\sigma}\partial_\rho h_{\sigma\alpha} - \Sigma^{k\ \beta}_{\alpha}\partial^\sigma h_{\beta\sigma}\right) \\ &= \epsilon^{ijk}\epsilon^{jks}\Sigma^{s\ \beta}_{\mu}(\partial_\beta h - \partial^\sigma h_{\beta\sigma}) \\ &\quad - (\Sigma^{i\ \rho}_{\mu}\eta^{\sigma\alpha} - \Sigma^{i\ \sigma}_{\mu}\eta^{\rho\alpha} - \Sigma^{i\alpha\rho}\eta^\sigma{}_\mu + \Sigma^{i\alpha\sigma}\eta^\rho{}_\mu)\partial_\rho h_{\sigma\alpha}. \end{aligned} \tag{8.151}$$

Here we have again used the identity (8.72). The expression (8.151) becomes

$$\Sigma^{i\ \nu}_{\mu}\partial_\nu h - \Sigma^{i\ \alpha}_{\mu}\partial^\beta h_{\alpha\beta} + \Sigma^{i\alpha\beta}\partial_\beta h_{\mu\alpha} + \Sigma^{i\alpha\beta}\partial_\mu h_{\alpha\beta}, \tag{8.152}$$

which gives

$$a^i_\mu = \frac{1}{4}(\Sigma^{i\alpha\beta}\partial_\beta(h_{\mu\alpha} + h_{\alpha\mu}) + \Sigma^{i\alpha\beta}\partial_\mu h_{\alpha\beta}). \tag{8.153}$$

As a check, we note that this expression is invariant under the action of diffeomorphisms given by (8.131). Also, under the gauge transformations (8.132) the connection undergoes gauge transformation $\delta_\phi a^i_\mu = \partial_\mu \phi^i$ as it should.

Given that we know that the antisymmetric part of $h_{\mu\nu}$ is not going to enter the second-order Lagrangian, the simplest way to compute it is to assume that $h_{\mu\nu}$ is symmetric from the start. With this assumption, the expression for the connection perturbation becomes

$$a^i_\mu = -\frac{1}{2}\Sigma^{i\alpha\beta}\partial_\alpha h_{\beta\mu}. \tag{8.154}$$

This is a remarkably simple expression for the linearised SD connection in terms of the metric perturbation.

8.5.8 Computation of the Linearised Action in Terms of $h_{\mu\nu}$

We now substitute the found solution (8.153) for the connection perturbation into the linearised Lagrangian to obtain the second-order linearised Lagrangian in terms of $h_{\mu\nu}$ only. Again, we can view this manipulation as a necessary step for the computation of the propagators in what follows.

We first perform the computation without making the assumption that $h_{\mu\nu}$ is symmetric. The result is that the second-order Lagrangian in terms of $h_{\mu\nu}$ can be written in terms of the symmetric part of $h_{\mu\nu}$ only, and so we could make an assumption that $h_{\mu\nu}$ is symmetric from the start. However, it is a useful exercise to perform the more complicated calculation without the symmetry assumption. The easiest way to do this calculation is to note that when a^i_μ is determined by $h_{\mu\nu}$ the two terms in (8.136) become multiples of each other, and so it is enough to compute only one of them. It is easier to compute the first term. So, we write the second-order Lagrangian as

$$\begin{aligned} L^{(2)} = \langle \tilde{a}, a\rangle = \frac{1}{4}(\Sigma^{i}_{\mu}{}^{\alpha}\partial_\alpha h + \Sigma^{i\alpha\beta}\partial_\alpha h_{\beta\mu} - \Sigma^{i}_{\mu}{}^{\alpha}\partial^\beta h_{\alpha\beta}) \\ (\Sigma^{i\rho\sigma}\partial_\sigma(h_{\nu\rho} + h_{\rho\nu}) + \Sigma^{i\rho\sigma}\partial_\nu h_{\rho\sigma})\eta^{\mu\nu}. \end{aligned} \tag{8.155}$$

We then use

$$\Sigma^i_{\mu\nu}\Sigma^i_{\rho\sigma} = \eta_{\mu\rho}\eta_{\nu\sigma} - \eta_{\mu\sigma}\eta_{\nu\rho} - \mathrm{i}\epsilon_{\mu\nu\rho\sigma} \tag{8.156}$$

to obtain

$$L^{(2)} = -\frac{1}{2}(\partial_\rho h_{(\mu\nu)})^2 + (\partial^\nu h_{(\mu\nu)})^2 + h_{\mu\nu}\partial^\mu\partial^\nu h + \frac{1}{2}(\partial_\mu h)^2, \tag{8.157}$$

which only depends on the symmetric part of $h_{\mu\nu}$ as expected.

An alternative way of doing the same computation is by making the symmetry assumption on $h_{\mu\nu}$ from the start, since we know the antisymmetric part cannot enter the final result. In this case we can use the simpler form of the connection (8.153) and compute the Lagrangian as a multiple of the second quadratic in the connection term. We then have

$$L^{(2)} = -2\epsilon^{ijk}\Sigma^{i\mu\nu}a^j_\mu a^k_\nu = \frac{1}{2}G^{\mu\nu\alpha\beta\rho\sigma}\partial_\alpha h_{\beta\mu}\partial_\rho h_{\sigma\nu}, \tag{8.158}$$

where we introduced

$$G^{\mu\nu\alpha\beta\rho\sigma} := -\epsilon^{ijk}\Sigma^{i\mu\nu}\Sigma^{j\alpha\beta}\Sigma^{k\rho\sigma} \tag{8.159}$$

This can be evaluated using the identity (8.72) as well as (8.156). This again gives (8.157) but with symmetric $h_{\mu\nu}$ assumed from the start. Note that we have obtained a remarkably simple expression (8.158) for the linearised metric Lagrangian. All of the complications of different possible tensor contractions have been hidden into the tensor (8.159).

Another important outcome of the previous calculation is that we have also represented the kinetic term of spin two particles using the first-order differential operator (8.153) giving a^i_μ from $h_{\mu\nu}$. Recalling the discussion of Section 8.2, we see that we are essentially dealing with the operator $\delta_{(2,2)}$ here; compare (8.71) with (8.153). This shows that indeed linearised chiral first-order perturbation theory for gravity realises the spin two particles kinetic term in a completely different way as compared to the usual metric formalism. The spin-preserving Dirac operator $\delta_{(2,2)}$ is used instead of the spin-increasing operator $d_{(2,2)}$ of the usual metric story. This has been made possible by the use of chiral objects such as $\Sigma^i_{\mu\nu}$.

8.5.9 Gauge-Fixing I: Symmetric $h_{\mu\nu}$

We now want to invert the kinetic term, adding gauge-fixing terms as necessary. We will consider two different gauge-fixing procedures. One is following the usual metric formalism route. Thus, we have seen that the Lorentz transformations act on the tensor $h_{\mu\nu}$ by shifting its antisymmetric part. This means that there exists a gauge in which this tensor is symmetric. Let us work out the details.

Since we are going to gauge-fix Lorentz transformations by demanding $h_{\mu\nu}$ to be symmetric, and impose the standard metric formalism gauge-fixing condition to gauge-fix diffeomorphisms, we will only need to fix the gauge after the connection has been integrated out. So, the first few steps can be carried out without worrying about the gauge-fixing. We add currents for both $h_{\mu\nu}$ and a^i_μ and attempt to integrate out these fields from the action obtaining a functional of the currents. So, we consider

$$L^{(2)} = 2\left(\Sigma^{i\mu\nu}h - \Sigma^{i\mu\alpha}h_\alpha{}^\nu + \Sigma^{i\nu\alpha}h_\alpha{}^\mu\right)\partial_\mu a^i_\nu + 2\epsilon^{ijk}\Sigma^{i\mu\nu}a^j_\mu a^k_\nu + J^{\mu\nu}h_{\mu\nu} + J^{\mu i}a^i_\mu. \tag{8.160}$$

Varying this with respect to the connection we get a modification of the equation (8.147)

$$\Sigma^{i}_\mu{}^\nu\partial_\nu h + \Sigma^{i\alpha\beta}\partial_\alpha h_{\beta\mu} - \Sigma^{i}_\mu{}^\alpha\partial^\beta h_{\alpha\beta} + \frac{1}{2}J^i_\mu = 2(J_\Sigma(a))^i_\mu. \tag{8.161}$$

This equation is solved as before with the solution being

$$a^i_\mu = \frac{1}{4}(\Sigma^{i\alpha\beta}\partial_\beta(h_{\mu\alpha} + h_{\alpha\mu}) + \Sigma^{i\alpha\beta}\partial_\mu h_{\alpha\beta} + (J_\Sigma^{-1}(J))^i_\mu), \tag{8.162}$$

where for now we do not make the symmetry assumption for $h_{\mu\nu}$. Substituting this back into the Lagrangian, we get

$$\begin{aligned} L^{(2)} = &-\frac{1}{2}(\partial_\rho h_{(\mu\nu)})^2 + (\partial^\nu h_{(\mu\nu)})^2 + h_{\mu\nu}\partial^\mu\partial^\nu h + \frac{1}{2}(\partial_\mu h)^2 \\ &- \frac{1}{4}h_{\mu\nu}(\Sigma^{i\mu\alpha}\partial_\alpha J^{i\nu} + \Sigma^{i\nu\alpha}\partial_\alpha J^{i\mu} + \Sigma^{i\mu\nu}\partial^\alpha J^i_\alpha) + J^{\mu\nu}h_{\mu\nu} \\ &- \frac{1}{16}((J^i_\mu)^2 + \epsilon^{ijk}\Sigma^{i\mu\nu}J^j_\mu J^k_\nu). \end{aligned} \tag{8.163}$$

The last step is to integrate out the metric perturbation. It is here that we need to make our gauge-fixing assumptions. First, we clearly need to gauge-fix the diffeomorphism symmetry. Because the arising metric Lagrangian is as standard, it is easiest to fix this gauge in the standard way as well, by adding to the Lagrangian the standard de Donder gauge-fixing term (8.79) for the symmetric part of $h_{\mu\nu}$. In fact, this term can be added in a first-order form, by adding a new auxiliary field ξ_μ with one spacetime index, as well as a quadratic term $(\xi_\mu)^2$. Indeed, we can instead add

$$L_{\text{g.f.}} = 2\xi^\mu(\partial^\nu h_{(\mu\nu)} - \frac{1}{2}\partial_\mu h) + (\xi_\mu)^2. \tag{8.164}$$

Integrating out ξ_μ results in (8.79). Moreover, we can think of ξ_μ as a new component of the connection a^i_μ that we need to add in order to gauge-fix the diffeomorphisms. This interpretation becomes particularly clear in the spinor formalism. But for now, it is sufficient for our purposes just to add the gauge-fixing term (8.79). The resulting (partially) gauge-fixed Lagrangian is

$$\begin{aligned} L^{(2)} + L_{\text{g.f.}} = &\frac{1}{2}h^{(\mu\nu)}\Box h_{(\mu\nu)} - \frac{1}{4}h\Box h - \frac{1}{2}h_{(\mu\nu)}\Sigma^{i\mu\alpha}\partial_\alpha J^{i\nu} + J^{\mu\nu}h_{\mu\nu} \\ &- \frac{1}{4}h_{\mu\nu}\Sigma^{i\mu\nu}\partial^\alpha J^i_\alpha - \frac{1}{16}((J^i_\mu)^2 + \epsilon^{ijk}\Sigma^{i\mu\nu}J^j_\mu J^k_\nu), \end{aligned}$$

where $\Box := \partial^\mu\partial_\mu$. For this Lagrangian we can find the symmetric part of the metric perturbation in terms of the currents. As for the antisymmetric part, there is clearly no kinetic term for it, and so it cannot be determined unless we fix a gauge. The simplest possibility is to just gauge-fix the antisymmetric part of $h_{\mu\nu}$ to be zero. In this case, going to the momentum space, the equation for $h_{\mu\nu}$ (assumed symmetric) is

$$-k^2 h_{\mu\nu} + \frac{k^2}{2}\eta_{\mu\nu}h - \frac{\mathrm{i}}{2}\Sigma^i_{(\mu}{}^\alpha k_\alpha J^i_{\nu)} + J_{\mu\nu} = 0, \tag{8.165}$$

where we assumed the current $J_{\mu\nu}$ to be symmetric. Taking the trace, one solves for the trace part of the metric perturbation

$$h = -\frac{1}{k^2}\left(\frac{\mathrm{i}}{2}\Sigma^{i\alpha\beta}k_\alpha J^i_\beta + \eta^{\alpha\beta}J_{\alpha\beta}\right), \tag{8.166}$$

so that

$$h_{\mu\nu} = \frac{1}{k^2}J_{\mu\nu} - \frac{\mathrm{i}}{2k^2}\Sigma^{i\ \alpha}_{(\mu}k_\alpha J^i_{\nu)} - \frac{1}{2k^2}\eta_{\mu\nu}\left(\frac{\mathrm{i}}{2}\Sigma^{i\alpha\beta}k_\alpha J^i_\beta + \eta^{\alpha\beta}J_{\alpha\beta}\right). \tag{8.167}$$

Substituting this back into the Lagrangian gives

$$\begin{aligned} L^{(2)} = &\frac{1}{2k^2}\left(J_{\mu\nu}(-k) + \frac{\mathrm{i}}{2}\Sigma^{i\ \alpha}_{(\mu}k_\alpha J^i_{\nu)}(-k)\right)\left(J_{\mu\nu}(k) - \frac{\mathrm{i}}{2}\Sigma^{i\ \alpha}_{(\mu}k_\alpha J^i_{\nu)}(k)\right)\\ &- \frac{1}{4k^2}\left(-\frac{\mathrm{i}}{2}\Sigma^{i\alpha\beta}k_\alpha J^i_\beta(-k) + \eta^{\alpha\beta}J_{\alpha\beta}(-k)\right)\left(\frac{\mathrm{i}}{2}\Sigma^{i\alpha\beta}k_\alpha J^i_\beta(k) + \eta^{\alpha\beta}J_{\alpha\beta}(k)\right)\\ &- \frac{1}{16}(J^i_\mu(-k)J^{i\mu}(k) + \epsilon^{ijk}\Sigma^{i\mu\nu}J^j_\mu(-k)J^k_\nu(k)), \end{aligned}$$

where we have explicitly indicated the momentum dependence. Expanding the squares here we get the sought propagators. The metric–metric propagator is represented by the following terms

$$\frac{1}{2k^2}J_{\mu\nu}(-k)J^{\mu\nu}(k) - \frac{1}{4k^2}\eta^{\alpha\beta}J_{\alpha\beta}(-k)\eta^{\mu\nu}J_{\mu\nu}(k). \tag{8.168}$$

This results in the usual metric propagator given by (8.82). The metric–connection terms are

$$-\frac{\mathrm{i}}{2k^2}J^{\mu\nu}(-k)\Sigma^{i\ \alpha}_{(\mu}k_\alpha J^i_{\nu)}(k) - \frac{\mathrm{i}}{4k^2}(\eta^{\alpha\beta}J_{\alpha\beta}(-k))(\Sigma^{i\mu\nu}k_\mu J^i_\nu(k)). \tag{8.169}$$

This results in the propagator

$$\langle h_{\mu\nu}a^i_\mu\rangle = -\frac{1}{2k^2}\Sigma^{i\ \alpha}_{(\mu}k_\alpha\eta_{\nu)\rho} + \frac{1}{4k^2}\eta_{\mu\nu}\Sigma^{i\ \alpha}_{\rho}k_\alpha. \tag{8.170}$$

Finally, the connection–connection terms are

$$\begin{aligned} &\frac{1}{16k^2}(\Sigma^{i\mu\alpha}k_\alpha J^{i\nu}(-k) + \Sigma^{i\nu\alpha}k_\alpha J^{i\mu}(-k))(\Sigma^{j\ \beta}_{\mu}k_\beta J^j_\nu(k))\\ &\qquad -\frac{1}{16k^2}(\Sigma^{i\alpha\beta}k_\alpha J^i_\beta(-k))(\Sigma^{i\mu\nu}k_\mu J^i_\nu(k))\\ &\qquad -\frac{1}{16}(J^i_\mu(-k)J^{i\mu}(k) + \epsilon^{ijk}\Sigma^{i\mu\nu}J^j_\mu(-k)J^k_\nu(k)). \end{aligned} \tag{8.171}$$

Let us simplify the second term in the first line. We have

$$\begin{aligned} \Sigma^{i\nu\alpha}k_\alpha J^{i\mu}(-k)\Sigma^{j\ \beta}_{\mu}k_\beta J^j_\nu(k) &= \Sigma^{j\nu\alpha}k_\alpha J^{i\mu}(-k)\Sigma^{i\ \beta}_{\mu}k_\beta J^j_\nu(k)\\ &\quad + \epsilon^{ijk}\epsilon^{klm}\Sigma^{l\nu\alpha}\Sigma^{m\mu\beta}k_\alpha J^i_\mu(-k)k_\beta J^j_\nu(k)\\ &= \Sigma^{i\alpha\beta}k_\alpha J^i_\beta(-k)\Sigma^{i\mu\nu}k_\mu J^i_\nu(k) + k^2\epsilon^{ijk}\Sigma^{k\mu\nu}J^i_\mu(-k)J^j_\nu(k)\\ &\quad + 2\epsilon^{ijk}k^\alpha\Sigma^{i\ \nu}_{\alpha}J^j_\nu(k)k^\beta J^k_\beta(-k). \end{aligned}$$

There are several cancellations and we get the following connection–connection current terms

$$\frac{1}{8k^2}\epsilon^{ijk}\Sigma^{j\mu\nu}k_\mu J^k_\nu(k)k^\alpha J^i_\alpha(-k). \tag{8.172}$$

This results in the connection–connection propagator

$$\langle a^i_\mu a^j_\nu\rangle = -\frac{\mathrm{i}}{8k^2}\epsilon^{ijk}(k^\alpha\Sigma^k_{\alpha\mu}k_\nu - k^\alpha\Sigma^k_{\alpha\nu}k_\mu). \tag{8.173}$$

To summarise, in this version of the perturbation theory we have two propagating fields: a symmetric tensor $h_{\mu\nu}$ as well as a connection a^i_μ. There are three different propagators, connecting all three different possible pairs of fields.

8.5.10 Gauge-Fixing II: Lorentz Gauge

We will now work out details of a different gauge-fixing procedure. Instead of demanding $h_{\mu\nu}$ to be symmetric to gauge-fix the Lorentz symmetry, we will use an asymmetric gauge-fixing condition. We continue to set the ASD part of the antisymmetric part of $h_{\mu\nu}$ to zero by a suitable chiral half of Lorentz transformation. But we will fix the SD chiral half of Lorentz in a different way, simply by adding the Lorentz gauge-fixing term $h^i\partial^\mu a^i_\mu$ to the action. Here h^i is a new field, varying with respect to which imposes the sharp gauge-fixing condition $\partial^\mu a^i_\mu = 0$. With this choice of gauge-fixing, both the diffeomorphism and (chiral half of) Lorentz symmetry are fixed with a condition that involves derivatives.

The procedure of integrating out the fields has to be repeated from scratch because we now have a new connection involving term in the Lagrangian. The Lagrangian to consider is now

$$\begin{aligned}L^{(2)} &= 2\left(\Sigma^{i\mu\nu}h - \Sigma^{i\mu\alpha}h_\alpha{}^\nu + \Sigma^{i\nu\alpha}h_\alpha{}^\mu\right)\partial_\mu a^i_\nu + 2\epsilon^{ijk}\Sigma^{i\mu\nu}a^j_\mu a^k_\nu \\ &\quad + 2h^i\partial^\mu a^i_\mu + J^{\mu\nu}h_{\mu\nu} + J^{\mu i}a^i_\mu,\end{aligned} \tag{8.174}$$

where the coefficient of two in the new term is for future convenience. The equation that we need to solve for a^i_μ is

$$\Sigma^{i\ \nu}_{\mu}\partial_\nu h + \Sigma^{i\alpha\beta}\partial_\alpha h_{\beta\mu} - \Sigma^{i\ \alpha}_{\mu}\partial^\beta h_{\alpha\beta} - \partial_\mu h^i + \frac{1}{2}J^i_\mu = 2(J_\Sigma(a))^i_\mu, \tag{8.175}$$

with the solution being

$$a^i_\mu = \frac{1}{4}\Big(\Sigma^{i\alpha\beta}\partial_\beta(h_{\mu\alpha} + h_{\alpha\mu}) + \Sigma^{i\alpha\beta}\partial_\mu h_{\alpha\beta} + \partial_\mu h^i - \epsilon^{ijk}\Sigma^{j\ \nu}_{\mu}\partial_\nu h^k + (J^{-1}_\Sigma(J))^i_\mu\Big). \tag{8.176}$$

Note that this only depends on the SD part of the antisymmetric part of $h_{\mu\nu}$ the ASD part drops out as before. We now substitute this back into the Lagrangian, which we write as

$$L^{(2)} = \left(\Sigma^{i\ \nu}_{\mu}\partial_\nu h + \Sigma^{i\alpha\beta}\partial_\alpha h_{\beta\mu} - \Sigma^{i\ \alpha}_{\mu}\partial^\beta h_{\alpha\beta} - \partial_\mu h^i + \frac{1}{2}J^i_\mu\right)a_{i\mu}. \tag{8.177}$$

Many of the terms are as before, and we only need to work out the terms involving the new h^i field. The terms quadratic in this field give simply $-(1/4)(\partial_\mu h^i)^2$. The terms linear in this field give

$$-\frac{1}{2}\Big(\Sigma^{i\alpha\beta}\partial_\beta(h_{\mu\alpha}+h_{\alpha\mu})+\Sigma^{i\alpha\beta}\partial_\mu h_{\alpha\beta}+\frac{1}{2}(-J^i_\mu+\epsilon^{ijk}\Sigma^{j\ \nu}_{\mu}J^k_\nu\Big)\partial^\mu h^i.$$

Thus, the action with the connection perturbation integrated out is

$$\begin{aligned}L^{(2)} &= -\frac{1}{4}(\partial_\rho h_{\mu\nu}+\partial_\rho h_{\nu\mu})\partial^\rho h^{\mu\nu}+\frac{1}{2}(\partial_\mu h)^2-\frac{1}{4}(\partial_\mu h^i)^2\\ &+\frac{1}{4}(\partial^\nu h_{\mu\nu}+\partial^\nu h_{\nu\mu})^2+h_{\mu\nu}\partial^\mu\partial^\nu h-\frac{1}{2}(\partial^\nu h_{\mu\nu}+\partial^\nu h_{\nu\mu})\Sigma^{i\mu\alpha}\partial_\alpha h^i\\ &-\frac{1}{2}\Sigma^{i\alpha\beta}\partial_\mu h_{\alpha\beta}\partial^\mu h^i+\frac{1}{4}(J^i_\mu-\epsilon^{ijk}\Sigma^{j\ \nu}_{\mu}J^k_\nu)\partial^\mu h^i\\ &-\frac{1}{4}h_{\mu\nu}(\Sigma^{i\mu\alpha}\partial_\alpha J^{i\nu}+\Sigma^{i\nu\alpha}\partial_\alpha J^{i\mu}+\Sigma^{i\mu\nu}\partial^\alpha J^i_\alpha)+J^{\mu\nu}h_{\mu\nu}\\ &-\frac{1}{16}((J^i_\mu)^2+\epsilon^{ijk}\Sigma^{i\mu\nu}J^j_\mu J^k_\nu),\end{aligned}$$

where we grouped terms in a suggesting way. We now add a gauge-fixing term that we choose to be a modification of the de Donder gauge

$$L_{\text{g.f.}} = -\left(\frac{1}{2}(\partial^\nu h_{\mu\nu}+\partial^\nu h_{\nu\mu})-\frac{1}{2}\partial_\mu h-\frac{1}{2}\Sigma^{i\ \nu}_{\mu}\partial_\nu h^i\right)^2. \tag{8.178}$$

The gauge-fixed Lagrangian becomes

$$\begin{aligned}L^{(2)}+L_{\text{g.f.}} &= -\frac{1}{4}(\partial_\rho h_{\mu\nu}+\partial_\rho h_{\nu\mu})\partial^\rho h^{\mu\nu}+\frac{1}{4}(\partial_\mu h)^2-\frac{1}{2}(\partial_\mu h^i)^2\\ &-\frac{1}{2}\Sigma^{i\alpha\beta}\partial_\mu h_{\alpha\beta}\partial^\mu h^i+\frac{1}{4}(J^i_\mu-\epsilon^{ijk}\Sigma^{j\ \nu}_{\mu}J^k_\nu)\partial^\mu h^i\\ &-\frac{1}{4}h_{\mu\nu}(\Sigma^{i\mu\alpha}\partial_\alpha J^{i\nu}+\Sigma^{i\nu\alpha}\partial_\alpha J^{i\mu}+\Sigma^{i\mu\nu}\partial^\alpha J^i_\alpha)+J^{\mu\nu}h_{\mu\nu}\\ &-\frac{1}{16}((J^i_\mu)^2+\epsilon^{ijk}\Sigma^{i\mu\nu}J^j_\mu J^k_\nu).\end{aligned}$$

The final step is to integrate out the metric perturbation, which is no longer assumed to be symmetric, but whose antisymmetric part only has the SD part. To integrate out $h_{\mu\nu}$ let us split the problem into two parts. We have already solved the problem of integrating out the symmetric part of $h_{\mu\nu}$, and this remains unmodified. So, we write

$$h_{\mu\nu} = h_{(\mu\nu)}+\frac{1}{4}\Sigma^i_{\mu\nu}\Sigma^{i\rho\sigma}h_{\rho\sigma}, \tag{8.179}$$

and also write the corresponding current as

$$J_{\mu\nu} = \tilde{J}_{\mu\nu}+\frac{1}{2}\Sigma^i_{\mu\nu}J^i, \tag{8.180}$$

where $\tilde{J}_{\mu\nu}$ is symmetric. Integrating out the symmetric part then gives the result as in the previous section with $\tilde{J}_{\mu\nu}$ in place of $J_{\mu\nu}$. The result is collected in

(8.168), (8.169), and (8.172). Let us work out the terms following from integrating out the antisymmetric part of $h_{\mu\nu}$. These terms are

$$-\frac{1}{2}(\partial_\mu h^i)^2 - \frac{1}{2}\Sigma^{i\mu\nu}\partial_\alpha h_{\mu\nu}\partial^\alpha h^i + \frac{1}{4}(J^i_\mu - \epsilon^{ijk}\Sigma^{j\,\nu}_{\mu} J^k_\nu)\partial^\mu h^i \\ -\frac{1}{4}h_{\mu\nu}\Sigma^{i\mu\nu}\partial^\alpha J^i_\alpha + \frac{1}{2}J^i\Sigma^{i\mu\nu}h_{\mu\nu}. \tag{8.181}$$

Varying with respect to $\Sigma^{i\mu\nu}h_{\mu\nu}$ we get

$$\Box h^i = \frac{1}{2}\partial^\mu J^i_\mu - J^i, \tag{8.182}$$

which we can solve for h^i

$$h^i = \frac{1}{k^2}\left(J^i - \frac{\mathrm{i}}{2}k^\mu J^i_\mu\right). \tag{8.183}$$

Varying with respect to h^i gives

$$\Box h^i + \frac{1}{2}\Sigma^{i\mu\nu}\Box h_{\mu\nu} - \frac{1}{4}\partial^\mu J^i_\mu + \frac{1}{4}\epsilon^{ijk}\Sigma^{j\mu\nu}\partial_\mu J^k_\nu = 0. \tag{8.184}$$

Using (8.182) we can write this equation as

$$\Sigma^{i\mu\nu}\Box h_{\mu\nu} = 2J^i - \frac{1}{2}\partial^\mu J^i_\mu - \frac{1}{4}\epsilon^{ijk}\Sigma^{j\mu\nu}\partial_\mu J^k_\nu, \tag{8.185}$$

which we can solve for $\Sigma^{i\mu\nu}h_{\mu\nu}$. Substituting everything into the Lagrangian we get the following current–current terms

$$-\frac{1}{2k^2}J^i(-k)J^i(k) + \frac{\mathrm{i}}{4k^2}J^i(-k)(k^\mu J^i_\mu(k) + \epsilon^{ijk}\Sigma^{j\mu\nu}k_\mu J^k_\nu(k)) \\ -\frac{1}{8k^2}\epsilon^{ijk}k^\alpha J^i_\alpha(-k)\Sigma^{j\mu\nu}k_\mu J^k_\nu(k).$$

We note that the last term here precisely cancels the similar term (8.172) obtained by integrating out the symmetric part of $h_{\mu\nu}$. Thus, in this version of the gauge-fixed theory, there is no connection–connection propagator, which simplifies calculations considerably. Thus, this version of the theory is preferable to the one where $h_{\mu\nu}$ is required to be symmetric.

Moreover, the terms containing J^i combine nicely with the already existing terms in (8.168) and (8.169). Let us work out how the propagator for $h_{\mu\nu}$ gets modified. Variation of the Lagrangian with respect to J^i inserts a factor of $(1/2)\Sigma^{i\mu\nu}h_{\mu\nu}$. This, together with the fact that the exponent of the generating function contains $-(\mathrm{i}/2)(J^i)^2$, implies the following two-point function

$$\left\langle \frac{1}{2}\Sigma^{i\mu\nu}h_{\mu\nu}\frac{1}{2}\Sigma^{j\rho\sigma}h_{\rho\sigma}\right\rangle = -\frac{\mathrm{i}}{k^2}\delta^{ij}, \tag{8.186}$$

which in turn implies

$$\langle h_{[\mu\nu]}h_{[\rho\sigma]}\rangle = -\frac{\mathrm{i}}{4k^2}\Sigma^i_{\mu\nu}\Sigma^i_{\rho\sigma}. \tag{8.187}$$

On the other hand we have

$$\Sigma^i_{\mu\nu}\Sigma^i_{\rho\sigma} = \eta_{\mu\rho}\eta_{\nu\sigma} - \eta_{\mu\sigma}\eta_{\nu\rho} - \mathrm{i}\epsilon_{\mu\nu\rho\sigma}. \tag{8.188}$$

Combining with (8.82), this gives

$$\langle h_{\mu\nu}h_{\rho\sigma}\rangle = -\frac{\mathrm{i}}{4k^2}\left(3\eta_{\mu\rho}\eta_{\nu\sigma} + \eta_{\mu\sigma}\eta_{\nu\rho} - 2\eta_{\mu\nu}\eta_{\rho\sigma} - \mathrm{i}\epsilon_{\mu\nu\rho\sigma}\right). \tag{8.189}$$

The appearing coefficients are hard to understand in tensor notations. This propagator takes a much simpler form in spinor notations.

8.5.11 Calculation of the Second-Order Perturbation of the 2-Form Field

Interaction vertices contain the second-order perturbation of the 2-form field, and so we need to find an expression for this in terms of the tensor $h_{\mu\nu}$ that we use to parametrise the first-order perturbation. This computation is somewhat laborious, even though the final answer is very simple. It seems to be most easily done using the spinor formalism. We use the definition (8.25). Replacing the frame with $e^{AA'} + \delta e^{AA'}$, we have to replace the 2-form field with $\Sigma^{AB} + \delta\Sigma^{AB} + (1/2)\delta^2\Sigma^{AB}$ where

$$\delta\Sigma^{AB} = \delta e^{(AA'}e^{B)}{}_{A'} \qquad \delta^2\Sigma^{AB} = \delta e^{AA'}\delta e^{B}{}_{A'}. \tag{8.190}$$

We can represent the perturbation of the frame as a linear combination of the frame 1-forms

$$\delta e^{AA'} = h^{AA'}{}_{BB'}e^{BB'}. \tag{8.191}$$

Also, whenever there is a wedge product of two copies of the frame, we can decompose this into the SD and ASD basic forms. The relevant expression is (8.30), with SD forms given by (8.25) and ASD forms given by (8.29). Using this we get

$$\delta\Sigma^{AB} = -\Sigma^{M(A}h^{B)A'}{}_{MA'} - \bar{\Sigma}_{A'B'}h^{(AA'B)B'} \tag{8.192}$$

and

$$\delta^2\Sigma^{AB} = -\Sigma^{MN}h^{AA'}{}_{M}{}^{B'}h^{B}{}_{A'NB'} + \bar{\Sigma}^{M'N'}h^{AA'N}{}_{M'}h^{B}{}_{A'NN'}. \tag{8.193}$$

Before we rewrite these expressions any further, let us discuss how these objects depend on the 'gauge' part of $h^{AA'}{}_{BB'}$. As we already know from the previous discussion, the antisymmetric part of this rank two tensor is pure gauge. In spinor notations, the SD part of this antisymmetric part is represented by $h^{(AA'B)}{}_{A'}$, while the ASD part is represented by $h^{A(A'}{}_{A}{}^{B')}$. It is clear that the first variation $\delta\Sigma^{AB}$ is independent of $h^{A(A'}{}_{A}{}^{B')}$. The second variation depends on this part of the perturbation. But we can always make an ASD SO(3) transformation to set this part to zero. We will always assume that this has been done.

The expression (8.192) can be seen to match (8.125). To see this we must parametrise $h^{AA'}{}_{BB'}$ by a tensor $h_{\mu\nu}$. We use the parametrisation

$$h^{AA'}{}_{MM'} = -\frac{1}{2} h_{\mu\nu} e^{\mu AA'} e^{\nu}_{MM'}, \tag{8.194}$$

where $e^{\mu AA'}$ is the inverse frame and the numerical factor is needed to match (8.125). Indeed, using the identity (8.31) we have

$$h^{BA'}{}_{MA'} = -\frac{1}{2} h_{\mu\nu} e^{\mu BA'} e^{\nu}_{MA'} = -\frac{1}{2} h_{\mu\nu} \left(\frac{1}{2} \eta^{\mu\nu} \epsilon^{B}{}_{M} - \frac{1}{2} \Sigma^{\mu\nu B}{}_{M} \right). \tag{8.195}$$

Therefore, the first term in $\Sigma^{M(A} h^{B)A'}{}_{MA'}$ is $(1/4)\Sigma^{AB} h$, which matches (8.125). The second term can be seen to be the one corresponding to the term with ϵ^{ijk} in (8.125), but developing the formalism further to allow the precise numerical matching is not important for us.

Let us now express the second variation of the 2-form field in terms of $h_{\mu\nu}$, which is not assumed to be symmetric. The parametrisation (8.194), together with the identity (8.31) gives

$$h^{AA'}{}_{(M}{}^{B'} h^{B}{}_{A'N)B'} = \frac{1}{4} h_{\mu\nu} h_{\rho\sigma} \Sigma^{\mu\rho AB} \Sigma^{\nu\sigma}{}_{MN}, \tag{8.196}$$

$$h^{AA'N}{}_{(M'} h^{B}{}_{|A'N|N')} = \frac{1}{4} h_{\mu\nu} h_{\rho\sigma} \Sigma^{\mu\rho AB} \bar{\Sigma}^{\nu\sigma}{}_{M'N'}.$$

This means that

$$\delta^2 \Sigma^{AB} = \frac{1}{4} h_{\mu\nu} h_{\rho\sigma} \Sigma^{\mu\rho AB} \left(\Sigma^{MN} \Sigma^{\nu\sigma}{}_{MN} - \bar{\Sigma}^{M'N'} \bar{\Sigma}^{\nu\sigma}{}_{M'N'} \right). \tag{8.197}$$

On the other hand, the object in brackets is just a multiple of the identity tensor in the space of antisymmetric matrices

$$\Sigma^{MN}_{\alpha\beta} \Sigma^{\nu\sigma}{}_{MN} - \bar{\Sigma}^{M'N'}_{\alpha\beta} \bar{\Sigma}^{\nu\sigma}{}_{M'N'} = 2\delta^{\nu}_{[\alpha} \delta^{\sigma}_{\beta]}. \tag{8.198}$$

The coefficient in this formula can be checked by, e.g., multiplying it by $\Sigma^{\alpha\beta}_{AB}$ and using

$$\Sigma^{MN}_{\alpha\beta} \Sigma^{\alpha\beta}_{AB} = 2\epsilon_{(A}{}^{M} \epsilon_{B)}{}^{N}, \tag{8.199}$$

which is easily derivable from the definition (8.24). Thus, overall we get

$$\delta^2 \Sigma^{AB}_{\alpha\beta} = \frac{1}{2} h_{\mu\alpha} h_{\nu\beta} \Sigma^{\mu\nu AB}. \tag{8.200}$$

This formula can be finally be converted to SO(3) notations that we are using in our perturbation theory

$$\delta^2 \Sigma^{i}_{\mu\nu} = \frac{1}{2} h_{\mu\alpha} h_{\nu\beta} \Sigma^{i\alpha\beta}. \tag{8.201}$$

This, together with the propagators computed previously, gives all the ingredients required to compute amplitudes using the perturbative expansion (8.121) of the action.

Fixing the numerical coefficient in (8.201) was somewhat painful. However, if one expects such a relation to be true, then the overall coefficient can be verified using that

$$\delta^2\Sigma^{(i}\Sigma^{j)} + \delta\Sigma^i\delta\Sigma^j \sim \delta^{ij} \tag{8.202}$$

must hold. This is much simpler calculation, using the already known identities for the 2-forms Σ^i, in particular (8.134), and confirms that the numerical coefficient in (8.201) is correct.

8.6 Chiral Connection Perturbation Theory

We now work out the chiral connection perturbative description of gravity. As we already know, this is obtained by starting with the Plebanski description and integrating out the 2-form field. There are two versions of this formalism. In one, we integrate out all the auxiliary fields, including the Lagrange multiplier field Ψ^{ij}. This results in the pure connection description (6.8). This action is non-polynomial in the curvature of the connection, and so its perturbative expansion produces an infinite number of terms, which are quite difficult to work out at increasing orders of perturbation theory.

In the other version we leave the field Ψ^{ij} in the game. This gives the first-order formalism with two independent fields, A^i_μ, Ψ^{ij} and the Lorentzian signature action given by

$$S[A,\Psi] = \frac{2}{\mathrm{i}}\int (M^{-1})^{ij} F^i F^j, \tag{8.203}$$

with the matrix M given by $M = \Psi + (\Lambda/3)\mathbb{I}$. We have again set $32\pi G = 1$. The matrix M^{-1} can then be expanded in powers of Ψ. This again produces a description with vertices of arbitrary valency. But in this version of the theory the vertices are at least straightforward to work out. We will develop the first-order version of the chiral connection perturbation theory, as this is also the version that exhibits direct similarities with the chiral Yang–Mills theory and chiral metric perturbation theory considered previously.

The main outcome of our analysis in this section is the realisation that the connection version of the perturbation theory is significantly simpler than that in terms of the metric. The main reason for this is that we necessarily have to develop the theory on a curved background. On a curved background, the transformation properties of the gauge field differ from those on a flat background. The new feature is that not only the metric, but also the connection field, transform nontrivially under diffeomorphisms. Moreover, the transformation rule for the connection can be rewritten so as to not involve the derivatives of the parameter of the transformation. It thus becomes trivial to gauge-fix the components of the connection that are pure gauge. We will see that this corresponds to the components taking values in $S_+ \otimes S_-$ subspace of $S_+^2 \otimes S_+ \otimes S_-$. This implies that there exists a gauge in which only the $S_+^3 \otimes S_-$ component of the connection propagates. This results in an extremely simple perturbation theory, directly analogous to the chiral perturbation theory for Yang–Mills theory.

8.6.1 Second-Order Lagrangian

We first work out the perturbation theory around a constant curvature background, and then make comments about the more general case. The constant curvature background is described by a field configuration in which $F^i = (\Lambda/3)\Sigma^i$, and we expand the connection Lagrangian (8.203) around this background. The background profile for the field Ψ^{ij} is trivial. The first variation of the Lagrangian in (8.203) is

$$\delta L = \frac{2}{\mathrm{i}}\mathrm{Tr}(-M^{-1}\delta M M^{-1} FF + M^{-1} 2F\delta F), \tag{8.204}$$

where $\delta M = \delta\Psi := \psi$. The second-order Lagrangian is the second variation divided by two. Using the fact that the background value of $M^{-1} = (3/\Lambda)\mathbb{I}$, we have

$$S^{(2)} = \frac{6}{\mathrm{i}\Lambda}\int \mathrm{Tr}(\psi^2\Sigma\Sigma - 2\psi\Sigma d_A a + d_A a d_A a + F[a,a]). \tag{8.205}$$

The last two terms here vanish by integration by parts

$$\int d_A a^i d_A a^i = \int a^i d_A d_A a^i = \int a^i \epsilon^{ijk} F^j a^k = -\int F^i \epsilon^{ijk} a^j a^k, \tag{8.206}$$

and so we have simply

$$L^{(2)} = \frac{6}{\Lambda}\left(2(\psi^{ij})^2 - 2\psi^{ij}\Sigma^{i\mu\nu} d^A_\mu a^j_\nu\right), \tag{8.207}$$

where we have used the self-duality of $\Sigma^i_{\mu\nu}$ and d^A_μ is the covariant derivative with respect to the background connection. Integrating out the auxiliary field ψ^{ij}, we immediately reproduce (5.183).

8.6.2 Gauge Transformations

Let us now spell out the action of gauge transformations. First, since the background value of the Lagrange multiplier field is zero, its perturbation does not transform under either SO(3) rotations or diffeomorphisms. The transformations of the connection are as follows

$$\delta_\phi a^i = d_A \phi^i, \qquad \delta_\xi a^i = i_\xi F^i, \tag{8.208}$$

where we corrected the diffeomorphism by a suitable gauge transformation to eliminate the derivative of the parameter ξ^μ from the transformation law. We see that the Lagrangian (8.207) is invariant under both transformations. Indeed, for gauge transformations we the commutator of two covariant derivatives, which produces a copy of the background curvature

$$\delta_\phi \Sigma^{i\mu\nu} d^A_\mu a^j_\nu = \Sigma^{i\mu\nu} d^A_\mu d^A_\nu \phi^j = \frac{\Lambda}{6}\Sigma^{i\mu\nu}\epsilon^{jkl}\Sigma^k_{\mu\nu}\phi^l \\ = \frac{2\Lambda}{3}\epsilon^{jil}\phi^l. \tag{8.209}$$

This gets contracted with ψ^{ij} that is symmetric, and so gives zero. For the variation under diffeomorphisms, we have analogously

$$\delta_\xi \Sigma^{i\mu\nu} d^A_\mu a^j_\nu = \Sigma^{i\mu\nu} d^A_\mu (\xi^\alpha \Sigma^j_{\alpha\nu}) = \Sigma^{i\mu\nu}\Sigma^j_{\alpha\nu}\nabla_\mu \xi^\alpha. \tag{8.210}$$

Here we have used the fact that the 'total' covariant derivative of the objects $\Sigma^i_{\mu\nu}$ vanishes

$$\partial_\rho \Sigma^i_{\mu\nu} + \epsilon^{ijk} A^j \Sigma^k_{\mu\nu} - \Gamma^\alpha{}_{\mu\rho}\Sigma^i_{\alpha\nu} - \Gamma^\alpha{}_{\nu\rho}\Sigma^i_{\mu\alpha} = 0, \tag{8.211}$$

where $\Gamma^\alpha{}_{\mu\nu}$ are the Christoffel symbols of the background metric. Using this, we can rewrite

$$d^A_\mu(\xi^\alpha\Sigma^i_{\alpha\nu}) = \Sigma^i_{\alpha\nu}\partial_\mu\xi^\alpha + \xi^\alpha(\partial_\mu\Sigma^i_{\alpha\nu} + \epsilon^{ijk}A^j_\mu\Sigma^i_{\alpha\nu}) \\ = \Sigma^i_{\alpha\nu}\partial_\mu\xi^\alpha + \xi^\alpha(\Gamma^\rho{}_{\alpha\mu}\Sigma^i_{\rho\nu} + \Gamma^\rho{}_{\nu\mu}\Sigma^i_{\alpha\rho}) = \Sigma^i_{\alpha\nu}\nabla_\mu\xi^\alpha + \Gamma^\rho{}_{\nu\mu}\xi^\alpha\Sigma^i_{\alpha\rho},$$

and the last term vanishes when contracted with $\Sigma^{i\mu\nu}$ in (8.210). But then according to (5.138), the contraction $\Sigma^{i\mu\nu}\Sigma^j_{\alpha\nu}$ of two Σ's is proportional to either the Kronecker δ^{ij} or the antisymmetric tensor ϵ^{ijk}, and so vanishes when contracted with ψ^{ij} in the Lagrangian.

8.6.3 Gauge-Fixing

Under diffeomorphisms the connection perturbation transforms as

$$\delta a^i_\mu = \frac{\Lambda}{3}\xi^\alpha\Sigma^i_{\alpha\mu}. \tag{8.212}$$

Let us now recall that in spinor terms the connection perturbation takes values in the space $S^2_+ \otimes S_+ \otimes S_-$. This has two irreducible components $S^3_+ \otimes S_-$ and $S_+ \otimes S_-$. The elements of $S_+ \otimes S_-$ are precisely of the form $\xi^\alpha\Sigma^i_{\alpha\mu}$. This means that the diffeomorphisms shift the $S_+ \otimes S_-$ component of the connection perturbation. A possible gauge-fix is to set this component of the connection to zero.

To see the most efficient way to gauge-fix SO(3) rotations we work out the Lagrangian with ψ^{ij} integrated out. This is given by

$$L = -\frac{3}{2\Lambda}\left(\delta_{ik}\delta_{jl} + \delta_{il}\delta_{jk} - \frac{2}{3}\delta_{ij}\delta_{kl}\right)(\Sigma^{i\mu\nu}d^A_\mu a^j_\nu)(\Sigma^{k\rho\sigma}d^A_\rho a^l_\sigma). \tag{8.213}$$

Let us assume that the connection is gauge-fixed to be in $S^3_+ \otimes S_-$. The derivative in $d^A_{[\mu} a^i_{\nu]}$ is the exterior covariant derivative with respect to the background connection. Because antisymmetrisation is taken, it can be replaced for free with the total covariant derivative with respect to both the background SL(2, $\mathbb{C}$) connection as well as the affine one. After this is done, the objects $\Sigma^{i\mu\nu}$ can be taken under the derivative sign for free, as they are killed by the total covariant derivative. For the terms resulting from $\delta_{ij}\delta_{kl}$ part of the projector on spin two this means that $\Sigma^{i\mu\nu} d^A_\mu a^i_\nu = \nabla^A_\mu (\Sigma^{i\mu\nu} a^i_\nu) = 0$, where ∇^A is the total derivative, as the combination in brackets extracts the $S_+ \otimes S_-$ part that vanishes.

To compute the term that is produced by the $\delta_{il}\delta_{jk}$ part of the projector we note that in general

$$\Sigma^{i\mu\nu} d^A_\mu a^j_\nu = \Sigma^{j\mu\nu} d^A_\mu a^i_\nu + \epsilon^{ijk}\epsilon^{klm}\Sigma^{l\mu\nu} d^A_\mu a^m_\nu . \tag{8.214}$$

But for connections in $S^3_+ \otimes S_-$ we have

$$\epsilon^{klm}\Sigma^{l\mu\nu} a^m_\nu = -a^k_\mu, \tag{8.215}$$

compare (8.70). This means that for such connections

$$\Sigma^{i\mu\nu} d^A_\mu a^j_\nu = \Sigma^{j\mu\nu} d^A_\mu a^i_\nu - \epsilon^{ijk}\nabla^{A\mu} a^k_\mu, \tag{8.216}$$

and so

$$\begin{aligned} L &= -\frac{3}{2\Lambda}(2(\Sigma^{i\mu\nu} d^A_\mu a^j_\nu)(\Sigma^{i\rho\sigma} d^A_\rho a^j_\sigma) - (\nabla^{A\mu} a^i_\mu)^2) \\ &= -\frac{3}{\Lambda}\left((\nabla^A_\mu a^i_\nu)^2 - (\nabla^{A\mu} a^{i\nu})(\nabla^A_\nu a^i_\mu) - \mathrm{i}\epsilon^{\mu\nu\rho\sigma} d^A_\mu a^i_\nu d^A_\rho a^i_\sigma - \frac{1}{2}(\nabla^{A\mu} a^i_\mu)^2 \right). \end{aligned}$$

We have again replaced the covariant derivatives d^A with the total derivatives ∇^A, for later convenience. We now integrate by parts in the second and third terms. For the second term

$$\begin{aligned} (\nabla^{A\mu} a^{i\nu})(\nabla^A_\nu a^i_\mu) &\hat{=} - a^{i\nu}\nabla^{A\mu}\nabla^A_\nu a^i_\mu \\ &= -a^{i\nu}\nabla^A_\nu \nabla^{A\mu} a^i_\mu - a^{i\nu}(-R^\rho{}_\mu{}^\mu{}_\nu a^i_\rho + \epsilon^{ijk} F^{j\mu}{}_\nu a^k_\mu) \\ &\hat{=} (\nabla^{A\mu} a^i_\mu)^2 - a^{i\nu}\left(R^\rho{}_\nu a^i_\rho - \frac{\Lambda}{3}\epsilon^{ijk}\Sigma^j_\nu{}^\mu a^k_\mu \right). \end{aligned} \tag{8.217}$$

where $\hat{=}$ means modulo surface terms. We now use that for our constant curvature background $R_{\mu\nu} = \Lambda g_{\mu\nu}$, where $g_{\mu\nu}$ is the background metric. We also use that for connections in $S^3_+ \otimes S_-$, we have $J_\Sigma(a) = -a$. This finally gives

$$(\nabla^{A\mu} a^{i\nu})(\nabla^A_\nu a^i_\mu) \hat{=} (\nabla^{A\mu} a^i_\mu)^2 - \frac{4\Lambda}{3}(a^i_\mu)^2. \tag{8.218}$$

For the third term in the Lagrangian we have

$$\begin{aligned} \mathrm{i}\epsilon^{\mu\nu\rho\sigma} d^A_\mu a^i_\nu d^A_\rho a^i_\sigma &\hat{=} - \mathrm{i}\epsilon^{\mu\nu\rho\sigma} a^i_\nu d^A_\mu d^A_\rho a^i_\sigma \\ &= -\frac{\Lambda}{6}\mathrm{i}\epsilon^{\mu\nu\rho\sigma} a^i_\nu \epsilon^{ijk}\Sigma^j_{\mu\rho} a^k_\sigma = -\frac{\Lambda}{3} a^i_\nu \epsilon^{ijk}\Sigma^{j\nu\sigma} a^k_\sigma = \frac{\Lambda}{3}(a^i_\mu)^2, \end{aligned} \tag{8.219}$$

where we have used the self-duality of $\Sigma^i_{\mu\nu}$ as well as the fact that the connection is in $S^3_+ \otimes S_-$ to get the last equality. Overall, this gives

$$L = -\frac{3}{\Lambda}((\nabla^A_\mu a^i_\nu)^2 - \frac{3}{2}(\nabla^{A\mu} a^i_\mu)^2 + \Lambda(a^i_\mu)^2). \tag{8.220}$$

This makes it clear that the most convenient gauge-fixing term is

$$L_{\text{g.f.}} = -\frac{9}{2\Lambda}(\nabla^{A\mu} a^i_\mu)^2, \tag{8.221}$$

where it is understood that the connection should be taken in $S^3_+ \otimes S_-$.

8.6.4 Spinor Interpretation

The previous gauge-fixing procedure can be understood most clearly using the language of spinors. To see this, let us start by converting the Lagrangian (8.207) into the spinor notations. The symmetric tracefree matrix ψ^{ij} in spinor notations is simply a generic element $\psi^{ABCD} \in S^4_+$. As we will verify later in this section, there is an overall factor of two that must be added in the process of spinor conversion. This has to do with the fact that there is a factor of $\sqrt{2}$ that appears when the objects $\Sigma^i_{\mu\nu}$ are converted to $\Sigma^{AB}_{MM'NN'}$. Overall, using (8.26) we get

$$L^{(2)} = \frac{12}{\Lambda}\left(2\psi^{ABCD}\psi_{ABCD} - 2\psi^{ABCD}\nabla_{AM'}a_{CDB}{}^{M'}\right), \tag{8.222}$$

where we replaced the covariant derivative with the total derivative with respect to both the Lorentz and affine connections. This is necessary, as in spinor notations, all indices are converted to spinor ones and are on the same footing. Integrating out the auxiliary field gives an extremely simple spinor form of the spin two kinetic term in the connection formalism

$$L^{(2)} = -\frac{6}{\Lambda}\left(\nabla_{(AM'}a_{BCD)}{}^{M'}\right)^2. \tag{8.223}$$

This form of the Lagrangian should be compared to that in the case of the chiral formulation of Yang–Mills theory, see (8.111). There is perfect analogy. The only difference is that there is no Lie algebra index in the case of gravity, and that the number of unprimed spinor indices has increased from one in the case of the Yang–Mills theory to three in the case of gravity, as is appropriate for a spin two field.

In order to do a check that the used normalisation is correct, we compute the leading term in the Lagrangian (8.223), the one that contains the $\Box$ operator. The coefficient in front can be computed by simply removing the symmetrisation from the indices $ABCD$ in (8.223). We then use that $\nabla_{AM'}\nabla^{AN'} = \frac{1}{2}\epsilon_{M'}{}^{N'}\Box + \cdots$ where dots stand for the curvature terms. The presence of the fact of $1/2$ in this relation shows that the overall coefficient is indeed correct as it reproduces $-3/\Lambda$ present in (8.220).

The spinor form of writing the Lagrangian makes all the gauge symmetries manifest. Indeed, the Lagrangian is explicitly independent of the $S_+ \otimes S_-$ component of the connection, as the symmetrisation on the three unprimed spinor indices of $a_{ABCM'}$ is taken to project the connection onto $S_+^3 \otimes S_-$. Second, the gauge invariance is also obvious in the spinor formalism due to the identity

$$2\nabla_{(AM'}\nabla_{B)}{}^{M'}\phi^{CD} = \left(\nabla_{AM'}\nabla_{B}{}^{M'} - \nabla_{B}{}^{M'}\nabla_{AM'}\right)\phi^{CD} \sim \Sigma^{(C|E|\ M'}_{ABM'}\phi_E{}^{D)} = 2\epsilon_A{}^{(C}\phi_B{}^{D)}, \tag{8.224}$$

where we replaced the commutator of covariant derivatives with the curvature and used the fact that on the background the curvature is proportional to the Σ^i 2-form. The precise proportionality factor in the relation on the second line is of no importance for us and so no attempt was made to fix it. Under the projection on S_+^4 the expression (8.224) vanishes, which shows that the action is also invariant under gauge rotations.

The form (8.223), together with the analogy with the chiral formulation of Yang–Mills theory suggests a natural way that the gauge symmetry must be fixed. Indeed, recall that in the case of the Yang–Mills Lagrangian (8.111), the gauge was fixed simply by removing the symmetrisation and converting the operator $S_+ \otimes S_- \to S_+^2$ present in the non gauge-fixed Lagrangian to the Dirac operator $S_+ \otimes S_- \to S_+ \otimes S_+$. It is clear that the same gauge-fixing can be carried out in the case of gravity. Indeed, the operator present in (8.223) is one that carries out the map $S_+^3 \otimes S_- \to S_+^4$. This map is degenerate because the dimensions of the spaces do not match, the dimension of the source is 8 while the dimension of the target is 5. The mismatch is precisely the number of gauge generators. The gauge-fixing can then be carried out by changing the operator to one $S_+^3 \otimes S_- \to S_+^3 \otimes S_+$. This is a version of the Dirac operator, as is also the case in the chiral Yang–Mills formalism.

Let us verify that this gives the required gauge-fixing. We start by replacing the auxiliary field $\psi^{ABCD} \in S_+^4$ in (8.222) with a new field

$$\psi^{ABCD} \to \Psi^{ABCD} := \psi^{ABCD} + \epsilon^{A(B}\phi^{CD)} \tag{8.225}$$

that takes values in $S_+^3 \otimes S_-$. Here ϕ^{AB} is the auxiliary field required for gauge-fixing in the first-order formalism. The additional terms generated by this replacing are

$$L_{\text{g.f.}} = \frac{12}{\Lambda}\left(\frac{8}{3}\phi^{AB}\phi_{AB} + 2\phi^{AB}\nabla^{M}{}_{M'}a_{ABM}{}^{M'}\right). \tag{8.226}$$

Integrating out ϕ^{AB} gives

$$L_{\text{g.f.}} = -\frac{9}{2\Lambda}\left(\nabla^{M}{}_{M'}a_{ABM}{}^{M'}\right)^2, \tag{8.227}$$

which is the correct gauge-fixing term (8.221). Thus, the gauge-fixed first-order Lagrangian is

$$L^{(2)} + L_{\text{g.f.}} = \frac{12}{\Lambda}\Big(2\Psi^{ABCD}\Psi_{ABCD} - 2\Psi^{ABCD}\nabla_{AM'}a_{CDB}{}^{M'}\Big), \tag{8.228}$$

where now the auxiliary field $\Psi^{ABCD} \in S_+^3 \otimes S_+$ is only symmetric in the last three indices.

8.6.5 Propagators

We now want to verify that as in all chiral formalisms considered before there is no propagator of the auxiliary field with itself. We now work on a (constantly) curved background and so cannot use the momentum space representation in this computation. Still, the required conclusion can be seen by formal manipulations in position space. There is no simple way of doing this computation in tensor notations, so we have to use spinors on this occasion. We start by adding to the Lagrangian the sources for both fields $\Psi^{ABCD}, a_{ABCC'}$.

$$\begin{aligned} L = {} & \frac{12}{\Lambda}\Big(2\Psi^{ABCD}\Psi_{ABCD} - 2\Psi^{ABCD}\nabla_{AM'}a_{CDB}{}^{M'}\Big) \\ & + J^{ABCD}\Psi_{ABCD} + J^{ABCC'}a_{ABCC'}. \end{aligned} \tag{8.229}$$

Extremising with respect to the auxiliary field Ψ^{ABCD} gives

$$\Psi_{ABCD} = \frac{1}{2}\nabla_{AM'}a_{BCD}{}^{M'} - \frac{\Lambda}{32}J_{ABCD}. \tag{8.230}$$

Substituting back into the Lagrangian gives

$$L = -\frac{6}{\Lambda}\left(\nabla_{AM'}a_{BCD}{}^{M'} - \frac{\Lambda}{16}J_{ABCD}\right)^2 + J^{ABCC'}a_{ABCC'}. \tag{8.231}$$

Opening the brackets and integrating by parts gives

$$\begin{aligned} L = {} & \frac{3}{\Lambda}a^{ABCC'}(-\Box + \Lambda)a_{ABCC'} + \frac{3}{4}a_{ABCC'}\nabla_D{}^{C'}J^{DABC} \\ & - \frac{3\Lambda}{128}(J_{ABCD})^2 + J^{ABCC'}a_{ABCC'}. \end{aligned} \tag{8.232}$$

Extremising with respect to the connection gives

$$(-\Box + \Lambda)a^{ABCC'} = -\frac{\Lambda}{6}\left(J^{ABCC'} + \frac{3}{4}\nabla_D{}^{C'}J^{DABC}\right). \tag{8.233}$$

Substituting this into the Lagrangian gives

$$L = -\frac{\Lambda}{12(-\Box + \Lambda)}\left(J^{ABCC'} + \frac{3}{4}\nabla_D{}^{C'}J^{DABC}\right)^2 - \frac{3\Lambda}{128}(J^{ABCD})^2,$$

where we formally inverted the operator $-\Box + \Lambda$. Expanding the brackets gives the propagators. There is clearly the propagator for the connection with itself, and connection to the auxiliary field. The auxiliary field to itself terms can be seen to cancel out, and so there is no $\langle \Psi^{ABCD} \Psi^{MNRS} \rangle$ propagator as could have been expected by analogy with the chiral Yang–Mills perturbation theory story.

8.6.6 Chiral Connection Perturbation Theory on an Arbitrary Einstein Background

In the previous sections, we have worked out the connection perturbation theory on a constant curvature background. Recall now that in (5.187) we have worked out the pure connection kinetic term on an arbitrary Einstein background. This is a significantly more complicated second-order action, which on a constant curvature background, by integration by parts manipulations, becomes the simple (5.183). However, no such simplification is possible on a general background and (5.187) is the simplest available form of the second-order action.

The transformation properties of the connection on a general background are as follows. First, we have the usual rule $\delta_\phi a^i = d^A \phi^i$ for the gauge rotations. For the diffeomorphisms, we again can write $\delta_\xi a^i = i_\xi F^i$, where F^i is the background curvature. Using the fact that on an Einstein background the curvature is SD as a 2-form we have

$$\delta_\xi a^i = \left(\Psi^{ij} + \frac{\Lambda}{3} \delta^{ij} \right) i_\xi \Sigma^j. \tag{8.234}$$

Let us now assume that the matrix $\Psi + (\Lambda/3)\mathbb{I}$ is nondegenerate. If this is the case, we can use the transformations (8.234) to set to zero the $S_+ \otimes S_-$ component of the connection perturbation. The diffeomorphisms now act in a more complicated fashion, changing both the $S_+^3 \otimes S_-$ and $S_+ \otimes S_-$ parts of the connection, while in the constant curvature case there was no action on the $S_+^3 \otimes S_-$ part. Nevertheless, the gauge in which the $S_+ \otimes S_-$ part is set to zero is possible.

In this gauge the scalar first-order operator

$$(d^A a) = \frac{1}{2} \Sigma^{i\mu\nu} d^A_\mu a^i_\nu = \frac{1}{2} \nabla_\mu (\Sigma^{i\mu\nu} a^i_\nu) = 0.$$

Thus, in this gauge the first term in (5.187) is zero. Also in this gauge, the combination

$$(d^A a)^i = \frac{1}{2} \epsilon^{ijk} \Sigma^{j\mu\nu} d^A_\mu a^k_\nu = \frac{1}{2} \nabla_\mu (\epsilon^{ijk} \Sigma^{j\mu\nu} a^k_\nu) = -\frac{1}{2} \nabla^\mu a^i_\nu$$

because for connections in $S_+^3 \otimes S_-$ we have $a = -J_\Sigma(a)$. This means that if we gauge-fix Lorentz rotations using the Lorentz gauge $\nabla^\mu a^i_\mu = 0$ the second term of (5.187) is also zero. This leaves only the the last two terms of (5.187) nonvanishing. It is also clear that the object $(d^A a)^{ij} = (1/2) \bar{\Sigma}^{j\,\mu\nu} d^A_\mu a^i_\nu$ is essentially the operator $\delta^*_{(2,2)}$ acting on the connection with the result lying in the space $S_+^2 \otimes S_-^2$. Indeed, we have $S_+^2 \otimes S_-^2 = \Lambda^+ \otimes \Lambda^-$, and this is precisely where the object $(d^A a)^{ij}$ takes values. Thus, on an arbitrary Einstein background we have an efficient representation of the spin two kinetic term, schematically of the form $\delta_{(2,0)} M^{-1} \delta^*_{(2,0)}$, where $M = \Psi + (\Lambda/3)\mathbb{I}$ is the matrix of the background curvature. There is also the 'mass term', which is the last term in (5.187). Note that this representation of the spin two second-order Lagrangian is possible even on $\Lambda = 0$ backgrounds, as long as $\Psi^{ij} \neq 0$. This is the case, for example, on the background of a Schwarzschild black hole. This means that there likely exists a rather simple chiral pure connection version of the Schwarzschild black hole perturbation theory, which is still to be worked out.

9

Higher-Dimensional Descriptions

This chapter develops what can be called higher-dimensional descriptions of 4D general relativity (GR). There are two considerations that motivate our constructions.

Recall from the discussion of the Kaluza–Klein mechanism that one can obtain 4D GR (coupled to other fields) by the dimensional reduction of a theory whose dynamics is described by the higher-dimensional version of the Einstein–Hilbert action. This mechanism can be anticipated to be much more general in that it can be expected that if one starts with a higher-dimensional theory that is diffeomoprhism-invariant and has local degrees of freedom, then dimensional reduction to 4D will generically give rise to a theory of massless spin two particles interacting with other fields. In other words, nothing forces us to fix the higher-dimensional theory to be one described by the Einstein–Hilbert action. There are other diffeomoprhism invariant theories, as we shall see in this chapter, and they generically lead to 4D theories possessing the essential features of GR.

The second consideration relates to conformal invariance. It is well known that the equations describing 4D massless particles of arbitrary spin are conformally invariant. For example, Maxwell's equations are

$$dF = 0, \qquad d^*F = 0,$$

where $^*F$ is the Hodge dual of the field strength F. As we have discussed in the chapter on chiral descriptions of GR, in four dimensions the Hodge star on 2-forms only depends on the conformal class of the metric, and is invariant under arbitrary conformal rescalings. This means that Maxwell's equations only depend on the conformal class of the metric. This implies that symmetries of Maxwell's equations are not just the isometries of the background metric, but the larger set of conformal isometries, i.e., transformations that may, in general, multiply the metric by an arbitrary conformal factor. Thus, the group of symmetries of Maxwell's equations in the Minkowski space is not the Poincare group of translations plus Lorentz transformations, but the larger **conformal group** of the Minkowski space.

These statements have a spinor translation. In the spinor language the 2-form F is described by its self-dual (SD) and anti-self-dual (ASD) parts, which are elements of S^2_+ and S^2_-, respectively; see the previous chapter for our spinor conventions. For a real 2-form in the Minkowski space, the ASD part of F is the complex conjugate of its SD part. Maxwell's equations are then coded into a single complex equation

$$\nabla^B{}_{A'}\phi_{AB} = 0, \tag{9.1}$$

where $\nabla_{AA'}$ is the covariant derivative operator and $\phi_{AB} \in S^2_+$ is the SD part of the field strength. The original Maxwell's equations arise as the real and imaginary part of this complex equation. Moreover, the equation (9.1) is conformally invariant because it only depends on the decomposition of the space of 2-forms into SD and ASD parts, which is conformally invariant.

The equation (9.1) has an immediate generalisation to the case of higher spins. For example, the spin two version of this equation is $\nabla^D{}_{A'}\psi_{ABCD} = 0$, where $\psi_{ABCD} \in S^4_+$ is the rank four spinor encoding the SD part of the Weyl curvature. This equation is also conformally invariant.

The group of conformal transformations of the Minkowski space acts on its coordinates x^4, x^1, x^2 and x^3 nonlinearly. This is similar to the action of the conformal group in two dimensions, which is most conveniently described as the group of fractional linear transformations

$$z \to \frac{az+b}{cz+d}, \tag{9.2}$$

where $z \in \mathbb{C} \sim \mathbb{R}^2$. The parameters a, b, c, and d are elements of a matrix $g \in \mathrm{SL}(2,\mathbb{C})$. The conformal group is the quotient subgroup $\mathrm{PSL}(2,\mathbb{C})$ in which the elements g and $-g$ are identified. Similarly, as we shall see in this Chapter, the conformal group in four dimensions can be realised as the group of fractional linear transformations in which all entries of the formula (9.2) are replaced by 2×2 matrices.

In two dimensions, the action of the conformal group $\mathrm{PSL}(2,\mathbb{C}) = \mathrm{SO}(1,3)$ can be linearly realised by considering a larger space, the four-dimensional Minkowski space. One realises the compactified complex plane as the space of future-directed null rays through the origin of $\mathbb{R}^{1,3}$. In other words, the compactified $\mathbb{R}^2$ is the two-sphere that arises as the projectivised light cone in $\mathbb{R}^{1,3}$. The action of the conformal group on $\mathbb{R}^{1,3}$ is the usual linear action of the Lorentz group in the Minkowski space. The nonlinearity present in (9.2) then has its origin in the projectivisation needed to pass from the light cone in $\mathbb{R}^{1,3}$ where the action is linear to its projective version.

One can similarly realise the action of the 4D conformal group as the group of linear transformations of a bigger space. Thus, the Minkowski space can be realised as the projectivised light cone in a six-dimensional space, where the conformal group acts linearly. However, the orthogonal groups in 6D are isomorphic to various real forms of the complex special linear group SL(4, $\mathbb{C}$) in four dimensions. This is the twistor isomorphism already discussed in (5.94). This implies that it is most convenient to think about the projectivised light-cone in 6D as the space of two-planes in an auxiliary four-dimensional space called the **twistor space**. The action of the conformal group is then just the natural action of a suitable real form of SL(4, $\mathbb{C}$) on $\mathbb{C}^4$.

We thus have two seemingly different considerations, indicating that it may be beneficial to introduce a higher-dimensional space to describe 4D gravity. In this chapter we shall see that they are not unrelated.

9.1 Twistor Space

We have motivated the twistor space as the geometric construction that gives the linear version of the action of the conformal group in the Minkowski space. With suitable modifications, this construction exists for all three possible signatures in four dimensions. For this reason, we start our description in the complexified setting, and then discuss relevant reality conditions that reduce everything to a space of desired signature.

9.1.1 The Twistor Space of $\mathbb{C}^4$

The complexified version of the twistor space is simplest to describe. The idea is to realise the four-dimensional complex space $M = \mathbb{C}^4$ as the space of certain geometric objects in some other space $\mathbb{T} = \mathbb{C}^4$. The geometric objects in question are two-planes through the origin.

It is customary in the twistor literature to use the capital letter Z to refer to coordinates in the twistor space $Z \in \mathbb{T} = \mathbb{C}^4$. Every two-plane through the origin of $\mathbb{T}$ can be characterised by two (non-colinear) vectors $Z_1, Z_2 \in \mathbb{C}^4$. Having two such vectors, one can form the **bi-vector** $Z_1 \wedge Z_2$. The space of bi-vectors is the six-dimensional space $\Lambda^2\mathbb{C}^4$. If $e_i, i = 1, \ldots, 4$ is a basis in $\mathbb{C}^4$, then the corresponding basis in the space of bi-vectors is $e_i \wedge e_j, i < j$, and a general bi-vector is of the form

$$Y = \frac{1}{2} y^{ij} e_i \wedge e_j, \tag{9.3}$$

with the summation implied. The coefficients $y^{ij} = y^{[ij]}$ form the so-called **Plücker coordinates** in the six-dimensional space of bi-vectors. The bi-vectors corresponding to two-planes are those that satisfy $Y \wedge Y = 0$, or, in terms of coordinates

$$y^{12}y^{34} + y^{31}y^{24} + y^{14}y^{23} = 0. \tag{9.4}$$

Bi-vectors satisfying this equation are called **simple**, or decomposable. This is an equation for a null surface (quadric)

$$Q = \{Y \in \Lambda^2\mathbb{C}^4 : Y \wedge Y = 0\} \tag{9.5}$$

in the space of bi-vectors $\mathbb{C}^6 = \Lambda^2\mathbb{C}^4$.

Two simple bi-vectors that differ by an overall scale correspond to the same two-plane in $\mathbb{T}$. This means that the space of two-planes through the origin of $\mathbb{T}$ can be described as the space of simple bi-vectors Y modulo rescaling, or equivalently as the projectivised quadric Q, the projective version PQ of Q. This is precisely analogous to the realisation of compactified $\mathbb{R}^2$ as the projectivised quadric (light cone) in four-dimensional Minkowksi space. The space of the planes through the origin of $\mathbb{T} = \mathbb{C}^4$ is called the **Grassmanian** $\mathrm{Gr}_2(\mathbb{C}^4)$. We thus see that using Plücker coordinates, the Grassmanian $\mathrm{Gr}_2(\mathbb{C}^4)$ of two-planes though the origin of $\mathbb{T}$ is the projective quadric of complex dimension 4 in $\mathbb{C}^6$, i.e., $\mathrm{Gr}_2(\mathbb{C}^4) = PQ$. We want to identify 'our' space with it

$$M = PQ = \mathrm{Gr}_2(\mathbb{C}^4). \tag{9.6}$$

9.1.2 Action of the Conformal Group

There is a natural action of the complex general linear group $\mathrm{GL}(4, \mathbb{C})$ on $\mathbb{T} = \mathbb{C}^4$. If $Z = z^i e_i$ is a vector in $\mathbb{T}$ then $g : z^i \to g^i{}_j z^j$. This induces the action on the space of bi-vectors $g : Y \to Y_g$. In coordinates, this reads

$$g : y^{ij} \to g^i{}_k g^j{}_l y^{kl}. \tag{9.7}$$

Because $Y_g \wedge Y_g = \det(g) Y \wedge Y$ the group $\mathrm{GL}(4, \mathbb{C})$ preserves the quadric (9.5). If we pass to the projective quadric PQ, one finds that it is the group $\mathrm{SL}(4, \mathbb{C})$ that acts on PQ effectively and transitively. This means that PQ is a group coset. Indeed, the stabiliser of the plane that is the span of, e.g., e_3, e_4 is the subgroup of matrices of the form

$$P = \left\{ \begin{pmatrix} * & * & 0 & 0 \\ * & * & 0 & 0 \\ * & * & * & * \\ * & * & * & * \end{pmatrix} \in \mathrm{SL}(4, \mathbb{C}) \right\}, \tag{9.8}$$

where a star denotes a nonzero entry. This means that the projective quadric PQ is the coset

$$PQ = \mathrm{Gr}_2(\mathbb{C}^4) = \mathrm{SL}(4, \mathbb{C})/P. \tag{9.9}$$

The group $\mathrm{SL}(4, \mathbb{C})$ acts on PQ by multiplication from the left. This is the conformal group of the complexified Minkowski space $PQ = \mathbb{C}^4$.

9.1.3 Coordinatisation of $\mathrm{Gr}_2(\mathbb{C}^4)$

We now describe an explicit set of coordinates on the Grassmanian $\mathrm{Gr}_2(\mathbb{C}^4)$ and verify that the action of the conformal group is a generalised version of the fractional linear transformations (9.2). We will also understand why it was natural, see (8.1), to put the coordinates of the Minkowski space together into a 2×2 matrix.

Consider a two-plane spanned by twistors $Z_1, Z_2 \in \mathbb{C}^4$. We can take arbitrary linear combinations of Z_1 and Z_2 without changing the plane. We thus have a group $\mathrm{GL}(2, \mathbb{C})$ at our disposal to put the twistors Z_1 and Z_2 into some convenient form. There are $4 + 4$ (complex) parameters needed to specify Z_1 and Z_2, and using $\mathrm{GL}(2, \mathbb{C})$ we can set four of them to desired values. One can then see that for a generic two-plane we can put Z_1 and Z_2 into the form

$$Z_1 = \alpha e_1 + \gamma e_2 + e_3, \qquad Z_2 = \beta e_1 + \delta e_2 + e_4. \tag{9.10}$$

In other words, using $\mathrm{GL}(2, \mathbb{C})$ we can set two of the 8 parameters of Z_1 and Z_2 to zero and two others to the identity. It is convenient to put the two columns Z_1 and Z_2 into a 2×4 matrix

$$Y = \begin{pmatrix} \alpha & \beta \\ \gamma & \delta \\ 1 & 0 \\ 0 & 1 \end{pmatrix}, \tag{9.11}$$

which we also denote by Y. The conformal group $\mathrm{SL}(4, \mathbb{C})$ acts on Y by matrix multiplication from the left. To describe this action explicitly we use the block matrix notation and rewrite

$$Y = \begin{pmatrix} \mathbf{x} \\ \mathbb{I} \end{pmatrix}, \qquad \mathbf{x} := \begin{pmatrix} \alpha & \beta \\ \gamma & \delta \end{pmatrix}, \tag{9.12}$$

and

$$g = \begin{pmatrix} A & B \\ C & D \end{pmatrix}, \tag{9.13}$$

where A, B, C, and D are 2×2 complex matrices. We then have

$$Y \to Y_g = gY = \begin{pmatrix} A\mathbf{x} + B \\ C\mathbf{x} + D \end{pmatrix} \sim \begin{pmatrix} \mathbf{x}_g \\ \mathbb{I} \end{pmatrix}, \tag{9.14}$$

where we have put the result into the form (9.11) by a $\mathrm{GL}(2, \mathbb{C})$ transformation and

$$\mathbf{x}_g = (A\mathbf{x} + B)(C\mathbf{x} + D)^{-1}. \tag{9.15}$$

This shows that the conformal group $\mathrm{SL}(4, \mathbb{C})$ acts on $M = \mathbb{C}^4 = \mathrm{Gr}_2(\mathbb{C}^4)$ by matrix fractional linear transformations.

It is instructive to describe various subgroups of the conformal group. First, the subgroup of transformations that preserve the origin $\mathbf{x} = 0$ is $B = 0$, which we have already seen in (9.8). Of these, the transformations with $C = 0$ are complexified Lorentz rotations combined with dilatations

$$\mathbf{x} \to A\mathbf{x}D^{-1}. \tag{9.16}$$

When $\det(A) = \det(D) = 1$ this is the already familiar to us action of the complexified Lorentz group. When $A = \lambda\mathbb{I}, D = \lambda^{-1}\mathbb{I}, \lambda \in \mathbb{C}$ this is a dilatation. More generally, the transformations with $C = 0, B \neq 0$ act as

$$\mathbf{x} \to A\mathbf{x}D^{-1} + BD^{-1}, \tag{9.17}$$

which is a complexified Lorentz rotation (together with dilatation) plus complexified translation. Finally, the transformations with $C \neq 0$ are the special conformal transformations. All this exactly parallels the usual conformal action (9.2) on $S^2 \sim \mathbb{C}$. We thus see that the generalisation required to go from 2D to 4D is to replace complex numbers by 2×2 matrices. In the following sections we will see that various different signatures that we can have in 4D correspond to various 'reality' conditions imposed on the complex matrices $\mathbf{x}$, exactly reproducing the already familiar story from Section 5.5. In particular, in the case of $\mathbb{R}^4$ the required 2×2 matrices will be those that correspond to quaternions. Thus, the generalisation that is required to go from $\mathbb{R}^2$ to $\mathbb{R}^4$ is to replace $\mathbb{C}$ by $\mathbb{H}$, and the conformal group $\mathrm{SL}(2, \mathbb{C})$ by $\mathrm{SL}(2, \mathbb{H})$, as we shall see in Section 9.2.

9.1.4 Twistor as Two Spinors

The subgroup of matrices with $B, C = 0$ and $\det(A) = \det(D) = 1$ is the complexified Lorentz group sitting inside the conformal group $\mathrm{SL}(4, \mathbb{C})$. With respect to this Lorentz group the fundamental representation $\mathbb{T}$ of $\mathrm{SL}(4, \mathbb{C})$ becomes reducible and splits into two $\mathbb{C}^2$ representations, the spinor representations of the Lorentz group. Let us therefore coordinatise $\mathbb{T} = \mathbb{C}^2 \oplus \mathbb{C}^2$ by two-component columns $\pi, \omega \in \mathbb{C}^2$, so that a twistor in $\mathbb{T}$ can be represented as a bi-spinor

$$Z = \begin{pmatrix} \pi \\ \omega \end{pmatrix}, \qquad \pi = \begin{pmatrix} \pi_1 \\ \pi_2 \end{pmatrix}, \qquad \omega = \begin{pmatrix} \omega_1 \\ \omega_2 \end{pmatrix}. \tag{9.18}$$

The action of the complexified Lorentz group on π, ω is

$$\pi \to A\pi, \qquad \omega \to D\omega. \tag{9.19}$$

The discussion of spinors in the previous chapter allows us to interpret π and ω as the two spinors of different types $\pi \equiv \pi_A, \omega \equiv \omega_{A'}$, with their spinor indices down.

We then have the spinor index–raising operation, which from a column π produces a row $(\epsilon\pi)^T$, and similarly from the column ω the row $(\epsilon\omega)^T$. The action of Lorentz group on the spinors with indices raised is

$$(\epsilon\pi)^T \to (\epsilon A\pi)^T = \pi^T A^T \epsilon^T = \pi^T \epsilon^T A^{-1} = (\epsilon\pi)^T A^{-1}, \tag{9.20}$$

where we have used the relation $A^T\epsilon = \epsilon A^{-1}$, which is true for any $A \in \mathrm{SL}(2,\mathbb{C})$. We similarly have $(\epsilon\omega)^T \to (\epsilon\omega)^T D^{-1}$. This means that the matrix $\mathbf{x}$ transforming as (9.16) should be interpreted as a bi-spinor with its primed index raised

$$\mathbf{x} \equiv \mathbf{x}_A{}^{A'}. \tag{9.21}$$

We will require this interpretation in the following sections.

9.1.5 Twistor Space as the Coset

The action of $\mathrm{SL}(4,\mathbb{C})$ on $\mathbb{T}$ is the simple action of a matrix group on columns. Given that to get to $M = PQ$ we pass to the projective version of the space of bi-vectors, it makes sense to consider also the projective version $P\mathbb{T}$ of the space $\mathbb{T}$. This is the space of lines through the origin of $\mathbb{T}$, which can also be described as the Grassmanian $\mathrm{Gr}_1(\mathbb{C}^4)$. This space is called the projective twistor space.

The projective version $P\mathbb{T}$ of the space $\mathbb{T}$ is also a group coset. Indeed, the stabiliser of the line in the direction of the vector e_4 is the subgroup of matrices of the form

$$R = \left\{ \begin{pmatrix} * & * & * & 0 \\ * & * & * & 0 \\ * & * & * & 0 \\ * & * & * & * \end{pmatrix} \in \mathrm{SL}(4,\mathbb{C}) \right\}. \tag{9.22}$$

This gives

$$P\mathbb{T} = \mathrm{Gr}_1(\mathbb{C}^4) = \mathrm{SL}(4,\mathbb{C})/R. \tag{9.23}$$

9.1.6 Twistor Double Fibration

Both PQ and $P\mathbb{T}$ arise as the quotients of the complex special linear group in four dimensions by the so-called **parabolic** subgroups P and R. There is a smaller parabolic subgroup that is the intersection of P and R. This is the subgroup of matrices of the form

$$Q = \left\{ \begin{pmatrix} * & * & 0 & 0 \\ * & * & 0 & 0 \\ * & * & * & 0 \\ * & * & * & * \end{pmatrix} \in \mathrm{SL}(4,\mathbb{C}) \right\}. \tag{9.24}$$

The group coset $\mathrm{SL}(4,\mathbb{C})/Q$ has complex dimension five, and is the so-called **flag** manifold. A point in this space is a two-dimensional projective plane in $\mathbb{T}$, together with a line in this plane. Both PQ and $P\mathbb{T}$ arise as bases of fibrations with $\mathrm{SL}(4,\mathbb{C})/Q$ as the total space of the bundle. We have the following double fibration playing central role in twistor theory:

$$\begin{array}{ccc} & \mathrm{SL}(4,\mathbb{C})/Q & \\ \swarrow^{\eta} & & \searrow^{\tau} \\ P\mathbb{T} = \mathrm{SL}(4,\mathbb{C})/R & & PQ = \mathrm{SL}(4,\mathbb{C})/P \end{array} \tag{9.25}$$

9.1.7 Geometric Interpretation

The realisation of complexified Minkowski space as the Grassmanian of two-planes in $\mathbb{C}^4$ can be phrased in geometric terms as follows. The projective twistor space is $P\mathbb{T} = \mathbb{C}P^3$, the complex projective space of dimension three. The two-planes through the origin in $\mathbb{T}$ are nothing else but the complex lines in the projective twistor space. Thus, we can say that points in complexified Minkowski space are lines in the projective twistor space.

What is then the Minkowski space interpretation of points in the projective twistor space? A point in $P\mathbb{T}$ is a line Z in $\mathbb{T}$ through the origin. There are different two-planes in $\mathbb{T}$ that share this line. To specify a two-plane sharing a given line Z we need to prescribe another twistor $\tilde{Z}$. Adding to $\tilde{Z}$ any multiple of Z does not change the plane, so we only need to specify three complex parameters, modulo an overall scale. This shows that the space of two-planes sharing a given line is two-dimensional. Thus, to every point in $P\mathbb{T}$, there corresponds a set of points in the complexified Minkowski space of complex dimension two. Let us call these sets of points **α-planes**. Thus, we can say that a point in the projective twistor space is an α-plane in the complexified Minkowski space.

Overall, we have the following correspondence

$$\begin{aligned} &\textbf{lines in projective twistor space} \Leftrightarrow \textbf{points in } M, \\ &\textbf{points in projective twistor space} \Leftrightarrow \alpha\textbf{-planes in } M. \end{aligned} \tag{9.26}$$

Let us work out a coordinate description of this. A line in $P\mathbb{T}$ is a two-plane in $\mathbb{T}$, and for a two-plane that is the span of vectors (9.10), this is a set of points

$$\omega_1 Z_1 + \omega_2 Z_2 = \begin{pmatrix} \mathbf{x}\omega \\ \omega \end{pmatrix}, \tag{9.27}$$

where $\omega_{1,2}$ are complex parameters. We thus see that a line in projective twistor space that corresponds to a 2×2 matrix $\mathbf{x}$ is the line where the coordinates π and ω on $\mathbb{T}$ satisfy

$$\pi = \mathbf{x}\,\omega. \tag{9.28}$$

We can also write this in spinor notations as

$$\pi_A = \mathbf{x}_A{}^{A'} \omega_{A'}. \tag{9.29}$$

In the twistor literature this is called the **incidence relation.**

In the opposite direction, let us start from a line in $\mathbb{T}$ in the direction of some twistor Z represented by a pair of spinors $(\pi,\omega)\in\mathbb{C}^2\oplus\mathbb{C}^2$. We are interested in all two-planes in $\mathbb{T}$ that share this line. Let us take some other twistor $\tilde{Z}=(\tilde{\pi},\tilde{\omega})$ not collinear with Z. We then form the 2×4 matrix representing the plane spanned by Z and $\tilde{Z}$

$$\begin{pmatrix} \pi & \tilde{\pi} \\ \omega & \tilde{\omega} \end{pmatrix}. \tag{9.30}$$

Mixing a and $\tilde{a}$ with an $\mathrm{GL}(2,\mathbb{C})$ transformation (that acts on Y by multiplication from the right) this can be put into the form (9.11) with

$$\mathbf{x}=\begin{pmatrix} \pi_1 & \tilde{\pi}_1 \\ \pi_2 & \tilde{\pi}_2 \end{pmatrix}\begin{pmatrix} \omega_1 & \tilde{\omega}_1 \\ \omega_2 & \tilde{\omega}_2 \end{pmatrix}^{-1}. \tag{9.31}$$

To see what this means, let us first work out the α-plane that corresponds to the line in $\mathbb{T}$, which in turn corresponds to the origin of the complexified Minkowski space. As we can see from (9.29), this is the line $\pi=0$ and ω arbitrary, which by rescaling can always be fixed to be

$$\omega=\begin{pmatrix} \xi \\ 1 \end{pmatrix}. \tag{9.32}$$

Here $\xi\in\mathbb{C}$. So we have

$$\begin{aligned}\mathbf{x}=\begin{pmatrix} 0 & \tilde{\pi}_1 \\ 0 & \tilde{\pi}_2 \end{pmatrix}\begin{pmatrix} \xi & \tilde{\omega}_1 \\ 1 & \tilde{\omega}_2 \end{pmatrix}^{-1} &= \frac{1}{\tilde{\omega}_1-\xi\tilde{\omega}_2}\begin{pmatrix} \tilde{\pi}_1 & -\tilde{\pi}_1\xi \\ \tilde{\pi}_2 & -\tilde{\pi}_2\xi \end{pmatrix} \\ &= \frac{1}{\tilde{\omega}_1-\xi\tilde{\omega}_2}\begin{pmatrix} \tilde{\pi}_1 \\ \tilde{\pi}_2 \end{pmatrix}\begin{pmatrix} 1 & -\xi \end{pmatrix}.\end{aligned} \tag{9.33}$$

The two-component row here is just what we defined in (8.6) to be the two-component spinor ω with its index raised

$$\begin{pmatrix} 1 & -\xi \end{pmatrix}=(\epsilon\omega)^T \tag{9.34}$$

This shows that the α-plane through the origin of the Minkowski space is the set of 2×2 matrices

$$\mathbf{x}_A{}^{A'}=\frac{\tilde{\pi}_A\omega^{A'}}{[\tilde{\omega}\omega]}, \tag{9.35}$$

where $[\tilde{\omega}\omega]:=\tilde{\omega}_{A'}\omega^{A'}$ is the spinor pairing. This should be interpreted as the α-plane corresponding to the twistor $(0,\omega)$ with the spinor ω thus being fixed. The twistor ($\tilde{\pi}$ and $\tilde{\omega}$) is changing and gives a two-parameter surface in the complexified Minkowski space.

For a general twistor (π and ω), we can always represent the matrix (9.31) as

$$\mathbf{x} = \begin{pmatrix} \pi_1 & \pi_1 \\ \pi_2 & \pi_2 \end{pmatrix} \begin{pmatrix} \omega_1 & \tilde{\omega}_1 \\ \omega_2 & \tilde{\omega}_2 \end{pmatrix}^{-1} + \begin{pmatrix} 0 & \tilde{\pi}_1 - \pi_1 \\ 0 & \tilde{\pi}_2 - \pi_2 \end{pmatrix} \begin{pmatrix} \omega_1 & \tilde{\omega}_1 \\ \omega_2 & \tilde{\omega}_2 \end{pmatrix}^{-1}.$$

In spinor notations this can be written as

$$\mathbf{x}_A{}^{A'} = \mathbf{x}_{0A}{}^{A'} + \frac{(\tilde{\pi}_A - \pi_A)\omega^{A'}}{[\tilde{\omega}\omega]}, \qquad \mathbf{x}_{0A}{}^{A'} := \frac{\pi_A(\omega^{A'} - \tilde{\omega}^{A'})}{[\tilde{\omega}\omega]}. \tag{9.36}$$

This represents a point $\mathbf{x}_{0A}{}^{A'}$ on the α-plane in question, plus a two-parameter set of vectors giving other points on the same plane, parametrised by the spinor $\tilde{\pi}_A - \pi_A$.

9.1.8 Conformal Metric on M

Generic two-planes in $\mathbb{T}$ only intersect at the origin. However, there are two-planes that intersect along a line in $\mathbb{T}$. As we have just described, these correspond to points of the Minkowski space that lie on the same α-plane, the α-plane that corresponds to the line in $\mathbb{T}$ in question. This means that we can introduce a natural conformal metric on the complexified Minkowski space. We define this metric so that points in the Minkowski space that lie on the same α-plane are null-separated.

Such a metric can be easily described using the Plücker coordinates on $\mathrm{Gr}_2(\mathbb{C}^4)$. Indeed, we have described $\mathrm{Gr}_2(\mathbb{C}^4)$ as the projective quadric PQ in $\mathbb{C}^6$. The tangent space to a point $Y \in Q$ consists of bi-vectors dY satisfying $dY \wedge Y = 0$. There is then a natural conformal metric given by

$$ds^2 \sim dY \wedge dY. \tag{9.37}$$

This gives a top form on $\mathbb{C}^4$, which we divide by an arbitrary volume form to obtain a number. Alternatively, using the index notation we have $ds^2 = \epsilon_{ijkl} dy^{ij} dy^{kl}$, where ϵ_{ijkl} is some completely antisymmetric tensor on $\mathbb{C}^4$.

The metric (9.37) has the desired properties. Indeed, let us consider a point in the Minkowski space that is represented by two-plane in $\mathbb{T}$ with Plücker coordinate $Y \in \Lambda^2\mathbb{C}^4$. Let us then consider a nearby two-plane $Y + dY$, with dY small. In order for this bi-vector to represent a two-plane, we must have $dY \wedge Y = 0$. There are then two possibilities. If dY is decomposable $dY \wedge dY = 0$, then $dY \wedge Y$ implies that bi-vectors dY and Y share a vector. This means that two-planes represented by Y and $Y + dY$ share a line and the two Minkowski points Y and $Y + dY$ are on the same α-plane. We wanted α-planes to be null, and $dY \wedge dY = 0$ guarantees that. On the other hand, dY does not have to be decomposable. In this case, the two-planes Y and $Y + dY$ do not share a line and the corresponding Minkowski space points do not lie on the same α-plane.

It is instructive to work out the metric (9.37) explicitly, using the parametrisation (9.12). We have

$$dY = (d\alpha e_1 + d\gamma e_2) \wedge e_4 + e_3 \wedge (d\beta e_1 + d\gamma e_2), \tag{9.38}$$

where $d\alpha, d\beta, d\gamma$, and $d\delta$ are components of the matrix $d\mathbf{x}$. Therefore,

$$dY \wedge dY \sim d\alpha d\delta - d\beta d\gamma = \det(d\mathbf{x}). \tag{9.39}$$

This is the familiar metric from Section 5.5 on $\mathbb{R}^4$ of various signatures expressed as the determinant of a 2×2 matrix $\mathbf{x}$. This immediately confirms that the α-planes of the origin (9.35), and general (9.36), are totally null. Indeed, the matrix $\mathbf{x}_A{}^{A'}$ in (9.35) has zero determinant, and so all points it represents are null-separated from the origin. In (9.36), the second term is a matrix with zero determinant, so all points on this α-plane are null-separated from $\mathbf{x}_0$.

In words, we see from (9.36) that a general α-plane in M is parametrised by a twistor (π, ω) and points on it can be represented as the sum of two null vectors, one given by the product of spinor π_A times an arbitrary primed spinor, and the other as the product of an arbitrary unprimed spinor times $\omega^{A'}$.

9.1.9 Split Signature Version

We now work out the different possible real versions of the previous description. The split signature case is easiest. Indeed, we know from Section 5.5 that the matrix $\mathbf{x}$ must be real. The most natural way to realise this setup is to have all spaces under consideration to be real. Thus, in this case, $\mathbb{T} = \mathbb{R}^4$ and the space of bi-vectors is $\Lambda^2\mathbb{R}^4$. The projective quadric PQ is real four-dimensional. The conformal metric (9.37) is real of a split signature. The conformal group is $\mathrm{SL}(4, \mathbb{R})$.

Let us note that in the present real setting, the projective twistor space $P\mathbb{T} = \mathbb{R}P^3$ is real three-dimensional. This, in particular, shows that the projective twistor space cannot be viewed as the total space of some fibre bundle over M, because $P\mathbb{T}$ is of lower dimension than M. We make this comment because in Section 9.2 we will see that in the case of Euclidean signature, it will be possible to interpret the projective twistor space as the total space of a bundle over M. The only natural fibre bundle over M that we have in the split signature setting is the five-dimensional bundle with fibres being all α-planes that pass through a given point of M. Such a fibre can be parametrised by a primed spinor ω up to scale, and thus by a copy of $\mathbb{R}P^1$.

9.1.10 Minkowski Space Version

We now work out the reality conditions to obtain the real Minkowski space version of the twistor space. We know from Section 5.5 that the relevant reality condition in this case and that the matrix $\mathbf{x}$ is anti-Hermitian. There is an involution on the space of two-planes in $\mathbb{T}$ that gives this, but this involution does not come from an involution on $\mathbb{T}$. This makes the Minkowski space version of twistor space somewhat harder to describe as compared to the other two signatures.

For an anti-Hermitian matrix $\mathbf{x}$, the matrix Y (9.11) is of the form

$$Y_{\mathbb{R}^{1,3}} = \begin{pmatrix} \mathrm{i}\alpha & \mathrm{i}\beta \\ \mathrm{i}\beta^* & \mathrm{i}\delta \\ 1 & 0 \\ 0 & 1 \end{pmatrix}, \qquad \alpha, \delta \in \mathbb{R}, \beta \in \mathbb{C}. \tag{9.40}$$

The corresponding bi-vector is

$$Y = (\mathrm{i}\alpha e_1 + \mathrm{i}\beta^* e_2 + e_3) \wedge (\mathrm{i}\beta e_1 + \mathrm{i}\delta e_2 + e_4), \tag{9.41}$$

and the corresponding Plücker coordinates are

$$\begin{aligned} y_{12} &= |\beta|^2 - \alpha\delta, \quad y_{31} = \mathrm{i}\beta, \quad y_{14} = \mathrm{i}\alpha, \\ y_{23} &= -\mathrm{i}\delta, \qquad\quad y_{24} = \mathrm{i}\beta^*, \quad y_{34} = 1. \end{aligned} \tag{9.42}$$

We thus see that

$$y_{12}, y_{34} \in \mathbb{R}, \qquad y_{14}, y_{23} \in \mathrm{i}\mathbb{R}, \qquad y_{24} = -y_{31}^*. \tag{9.43}$$

Our task now is to find a subgroup of $\mathrm{SL}(4,\mathbb{C})$ that preserves these conditions.

Let us consider a Lie algebra matrix $X \in \mathfrak{sl}(4,\mathbb{C})$. Its action on a bi-vector $e_i \wedge e_j$ is

$$e_i \wedge e_j \to X_i{}^k e_k \wedge e_j + e_i \wedge X_j{}^k e_k. \tag{9.44}$$

This corresponds to the following 6×6 matrix acting on columns of Plücker coordinates

$$\rho(X) = \begin{pmatrix} z_1{}^1 + z_2{}^2 & 0 & z_2{}^4 & -z_1{}^3 & -z_2{}^3 & -z_1{}^4 \\ 0 & z_3{}^3 + z_4{}^4 & z_3{}^1 & -z_4{}^2 & z_4{}^1 & z_3{}^2 \\ z_4{}^2 & z_1{}^3 & z_1{}^1 + z_4{}^4 & 0 & -z_4{}^3 & z_1{}^2 \\ -z_3{}^1 & -z_2{}^4 & 0 & z_2{}^2 + z_3{}^3 & -z_2{}^1 & z_3{}^4 \\ -z_3{}^2 & z_1{}^4 & -z_3{}^4 & -z_1{}^2 & z_3{}^3 + z_1{}^1 & 0 \\ -z_4{}^1 & z_2{}^3 & z_2{}^1 & z_4{}^3 & 0 & z_2{}^2 + z_4{}^4 \end{pmatrix}.$$

Here $\rho(X)$ is the representation of the $\mathfrak{sl}(4,\mathbb{C})$ matrix X on bi-vectors, and we have ordered the basis as $e_{12}, e_{34}, e_{14}, e_{23}, e_{31}$, and e_{24}. For example, the first row of this matrix follows from

$$\begin{aligned} e_1 \wedge e_2 &\to (z_1{}^1 e_1 + z_1{}^3 e_3 + z_1{}^4 e_4) \wedge e_2 + e_1 \wedge (z_2{}^2 z_2 + z_2{}^3 e_3 + z_2{}^4 e_4) \\ &= (z_1{}^1 + z_2{}^2) e_{12} + z_2{}^4 e_{14} - z_1{}^3 e_{23} - z_2{}^3 e_{31} - z_1{}^4 e_{24}. \end{aligned}$$

The other rows are obtained similarly. On the other hand, we can rewrite the reality conditions in (9.43) by introducing a matrix

$$\theta = \begin{pmatrix} \mathbb{I} & 0 & 0 \\ 0 & -\mathbb{I} & 0 \\ 0 & 0 & -\sigma_1 \end{pmatrix}, \qquad \sigma_1 = \begin{pmatrix} 0 & 1 \\ 1 & 0 \end{pmatrix}, \tag{9.45}$$

so that

$$y^* = \theta y. \tag{9.46}$$

The matrices $\rho(X)$ that commute with this involution are those that satisfy $(\rho(X)y)^* = \theta\rho(X)y$, which gives

$$\theta\rho(X) = \rho(X)^*\theta. \tag{9.47}$$

An explicit calculation shows that this is equivalent to the following conditions

$$\begin{gathered} {z_1}^1 + {z_4}^4 \in \mathbb{R}, \qquad {z_2}^2 + {z_3}^3 \in \mathbb{R}, \\ {z_1}^1 + {z_2}^2 \in \mathbb{R}, \qquad {z_3}^3 + {z_4}^4 \in \mathbb{R}, \\ ({z_1}^1 + {z_3}^3)^* = {z_2}^2 + {z_4}^4, \end{gathered}$$

as well as

$$\begin{gathered} {z_1}^3, {z_2}^4, {z_3}^1, {z_4}^2 \in \mathrm{i}\mathbb{R}, \\ ({z_1}^2)^* = -{z_4}^3, \quad ({z_2}^1)^* = -{z_3}^4, \quad ({z_2}^3)^* = -{z_1}^4, \quad ({z_3}^2)^* = -{z_4}^1. \end{gathered}$$

It is not hard to see that this implies that the matrix $X \in \mathfrak{sl}(4, \mathbb{C})$ is of the form

$$X = \begin{pmatrix} A & B \\ C & -A^\dagger \end{pmatrix}, \tag{9.48}$$

where B and C are arbitrary anti-Hermitian 2×2 matrices and A is arbitrary, but satisfying $\mathrm{Tr}(A) \in \mathbb{R}$ in order to have $\mathrm{Tr}(X) = 0$.

Let us now understand what kind of condition on X can produce matrices of this type. It is not hard to see that (9.48) are precisely those matrices that satisfy

$$FX + X^\dagger F = 0, \tag{9.49}$$

where

$$F = \begin{pmatrix} 0 & \mathbb{I} \\ \mathbb{I} & 0 \end{pmatrix}, \tag{9.50}$$

with $\mathbb{I}$ being the 2×2 identity matrix. On the other hand, the matrices satisfying (9.49) are precisely those that leave invariant the following quadratic form on $\mathbb{T}$

$$|Z|^2_{Mink} := Z^\dagger F Z. \tag{9.51}$$

Indeed, the action of Lie algebra of SL(4, $\mathbb{C}$) on this inner product is

$$Z^\dagger FZ \to Z^\dagger FXZ + (XZ)^\dagger FZ = Z^\dagger(FX + X^\dagger F)Z, \tag{9.52}$$

which equals zero in view of (9.49). Thus, the real quadratic form (9.51) remains invariant under transformations of the form (9.48), and in turn, transformations of this form are precisely those that leave invariant the quadratic form (9.51). The eigenvalues of F are ± 1, which shows that the signature of the Hermitian quadratic form (9.51) is (2, 2). Thus, the conformal group of Minkowski space is SU(2, 2), the group of transformations of $\mathbb{C}^4$ preserving the Hermitian form (9.51), which is of signature (2, 2).

It is also interesting to give the coordinate description of the twistor space in this case. If we consider the action of X of the form (9.48) on columns (9.18) we see that $\pi \to A\pi, \omega \to -A^\dagger\omega$, where $A \in \mathfrak{sl}(2, \mathbb{C})$. According to our discussion in (8.4), (8.5) the two-component column π should be given the interpretation of an unprimed spinor with lower index, while ω should be given interpretation of a primed spinor with an upper index; see (8.16). Thus, in the Minkowski case, the spinor interpretation of a twistor four-component column is

$$Z = \begin{pmatrix} \pi_A \\ \omega^{A'} \end{pmatrix}. \tag{9.53}$$

The quadratic form (9.51) is then

$$|Z|^2_{Mink} = \pi^*_{A'}\omega^{A'} + \pi_A\omega^{*A}, \tag{9.54}$$

where we have used the fact that the complex conjugate of a spinor is the spinor of opposite chirality. This in particular shows that the quadratic form (9.51) is zero on two-planes in $\mathbb{T}$ that represent points in M. Indeed, on such planes $\pi_A = \mathbf{x}_{AA'}\omega^{A'}$, and, therefore, $\pi^*_{A'} = -\mathbf{x}_{AA'}\omega^{*A}$, where we have used the anti-Hermiticity of $\mathbf{x}_{AA'}$. This shows that $|a|^2_{Mink}$ vanishes on the two-planes that corresponds to points in the real Minkowski space. This can be used as an alternative characterisation of such two-planes. The 'real' points in the Grassmanian of two-planes in $\mathbb{T}$ are those that are null with respect to (9.51), with the meaning of 'real' being that they correspond to points in the real Minkowski space.

We also note that, like in the split signature case, the (projective) twistor space of the Minkowski space M is not a fibre bundle over M. Points of the projective twistor space–lines in $\mathbb{T}$–corresponding to a given point in M–2-plane in $\mathbb{T}$–are those contained in the given two-plane. Because in the Minkowski case the real two-planes in question are totally null with respect to the Hermitian inner product on $\mathbb{T}$; these are the null lines in $\mathbb{T}$. However, a given null line in $\mathbb{T}$ is contained in more than one null two-plane. This means that there is no well-defined projection from $P\mathbb{T}$ to M and $P\mathbb{T}$ is not a fibre bundle with M as the base.

There is, however, a natural bundle over M even in this setting, which is the Minkowski signature version of the fibration τ in (9.25). Indeed, we can consider the bundle whose fibres are α-planes though a given point in M. The fibre can be coordinatised by primed spinors up to scale, and is thus a copy of $\mathbb{C}P^1$ in this case. This is the Minkowski signature version of the coset $\mathrm{SL}(4,\mathbb{C})/Q$ from (9.25), which in this signature has real dimension six.

9.2 Euclidean Twistors

We now develop the Euclidean version of the twistor story. We dedicate a separate section to this material, because there are many aspects that are not shared by the split and Minkowski cases. The main distinguishing feature of the Euclidean case is that the twistor space turns out to have the fibre bundle structure–it is the total space of the primed spinor bundle over M, as we shall soon see. Another specialty of the Euclidean setting is that many constructions are obtained by directly generalising those that are already familiar from the case of $S^2 \sim \mathbb{C}P^1$ by replacing complex numbers $\mathbb{C}$ with quaternions $\mathbb{H}$.

9.2.1 Euclidean Signature Conformal Group

We know from Section 5.5 that that matrix $\mathbf{x}$ parametrising two-planes in $\mathbb{T}$ must be of the form

$$\mathbf{x} = \begin{pmatrix} \alpha & \beta \\ -\beta^* & \alpha^* \end{pmatrix}. \tag{9.55}$$

Equivalently, we can say that the matrix $\mathbf{x}$ must take values in the space of quaternions $\mathbf{x} \in \mathbb{H}$. Then the conformal group is just the group $\mathrm{SL}(2,\mathbb{H})$ of 2×2 matrices with quaternionic entries.

To define the group $\mathrm{SL}(2,\mathbb{H})$, let us consider a 2×2 quaternionic matrix

$$g = \begin{pmatrix} A & B \\ C & D \end{pmatrix}, \qquad A, B, C, D \in \mathbb{H}. \tag{9.56}$$

We can then view A, B, C, and D either as unitary 2×2 matrices or as quaternions. In the first interpretation we get a 4×4 matrix g, whose determinant can be expressed as

$$\det(g) = \det(A)\det(D - CA^{-1}B), \tag{9.57}$$

where on the right-hand side we interpret A, B, C, and D as 2×2 matrices. On the other hand, the determinant of a unitary 2×2 matrix is the norm squared of the corresponding quaternion. This means that we can define the determinant of a 2×2 quaternionic matrix as

$$\det(g) = |A|^2 |D - CA^{-1}B|^2, \tag{9.58}$$

where now A, B, C, and D are viewed as quaternions. This makes it clear that for quaternionic 2×2 matrices $\det(g) \geq 0$. Having defined the determinant we can define the group $SL(2, \mathbb{H})$ as the group of unimodular 2×2 matrices with quaternionic entries. Such a group acts on $\mathbb{H}$ by fractional linear transformations, and this is the realisation of the conformal group of $\mathbb{R}^4$. As already mentioned, this is the direct generalisation of the situation in 2D, with complex numbers replaced by quaternions, and complex unimodular 2×2 matrices replaced by quaternionic such matrices.

The group $SL(2, \mathbb{H})$ can be interpreted as the subgroup of $SL(4, \mathbb{C})$ that commutes with some involution of the twistor space $\mathbb{T} = \mathbb{C}^4$. To see this interpretation, let us think about twistors as a pair of spinors $Z = (\pi$ and $\omega)$. We then have the following involution on Euclidean signature spinors

$$\hat{\pi} := \epsilon \pi^*, \qquad \hat{\omega} := \epsilon \omega^*, \tag{9.59}$$

where ϵ is the antisymmetric matrix (8.7) and star denotes the complex conjugation. We note that because $\epsilon^2 = -\mathbb{I}$ the involution squares to minus the identity.

$$\hat{}\hat{} = -\text{id}. \tag{9.60}$$

Then, using the fact that for unitary matrices $A\epsilon = \epsilon A^*$ we can see that unitary transformations commute with the involutions defined

$$A\hat{\pi} = (A\omega)^\wedge, \qquad D\hat{\omega} = (D\omega)^\wedge, \qquad A, D \in SU(2). \tag{9.61}$$

We now define an involution on $\mathbb{T}$

$$\hat{Z} = \begin{pmatrix} \pi \\ \omega \end{pmatrix}^\wedge := \begin{pmatrix} \hat{\pi} \\ \hat{\omega} \end{pmatrix}. \tag{9.62}$$

It is easy to check that $SL(2, \mathbb{H})$ transformations commute with the involution defined

$$g \begin{pmatrix} \xi \\ \eta \end{pmatrix}^\wedge = \left(g \begin{pmatrix} \xi \\ \eta \end{pmatrix} \right)^\wedge, \qquad g \in SL(2, \mathbb{H}). \tag{9.63}$$

Thus, we see that it is the $\hat{}$-involution (9.62) on $\mathbb{T}$ that gives us the desired Euclidean signature real form of the conformal group.

Because the involution we have defined squares to minus the identity there are no 'real' vectors in $\mathbb{T}$. However, there are real bi-vectors. Indeed, any bi-vector of the type

$$Y = \hat{Z} \wedge Z \tag{9.64}$$

is real. Indeed, we have

$$\hat{Y} = -Z \wedge \hat{Z} = \hat{Z} \wedge Z = Y. \tag{9.65}$$

Thus, while there are no 'real' lines in $\mathbb{T}$, there are real two-planes and thus real points in M. In particular, we note that the matrix Y (9.11) that is constructed from a unitary matrix $\mathbf{x}$ is

$$\begin{pmatrix} \alpha & \beta \\ -\beta^* & \alpha^* \\ 1 & 0 \\ 0 & 1 \end{pmatrix} = \begin{pmatrix} \pi & -\hat{\pi} \\ \omega & -\hat{\omega} \end{pmatrix}, \tag{9.66}$$

where

$$\pi = \begin{pmatrix} \alpha \\ -\beta^* \end{pmatrix}, \qquad \omega = \begin{pmatrix} 1 \\ 0 \end{pmatrix}, \tag{9.67}$$

and is thus of the type (9.64) and corresponds to a real two-plane.

The form (9.64) of real two-planes implies that every twistor $Z \in \mathbb{T}$ corresponds to a unique real point $\hat{Z} \wedge Z$ in the space of two-planes. This means that in the case of Euclidean signature, we have a well-defined projection $P\mathbb{T} \to M = \mathbb{R}^4$. In other words, the twistor space $\mathbb{T}$ is the total space of a bundle over the space of real two-planes, which coincides with 'our' space M. Similarly, the projective twistor space $P\mathbb{T}$ is the total space of a bundle over M. Each fibre of $P\mathbb{T} \to M$ is a copy of $S^2 \sim \mathbb{C}$, and consists of all lines in $\mathbb{T}$ that form the given real two-plane. Explicitly, as we saw in (9.29), this is the set of twistors (π and ω) satisfying $\pi_A = \mathbf{x}_A{}^{A'} \omega_{A'}$, which is parametrised by the primed spinor ω up to scale, i.e., by a copy of $\mathbb{C}$. Thus, we can say that the Euclidean signature twistor space is the total space of the primed spinor bundle over $M = \mathbb{R}^4$. This fibred structure of the twistor space is not a general feature, and holds only in Euclidean signature, as we have seen from our previous discussion of the other two signatures.

9.2.2 Euclidean Spinors

Before we can continue with further developments related to Euclidean signature twistors, we need to establish some facts about Euclidean spinors. This is done in complete analogy with the treatment in Section 8.1, but with the matrix $\mathbf{x}$ changed accordingly. We have already encountered the matrix $\mathbf{x}_E$ in (5.76). We repeat it here for convenience

$$\mathbf{x}_E = \begin{pmatrix} -x^4 + \mathrm{i}x^3 & \mathrm{i}x^1 + x^2 \\ \mathrm{i}x^1 - x^2 & -x^4 - \mathrm{i}x^3 \end{pmatrix}. \tag{9.68}$$

Under Lorentz transformations, this matrix transforms as $\mathbf{x} \to g_L \mathbf{x} g_R^\dagger, g_L, g_R \in \mathrm{SU}(2)$. Given that in general the spinors with their index raised transform by multiplying them with g^{-1} from the right, this transformation property of $\mathbf{x}$ implies that it should be interpreted as $\mathbf{x}_A{}^{A'}$, i.e., a bi-spinor with the primed index raised.

We can write the Euclidean norm squared as

$$|\mathbf{x}|^2 := \det(\mathbf{x}) = x_1^2 + x_2^2 + x_3^2 + x_4^2. \tag{9.69}$$

We can alternatively write the metric on $\mathbb{R}^4$ as

$$ds^2 = \frac{1}{2}\text{Tr}(d\mathbf{x}d\mathbf{x}^\dagger), \tag{9.70}$$

where $\mathbf{x}^\dagger = (\mathbf{x}^T)^*$ is the Hermitian conjugation. For future use, we note the useful property

$$\mathbf{x}\mathbf{x}^\dagger = \mathbb{I}|x|^2. \tag{9.71}$$

For any unitary matrix we have $\epsilon\mathbf{x}\epsilon^T = \mathbf{x}^*$. This implies that the matrix $\mathbf{x}_A{}^{A'}$ with its index A raised and A' lowered, and then the two interchanged, is the matrix $\mathbf{x}^\dagger$. In other words, we can write (9.70) as

$$ds^2 = \frac{1}{2}dx_A{}^{A'}dx_{A'}{}^A. \tag{9.72}$$

This means that if we introduce the soldering form via

$$\mathbf{x}_A{}^{A'} = \sqrt{2}e_{\mu A}{}^{A'}x^\mu \tag{9.73}$$

then the metric of $\mathbb{R}^4$ takes the form

$$\delta_{\mu\nu} = e_{\mu A}{}^{A'}e_{\nu A'}{}^A. \tag{9.74}$$

The components of the soldering form are given by

$$e_4 = \frac{1}{\sqrt{2}}\begin{pmatrix} -1 & 0 \\ 0 & -1 \end{pmatrix}, \qquad e_3 = \frac{1}{\sqrt{2}}\begin{pmatrix} \mathrm{i} & 0 \\ 0 & -\mathrm{i} \end{pmatrix}, \tag{9.75}$$
$$e_1 = \frac{1}{\sqrt{2}}\begin{pmatrix} 0 & \mathrm{i} \\ \mathrm{i} & 0 \end{pmatrix}, \qquad e_2 = \frac{1}{\sqrt{2}}\begin{pmatrix} 0 & 1 \\ -1 & 0 \end{pmatrix}.$$

We now compute the components of the ASD 2-forms given by

$$\bar{\Sigma}^{A'B'}{}_{\mu\nu} = e_{[\mu}^{AA'}e_{\nu]A}{}^{B'}. \tag{9.76}$$

We have

$$\bar{\Sigma}_{41} = \frac{1}{2}((\epsilon e_4)^T e_1 - (\epsilon e_1)^T e_4) = \frac{1}{2}\begin{pmatrix} \mathrm{i} & 0 \\ 0 & -\mathrm{i} \end{pmatrix}, \tag{9.77}$$
$$\bar{\Sigma}_{42} = \frac{1}{2}((\epsilon e_4)^T e_2 - (\epsilon e_2)^T e_4) = \frac{1}{2}\begin{pmatrix} -1 & 0 \\ 0 & -1 \end{pmatrix},$$
$$\bar{\Sigma}_{43} = \frac{1}{2}((\epsilon e_4)^T e_3 - (\epsilon e_3)^T e_4) = \frac{1}{2}\begin{pmatrix} 0 & -\mathrm{i} \\ -\mathrm{i} & 0 \end{pmatrix}.$$

The other components follow from anti–self-duality (ASD), e.g., $\bar{\Sigma}_{23} = -\bar{\Sigma}_{41}$, which can of course be also checked explicitly. Finally, it can be checked that all $\bar{\Sigma}$'s are real with respect to the involution (9.59)

$$\epsilon(\bar{\Sigma}_{\mu\nu})^*\epsilon^T = \bar{\Sigma}_{\mu\nu}. \tag{9.78}$$

All matrices $\bar{\Sigma}^{A'B'}$ are also symmetric $(\bar{\Sigma}_{\mu\nu})^T = \bar{\Sigma}_{\mu\nu}$. We will need these objects in the following subsection when we describe almost complex structures on $\mathbb{R}^4$.

9.2.3 Euclidean Twistors and Almost Complex Structures

As we have just seen, the Euclidean (projective) twistor space is naturally the total space of a two-sphere bundle over M. This allows us to interpret the projective twistor space $P\mathbb{T}$ as the bundle of almost complex structures over M. This interpretation was first given in Atiyah et al. (1978), and we give some details here.

Recall that an **almost complex structure** on M is an endomorphism $J : TM \to TM$ of the tangent space that squares to minus the identity $J^2 = -\mathbb{I}$. Now, the twistor space of $M = \mathbb{R}^4$ is the total space of the primed spinor bundle over M. Given a primed spinor $\omega^{A'}$, we can construct an endomorphism of the tangent space from the ASD 2-forms $\bar{\Sigma}^{A'B'}$ contracted with $\omega^{A'}$ as well as $\hat{\omega}^{A'}$. Thus, we first raise one of the indices of $\bar{\Sigma}^{A'B'}_{\mu\nu}$ to convert it into an S^2_--valued endomorphism $\bar{\Sigma}_\mu^{A'B'\nu}$. We then contract with $\hat{\omega}_{A'}\omega_{B'}$

$$(J_\omega)_\mu{}^\nu := \frac{2}{\mathrm{i}} \frac{\bar{\Sigma}_\mu^{A'B'\nu}\hat{\omega}_{A'}\omega_{B'}}{[\hat{\omega}\omega]}, \tag{9.79}$$

where $[\hat{\omega}\omega] := \hat{\omega}_{A'}\omega^{A'}$, and $\hat{\omega}_{A'}$ is the Euclidean conjugation (9.59) that maps primed spinors to the same type spinors. We have $[\hat{\omega}\omega] := (\epsilon\omega)^T\epsilon\omega^* = \omega^T\omega^*$, and so this object is real. The factor of two in front is needed for the correct normalisation, and the factor of imaginary unit is needed to make the endomorphism real. Indeed, we can rewrite (9.79) as

$$J_\omega = \frac{2}{\mathrm{i}[\hat{\omega}\omega]}(\epsilon\omega^*)^T\bar{\Sigma}\omega, \tag{9.80}$$

where here the object $\bar{\Sigma}$ is with its first spacetime index down and second up to create an endomorphism of TM. We then have

$$(J_\omega)^* = J_\omega, \tag{9.81}$$

where we have used that $[\hat{\omega}\omega]$ is real and

$$\left((\epsilon\omega^*)^T\bar{\Sigma}\omega\right)^* = -\omega^T\epsilon(\bar{\Sigma})^*\epsilon^T\epsilon\omega^* = -\omega^T\bar{\Sigma}\epsilon\omega^* = -(\epsilon\omega^*)^T\bar{\Sigma}\omega,$$

and we used (9.78) as well as the symmetry of $\bar{\Sigma}$. We also note that (9.79) only depends on the spinor ω up to scale, and so it is parametrised by a copy of $\mathbb{C}P^1$.

We now use (8.33), which is also valid in Euclidean signature, to obtain

$$J_\omega J_\omega = -\frac{\mathbb{I}}{[\hat{\omega}\omega]^2}(\epsilon^{A'C'}\epsilon^{D'B'} + \epsilon^{A'D'}\epsilon^{C'B'})\hat{\omega}_{A'}\omega_{B'}\hat{\omega}_{C'}\omega_{D'} = -\mathbb{I}. \tag{9.82}$$

Thus, J_ω is indeed an almost complex structure on $\mathbb{R}^4$. Unlike the case of two dimensions, in 4D there is no longer a unique almost complex structure.

The bundle of such almost complex structures over $\mathbb{R}^4$ is the (projective) primed spinor bundle.

We can explicitly describe the eigenspaces of J_ω as follows. First, we need the following identity

$$\bar{\Sigma}^{A'B'}{}_\mu{}^\nu e_\nu^{CC'} = -e_\mu^{C(A'}\epsilon^{B')C'}, \tag{9.83}$$

which can be checked by an explicit computation using the definition of $\bar{\Sigma}^{A'B'}_{\mu\nu}$ as well as $e_{\mu AA'}e^{\mu BB'} = -\epsilon_A{}^B\epsilon_{A'}{}^{B'}$, the latter following from (9.74). We then act with J_ω on a 1-form of the type $e_\nu^{AA'}\pi_A\omega_{A'}$, where π_A is an arbitrary spinor. We have

$$\begin{aligned}(J_\omega)_\mu{}^\nu e_\nu^{CC'}\pi_C\omega_{C'} &= \frac{\mathrm{i}}{[\hat{\omega}\omega]}(e_\mu^{CA'}\epsilon^{B'C'} + e_\mu^{CB'}\epsilon^{A'C'})\hat{\omega}_{A'}\omega_{B'}\pi_C\omega_{C'} \\ &= \mathrm{i}e_\mu^{CC'}\pi_C\omega_{C'}.\end{aligned} \tag{9.84}$$

Thus, the 1-forms of the type $e_\nu^{AA'}\pi_A\omega_{A'}$ are the $(1,0)$ forms. It can similarly be checked that $e_\nu^{AA'}\pi_A\hat{\omega}_{A'}$ are the $(0,1)$ forms.

Given a real vector field (or a 1-form) on M, one often needs to decompose it into its $(1,0)$ and $(0,1)$ parts. The following identity is the most useful way of obtaining such a decomposition

$$\epsilon_{A'}{}^{B'} = \frac{\hat{\omega}_{A'}\omega^{B'} - \omega_{A'}\hat{\omega}^{B'}}{[\hat{\omega}\omega]}. \tag{9.85}$$

This identity can be verified by checking that $\epsilon_{A'}{}^{B'}$ as given in (9.85) returns $\omega_{A'}, \hat{\omega}_{A'}$ when the contractions $\epsilon_{A'}{}^{B'}\omega_{B'}$ and $\epsilon_{A'}{}^{B'}\hat{\omega}_{B'}$ are computed. Since arbitrary spinor $\eta_{A'}$ can be decomposed into $\omega_{A'}, \hat{\omega}_{A'}$, this means that $\epsilon_{A'}{}^{B'}\eta_{B'} = \eta_{A'}$, which is the defining property of the $\epsilon_{A'}{}^{B'}$. We can then insert the identity (9.85) into the object $e_\mu^{AA'}$ to obtain

$$e_\mu^{AA'} = \frac{1}{[\hat{\omega}\omega]}\left(e_\mu^{AB'}\hat{\omega}_{B'}\omega^{A'} - e_\mu^{AB'}\omega_{B'}\hat{\omega}^{A'}\right), \tag{9.86}$$

which is the desired decomposition of the 1-forms $e_\mu^{AA'}$ into $(0,1)$ and $(1,0)$ parts. Then, using $\xi_\mu = e_\mu^{AA'}\xi_{AA'}$ any 1-form can be decomposed.

9.2.4 Fubini-Study Metric

It turns out that there is a natural metric as well as a compatible almost complex structure in the Euclidean twistor space. As we shall soon see, in particular, the existence of the latter is related to the fact that the twistor space has the interpretation of the bundle of almost complex structures over $\mathbb{R}^4$. The reason why a natural metric exists on the twistor space is more subtle. We will give a 7D explanation in the following sections.

To see how a metric and an almost complex structure arise on the twistor space of the complexified Minkowski space, we note that this twistor space is $\mathbb{T} = \mathbb{C}^4$, and its projective version is $\mathbb{C}P^3$. This space is a complex manifold, i.e., has an integrable almost complex structure J. It also has a compatible Hermitian metric, i.e., a metric that has the property $g(J\cdot, J\cdot) = g(\cdot,\cdot)$. The metric in question is the so-called **Fubini–Study** metric of the complex projective space. Let us describe this metric in coordinates.

The space $\mathbb{T} = \mathbb{C}^4$ comes with its natural Hermitian metric

$$ds^2_{\mathbb{C}^4} = \sum_{\alpha=1}^{4} |dZ_\alpha|^2. \tag{9.87}$$

The projective twistor space $P\mathbb{T} = \mathbb{C}P^3$ can be coordinatised by projective coordinates $[Z_1, Z_1, Z_3, Z_4] \sim [z_1, z_2, z_3, 1]$. In other words, let us introduce the following set of coordinates on $\mathbb{C}^4$

$$Z_i = z_i t, \quad i = 1, 2, 3, \quad Z_4 = t \tag{9.88}$$

It is then an exercise to check that the flat Hermitian metric (9.87) on $\mathbb{C}^4$ takes the following form in the previous coordinates

$$ds^2_{\mathbb{C}^4} = (1 + \sum_i |z_i|^2)|t|^2 \left[\left| \frac{dt}{t} + \frac{\sum_i \bar{z}_i dz_i}{1 + \sum_i |z_i|^2} \right|^2 + ds^2_{FS} \right], \tag{9.89}$$

where

$$ds^2_{FS} := \frac{\sum_i |dz_i|^2 (1 + \sum_j |z_j|^2) - \sum_{i,j} \bar{z}_i z_j dz_i d\bar{z}_j}{(1 + \sum_i |z_i|^2)^2} \tag{9.90}$$

is the Fubini–Study metric on $\mathbb{C}P^3$.

The calculation we just performed allows us to view the sphere $S^7 \subset \mathbb{R}^8 = \mathbb{C}^4$ given by the equation $\sum_\alpha |Z_\alpha|^2 = 1$ as a S^1 bundle over $\mathbb{C}P^3$. Indeed, we can pull back the metric (9.87) to the sphere S^7 by choosing

$$t = \frac{e^{\mathrm{i}\psi}}{\sqrt{1 + \sum_i |z_i|^2}}, \tag{9.91}$$

where ψ is a coordinate on S^1. In this coordinates the metric (9.89) gives the following form of the metric on S^7

$$ds^2_{S^7} = (d\psi + a)^2 + ds^2_{FS}, \tag{9.92}$$

where

$$a := \frac{\mathrm{i}}{2} \frac{\sum_i (z_i d\bar{z}_i - \bar{z}_i dz_i)}{1 + \sum_i |z_i|^2} \tag{9.93}$$

is a $\mathrm{U}(1)$ connection on $\mathbb{C}P^3$. This explicitly realises S^7 as the total space of an S^1 bundle over $\mathbb{C}P^3$. Note that the described construction is the direct generalisation of the construction of the Hopf fibration in Section 1.13. Indeed, if there is just one complex coordinate $z_1 = z$, the described computation is identical to the one performed in Section 1.13, and the Fubini–Study metric coincides with (quarter of) the metric on the unit $S^2 = \mathbb{C}P^1$.

9.2.5 $\mathbb{C}P^3$ as an S^2 Bundle over S^4

We now perform a similar computation, but this time parametrise $\mathbb{C}^4$ using the previously developed twistor interpretation. Thus, we view $\mathbb{C}^4 = \mathbb{T}$ as a $\mathbb{C}^2$ bundle over S^4, viewing the latter as another copy of $\mathbb{C}^2$. Indeed, recall that the (projective) twistor space is the bundle of α-planes through real points in M, and all twistors that lie in a given real two-plane in $\mathbb{T}$ are of the form $\pi = \mathbf{x}\omega$. This gives a parametrisation of $\mathbb{T} = \mathbb{C}^4$ by two complex coordinates α and β in $\mathbf{x}$ and two complex coordinates components of the column ω.

Thus, we now have $Z = (\pi, \omega)$. In terms of π and ω the Hermitian metric (9.87) reads

$$ds^2_{\mathbb{C}^4} = d\pi^\dagger d\pi + d\omega^\dagger d\omega, \tag{9.94}$$

where we view π and ω as two-component columns. Parametrising π as $\mathbf{x}\omega$ we get

$$ds^2_{\mathbb{C}^4} = (\omega^\dagger d\mathbf{x}^\dagger + d\omega^\dagger \mathbf{x}^\dagger)(d\mathbf{x}\omega + \mathbf{x}d\omega) + d\omega^\dagger d\omega. \tag{9.95}$$

Using the fact that $\mathbf{x}^\dagger\mathbf{x} = |\mathbf{x}|^2\mathbb{I}$ we can rewrite the previous equation as

$$ds^2_{\mathbb{C}^4} = (1+|\mathbf{x}|^2)\left[\left(d\omega^\dagger + \omega^\dagger \frac{d\mathbf{x}^\dagger \mathbf{x}}{1+|\mathbf{x}|^2}\right)\left(d\omega + \frac{\mathbf{x}^\dagger d\mathbf{x}}{1+|\mathbf{x}|^2}\omega\right) + \omega^\dagger \frac{d\mathbf{x}^\dagger d\mathbf{x}}{(1+|\mathbf{x}|^2)^2}\omega\right]. \tag{9.96}$$

We now parametrise $\omega \in \mathbb{C}^2$ projectively

$$\omega = t\eta, \qquad \eta := \begin{pmatrix} z \\ 1 \end{pmatrix}. \tag{9.97}$$

We also restrict to the sphere $S^7 \subset \mathbb{C}^4$ given by the equation $\pi^\dagger\pi + \omega^\dagger\omega = 1$. This gives $|t|^2(1+|z|^2)(1+|\mathbf{x}|^2) = 1$, which allows to parametrise

$$t = \frac{e^{i\psi}}{\sqrt{(1+|z|^2)(1+|\mathbf{x}|^2)}}. \tag{9.98}$$

A straightforward computation then gives

$$d\omega + \frac{\mathbf{x}^\dagger d\mathbf{x}}{1+|\mathbf{x}|^2}\omega = t\left(\left(id\psi - \frac{d|z|^2}{2(1+|z|^2)}\right)\eta + D\eta\right), \tag{9.99}$$

where

$$D\eta := (d+\mathbf{A})\eta, \qquad \mathbf{A} := \frac{\mathbf{x}^\dagger d\mathbf{x} - d\mathbf{x}^\dagger \mathbf{x}}{2(1+|\mathbf{x}|^2)}. \tag{9.100}$$

The 2×2 matrix A is anti-Hermitian, and will later be identified with a chiral half of the spin connection on S^4. This gives

$$\frac{1}{|t|^2(1+|z|^2)}\left(d\omega^\dagger + \omega^\dagger \frac{d\mathbf{x}^\dagger \mathbf{x}}{1+|\mathbf{x}|^2}\right)\left(d\omega + \frac{\mathbf{x}^\dagger d\mathbf{x}}{1+|\mathbf{x}|^2}\omega\right) \\ = d\psi^2 + \mathrm{i} d\psi \frac{(D\eta)^\dagger \eta - \eta^\dagger D\eta}{\eta^\dagger \eta} + \frac{(D\eta)^\dagger D\eta}{\eta^\dagger \eta} - \frac{((D\eta)^\dagger \eta + \eta^\dagger D\eta)^2}{4(1+|z|^2)^2}, \tag{9.101}$$

where we have used

$$(D\eta)^\dagger \eta + \eta^\dagger D\eta = d|z|^2. \tag{9.102}$$

The terms on the right-hand side of (9.101) can be rewritten as

$$\left(d\psi + \frac{\mathrm{i}}{2}\frac{(D\eta)^\dagger \eta - \eta^\dagger D\eta}{\eta^\dagger \eta}\right)^2 + \frac{(D\eta)^\dagger ((\eta^\dagger \eta)\mathbb{I} - \eta\eta^\dagger)D\eta}{(\eta^\dagger \eta)^2}. \tag{9.103}$$

Overall, using the fact that $d\mathbf{x}^\dagger d\mathbf{x} = |d\mathbf{x}|^2\mathbb{I}$ we can write the metric on S^7 in these coordinates in the form (9.92) with

$$a = \frac{\mathrm{i}}{2}\frac{(D\eta)^\dagger \eta - \eta^\dagger D\eta}{\eta^\dagger \eta}, \tag{9.104}$$

and

$$ds^2_{\mathbb{C}P^3} = \frac{(D\eta)^\dagger ((\eta^\dagger \eta)\mathbb{I} - \eta\eta^\dagger)D\eta}{(\eta^\dagger \eta)^2} + \frac{|d\mathbf{x}|^2}{(1+|\mathbf{x}|^2)^2}. \tag{9.105}$$

The last term here is (quarter of) the metric on S^4 of unit radius. This represents $\mathbb{C}P^3$ as the total space of an $S^2 \sim \mathbb{C}P^1$ bundle over S^4.

We can further rewrite the fibre part of the metric in (9.105) by introducing the conjugate spinor $\hat{\eta}$ we have

$$\hat{\eta} = \epsilon\eta^* = \begin{pmatrix} 1 \\ -\bar{z} \end{pmatrix}. \tag{9.106}$$

We then have

$$(\eta^\dagger \eta)\mathbb{I} - \eta\eta^\dagger = \hat{\eta}\hat{\eta}^\dagger. \tag{9.107}$$

The spinor contraction appearing in the metric is then

$$\hat{\eta}^\dagger D\eta = (\epsilon\eta^*)^\dagger D\eta = (\epsilon\eta)^T D\eta = \eta^{A'}(D\eta)_{A'}, \tag{9.108}$$

where we rewrote the result in spinor notations. Also, using $\hat{\eta}^{A'}\hat{\eta}_{A'} = 0$ we can rewrite

$$(D\eta)^\dagger \hat{\eta} = \hat{\eta}^{A'}(D\hat{\eta})_{A'}. \tag{9.109}$$

We also have $[\hat{\eta}\eta] = \eta^\dagger \eta$. Thus, we can rewrite the metric (9.105) more compactly as

$$ds^2_{\mathbb{C}P^3} = \frac{\tau\bar{\tau}}{[\hat{\eta}\eta]^2} + \frac{|d\mathbf{x}|^2}{(1+|\mathbf{x}|^2)^2}, \tag{9.110}$$

where

$$\tau = \eta^{A'}(D\eta)_{A'}, \qquad \bar{\tau} = \hat{\eta}^{A'}(D\hat{\eta})_{A'} \tag{9.111}$$

are complex-valued 1-forms on the total space of the $\mathbb{C}P^3 \to S^4$ bundle, which are complex conjugate to each other. We note that we could have replaced $\eta_{A'} \to \omega_{A'} = t\eta_{A'}$ in (9.110). Indeed, we have

$$\omega^{A'}(D\omega)_{A'} = t\eta^{A'}(d(t\eta_{A'}) + A_{A'}{}^{B'}t\eta_{B'}) = t^2\eta^{A'}(D\eta)_{A'}, \tag{9.112}$$

where we have used that the spinor contraction of $\eta^{A'}$ with itself is zero, and so there is no dt term. Then the numerator of the first term in (9.110) is homogeneous of degree two in t, and the denominator is similarly homogeneous. Thus, the first term in the metric can be written in terms of the spinor $\omega_{A'} \in \mathbb{C}^2$ but descends to a well-defined quadratic form on the projective space $\mathbb{C}P^1$.

9.2.6 Almost Complex Structures on the Twistor Space

We can now connect the calculation of the previous subsection culminating in (9.110) with the description of the Euclidean twistor space as the total space of the bundle of almost complex structures over M.

First, the complex projective space $\mathbb{C}P^3$ is a complex manifold in that there is an integrable almost complex structure that is also metric compatible. This almost complex structure is easiest to describe in terms of the $(1,0)$ and $(0,1)$ decomposition that it induces. These can be read off the metric, which must be of the form that is a sum of $(1,0)$ forms times their complex conjugates. The $(1,0)$ form in the fibre direction is readily read off from (9.110), and is given by $\tau = \eta^{A'}D\eta_{A'}$. As we already discussed, this is not really a 1-form on $\mathbb{C}P^3$, but rather 1-form transforming homogeneously under rescalings used to pass from $\mathbb{C}^4$ to $\mathbb{C}P^3$. In other words, it is a section of an appropriate line bundle over $\mathbb{C}P^3$. It is essentially the 1-form dz on $\mathbb{C}P^1$ corrected by basic terms coming from the connection A.

The basic $(1,0)$ forms can be identified from the second term of the metric (9.110). Indeed, one can write $|d\mathbf{x}|^2$ as a multiple of $\mathrm{Tr}(d\mathbf{x}d\mathbf{x}^\dagger)$. We can then use $d\mathbf{x}^\dagger = \epsilon\mathbf{x}^T\epsilon^T$, and insert the decomposition of identity (9.85). This shows that the numerator in the second term in (9.110) is a multiple of $\mathrm{Tr}(\epsilon d\mathbf{x}\omega(d\mathbf{x}\hat{\omega})^T)$. However, the object $d\mathbf{x}\omega$ is a multiple of 1-form $e_{\mu A}{}^{A'}\omega_{A'}$, which we identified in (9.84) as $(1,0)$ forms with respect to the almost complex structure J_ω given by (9.79). Thus, the basic $(1,0)$ 1-forms as read off from the metric (9.110) coincide with those determined by J_ω.

This can be reinterpreted as follows. The Euclidean (projective) twistor space is the total space of the $\mathbb{C}P^1$ bundle over $\mathbb{R}^4$, which we interpreted as the bundle of almost complex structures. Given that every point in the fibre $\mathbb{C}P^1$ defines an almost complex structure on $\mathbb{R}^4$, and that the fibre itself is a complex manifold, there is a natural almost complex structure that can be given to the twistor space as the whole. Indeed, at every point of $P\mathbb{T}$, we define the action of J as the corresponding J_ω on basic vector fields, and the unique J on $\mathbb{C}P^1$ on vertical vector fields. As we just saw, this is precisely the construction that produces the natural integrable almost complex structure on $\mathbb{C}P^3$ viewed as the total space of the primed spinor bundle over S^4.

Any almost complex structure on the twistor space can be alternatively described by the corresponding $(3,0)$ form. The $(1,0)$ forms wedged against this must then produce zero, which serves to define the space of $(1,0)$ forms. Let us see what the $(3,0)$ form is for the already described complex structure on $\mathbb{C}P^3$. It is given by the wedge product of three different $(1,0)$ forms, i.e., by

$$\Omega^{(3,0)} \sim \tau d\mathbf{x}^{AA'}\omega_{A'} d\mathbf{x}_A{}^{B'}\omega_{B'}. \tag{9.113}$$

Given that $d\mathbf{x}^{AA'}$ is a multiple of the soldering 1-form $e^{AA'}$, the object $d\mathbf{x}^{AA'}d\mathbf{x}_A{}^{B'}$ is just a multiple of the ASD 2-form $\bar{\Sigma}^{A'B'}$. Overall, we see that the $(3,0)$ form our our complex structure on $\mathbb{C}P^3$ is given by

$$\Omega^{(3,0)} = \omega^{A'} D\omega_{A'} \bar{\Sigma}^{B'C'} \omega_{B'}\omega_{C'}, \tag{9.114}$$

where we rewrote everything in terms of $\omega \in \mathbb{C}^2$. This shows that $\Omega^{(3,0)}$ is of homogeneity degree four in ω, and similarly to τ only exists as a section of a line bundle over $\mathbb{C}P^3$.

It turns out that there exists another natural almost complex structure on the twistor space. Indeed, there is a choice that is made in the previous construction. Thus, to construct the integrable almost complex structure on the projective twistor space, we combined the unique almost complex structure in the fibre $\mathbb{C}P^1$ with the almost complex structure J_ω on the base. However, there are two possible relative signs for this combination. The other possible choice is to say that the $(1,0)$ forms in the fibres are put together with the $(0,1)$ basic forms to form the space of $(1,0)$ forms. This corresponds to reversing the sign of J_ω when putting it together with the almost complex structure in the fibre. The almost complex structure on the projective twistor space obtained this way is not integrable. However, it is interesting, due to reasons that will become clear in the following sections.

The $(3,0)$ form defining this almost complex structure is given by

$$\tilde{\Omega}^{(3,0)} = \frac{1}{[\hat{\omega}\omega]^2} \hat{\omega}^{A'} D\hat{\omega}_{A'} \bar{\Sigma}^{B'C'} \omega_{B'}\omega_{C'}. \tag{9.115}$$

The corresponding $(3,0)$ form exists as an actual 3-form on $\mathbb{C}P^3$. Indeed, both numerator and denominator here are of homogeneity 2 in both ω and $\hat{\omega}$, which produces an actual 3-form on $\mathbb{C}P^3$. In Section 9.6 we shall see that both the integrable (9.114) and non-integrable (9.115) almost complex structures arise naturally from a geometric construction based on 3-form in seven dimensions.

9.3 Quaternionic Hopf Fibration

The unprojectivised twistor space is $\mathbb{C}^4 = \mathbb{R}^8$. There is naturally a seven-sphere $S^7 \subset \mathbb{R}^8$ obtained by setting the radial coordinate in $\mathbb{R}^8$ to unity. We have already seen that we can view the seven-sphere as the total space of the circle bundle over the projective twistor space $\mathbb{C}P^3$. In other words, because we can identify $\mathbb{C}^4 = \mathbb{R}^8$, there is a natural circle action on $S^7 \subset \mathbb{R}^8$, the seven-sphere is fibred by copies of S^1, and the space of such fibres is the complex projective space $\mathbb{C}P^3$.

In this section, we will see that there is another natural way to think about the twistor space $\mathbb{C}^4$. This arises because we can instead identify $\mathbb{R}^8 = \mathbb{H}^2$, viewing a point in $\mathbb{R}^8$ as a pair of quaternions. We then have a natural action of the group of unimodular quaternions on S^7. Unimodular quaternions form the group SU(2). The orbits of this action are thus copies of three-sphere S^3, and so S^7 gets fibred by copies of S^3. The space of such orbits turns out to be a copy of S^4. This is the quaternionic Hopf fibration. It is the precise analog of the Hopf fibration $S^3 \to S^2$ as was described in Section 1.13, with $\mathbb{C}$ replaced by $\mathbb{H}$ everywhere. It is of interest to us because it gives yet another viewpoint on the Euclidean twistor space, in particular the twistor space of the four-dimensional sphere S^4.

9.3.1 The Hopf Projection

We now take $\mathbb{R}^8 = \mathbb{R}^4 \oplus \mathbb{R}^4$, and identify both copies of $\mathbb{R}^4$ with 2×2 matrices of the type (9.68). We consider the following codimension one surface

$$|\mathbf{q}|^2 + |\mathbf{p}|^2 = 1. \tag{9.116}$$

Here $\mathbf{q}, \mathbf{p}$ are matrices of the type (9.68), and the norm is as given in (9.69). This surface is clearly the round seven-sphere S^7.

We now consider a map from this S^7 to the sphere $S^4 \subset \mathbb{R}^5$. This projection, explicitly, is

$$(\mathbf{q}, \mathbf{p}) \to (2\mathbf{q}\mathbf{p}^\dagger, |\mathbf{p}|^2 - |\mathbf{q}|^2) \in \mathbb{R}^5. \tag{9.117}$$

However, in view of (9.116) it is clear that the right-hand side in (9.117) actually lies on the surface $S^4 \subset \mathbb{R}^5$. It is also clear that the fibres of this projection are copies of $\mathrm{SU}(2) = S^3$. Indeed, points

$$(\mathbf{q}, \mathbf{p}) \sim (\mathbf{q}g, \mathbf{p}g), \quad g \in \mathrm{SU}(2) \tag{9.118}$$

are mapped to the same point on the base S^4.

The following parametrisation of the projection gives a convenient set of coordinates on the total space of this fibration

$$\mathbf{q} = \frac{\mathbf{x}h}{\sqrt{1+|\mathbf{x}|^2}}, \qquad \mathbf{p} = \frac{h}{\sqrt{1+|\mathbf{x}|^2}}. \tag{9.119}$$

Here $h \in \mathrm{SU}(2)$.

9.3.2 Compatibility with the Twistor Description

An important point is that a choice has been made in writing formulas (9.117) and (9.119). Indeed, it is clear that we can achieve the parametrisation (9.119) by using the action (9.118) of SU(2) on $\mathbb{R}^8 = \mathbb{H}^2$. The action that is used here is multiplication from the right. We could have used the multiplication from the left instead, with appropriate changes in (9.117). However, it is the action (9.118) that is compatible with viewing the twistor space $\mathbb{R}^8$ as the total space of the primed spinor bundle over S^4.

Let us see this. The Euclidean Lorentz group sits in the conformal group $\mathrm{SL}(4,\mathbb{H})$ given by matrices of the form (9.56) as the subgroup $B, C = 0$ and A, D being unit quaternions. This subgroup acts on two-columns with quaternionic entries $\mathbf{q}, \mathbf{p}$ as $\mathbf{q} \to A\mathbf{q}, \mathbf{p} \to D\mathbf{p}$. This allows us to parametrise the 2×2 unitary matrices corresponding to $\mathbf{q}, \mathbf{p}$ as

$$\mathbf{q} = (\pi, -\hat{\pi}), \qquad \mathbf{p} = (\omega, -\hat{\omega}), \tag{9.120}$$

where π and $\omega \in \mathbb{C}^2$ are two-dimensional complex columns. Then the action of the Euclidean Lorentz group on π and ω is just that on unprimed π and primed ω spinors.

Note now that $\mathbf{q}$ and $\mathbf{p}$ in (9.119) are related as $\mathbf{q} = \mathbf{x}\mathbf{p}$. In terms of the spinors π and ω the relation $\mathbf{q} = \mathbf{x}\mathbf{p}$ becomes $\pi = \mathbf{x}\omega$, as well as $\hat{\pi} = \mathbf{x}\hat{\omega}$. But the second relation follows from the first using the unitarity of $\mathbf{x}$. On the other hand, the relation $\pi = \mathbf{x}\omega$ is already familiar, describing the Euclidean twistor space as the total space of the primed spinor bundle over M. So, the parametrisation of S^7 by $\mathbf{x}, h$ used in deriving the Hopf fibration is compatible with the description of the twistor space $\mathbb{R}^8$ as the total space of an $\mathbb{R}^4 = \mathbb{C}^2$ bundle over S^4. This would not be so had we used the left multiplication by unit quaternions on (9.118) to get the projection to S^4. Indeed, in that case, we would have $\mathbf{q} = \mathbf{p}\mathbf{x}$, which does not lead to the right relation between π and ω. This discussion is important because it shows that the quaternionic Hopf fibration gives an alternative description of the same twistor space as we considered before.

To make everything completely explicit, we take, as in (9.97), $\omega = t\eta$ with

$$t = \frac{re^{i\psi}}{\sqrt{(1+|\mathbf{x}|^2)(1+|z|^2)}}. \tag{9.121}$$

Then $\mathbf{p} = (\omega, -\hat{\omega})$ is of the form as in (9.218) with

$$h = \frac{1}{\sqrt{1+|z|^2}} \begin{pmatrix} z & -1 \\ 1 & z^* \end{pmatrix} \begin{pmatrix} e^{i\psi} & 0 \\ 0 & e^{-i\psi} \end{pmatrix}, \tag{9.122}$$

which is unitary unimodular as required. This makes the relation between parametrisations of S^7 by $\mathbf{x}, h$ as in Hopf fibration and $\mathbf{x}, z, \psi$ as in its description as $S^7 \to \mathbb{C}P^3$ explicit.

Note that in (9.122) we see the right action of $\mathrm{U}(1) \subset \mathrm{SU}(2)$ on quaternions. We note that this right action on quaternions $\mathbf{p}, \mathbf{q}$ parametrised by spinors π

and ω as in (9.120) is compatible with the action of U(1) on π and ω via π and $\omega \to e^{i\psi}\pi, e^{i\psi}\omega$. Indeed, we have

$$\begin{pmatrix} \pi & -\hat{\pi} \end{pmatrix} \begin{pmatrix} e^{i\psi} & 0 \\ 0 & e^{-i\psi} \end{pmatrix} = \begin{pmatrix} e^{i\psi}\pi & -(e^{i\psi}\pi)^\wedge \end{pmatrix}. \tag{9.123}$$

This discussion shows that the two projections we have constructed, namely $S^7 \to \mathbb{C}P^3$ and $S^7 \to S^4$, are compatible in the sense that the action of U(1) that fibres S^7 by copies of S^1 with the space of such fibres being $\mathbb{C}P^3$ is the same as the action of U(1) $\subset$ SU(2), with SU(2) acting on quaternions $\mathbf{q}$ and $\mathbf{p}$ via right multiplication.

In other words, this means that the two described ways of realising S^7 as the total space of a circle bundle over the projective twistor space of S^4 coincide. Indeed, on one hand, we have the already described construction of the total space of an S^1 bundle over $\mathbb{C}P^3$, with the later viewed as an S^2 bundle over S^4. On the other hand, we have the Hopf fibration, which is an S^3 bundle over S^4. But S^3 can be viewed as an S^1 bundle over S^2, using the usual Hopf fibration. Combining the two, we have another description of S^7 as a circle bundle over an S^2 bundle over S^4. This gives two descriptions of the projective twistor space of S^4. We have realised them so that they coincide.

However, we also note that the right action of SU(2) on $\mathbb{H}^2$ is *not* compatible with the natural complex structure on this space that comes from parametrising the quaternions as in (9.120) and then interpreting π and ω as holomorphic coordinates. Indeed, we see from (9.120) that the right action of SU(2) mixes π with $\hat{\pi}$ and ω with $\hat{\omega}$. Thus, it does not commute with the natural complex structure on the space of pairs $(\pi, \omega) \in \mathbb{C}^4$. This remark is going to be important when we consider G_2 structures on S^7 in Section 9.6, because it explains why non-integrable almost complex structure on the twistor space is related to the quaternionic Hopf fibration.

9.3.3 Metric on the Total Space

We now compute the metric on S^7 in terms of the coordinates $\mathbf{x}$ and h of the Hopf fibration. The computation is straightforward, even though somewhat lengthy. We have

$$d\mathbf{p} = -\frac{1}{2}\frac{1}{(1+|\mathbf{x}|^2)^{3/2}} d|\mathbf{x}|^2 h + \frac{1}{(1+|\mathbf{x}|^2)^{1/2}} dh,$$
$$d\mathbf{q} = -\frac{1}{2}\frac{1}{(1+|\mathbf{x}|^2)^{3/2}} d|\mathbf{x}|^2 \mathbf{x}h + \frac{1}{(1+|\mathbf{x}|^2)^{1/2}} (d\mathbf{x}h + \mathbf{x}dh).$$

Let us also write the corresponding Hermitian conjugates

$$d\mathbf{p}^\dagger = -\frac{1}{2}\frac{1}{(1+|\mathbf{x}|^2)^{3/2}} d|\mathbf{x}|^2 h^{-1} - \frac{1}{(1+|\mathbf{x}|^2)^{1/2}} h^{-1}dhh^{-1},$$
$$d\mathbf{q}^\dagger = -\frac{1}{2}\frac{1}{(1+|\mathbf{x}|^2)^{3/2}} d|\mathbf{x}|^2 h^{-1}\mathbf{x}^\dagger + \frac{1}{(1+|\mathbf{x}|^2)^{1/2}} (h^{-1}d\mathbf{x}^\dagger - h^{-1}dhh^{-1}\mathbf{x}^\dagger).$$

Here we have used the fact that $h^\dagger = h^{-1}$ and that $dh^{-1} = -h^{-1}dhh^{-1}$. We now compute the pieces of the flat metric on $\mathbb{R}^8$

$$
\begin{aligned}
\frac{1}{2}\mathrm{Tr}(d\mathbf{p}d\mathbf{p}^\dagger) &= \frac{1}{4}\frac{1}{(1+|\mathbf{x}|^2)^3}d|\mathbf{x}|^2 d|\mathbf{x}|^2 - \frac{1}{2}\frac{1}{(1+|\mathbf{x}|^2)}\mathrm{Tr}(h^{-1}dhh^{-1}dh),\\
\frac{1}{2}\mathrm{Tr}(d\mathbf{q}d\mathbf{q}^\dagger) &= \frac{1}{4}\frac{|x|^2}{(1+|\mathbf{x}|^2)^3}d|\mathbf{x}|^2 d|\mathbf{x}|^2 - \frac{1}{2}\frac{1}{(1+|\mathbf{x}|^2)^2}d|\mathbf{x}|^2 d|\mathbf{x}|^2\\
&\quad + \frac{1}{2}\frac{1}{(1+|\mathbf{x}|^2)}\mathrm{Tr}(d\mathbf{x}d\mathbf{x}^\dagger) + \frac{1}{2}\frac{1}{(1+|\mathbf{x}|^2)}\mathrm{Tr}(\mathbf{x}dhh^{-1}d\mathbf{x}^\dagger - d\mathbf{x}dhg^{-1}\mathbf{x}^\dagger)\\
&\quad - \frac{1}{2}\frac{|\mathbf{x}|^2}{(1+|\mathbf{x}|^2)}\mathrm{Tr}(h^{-1}dhh^{-1}dh),
\end{aligned}
$$

where we have taken into account some obvious cancellations. We now add these two quantities to obtain the metric on S^7, taking into account some obvious simplifications

$$
\begin{aligned}
ds^2_{S^7} &= -\frac{1}{4}\frac{1}{(1+|\mathbf{x}|^2)^2}d|\mathbf{x}|^2 d|\mathbf{x}|^2 + \frac{1}{2}\frac{1}{(1+|\mathbf{x}|^2)}\mathrm{Tr}(d\mathbf{x}d\mathbf{x}^\dagger)\\
&\quad + \frac{1}{2}\frac{1}{(1+|\mathbf{x}|^2)}\mathrm{Tr}(\mathbf{x}dhh^{-1}d\mathbf{x}^\dagger - d\mathbf{x}dhg^{-1}\mathbf{x}^\dagger) - \frac{1}{2}\mathrm{Tr}(h^{-1}dhh^{-1}dh).
\end{aligned} \tag{9.124}
$$

We now complete the square in the terms on the second line

$$
\begin{aligned}
&\frac{1}{2}\frac{1}{(1+|\mathbf{x}|^2)}\mathrm{Tr}(\mathbf{x}dhh^{-1}d\mathbf{x}^\dagger - d\mathbf{x}dhg^{-1}\mathbf{x}^\dagger) - \frac{1}{2}\mathrm{Tr}(h^{-1}dhh^{-1}dh)\\
&\quad = -\frac{1}{2}\mathrm{Tr}\left(h^{-1}dh - \frac{1}{2}\frac{1}{(1+|\mathbf{x}|^2)}h^{-1}(d\mathbf{x}^\dagger\mathbf{x} - \mathbf{x}^\dagger d\mathbf{x})h\right)^2\\
&\qquad + \frac{1}{8}\frac{1}{(1+|\mathbf{x}|^2)^2}\mathrm{Tr}(d\mathbf{x}^\dagger\mathbf{x} - \mathbf{x}^\dagger d\mathbf{x})^2.
\end{aligned}
$$

The first term here is the desired metric together with the connection in the fibre. The last term here is to be combined with the terms in the first line of (9.124). To do this, it needs some rewriting. We have

$$
\begin{aligned}
\mathrm{Tr}(d\mathbf{x}^\dagger\mathbf{x} - \mathbf{x}^\dagger d\mathbf{x})^2 &= -4\mathrm{Tr}\left(d\mathbf{x}^\dagger\mathbf{x}\mathbf{x}^\dagger d\mathbf{x}\right) + d|\mathbf{x}|^2\mathrm{Tr}(d\mathbf{x}^\dagger\mathbf{x} + \mathbf{x}^\dagger d\mathbf{x})\\
&= -4|\mathbf{x}|^2\mathrm{Tr}(d\mathbf{x}^\dagger d\mathbf{x}) + 2d|\mathbf{x}|^2 d|\mathbf{x}|^2.
\end{aligned} \tag{9.125}
$$

We have used $d\mathbf{x}^\dagger\mathbf{x} = -\mathbf{x}^\dagger d\mathbf{x} + \mathbb{I}d|\mathbf{x}|^2$ to get the first relation. It is clear that the last term here cancels with the first term in (9.124). The other terms combine into the final result

$$
\begin{aligned}
ds^2_{S^7} &= \frac{1}{2}\frac{1}{(1+|\mathbf{x}|^2)^2}\mathrm{Tr}(d\mathbf{x}d\mathbf{x}^\dagger)\\
&\quad - \frac{1}{2}\mathrm{Tr}\left(h^{-1}dh + \frac{1}{2}\frac{1}{(1+|\mathbf{x}|^2)}h^{-1}(\mathbf{x}^\dagger d\mathbf{x} - d\mathbf{x}^\dagger\mathbf{x})h\right)^2.
\end{aligned} \tag{9.126}
$$

The first term here is (a quarter of) the usual metric on S^4 in conformally flat parametrisation. We can rewrite the second term more compactly by using the previously encountered connection (9.100), which is an anti-Hermitian matrix

$\mathbf{A}^\dagger = -\mathbf{A}$. We also introduce the Maurer–Cartan 1-form $\mathbf{m}$ and the connection 1-form W in the total space of the bundle

$$\mathbf{m} := h^{-1}dh, \qquad W := \mathbf{m} + h^{-1}\mathbf{A}h. \tag{9.127}$$

We can then write the metric on the total space of the $S^7 \to S^4$ Hopf fibration as

$$ds^2_{S^7} = \frac{1}{4}\left(ds^2_{S^4} + \sum_i (W^i)^2\right), \tag{9.128}$$

where

$$ds^2_{S^4} = \frac{4\sum_\mu (dx^\mu)^2}{(1+|\mathbf{x}|^2)^2} \tag{9.129}$$

is the usual metric on the four-sphere. The objects $W = W^i\tau^i$ with $\tau^i = (-\mathrm{i}/2)\sigma^i$ are the generators of the Lie algebra of SU(2). The coordinates $x^\mu, \mu = 1, \ldots, 4$ are those on S^4 in conformally flat parametrisation.

9.3.4 Checking the Connection

We have seen the connection (9.100) appearing in two different constructions. One was the description of $\mathbb{C}P^3$ as a two-sphere bundle over S^4, the other description of S^7 as a three-sphere bundle over S^4. These constructions are of course related, because S^3 is itself an S^1 bundle over S^2, and so our description of $\mathbb{C}P^3 \to S^4$ is in fact inside the description of the Hopf fibration $S^7 \to S^4$.

Let us now interpret the connection (9.100) as the chiral half of the spin connection for the metric on S^4. This connection acts naturally on the primed spinors, and so it is the ASD part of the spin connection. To see this, we introduce the matrix of ASD 2-forms

$$\Sigma := \frac{d\mathbf{x}^\dagger \wedge d\mathbf{x}}{(1+|\mathbf{x}|^2)^2}, \tag{9.130}$$

where we indicated the wedge product explicitly. To see that this is the correct normalisation we note that the soldering form for the metric (9.129) is $e^I = 2dx^I/(1+|\mathbf{x}|^2)$. On the other hand, computing (9.130) with $\mathbf{x}$ given by (9.68) one gets $\Sigma = \tau^i\Sigma^i$ with, e.g., $\Sigma^1 = 4(dx^4dx^3 - dx^1dx^2)/(1+|\mathbf{x}|^2)^2$. This is the correctly normalised chiral 2-form for the metric (9.129). However, the sign here is what in the previous chapters we called SD. At the same time, it is the ASD 2-form in the orientation (4123), and so we will continue to refer to it as ASD for the remainder of this chapter. We will not be writing a bar over Σ from now on to de-clatter notations.

The exterior derivative of (9.130) is given by

$$d\Sigma = -\frac{2}{(1+|\mathbf{x}|^2)^3}d|\mathbf{x}|^2 d\mathbf{x}^\dagger d\mathbf{x}. \tag{9.131}$$

We then compute

$$\mathbf{A}\Sigma - \Sigma\mathbf{A} = \frac{1}{2}\frac{1}{(1+|\mathbf{x}|^2)^3}\left((\mathbf{x}^\dagger d\mathbf{x} - d\mathbf{x}^\dagger\mathbf{x})d\mathbf{x}^\dagger d\mathbf{x} - d\mathbf{x}^\dagger d\mathbf{x}(\mathbf{x}^\dagger d\mathbf{x} - d\mathbf{x}^\dagger\mathbf{x})\right).$$

We need to do some massaging of the right-hand side using the identity (9.71). First, using this identity we can rewrite the expression in brackets here as

$$(\mathbb{I}d|\mathbf{x}|^2 - 2d\mathbf{x}^\dagger\mathbf{x})d\mathbf{x}^\dagger d\mathbf{x} - d\mathbf{x}^\dagger d\mathbf{x}(2\mathbf{x}^\dagger d\mathbf{x} - \mathbb{I}d|\mathbf{x}|^2). \tag{9.132}$$

We now group the second and third terms here, and again use the same identity. This gives for (9.132)

$$4d|\mathbf{x}|^2 d\mathbf{x}^\dagger d\mathbf{x}, \tag{9.133}$$

and comparing with (9.131) we have

$$d\Sigma + \mathbf{A}\wedge\Sigma - \Sigma\wedge\mathbf{A} = 0. \tag{9.134}$$

9.3.5 Checking the Einstein Condition

It is clear that we are in the Plebanski formalism setting for the four-sphere, and thus we also expect to be able to recover the Plebanski version of the Einstein equations, which is the statement that the curvature of the ASD connection $\mathbf{A}$ is ASD as a 2-form. This is an instructive calculation because we are now using 2×2 matrix notations doing curvature calculations in 4D. This is similar to the index-free formalism we have developed for 3D in Chapter 4.

Let us compute the curvature, which is given by $\mathbf{F} = d\mathbf{A} + \mathbf{A}\mathbf{A}$. The first term gives

$$\begin{aligned} d\mathbf{A} &= -\frac{1}{2}\frac{1}{(1+|\mathbf{x}|^2)^2}d|\mathbf{x}|^2(\mathbf{x}^\dagger d\mathbf{x} - d\mathbf{x}^\dagger\mathbf{x}) + \frac{1}{(1+|\mathbf{x}|^2)}d\mathbf{x}^\dagger d\mathbf{x} \\ &= -\frac{1}{(1+|\mathbf{x}|^2)}d|\mathbf{x}|^2\mathbf{A} + (1+|\mathbf{x}|^2)\Sigma. \end{aligned} \tag{9.135}$$

The $\mathbf{A}\mathbf{A}$ computation is again made simple by using the identity (9.71) to rewrite the connection in a convenient form. We have

$$\mathbf{A}\mathbf{A} = \frac{1}{4}\frac{1}{(1+|\mathbf{x}|^2)^2}(\mathbb{I}d|\mathbf{x}|^2 - 2d\mathbf{x}^\dagger\mathbf{x})(2\mathbf{x}^\dagger d\mathbf{x} - \mathbb{I}d|\mathbf{x}|^2) \tag{9.136}$$

$$= \frac{1}{4}\frac{1}{(1+|\mathbf{x}|^2)^2}\left(2d|\mathbf{x}|^2(\mathbf{x}^\dagger d\mathbf{x} - d\mathbf{x}^\dagger\mathbf{x}) - 4|\mathbf{x}|^2 d\mathbf{x}^\dagger d\mathbf{x}\right)$$

$$= \frac{1}{(1+|\mathbf{x}|^2)}d|\mathbf{x}|^2\mathbf{A} - |\mathbf{x}|^2\Sigma. \tag{9.137}$$

This immediately gives the expected Einstein equation

$$\mathbf{F} = \Sigma. \tag{9.138}$$

9.4 Twistor Description of Gravitational Instantons

Previously we have described the twistor space of flat $\mathbb{R}^4$ and conformally flat S^4 manifolds. We have seen that the arising in this case twistor space is $\mathbb{C}P^3$, a complex manifold. The purpose of this section is to explain that this can be generalised to the more nontrivial setting of half-flat geometries in which one of the two chiral halves of the Weyl curvature vanishes.

9.4.1 The Curved Twistor Space

We will only present the version of the story that works for Euclidean signature metrics. As we have seen, the twistor space in this case is the total space of the projective spinor bundle over M. Such a bundle can also be constructed for a general Euclidean metric on M. Let $\Sigma^{A'B'}$ be the associated ASD 2-forms, and $A^{A'B'}$ the the ASD chiral half of the spin connection. We can then construct the 3-form (9.114). This is a 3-form on the total space of the primed spinor bundle over M, and descends to a 3-form of homogeneity degree four in $\omega^{A'}$ on the projectivised spinor bundle.

9.4.2 Twistor Space of an Instanton Is a Complex Manifold

We can declare the 3-form (9.114) to be $(3,0)$, which then defines an almost complex structure on the twistor space. The corresponding $(1,0)$ forms are those whose wedge product with $\Omega^{(3,0)}$ is zero. These are the projective versions of the 1-forms

$$\tau = \omega^{A'} D\omega_{A'}, \qquad e^{AA'}\omega_{A'}, \tag{9.139}$$

where the last expression gives two different 1-forms for $A = 1, 2$.

A natural question is then whether the almost complex structure so defined is integrable. As the criterion of integrability we use the following statement: An almost complex structure is integrable if and only if the restriction of the exterior derivative on $(1,0)$ forms to the space of $(0,2)$ forms vanishes. This statement is one of the alternative ways to state the **Newlander–Niernberg theorem**. This then implies that $d = \partial + \bar{\partial}$. Thus, we need to compute the exterior derivative of the previous $(1,0)$ forms and project into the subspace of $(0,2)$ forms. We have

$$d(\omega^{A'} D\omega_{A'}) = D\omega^{A'} D\omega_{A'} + \omega^{A'} DD\omega_{A'}. \tag{9.140}$$

The first term here is a $(1,1)$ form, which can be seen by inserting in it the decomposition (9.85) of the identity. To compute the second term, we use $DD\omega_{A'} = F_{A'}{}^{B'}\omega_{B'}$, where $F_{A'B'}$ is the spinorial version of the curvature of the ASD connection. On an Einstein background we have

$$F^{A'B'} = \left(\Psi^{A'B'C'D'} - \frac{\Lambda}{3}\epsilon^{A'(C'}\epsilon^{D')B'}\right)\Sigma_{C'D'}, \tag{9.141}$$

which is the Plebanski second equation and $\Psi^{A'B'C'D'}$ is the spinor version of the matrix Ψ^{ij}. Now, the $(0,2)$ part of the basic 2-form $\Sigma^{A'B'}$ is

$$\Sigma^{A'B'}\Big|_{(0,2)} = \frac{\omega^{A'}\omega^{B'}}{[\hat{\omega}\omega]^2}\Sigma^{C'D'}\hat{\omega}_{C'}\hat{\omega}_{D'}. \tag{9.142}$$

This means that the $(0,2)$ part of $d\tau$ is a multiple of

$$\Psi^{A'B'C'D'}\omega_{A'}\omega_{B'}\omega_{C'}\omega_{D'}, \tag{9.143}$$

which vanishes for all $\omega^{A'}$ if and only if $\Psi^{A'B'C'D'} = 0$, which is the half-flatness condition.

For the basic $(1,0)$ forms we can use the covariant derivative and then the torsion-free condition $De^{AA'} = 0$. This gives

$$D(e^{AA'}\omega_{A'}) = -e^{AA'}D\omega_{A'}. \tag{9.144}$$

Inserting here the decomposition of the identity (9.85) we see that there is no $(0,2)$ component. Thus, the almost complex structure defined by $\Omega^{(3,0)}$ given by (9.114) is integrable if and only if the ASD chiral half of the Weyl curvature vanishes $\Psi^{A'B'C'D'} = 0$. This means that the twistor space of a gravitational instanton is a complex manifold. This fact has been used to construct new gravitational instantons using deformation theory of complex manifolds; see Ward (1980).

9.4.3 Generalising Twistors

We have already understood that the projective twistor space of S^4 is $\mathbb{C}P^3$, which is a complex manifold, and the complex structure on $\mathbb{C}P^3$ can be understood from the fact that $\mathbb{C}P^3$ is naturally an S^2 fibre bundle over S^4, with fibres parametrising different almost complex structures on S^4. But we have also seen that $\mathbb{C}P^3$ is itself naturally a base of the fibre bundle $S^7 \to \mathbb{C}P^3$ with circles as fibres. The two constructions intersect via the quaternionic Hopf fibration $S^7 \to S^4$, which can be thought of as either S^3 bundle over S^4 or as a circle bundle over the projective twistor space $\mathbb{C}P^3$ of S^4.

Thus, the twistor space $\mathbb{C}P^3$ of S^4 can be viewed as sitting inside $S^7 \subset \mathbb{C}^4$. This suggest that we can contemplate generalising the twistor theory. The usual twistor space of a Euclidean space M is the total space of the (projective) primed spinor bundle over M. It is interesting to consider a larger space, which is the total space of an S^1 bundle over the usual twistor space.

The usual twistor theory puts emphasis on the complex analytic aspects of the twistor construction, and also allows to use powerful theory of complex manifolds and their deformations to produce new examples of ASD Einstein manifolds. These aspects of twistor theory are well-described in, e.g., *Introduction to Twistor Theory* in, e.g., Haggett and Tod (1994). See also Atiyah et al. (2017) for a more recent account.

When we instead consider the total space of an S^1 bundle over the usual twistor space, the complex analytic aspects of the usual twistor story are no longer at the forefront. In particular, the seven-dimensional total space of the bundle cannot be a complex manifold. However, since the seven-dimensional space in question is fibred by S^1 over the usual twistor space, all of the twistor constructions are still relevant. Enlarging the space one just gets access to richer geometry. In particular, as we shall see in the following sections, there is a beautiful geometry of 3-forms in 7D, and bringing into play the total space of an S^1 bundle over the twistor space gives access to this geometry. This geometry in particular explains why the projective twistor space can naturally be endowed with a metric, something that remains a puzzle if one stays in the context of usual 6D twistor theory. The explanation of this is that there is a natural 3-form on the S^1 bundle over the twistor space of a Euclidean 4-manifold M, and generic 3-forms in seven dimensions define a metric.

Thus, going to 7D allows for geometric constructions not possible in the setting of the usual 6D twistor theory, and also emphasises different geometric aspects in the sense that the theory of complex manifolds no longer plays the dominating role. In particular, we shall see the first order Cauchy–Riemann equations guaranteeing integrability of the almost complex structure on the twistor space are replaced by certain other natural first-order differential equations in seven dimensions.

9.5 Geometry of 3-Forms in Seven Dimensions

The purpose of this section is to describe the geometry of 3-forms in 7D. The ultimate goal is to relate this geometry to the previous twistor constructions.

9.5.1 Stable 3-Forms

The beautiful geometry reviewed in this section has been known for more than a century; see Agricola (2008) for the history. In particular the characterisation of G_2 via 3-forms is a result due to Engel from 1900.

Let us start with some linear algebra in $\mathbb{R}^7$. A 3-form $C \in \Lambda^3\mathbb{R}^7$ is called *stable* if it lies in a open orbit under the action of GL(7); see Hitchin (2000). This notion gives a generalisation of nondegeneracy of forms and implies that any nearby form can be reached by a GL(7) transformation. Thus, stable 3-forms can also be called generic or nondegenerate.

For real 3-forms, there are exactly two distinct open orbits, characterised by the sign of a certain invariant, see Section 9.5.4, each of which is related to a real form of $G_2^{\mathbb{C}}$. The open orbit corresponding to the compact real form G_2 is what plays role in relation to the quaternionic Hopf fibration $S^7 \to S^4$. For every such C, there exists a set $e^1, \dots, e^7$ of 1-forms in which C is expanded in the following canonical form:

$$C = e^{567} + e^5\Sigma^1 + e^6\Sigma^2 + e^7\Sigma^3, \tag{9.145}$$

where $\Sigma^i, i = 1, 2, 3$ are already familiar to us as Euclidean chiral 2-forms (5.31).

The fact that is of central importance about stable 3-forms in seven dimensions is that a stabiliser of such a form in GL(7) is isomorphic to the exceptional Lie group G_2. This group has dimension 14, and this number arises as the dimension 49 of GL(7) minus the dimension 35 of $\Lambda^3\mathbb{R}^7$. Thus, the space of stable 3-forms is the homogeneous group manifold $\mathrm{GL}(7)/G_2$.

We can then generalise the notion of stable forms to 3-forms on a seven-dimensional differentiable manifold M. These are forms that are stable at every point.

9.5.2 The Metric

The most fundamental fact about stable 3-forms in 7D is that they define a metric. The latter is obtained as follows

$$g_C(\xi,\eta)v_C = \frac{1}{3} i_\xi C \wedge i_\eta C \wedge C, \tag{9.146}$$

where we explicitly indicated the wedge product. The right-hand side here is the top form, which is moreover ξ, η symmetric. This gives a symmetric pairing of two vector fields up to scaling. The scale factor is then completely determined by the requirement that v_C on the left-hand side is the volume form of g_C. Moreover, the sign of the volume form v_C is uniquely fixed by the requirement that the metric defined by (9.146) has specific (say, all plus) signature. In this way, a 3-form C defines both the metric g_C and an orientation.

It is then a simple computation that, for a 3-form presented in the canonical form (9.145), the arising metric is

$$g_C = \sum_{I=1}^{7} e^I e^I, \tag{9.147}$$

and the orientation is given by $e^{1\dots 7}$. Given that G_2 is the stabiliser of (9.145), it also stabilises the metric (9.147). This gives an embedding $G_2 \subset \mathrm{O}(7)$.

The form (9.145) corresponds to the compact real form G_2 of $G_2^{\mathbb{C}}$. The orbit corresponding to the non-compact real form $G_2^* \subset \mathrm{O}(3,4)$ is the orbit of a 3-form similar to (9.145) but with the signs in all three terms containing Σ^i changed. The formula (9.146) still defines a metric and an orientation at this time of signature $(3,4)$.

9.5.3 Relation to Urbantke Formula

The formula (9.146) is remarkable in particular because it provides an explanation for why the Urbantke (5.47) formula in 4D exists. To see this, consider the bundle of ASD 2-forms over a 4D Riemannian manifold. A general ASD 2-form can be parametrised as $\Sigma^i y^i$, and so y^i are the coordinates along the fibre. Here Σ^i are the canonical ASD 2-forms (5.31). We then form the following 3-form in the total space of this bundle

$$C_\Sigma = dy^1 dy^2 dy^3 + dy^1 \Sigma^1 + dy^2 \Sigma^2 + dy^3 \Sigma^3. \tag{9.148}$$

It is then easy to check that the metric (9.146) reduces on the fibres to a multiple of the metric δ^{ij}. On the other hand, on the base, the formula (9.146) reduces to the Urbantke formula (5.47).

The described 7D explanation of the 4D Urbantke formula makes one suspect that there should be a relation between the 4D Plebanski formalism with its SO(3) bundle of ASD 2-forms and the geometry of 3-forms in seven dimensions. The rest of this chapter is devoted to exhibiting aspects of this relation.

9.5.4 The Volume Functional

Given a stable 3-form, we construct the metric and the corresponding volume form as described in (9.146). The volume form can be computed in two different ways. First, one can compute the metric g_C times its volume form from (9.145). One can then take the determinant of the right-hand side, which results in a quantity of homogeneity degree 21 in C. The left-hand side gives $(\det(g))^{9/2}$. This means that the volume form $\sqrt{\det(g)}$ is a quantity of homogeneity degree $7/3$ in C. On the other hand, there is also an explicit formula for this quantity

$$v_C \sim \left(\tilde{\epsilon}^{a_1\ldots a_7}\tilde{\epsilon}^{b_1\ldots b_7}\tilde{\epsilon}^{c_1\ldots c_7}C_{a_1b_1c_1}\ldots C_{a_7b_7c_7}\right)^{1/3}. \tag{9.149}$$

The number appearing as the proportionality coefficient in this formula is unimportant to us. The quantity $\tilde{\epsilon}^{a_1\ldots a_7}$ is the completely antisymmetric densitiesed tensor available on any manifold without any additional structure such as a metric. The invariant in brackets in the formula (9.149), of degree seven in C, has been known since 1900, see Agricola (2008), and gives the stability criterion. Thus, the 3-form C is stable if and only if this invariant is different from zero. The sign of this invariant determines whether the form belongs to the compact or non-compact real orbit.

One can integrate the volume form constructed from C over M to get the volume functional

$$S[C] = \int_M v_C. \tag{9.150}$$

As is explained in particular in Hitchin (2000), the first variation of the functional (9.150) in C has a simple form

$$\delta S[C] \sim \int_M {}^*C \wedge \delta C. \tag{9.151}$$

The precise numerical coefficient in this equation is of no importance for us. The 4-form ${}^*C$ can be shown to be given by the Hodge dual of C computed with respect to the metric defined by C. This means that for the 3-form (9.145) the 4-form ${}^*C$ is given by

$${}^*C = e^{1234} + e^{67}\Sigma^1 + e^{75}\Sigma^2 + e^{56}\Sigma^3. \tag{9.152}$$

9.5.5 Complex Parametrisation

Let us give another form of the canonical expression for the 3-form (9.145). We first rewrite it as

$$C = -e^7(e^{12} + e^{34} + e^{65}) + e^5(e^{41} - e^{23}) + e^6(e^{42} - e^{31}). \tag{9.153}$$

We then notice that if we introduce complex-valued 1-forms

$$\theta^1 := e^1 + \mathrm{i}e^2, \quad \theta^2 = e^3 + \mathrm{i}e^4, \quad \theta^3 = e^6 + \mathrm{i}e^5 \tag{9.154}$$

then

$$e^{12} + e^{34} + e^{65} := \omega = \frac{1}{2\mathrm{i}}(\bar{\theta}^1\theta^1 + \bar{\theta}^2\theta^2 + \bar{\theta}^3\theta^3) \tag{9.155}$$

and

$$\begin{aligned} \mathrm{Re}(\theta^1\theta^2\theta^3) &= e^5(e^{41} - e^{23}) + e^6(e^{42} - e^{31}), \\ \mathrm{Im}(\theta^1\theta^2\theta^3) &= e^5(e^{42} - e^{31}) - e^6(e^{41} - e^{23}). \end{aligned} \tag{9.156}$$

This means that we can rewrite

$$\begin{aligned} C &= -e^7\omega + \mathrm{Re}(\theta^1\theta^2\theta^3), \\ {}^*C &= \frac{1}{2}\omega\omega + e^7\mathrm{Im}(\theta^1\theta^2\theta^3). \end{aligned} \tag{9.157}$$

While the form (9.145) makes manifest the SO(4) subgroup of G_2 preserving C, the form (9.157) makes manifest the SU(3) subgroup. Both forms will be useful in the calculations that follow.

9.5.6 Holonomy Reduction

The fundamental result Gray (1969) states: Let $C \in \Lambda^3 M$ be a 3-form on a 7-manifold. Then C is parallel with respect to the Levi–Civita connection of g_C if and only if $dC = 0$ and $d^*C = 0$. In other words, the condition of C being parallel with respect to the metric it defines is equivalent to the conditions of C being closed and co-closed, where co-closedness is again with respect to the metric it defines.

The next basic fact is that if a Riemannian manifold (M, g) has a parallel 3-form C, then the holonomy group of M is contained in G_2. In particular, this implies that the (M, g) is Ricci-flat. This is very interesting, because this means that we can code Einstein's equations in 7D as differential equations on an object of a completely different nature from the metric, i.e., on a 3-form. The 3-form C that is closed and co-closed then defines a metric algebraically, and this metric is guaranteed to be Ricci-flat. Actually, having a C that is parallel constraints the Riemann curvature stronger than just requiring the Ricci part to be zero, but in particular Ricci flatness is guaranteed.

Combining the result from Gray (1969) with the formula (9.151) for the first variation of the functional $S[C]$, we see that manifolds with holonomy contained in G_2 are critical points of $S[C]$, provided one varies C in a fixed cohomology class $\delta C = dB$, $B \in \Lambda^2 M$. This variational characterisation is explored in depth in Hitchin (2000).

9.5.7 Nearly Parallel G_2 Structures

Let us now instead assume that we have a 7-manifold M with a stable 3-form of positive type and satisfying

$$dC = \lambda {}^*C, \tag{9.158}$$

where ${}^*C$ is the 4-form that is the Hodge dual of C computed using the metric defined by C itself, and λ is a constant. In this case, the 3-form C is not closed, but ${}^*C$ is. Thus, the 3-form is not parallel in the sense of previous subsection. It is instead called **nearly parallel**, because the departure of C from being closed is as small as possible.

Nearly parallel G_2 structures have in particular been studied in Friedrich et al. (1997). The canonical example of such a structure is one on the seven-sphere; see the next section. What is important for us is that the metric defined by a nearly parallel G_2 structure is automatically Einstein with a nonzero (and positive) scalar curvature. Thus, the equation (9.158) can be viewed as encoding the Einstein equations in seven dimensions (but similarly to the case of parallel structures, giving in fact stronger equations). It is also important for us that these equations can be obtained from a variational principle. Thus, we write the following action Krasnov (2017a)

$$S[C] = \frac{1}{2}\int CdC + 6\lambda v_C. \tag{9.159}$$

Its critical points are precisely the 3-forms satisfying (9.158). This action can be viewed as a 7D analog of the 3D Chern–Simons theory. The difference is that it is not possible to write an interacting Abelian Chern–Simons theory in 3D, while this is possible in 7D due to the availability of the degree seven invariant whose cube root can be integrated over the manifold.

We will return to the equations (9.158) in the next section. We view (9.158) as the natural set of first-order partial differential equations that can be written for a 3-form in 7D. In Section 9.7 we will see that these equations can be thought of as generalising the integrability of certain almost complex structure in 6D. So, they are the main player in our envisaged $6D \to 7D$ generalisation of twistor theory.

9.5.8 3-Forms That Correspond to the Same Metric

The counting of components shows that 3-forms contain more information than just that of a metric. Indeed, to specify a metric in seven dimensions, we need $7 \times 8/2 = 28$ numbers, while the dimension of the space of 3-forms is 35. Thus, there are seven more components in a 3-form. It can be shown that these correspond to components of a unit spinor $\Psi : |\Psi|^2 = 1$ so that the 3-form in question can be represented as $C_{abc} = \Psi^T \gamma_a \gamma_b \gamma_c \Psi$. Here $\gamma_a, a = 1, \ldots, 7$ are the 8×8 γ-matrices for the metric g_C, the spinor Ψ is a real spinor of Spin(7), and $|\Psi|^2$ is a symmetric bilinear form on spinors available for Spin(7).

There is an alternative, very useful, characterisation of 3-forms that correspond to the same metric. This uses a vector field rather than a unit spinor. The expression we are after can be obtained by considering a rotation that mixes directions 4, 3, i.e., let

$$\begin{pmatrix} e^3 \\ e^4 \end{pmatrix} \to \begin{pmatrix} \cos(2\theta) & \sin(2\theta) \\ -\sin(2\theta) & \cos(2\theta) \end{pmatrix} \begin{pmatrix} e^3 \\ e^4 \end{pmatrix}. \tag{9.160}$$

This rotation does not change the metric, but mixes the 2-forms Σ^1 and Σ^2

$$\begin{pmatrix} \Sigma^1 \\ \Sigma^2 \end{pmatrix} \to \begin{pmatrix} \cos(2\theta) & \sin(2\theta) \\ -\sin(2\theta) & \cos(2\theta) \end{pmatrix} \begin{pmatrix} \Sigma^1 \\ \Sigma^2 \end{pmatrix}. \tag{9.161}$$

So, under this rotation the 3-form C goes to

$$\begin{aligned} C &\to e^{567} + \cos(2\theta)(e^5\Sigma^1 + e^6\Sigma^2) + \sin(2\theta)(e^5\Sigma^2 - e^6\Sigma^1) + e^7\Sigma^3 \\ &= e^{567} + (1 - 2\sin^2(\theta))(e^5\Sigma^1 + e^6\Sigma^2) + 2\sin(\theta)\cos(\theta)(e^5\Sigma^2 - e^6\Sigma^1) + e^7\Sigma^3 \\ &= (1 - 2\sin^2(\theta))C + 2\sin^2(\theta)(e^{567} + e^7\Sigma^3) + 2\sin(\theta)\cos(\theta)(e^5\Sigma^2 - e^6\Sigma^1), \end{aligned}$$

where we wrote the result in a suggestive form. We now notice that the 3-forms in the last two terms can be obtained as

$$e^{567} + e^7\Sigma^3 = e^7 \wedge i_7 C, \qquad e^5\Sigma^2 - e^6\Sigma^1 = i_7{}^*C, \tag{9.162}$$

where ${}^*C$ is given by (9.152). Thus, if we take

$$\alpha := \sin(\theta)e^7 \tag{9.163}$$

we see that the 3-form

$$\tilde{C} = (1 - 2|\alpha|^2)C + 2\alpha \wedge i_\alpha C + 2\sqrt{1 - |\alpha|^2}\, i_\alpha{}^*C \tag{9.164}$$

corresponds to the same metric. Here i_α is the operation of insertion of the vector field dual (with respect to the metric defined by C) to the 1-form α. Even though the fact that $\tilde{C}$ in (9.164) and C define the same metric was shown only for the 1-forms of the special type (9.163), this fact holds in general, because any 1-form can be aligned with e^7 by rotation.

In the similar way, the transformation rule of the dual 4-form is shown to be

$${}^*\tilde{C} = {}^*C - 2\alpha \wedge i_\alpha{}^*C - 2\sqrt{1 - |\alpha|^2}\, \alpha \wedge C. \tag{9.165}$$

The presence of the square root in these formulas signifies the fact that in the transformation by a 1-form, the 1-form α cannot be taken with norm larger than one, because this takes one out of the space of real 3-forms.

9.6 G_2-Structures on S^7

The purpose of this section is to study G_2 structures on the seven-sphere S^7. There is a unique G_2 structure on S^7 that comes from the embedding $S^7 \subset \mathbb{R}^8$. There is then a unique so-called Spin(7) structure on $\mathbb{R}^8$, which projecting against the radial vector field gives the desired G_2 structure on S^7. On the other hand, S^7 can be realised as the total space of the circle bundle over $\mathbb{C}P^3$. We shall see that there are two natural ways that the circle bundle over $\mathbb{C}P^3$ can be mapped into the round S^7 with its canonical G_2 structure. This leads to two different G_2 structures on the circle bundle over $\mathbb{C}P^3$, or we can say, two different G_2 structures on S^7. One of these will be related to the integrable almost complex structure (ACS) on $\mathbb{C}P^3$, while the other one will give rise to the non-integrable ACS.

9.6.1 Spin(7) Structure on $\mathbb{R}^8$

There is a 4-form in $\mathbb{R}^8$ whose stabiliser in GL(8) is the group Spin(7). The group Spin(7) acts naturally on $\mathbb{R}^8$ as its spinor representation. The 4-form in question is given by

$$\begin{aligned}\Theta &= \sum_{i=1}^{3} \Sigma^i \tilde{\Sigma}^i - \frac{1}{6}\Sigma^i\Sigma^i - \frac{1}{6}\tilde{\Sigma}^i\tilde{\Sigma}^i \qquad (9.166)\\ &= \sum_{i=1}^{3} \Sigma^i \tilde{\Sigma}^i + dx^4dx^1dx^2dx^3 + dx^8dx^5dx^6dx^7,\end{aligned}$$

where

$$\Sigma^1 = dx^4dx^1 - dx^2dx^3, \quad \Sigma^2 = dx^4dx^2 - dx^3dx^1, \quad \Sigma^3 = dx^4dx^3 - dx^1dx^2,$$
$$\tilde{\Sigma}^1 = dx^8dx^5 - dx^6dx^7, \quad \tilde{\Sigma}^2 = dx^8dx^6 - dx^7dx^5, \quad \tilde{\Sigma}^3 = dx^8dx^7 - dx^5dx^6.$$

We note that

$$\Theta = dx^8 \wedge C - {}^*C, \qquad (9.167)$$

where C is the canonical (9.145) 3-form on $\mathbb{R}^7$ and ${}^*C$ is its dual (9.152). We have indicated the wedge product explicitly to have a nicer looking expression. This in particular shows that Θ is SD (in the orientation 12345678) with respect to the standard flat metric on $\mathbb{R}^8$

$${}^*\Theta = \Theta. \qquad (9.168)$$

9.6.2 Canonical Nearly Parallel G_2 Structure on S^7

We now introduce spherical coordinates on $\mathbb{R}^8$. Using homogeneity and the fact that Θ is SD we have

$$\Theta = r^3 dr \wedge C + r^4\, {}^*C, \qquad (9.169)$$

where C is defined as $i_{\partial/\partial r}\Theta$ evaluated at $r = 1$, and as before ${}^*C$ is the Hodge dual of C computed using the metric defined by C. Using the fact that Θ is a closed form we deduce

$$dC = 4^*C. \tag{9.170}$$

As we have already mentioned, G_2 structures satisfying $dC = \lambda^*C$ for some constant λ are called **nearly parallel**, and so we have obtained the canonical nearly parallel G_2 structure on S^7.

9.6.3 Two Different Maps $S^7 \to \mathbb{C}P^3$ into the Canonical S^7

We now describe two distinct ways of mapping the seven-sphere viewed as the circle bundle over the projective twistor space $\mathbb{C}P^3$ into the previously desrcibed canonical S^7 with its canonical G_2 structure. Pulling back the canonical G_2 structure on S^7 via this map gives two different G_2 structures on S^7.

To describe both maps, we realise S^7 as the surface $\pi^\dagger\pi + \omega^\dagger\omega = 1$ in $\mathbb{C}^4$, where π and ω are both two-component spinors. We then need to describe a map from a pair (π and ω) into $\mathbb{R}^8$ with its canonical Spin(7) structure (9.166). There are two such natural maps that are of importance for us.

In the first case we set

$$\pi = \begin{pmatrix} x^1 + \mathrm{i}x^5 \\ x^2 + \mathrm{i}x^6 \end{pmatrix}, \qquad \omega = \begin{pmatrix} x^3 + \mathrm{i}x^7 \\ x^4 + \mathrm{i}x^8 \end{pmatrix}. \tag{9.171}$$

In other words, for this map we have

$$\begin{pmatrix} x^1 \\ x^2 \\ x^3 \\ x^4 \end{pmatrix} = \begin{pmatrix} \mathrm{Re}(\pi) \\ \mathrm{Re}(\omega) \end{pmatrix}, \qquad \begin{pmatrix} x^8 \\ x^6 \\ x^7 \\ x^8 \end{pmatrix} = \begin{pmatrix} \mathrm{Im}(\pi) \\ \mathrm{Im}(\omega) \end{pmatrix}. \tag{9.172}$$

In the second map we instead make the real and imaginary parts of π to be the coordinates of the first copy of $\mathbb{R}^4$, and those of ω of the second copy of $\mathbb{R}^4$ in $\mathbb{R}^8$. We put

$$\pi = \begin{pmatrix} -x^4 + \mathrm{i}x^3 \\ \mathrm{i}x^1 - x^2 \end{pmatrix}, \qquad \omega = \begin{pmatrix} -x^8 + \mathrm{i}x^7 \\ \mathrm{i}x^5 - x^6 \end{pmatrix}. \tag{9.173}$$

Here the specific complex linear combinations are motivated by the desire to have the quaternion describing the first copy of $\mathbb{R}^4$, i.e., the Euclidean matrix $\mathbf{x}$ given by (9.68) to be representable as $\mathbf{x} = (\pi, -\hat{\pi})$, and similarly for the quaternion for the second copy of $\mathbb{R}^4$.

It is clear that the seven-sphere $\pi^\dagger\pi + \omega^\dagger\omega = 1$ goes into the sphere $\sum_I (x^I)^2 = 1$ in both cases. Pulling back the 4-form (9.166) via these two different maps, and evaluating the insertion of $\partial/\partial r$ into Θ we get two different G_2 structures on the seven-sphere.

9.6.4 *First* Spin(7) *Structure on* $\mathbb{R}^8$

Let us compute Θ pulled back to (π and ω) $\in \mathbb{C}^4$ via the first map. We define

$$\omega := \frac{1}{2\mathrm{i}}\sum_{\alpha} d\pi^\dagger d\pi + d\omega^\dagger d\omega = dx^1dx^5 + dx^2dx^6 + dx^3dx^7 + dx^4dx^8,$$

as well as

$$\Omega := \frac{1}{4}d\pi^T\epsilon d\pi d\omega^T\epsilon d\omega = (dx^1 + \mathrm{i}dx^5)(dx^2 + \mathrm{i}dx^6)(dx^3 + \mathrm{i}dx^7)(dx^4 + \mathrm{i}dx^8).$$

In these coordinates, the 4-form (9.166) can be checked to be

$$\Theta = -\frac{1}{2}\omega\omega + \mathrm{Re}(\Omega). \tag{9.174}$$

For future reference, we also rewrite ω and Ω in spinor notations

$$\omega = -\frac{1}{2\mathrm{i}}(d\hat{\pi}^A d\pi_A + d\hat{\omega}^{A'}d\omega_{A'}), \qquad \Omega = \frac{1}{4}d\pi^A d\pi_A d\omega^{A'}d\omega_{A'}. \tag{9.175}$$

9.6.5 *Second* Spin(7) *Structure on* $\mathbb{R}^8$

To compute the second G_2 structure we parametrise

$$\mathbf{q} := \begin{pmatrix} -x^4 + \mathrm{i}x^3 & \mathrm{i}x^1 + x^2 \\ \mathrm{i}x^1 - x^2 & -x^4 - \mathrm{i}x^3 \end{pmatrix}, \quad \mathbf{q} := \begin{pmatrix} -x^8 + \mathrm{i}x^7 & \mathrm{i}x^5 + x^6 \\ \mathrm{i}x^5 - x^6 & -x^8 - \mathrm{i}x^7 \end{pmatrix},$$

so that $\mathbf{q} = (\pi, -\hat{\pi}), \mathbf{p} = (\omega, -\hat{\omega})$. We then have

$$\Sigma^i\tau^i := \Sigma^+ = \frac{1}{4}d\mathbf{q}^\dagger d\mathbf{q} = \frac{1}{4}\begin{pmatrix} d\pi^\dagger d\pi & -(d\pi^T\epsilon d\pi)^* \\ d\pi^T\epsilon d\pi & -d\pi^\dagger d\pi \end{pmatrix}, \tag{9.176}$$

$$\tilde{\Sigma}^i := \tau^i\Sigma^- = \frac{1}{4}d\mathbf{p}^\dagger d\mathbf{p} = \frac{1}{4}\begin{pmatrix} d\omega^\dagger d\omega & -(d\omega^T\epsilon d\omega)^* \\ d\omega^T\epsilon d\omega & -d\omega^\dagger d\omega \end{pmatrix}.$$

On the other hand, the 4-form (9.166) can be written as

$$\Theta = -2\mathrm{Tr}(-\frac{1}{6}\Sigma^+\Sigma^+ - \frac{1}{6}\Sigma^-\Sigma^- + \Sigma^+\Sigma^-). \tag{9.177}$$

Let us compute it in terms of π and ω coordinates. We have

$$\frac{1}{2}\mathrm{Tr}(d\mathbf{q}^\dagger d\mathbf{q}\, d\mathbf{p}^\dagger d\mathbf{p}) = d\pi^\dagger d\pi d\omega^\dagger d\omega - \mathrm{Re}(d\pi^T\epsilon d\pi(d\omega^T\epsilon d\omega)^*), \tag{9.178}$$

$$\frac{1}{2}\mathrm{Tr}(d\mathbf{q}^\dagger d\mathbf{q}\, d\mathbf{q}^\dagger d\mathbf{q}) = d\pi^\dagger d\pi d\pi^\dagger d\pi - d\pi^T\epsilon d\pi(d\pi^T\epsilon d\pi)^*.$$

The second result can be further simplified by noting that the second term is a multiple of the first. To see this, we first rewrite things in spinor notation. We have

$$d\pi^\dagger d\pi = -d\hat{\pi}^A d\pi_A, \quad d\pi^T\epsilon d\pi = -d\pi^A d\pi_A, \quad (d\pi^T\epsilon d\pi)^* = -d\hat{\pi}^A d\hat{\pi}_A.$$

Thus,

$$d\pi^T \epsilon d\pi (d\pi^T \epsilon d\pi)^* = d\pi^A d\pi_A d\hat{\pi}^B d\hat{\pi}_B = -d\hat{\pi}^B d\pi^A d\hat{\pi}_B d\pi_A. \tag{9.179}$$

On the other hand

$$(d\hat{\pi}^A d\pi^B - d\hat{\pi}^B d\pi^A) d\hat{\pi}_B d\pi_A = d\hat{\pi}^A d\pi_A d\hat{\pi}^B d\pi_B - d\hat{\pi}^B d\pi^A d\hat{\pi}_B d\pi_A.$$

But the left-hand side here is

$$(d\hat{\pi}^A d\pi^B - d\hat{\pi}^B d\pi^A) d\hat{\pi}_B d\pi_A = \epsilon^{BA} d\hat{\pi}^C d\pi_C d\hat{\pi}_B d\pi_A = -d\hat{\pi}^A d\pi_A d\hat{\pi}^B d\pi_B.$$

This gives

$$d\hat{\pi}^B d\pi^A d\hat{\pi}_B d\pi_A = 2 d\hat{\pi}^A d\pi_A d\hat{\pi}^B d\pi_B, \tag{9.180}$$

and thus

$$\frac{1}{2}\mathrm{Tr}(d\mathbf{q}^\dagger d\mathbf{q}\, d\mathbf{q}^\dagger d\mathbf{q}) = 3 d\hat{\pi}^A d\pi_A d\hat{\pi}^B d\pi_B. \tag{9.181}$$

This means that we can write the 4-form (9.166) as

$$4\tilde{\Theta} = \frac{1}{2}(d\pi^A d\hat{\pi}_A - d\omega^{A'} d\hat{\omega}_{A'})^2 + \mathrm{Re}(d\pi^A d\pi_A d\hat{\omega}^{A'} d\hat{\omega}_{A'}). \tag{9.182}$$

We now note that $\tilde{\Theta}$ is of the already familiar form (9.174)

$$\tilde{\Theta} = -\frac{1}{2}\tilde{\omega}\tilde{\omega} + \mathrm{Re}(\tilde{\Omega}) \tag{9.183}$$

with

$$\tilde{\omega} = \frac{1}{2\mathrm{i}}(d\hat{\pi}^A d\pi_A - d\hat{\omega}^{A'} d\omega_{A'}), \qquad \Omega = \frac{1}{4} d\pi^A d\pi_A d\hat{\omega}^{A'} d\hat{\omega}_{A'}. \tag{9.184}$$

This differs from (9.175) in the relative sign in the 2-form ω and the use of $\hat{\omega}$ rather than ω in Ω.

9.6.6 Two Different Almost Complex Structures on $\mathbb{R}^8$

It is now clear that the difference between the two Spin(7) structures (9.175), (9.184) stems from using two different almost complex structures on $\mathbb{R}^8$. Indeed, in both cases the 4-form on $\mathbb{R}^8$ is given by (minus half) the wedge product of the Kähler form squared plus the real part of the $(4,0)$ form. Both of these are fixed once the decomposition of $\mathbb{R}^8$ into $(1,0)$ and $(0,1)$ forms is given, which is equivalent to specifying an almost complex structure.

In the case of (9.175), the almost complex structure is the standard one on (π and ω) $\in \mathbb{C}^4$ that views π, ω as holomorphic coordinates. In the case of (9.184) the almost complex structure is instead the one with π and $\hat{\omega}$ as the holomorphic coordinates. Thus, the difference between these two cases is in the relative sign with which the two almost complex structures on $\mathbb{R}^4$ are put together. It can then be expected that upon taking the projectivisation (9.175) should give rise to the integrable (9.114) almost complex structure on $\mathbb{C}P^3$, while (9.184) will produce the non-integrable one (9.115). Calculations that follow will confirm this expectation.

9.6.7 First G_2 Structure on $S^7 \to \mathbb{C}P^3$

We now compute what (9.174) gives when we pass to the projective version of $\mathbb{C}^4$. Thus, we parametrise

$$Z^i = tz^i, \qquad Z^4 = t \tag{9.185}$$

to pass to the projective space $\mathbb{C}P^3$. In these coordinates the Kähler form becomes

$$\omega = \frac{\mathrm{i}}{2}\left(dtd\bar{t}(1+\sum_i |z^i|^2) + \bar{t}dt\sum_i z^i d\bar{z}^i - td\bar{t}\sum_i \bar{z}^i dz^i + |t|^2 \sum_i dz^i d\bar{z}^i\right).$$

We now parametrise

$$t = r\frac{e^{\mathrm{i}\psi}}{\sqrt{1+\sum_i |z^i|^2}}, \tag{9.186}$$

so that $r=1$ is the seven-sphere $\sum_\alpha |Z^\alpha|^2 = 1$. We have

$$\frac{dt}{t} = \frac{dr}{r} + \mathrm{i}d\psi - \frac{1}{2}\frac{\sum_i d|z^i|^2}{1+\sum_i |z^i|^2}. \tag{9.187}$$

A straightforward computation then gives

$$\frac{1}{r^2}\omega = \frac{dr}{r}(d\psi + a) + \omega_{\rm FS}, \tag{9.188}$$

where the $U(1)$ connection a is given by (9.93) and $\omega_{\rm FS}$ is the Kähler form for the Fubini–Study metric (9.90)

$$\omega_{\rm FS} = \frac{\mathrm{i}}{2}\frac{\sum_i dz^i dz^i (1+\sum_j |z^j|^2) - \sum_{i,j} \bar{z}^i z^j dz^i d\bar{z}^j}{(1+\sum_i |z^i|^2)^2}. \tag{9.189}$$

On the other hand, we have

$$\frac{1}{r^4}\Omega = \frac{e^{4\mathrm{i}\psi}}{(1+\sum_i |z^i|^2)^2}\left(\frac{dr}{r} + \mathrm{i}d\psi - \frac{1}{2}\frac{\sum_i d|z^i|^2}{1+\sum_i |z^i|^2}\right) dz^1 dz^2 dz^3.$$

Given that there is the wedge product with $dz^1dz^2dz^3$ here, we can rewrite the terms in the brackets as

$$\frac{1}{r^4}\Omega = \frac{e^{4\mathrm{i}\psi}}{(1+\sum_i |z^i|^2)^2}\left(\frac{dr}{r} + \mathrm{i}(d\psi + a)\right) dz^1 dz^2 dz^3, \tag{9.190}$$

where a is again the connection (9.93).

Combining these blocks into (9.174) and taking the interior product with the vector field $\partial/\partial r$ we get the sought 3-form on S^7 of unit radius $r = 1$

$$C_{S^7} = -(d\psi + a)\omega_{\rm FS} + {\rm Re}\frac{e^{4i\psi}dz^1dz^2dz^3}{(1+\sum_i |z^i|^2)^2}. \tag{9.191}$$

We can also restrict Θ to the seven-sphere and get the dual form

$$^*C_{S^7} = -\frac{1}{2}\omega_{\rm FS}\omega_{\rm FS} - (d\psi + a){\rm Im}\frac{e^{4i\psi}dz^1dz^2dz^3}{(1+\sum_i |z^i|^2)^2}. \tag{9.192}$$

It is instructive to compute dC explicitly and check that it is a multiple of $^*C$. To do this we need the following relation

$$da = 2\omega_{\rm FS}, \tag{9.193}$$

which in particular shows that $\omega_{\rm FS}$ is closed. We also have

$$d{\rm Re}\frac{e^{4i\psi}dz^1dz^2dz^3}{(1+\sum_i |z^i|^2)^2} = -4(d\psi + a){\rm Im}\frac{e^{4i\psi}dz^1dz^2dz^3}{(1+\sum_i |z^i|^2)^2}. \tag{9.194}$$

This immediately gives

$$dC = 4^*C. \tag{9.195}$$

9.6.8 Twistor Space Description

We now take the (9.174) and parametrise $\mathbb{C}^4$ as the total space of the $\mathbb{C}^2$ bundle over S^4 via $\pi = \mathbf{x}\omega$. Let us first compute the Kähler form in these coordinates. We have

$$\begin{aligned} d\pi^\dagger d\pi + d\omega^\dagger d\omega &= \omega^\dagger d\mathbf{x}^\dagger d\mathbf{x}\omega + d\omega^\dagger \mathbf{x}^\dagger d\mathbf{x}\omega + \omega^\dagger d\mathbf{x}^\dagger \mathbf{x} d\omega + (1+|\mathbf{x}|^2)d\omega^\dagger d\omega \\ &= (1+|\mathbf{x}|^2)\left[\left(d\omega^\dagger + \omega^\dagger\frac{d\mathbf{x}^\dagger\mathbf{x}}{1+|\mathbf{x}|^2}\right)\left(d\omega + \frac{\mathbf{x}^\dagger d\mathbf{x}}{1+|\mathbf{x}|^2}\omega\right)\right. \\ &\qquad \left. + \omega^\dagger\frac{d\mathbf{x}^\dagger d\mathbf{x}}{(1+|\mathbf{x}|^2)^2}\omega\right]. \end{aligned}$$

We then go to the projectivised version parametrising $\omega = t\eta$ with η as in (9.97) and

$$t = \frac{re^{i\psi}}{\sqrt{(1+|\mathbf{x}|^2)(1+|z|^2)}}. \tag{9.196}$$

This is the parametrisation that gives $\pi^\dagger\pi + \omega^\dagger\omega = r^2$.

The analog of (9.99) becomes

$$d\omega + \frac{\mathbf{x}^\dagger d\mathbf{x}}{1+|\mathbf{x}|^2}\omega = t\left(\left(\frac{dr}{r} + {\rm i}d\psi - \frac{d|z|^2}{2(1+|z|^2)}\right)\eta + D\eta\right). \tag{9.197}$$

Using this, after some algebra we get

$$\omega = \frac{1}{2{\rm i}}((d\pi^\dagger d\pi + d\omega^\dagger d\omega) = rdr(d\psi + a) + r^2\omega_{\mathbb{C}P^3},$$

where

$$\omega_{\mathbb{C}P^3} = \frac{1}{2\mathrm{i}}\left(\frac{\bar{\tau}\tau}{(1+|z|^2)^2} + \frac{\eta^\dagger \Sigma \eta}{1+|z|^2}\right) \tag{9.198}$$

is the Kähler form corresponding to the metric (9.110) and

$$a = \frac{\eta^\dagger D\eta - (D\eta)^\dagger \eta}{2\mathrm{i}(1+|z|^2)} \tag{9.199}$$

is the already familiar U(1) connection.

Let us now compute the (4, 0) form Ω. Since $d\omega^T \epsilon d\omega$ is proportional to $dtdz$ and this is wedged with $d\pi^T \epsilon d\pi$, only the terms involving $d\mathbf{x}$ must be kept in $d\pi^T \epsilon d\pi$. In more details, we have

$$d\omega^T \epsilon d\omega = 2t dt \eta^T \epsilon d\eta \tag{9.200}$$

and

$$d\pi^T \epsilon d\pi d\omega^T \epsilon d\omega = \omega^T d\mathbf{x}^T \epsilon d\mathbf{x} \omega 2t dt \eta^T \epsilon d\eta. \tag{9.201}$$

Using $\mathbf{x}^T \epsilon = \epsilon \mathbf{x}^\dagger$, as well as $d\mathbf{x}^\dagger d\mathbf{x} = (1+|\mathbf{x}|^2)^2 \Sigma$ we get

$$\Omega = \frac{dt}{2t}\frac{r^4 e^{4\mathrm{i}\psi}}{(1+|z|^2)^2}\eta^T \epsilon d\eta\, \eta^T \epsilon \Sigma \eta, \tag{9.202}$$

with

$$\frac{dt}{t} = \frac{dr}{r} + \mathrm{i}d\psi - \frac{d|\mathbf{x}|^2}{2(1+|\mathbf{x}|^2)} - \frac{d|z|^2}{2(1+|z|^2)}. \tag{9.203}$$

We now note that we can replace $d\eta$ with $D\eta$ in (9.202). Indeed, we have

$$\eta^T \epsilon \Sigma \eta = \eta^T \epsilon \frac{d\mathbf{x}^\dagger d\mathbf{x}}{(1+|\mathbf{x}|^2)^2}\eta = \frac{(d\mathbf{x}\eta)^T \epsilon d\mathbf{x}\eta}{(1+|\mathbf{x}|^2)^2}, \tag{9.204}$$

where we have used $\epsilon \mathbf{x}^\dagger = \mathbf{x}^T \epsilon$. We also have

$$\eta^T \epsilon A \eta = \eta^T \epsilon \frac{\mathbf{x}^\dagger d\mathbf{x} - d\mathbf{x}^\dagger \mathbf{x}}{2(1+|\mathbf{x}|^2)}\eta = \frac{(\mathbf{x}\eta)^T \epsilon d\mathbf{x}\eta}{1+|\mathbf{x}|^2}. \tag{9.205}$$

Thus, the 1-form $\eta^T \epsilon A \eta$ is a linear combination of two 1-forms $d\mathbf{x}\eta \equiv d\mathbf{x}_A{}^{A'} \eta_{A'}$, while the 2-form $\eta^T \epsilon \Sigma \eta$ is the wedge product of these two 1-forms. This means that the wedge product of $\eta^T \epsilon A \eta$ with $\eta^T \epsilon \Sigma \eta$ vanishes

$$\eta^T \epsilon A \eta \wedge \eta^T \epsilon \Sigma \eta = 0 \tag{9.206}$$

and so we can extend the exterior derivative in (9.202) into a covariant exterior derivative for free. Thus, we have

$$\Omega = \frac{dt}{2t}\frac{r^4 e^{4\mathrm{i}\psi}}{(1+|z|^2)^2}\eta^T \epsilon D\eta\, \eta^T \epsilon \Sigma \eta. \tag{9.207}$$

Projecting Θ on $\partial/\partial r$ vector field we get $C = i_{\partial/\partial r}\Theta$

$$C = -(d\psi + a)\omega_{\mathbb{C}P^3} + \frac{1}{2(1+|z|^2)^2}\mathrm{Re}\left(e^{4\mathrm{i}\psi}\eta^T\epsilon D\eta\,\eta^T\epsilon\Sigma\eta\right). \tag{9.208}$$

We note that the 3-form (9.114) that corresponds to an integrable almost complex structure on the twistor space has made its appearance here. Indeed, the real part of $e^{4\mathrm{i}\psi}\Omega^{3,0}$ appears in the second term in C.

To get the dual 4-form, we note that we can rewrite

$$\left(\mathrm{i}d\psi - \frac{d|\mathbf{x}|^2}{2(1+|\mathbf{x}|^2)} - \frac{d|z|^2}{2(1+|z|^2)}\right)\eta^T\epsilon D\eta\,\eta^T\epsilon\Sigma\eta = \mathrm{i}(d\psi + a)\eta^T\epsilon D\eta\,\eta^T\epsilon\Sigma\eta,$$

where a is the connection (9.199). To see this, let us spell out the connection a. We have

$$a = \frac{\eta^\dagger d\eta - d\eta^\dagger\eta + 2\eta^\dagger A\eta}{2\mathrm{i}\eta^\dagger\eta}, \tag{9.209}$$

with

$$\eta^\dagger A\eta = \frac{(\mathbf{x}\eta)^\dagger d\mathbf{x}\eta - (d\mathbf{x}\eta)^\dagger\mathbf{x}\eta}{2(1+|\mathbf{x}|^2)}. \tag{9.210}$$

When we wedge this with $\eta^T\epsilon\Sigma\eta$ the term with $d\mathbf{x}\eta$ does not contribute. Therefore, we can flip the sign in front of this term and write

$$\left(-\frac{d|\mathbf{x}|^2}{2(1+|\mathbf{x}|^2)}\right)\eta^T\epsilon\Sigma\eta = \frac{\eta^\dagger A\eta}{\eta^\dagger\eta}\eta^T\epsilon\Sigma\eta. \tag{9.211}$$

Similarly, we have

$$\left(-\frac{d|z|^2}{2(1+|z|^2)}\right)\eta^T\epsilon d\eta = \frac{\eta^\dagger d\eta - d\eta^\dagger\eta}{2\eta^\dagger\eta}\eta^T\epsilon d\eta \tag{9.212}$$

because only $d\bar{z}$ term contributes when multiplied by $dz \sim \eta^T\epsilon d\eta$.

These considerations shows that the dual 4-form given by the restriction of Θ to $r = 1$ is given by

$$^*C = -\frac{1}{2}\omega_{\mathbb{C}P^3}\omega_{\mathbb{C}P^3} + (d\psi + a)\mathrm{Im}\left(\frac{e^{4\mathrm{i}\psi}\eta^T\epsilon D\eta\,\eta^T\epsilon\Sigma\eta}{2(1+|z|^2)^2}\right). \tag{9.213}$$

9.6.9 Calculation

It is instructive to compute the exterior derivative of the 3-form (9.208) to check that the form is nearly parallel. One first checks that

$$da = 2\omega_{\mathbb{C}P^3} \tag{9.214}$$

as expected.

We then compute

$$d(\eta^T\epsilon D\eta\,\eta^T\epsilon\Sigma\eta) = (D\eta)^T\epsilon D\eta\,\eta^T\epsilon\Sigma\eta - \eta^T\epsilon D\eta\,(D\eta)^T\epsilon\Sigma\eta - \eta^T\epsilon D\eta\,\eta^T\epsilon\Sigma D\eta,$$

where we have used that $DD\eta = F\eta$ and $F = \Sigma$, as well as the fact that $\Sigma\Sigma$ contracted with four copies of spinor η vanishes. This is because we have $\Sigma^{A'B'}\Sigma^{C'D'} \sim \epsilon^{A'(C'}\epsilon^{D')B'}$, and copies of the spinor metric cause the spinor η to contract with itself, which vanishes. Finally, we have used $D\Sigma = 0$.

The previous expression can be simplified by rewriting it in spinor notations. We have

$$\begin{aligned} &D\eta^{A'}D\eta_{A'}\eta^{B'}\Sigma_{B'}{}^{C'}\eta_{C'} - \eta^{A'}D\eta_{A'}D\eta^{B'}\Sigma_{B'}{}^{C'}\eta_{C'} - \eta^{A'}D\eta_{A'}\eta^{B'}\Sigma_{B'}{}^{C'}D\eta_{C'} \\ &= D\eta^{A'}D\eta_{A'}\eta^{B'}\Sigma_{B'}{}^{C'}\eta_{C'} - 2\eta^{A'}D\eta_{A'}D\eta^{B'}\Sigma_{B'}{}^{C'}\eta_{C'}, \end{aligned}$$

and then using

$$D\eta_{A'}D\eta^{B'} = -\frac{1}{2}\epsilon_{A'}{}^{B'}D\eta^{C'}D\eta_{C'}, \tag{9.215}$$

which holds due to antisymmetry in $A'B'$, finally gives

$$d(\eta^T\epsilon D\eta\,\eta^T\epsilon\Sigma\eta) = 2(D\eta)^T\epsilon D\eta\,\eta^T\epsilon\Sigma\eta. \tag{9.216}$$

We then have

$$\begin{aligned} &d\frac{1}{2(1+|z|^2)^2}\mathrm{Re}\left(e^{4i\psi}\eta^T\epsilon D\eta\,\eta^T\epsilon\Sigma\eta\right) \\ &= \frac{1}{(1+|z|^2)^2}\mathrm{Re}\left(e^{4i\psi}\left(2id\psi - \frac{d|z|^2}{1+|z|^2}\right)\eta^T\epsilon D\eta\,\eta^T\epsilon\Sigma\eta + e^{4i\psi}(D\eta)^T\epsilon D\eta\,\eta^T\epsilon\Sigma\eta\right). \end{aligned}$$

A spinor index notation calculation in Section 9.7 shows that the terms in brackets can be rewritten as

$$e^{4i\psi}2i(d\psi + a)\eta^T\epsilon D\eta\,\eta^T\epsilon\Sigma\eta, \tag{9.217}$$

and so we have

$$d\frac{1}{2(1+|z|^2)^2}\mathrm{Re}\left(e^{4i\psi}\eta^T\epsilon D\eta\,\eta^T\epsilon\Sigma\eta\right) = 4(d\psi + a)\mathrm{Im}\left(\frac{e^{4i\psi}\eta^T\epsilon D\eta\,\eta^T\epsilon\Sigma\eta}{2(1+|z|^2)^2}\right).$$

This shows that we indeed have $dC = 4{}^*C$ with ${}^*C$ given by (9.213).

9.6.10 Second G_2 Structure on S^7

We now compute the G_2 structure on S^7 that is associated with the Hopf fibration, in Hopf fibration coordinates. We first compute the 4-form Θ (9.184), in Hopf coordinates. The arising G_2 structure is more involved than the first one we studied, in particular due to the fact that it is not compatible with the natural (integrable) almost complex structure on $\mathbb{C}P^3$. We present the computation for completeness. We will not compute it here in the twistor space coordinates, as the expression can be expected to be significantly more complicated than (9.208).

We need to change our parametrisation of the Hopf projection (9.119) slightly, to take into account the possibility of changing the radius of S^7. Thus, we take

$$\mathbf{q} = r\frac{\mathbf{x}h}{\sqrt{1+|\mathbf{x}|^2}}, \qquad \mathbf{p} = r\frac{h}{\sqrt{1+|\mathbf{x}|^2}}, \tag{9.218}$$

so that r^2 is the radius squared of S^7. In these coordinates we have

$$\begin{aligned} d\mathbf{p} &= \frac{h}{(1+|\mathbf{x}|^2)^{1/2}}dr - \frac{r}{2}\frac{1}{(1+|\mathbf{x}|^2)^{3/2}}d|\mathbf{x}|^2 h + \frac{r}{(1+|\mathbf{x}|^2)^{1/2}}dh, \\ d\mathbf{q} &= \frac{\mathbf{x}h}{(1+|\mathbf{x}|^2)^{1/2}}dr - \frac{r}{2}\frac{1}{(1+|\mathbf{x}|^2)^{3/2}}d|\mathbf{x}|^2\mathbf{x}h + \frac{r}{(1+|\mathbf{x}|^2)^{1/2}}(d\mathbf{x}h + \mathbf{x}dh). \end{aligned}$$

From this we get

$$4\Sigma^+ = \frac{1}{(1+|\mathbf{x}|^2)}dr^2\mathbf{m} - \frac{r^2}{(1+|\mathbf{x}|^2)^2}d|\mathbf{x}|^2\mathbf{m} - \frac{r^2}{(1+|\mathbf{x}|^2)}\mathbf{m}\mathbf{m},$$

where $\mathbf{m} = h^{-1}dh$ and

$$\begin{aligned} 4\Sigma^- &= dr^2 h^{-1}\mathbf{A}h + \frac{|\mathbf{x}|^2 dr^2}{(1+|\mathbf{x}|^2)}\mathbf{m} - \frac{r^2 d|\mathbf{x}|^2}{(1+|\mathbf{x}|^2)}h^{-1}\mathbf{A}h - \frac{r^2|\mathbf{x}|^2}{(1+|\mathbf{x}|^2)}\mathbf{m}\mathbf{m} \\ &+ \frac{r^2}{(1+|\mathbf{x}|^2)^2}d|\mathbf{x}|^2\mathbf{m} + r^2(1+|\mathbf{x}|^2)h^{-1}\Sigma h - r^2(h^{-1}\mathbf{A}h\mathbf{m} + \mathbf{m}h^{-1}\mathbf{A}h), \end{aligned}$$

where $\mathbf{A}$ is given by (9.100) and Σ is as in (9.130). We can simplify the result for Σ^- somewhat by using the relation

$$\frac{1}{(1+|\mathbf{x}|^2)}d|\mathbf{x}|^2\mathbf{A} = \mathbf{A}\mathbf{A} + |\mathbf{x}|^2\Sigma. \tag{9.219}$$

The last term here cancels a part of the second term on the second line in the expression for Σ^- and further allows the connections 1-forms $\mathbf{A}$ and $\mathbf{m}$ to be combined into the connection W; see (9.127). We get

$$4\Sigma^- = dr^2 W + r^2(S - WW) - 4\Sigma^+, \tag{9.220}$$

where we have have introduced a convenient notation

$$S := h^{-1}\Sigma h \tag{9.221}$$

for the lift of Σ to the total space of the bundle. As a check of the previous computations, it is not hard to check that both forms $\Sigma^\pm$ are closed

$$d\Sigma^\pm = 0. \tag{9.222}$$

We now compute the contraction of Θ given by (9.177) with the vector field $\partial/\partial r$, and set $r = 1$. This gives the sought-after G_2 structure on S^7 in Hopf coordinates. To compute it, we first rewrite (9.177) as

$$\Theta := \frac{1}{3}\mathrm{Tr}(\Sigma^+ + \Sigma^-)^2 - \frac{8}{3}\mathrm{Tr}(\Sigma^+\Sigma^-). \tag{9.223}$$

The contraction with the first term is very easy to compute. We have

$$\frac{16}{3r^2} i_{\partial/\partial r^2} \mathrm{Tr}(\Sigma^+ + \Sigma^-)^2 = -\frac{2}{3} \mathrm{Tr}\,(W^3 - WS)\,. \tag{9.224}$$

The other term gives

$$-\frac{8 \cdot 16}{3r^2} i_{\partial/\partial r^2} \mathrm{Tr}(\Sigma^+ \wedge \Sigma^-) = \mathrm{Tr}\left(\frac{8A\mathbf{m}^2}{(1+|\mathbf{x}|^2)} - \frac{8}{3}\mathbf{m}S + \frac{16}{3}\frac{|\mathbf{x}|^2}{(1+|\mathbf{x}|^2)^2}\mathbf{m}^3 \right),$$

where $A = h^{-1}\mathbf{A}h$ is the lift of the connection on the base to the total space of the bundle, and terms containing $d|x|^2$ drop out because $\mathrm{Tr}(A\mathbf{m}) = 0$ and $\mathrm{Tr}(\mathbf{mm}) = 0$ as a contraction of a symmetric and antisymmetric tensors. We now set $r = 1$ and rewrite everything in terms of W, A, thus eliminating $\mathbf{m}$. We have for our sought-after 3-form

$$\begin{aligned} 8C = \mathrm{Tr}\Bigg[& -\frac{2}{3}W^3 + \frac{2}{3}WS + \frac{8}{(1+|\mathbf{x}|^2)}A(W-A)^2 \\ & -\frac{8}{3}(W-A)S + \frac{16}{3}\frac{|\mathbf{x}|^2}{(1+|\mathbf{x}|^2)^2}(W-A)^3 \Bigg]. \end{aligned} \tag{9.225}$$

9.6.11 Simplification

The found expression (9.225) for the 3-form on S^7 can be simplified further. To this end, we will first write a different 3-form on S^7, obtained by taking the frame 1-forms corresponding to the metric (9.128), and writing the canonical 3-form (9.145). This gives

$$8C_0 = -2\,\mathrm{Tr}\left(\frac{1}{3}W^3 + WS\right). \tag{9.226}$$

The factor of eight on the left-hand side is due to the frame for the metric (9.128) containing an additional factor of half as compared to the frame $W^i, i = 1,2,3$ for S^3 and e^1, e^2, e^3, and e^4, which is the frame for the metric (9.129). The corresponding 4-form is

$$16^*C_0 = -2\,\mathrm{Tr}\left(\frac{1}{6}S^2 + W^2S\right). \tag{9.227}$$

These are, however, not the forms we are after, because dC_0 is not proportional to $^*C_0$. Indeed, we have

$$\begin{aligned} 8dC_0 &= -2\mathrm{Tr}\,((S-W^2)W^2 + (S-W^2)S - W(SW - WS)) \\ &= -2\mathrm{Tr}\,(SS + 2W^2S)\,. \end{aligned} \tag{9.228}$$

Here we have used the cyclicity of the trace, as well as

$$dW = S - WW, \qquad dS = SW - WS, \tag{9.229}$$

which are the Plebanski field equations for objects on S^4, lifted to the total space of the bundle. Both can be easily checked using the definitions of these objects. It is clear that the relative coefficient in the last expression in (9.228) is not correct for the equation $dC_0 \sim {}^*C_0$ to be satisfied.

The 3-form on S^7 for which this equation is satisfied has been found in the previous section. Our goal is now to relate C to C_0 via (9.164). By symmetry, the only 1-form that can be used for this purpose is a multiple of the form

$$\alpha = \frac{d|\mathbf{x}|^2}{(1+|\mathbf{x}|^2)} = \frac{2x_\mu dx^\mu}{(1+|\mathbf{x}|^2)}. \tag{9.230}$$

With the relevant metric being the quarter of (9.129), the dual vector field is equal to

$$\alpha^\sharp = (1+|\mathbf{x}|^2) 2x_\mu \frac{\partial}{\partial x_\mu}. \tag{9.231}$$

We need to compute the insertions of $\alpha^\sharp$ into $\mathbf{A}$ and Σ. To this end, let us write $\mathbf{x} = x^\mu \sigma_\mu$, where σ_μ are the 2×2 matrices that can be read off from (9.68). We then have

$$\mathbf{A} = \frac{1}{2(1+|\mathbf{x}|^2)} x^\mu dx^\nu \left(\sigma_\mu^\dagger \sigma_\nu - \sigma_\nu^\dagger \sigma_\mu\right), \tag{9.232}$$

and

$$\Sigma = \frac{1}{2(1+|\mathbf{x}|^2)^2} dx^\mu dx^\nu \left(\sigma_\mu^\dagger \sigma_\nu - \sigma_\nu^\dagger \sigma_\mu\right). \tag{9.233}$$

This immediately gives

$$i_{\alpha^\sharp} \Sigma = 4\mathbf{A}, \qquad i_{\alpha^\sharp} \mathbf{A} = 0. \tag{9.234}$$

This means that

$$c\alpha \wedge i_{c\alpha^\sharp} 8C_0 = 8c^2 \frac{d|\mathbf{x}|^2}{(1+|\mathbf{x}|^2)} \mathrm{Tr}(WA) = -8c^2\, \mathrm{Tr}(W(A^2 + |\mathbf{x}|^2 S),$$

where c is some coefficient to be worked out later, and we used (9.137) to write the second equality. We also have

$$i_{c\alpha^\sharp} 16{}^*C_0 = -8c\, \mathrm{Tr}\left(\frac{1}{3} AS + W^2 A\right).$$

The formula (9.164) then gives

$$\begin{aligned} 8\tilde{C}_0 = \mathrm{Tr}\Bigg[&\left(-\frac{2}{3}W^3 - 2WS\right)(1 - 8|\mathbf{x}|^2 c^2) \\ &- 16c^2 W(AA + |\mathbf{x}|^2 S) - \sqrt{1 - 4c^2|\mathbf{x}|^2}\, 4c\left(\frac{2}{3}AS + 2W^2 A\right)\Bigg]. \end{aligned} \tag{9.235}$$

This is to be compared to (9.225). We first note that the WS terms work out correctly. We now compare the W^3 terms. Demanding equality gives

$$-\frac{2}{3}(1-8|\mathbf{x}|^2c^2) = -\frac{2}{3}+\frac{16}{3}\frac{|\mathbf{x}|^2}{(1+|\mathbf{x}|^2)^2} \quad \Rightarrow \quad |c| = \frac{1}{1+|\mathbf{x}|^2}. \tag{9.236}$$

To match things further we note that there is no A^3 term in (9.235), while there are such terms in (9.225). This can be matched by once again using the identity (9.137). We can multiply this identity by A and take the trace. This gives

$$0 = \mathrm{Tr}\left(\frac{d|\mathbf{x}|^2}{(1+|\mathbf{x}|^2)}AA\right) = \mathrm{Tr}\left(A^3+|\mathbf{x}|^2 SA\right). \tag{9.237}$$

Using this identity, one can check that all the terms match provided we choose the minus sign for c in (9.236). Thus, finally, the 1-form to be used to write (9.225) in the form (9.164) is

$$-\frac{d|\mathbf{x}|^2}{(1+|\mathbf{x}|^2)^2} = d\phi, \tag{9.238}$$

where ϕ is the conformal factor

$$\phi := \frac{1}{(1+|\mathbf{x}|^2)}. \tag{9.239}$$

All in all, the sought G_2 structure on S^7 in Hopf coordinates is

$$\begin{aligned} 8C = \mathrm{Tr}\Big[&-\frac{2}{3}(1-8|\mathbf{x}|^2\phi^2)W^3 + 8(1-|\mathbf{x}|^2)\phi^2 W^2 A \\ &-16\phi^2 WA^2 - 2WS + \frac{8}{3}(1-|\mathbf{x}|^2)\phi^2 AS\Big]. \end{aligned} \tag{9.240}$$

Only at the origin $\mathbf{x}=0$ where $A=0$ this matches the 'canonical' form (9.226). At a general point the metric that this 3-form defines is the same as the metric of C_0 given by (9.226), but the forms do not coincide. They are related via (9.164), with the 1-form that needs to be used given by $d\phi$.

9.7 3-Form Version of the Twistor Construction

The purpose of this section is to put everything we have learned together and describe a generalisation of G_2 structure (9.208) on the circle bundle over $\mathbb{C}P^3$ to the circle bundle over the projective twistor space of a general gravitational instanton M. We will then see that imposing the nearly parallel condition (9.158) on the corresponding G_2 structure is equivalent to the condition of integrability of the corresponding almost complex structure.

9.7.1 G_2 Structure on the Circle Bundle Over the Twistor Space

We start with the G_2 structure inspired by (9.208), but with some changes. Thus, we set

$$C = -(d\psi + a)\omega + \mathrm{Re}(e^{4i\psi}\Omega), \tag{9.241}$$

where

$$\Omega := \frac{1}{2(\eta^\dagger\eta)^2}\eta^T\epsilon D\eta\,\eta^T\epsilon\Sigma\eta, \tag{9.242}$$

and

$$\omega := \frac{1}{2\mathrm{i}}\left(\frac{\bar\tau\tau}{(\eta^\dagger\eta)^2} + \frac{\eta^\dagger F\eta}{\eta^\dagger\eta}\right), \quad a := \frac{\eta^\dagger D\eta - (D\eta)^\dagger\eta}{2\mathrm{i}(\eta^\dagger\eta)}, \tag{9.243}$$

with $\tau = \eta^T\epsilon D\eta$ are all objects on the projective twistor space. Thus, C is a 3-form on a circle bundle over the projective twistor space of a general Riemannian 4-manifold, with ψ being the coordinate on S^1, while $z \in \mathbb{C}P^1$ is the coordinate on the fibres of the projective twistor space. The covariant derivative D is with respect to an SU(2) connection on M and F is its curvature so that $DD\eta = F\eta$. The object Σ is a 2-form with values in the space of anti-Hermitian 2×2 matrices, which is assumed to be metric in the sense that the simplicity condition (5.160) is satisfied. It is also assumed that the connection A is torsion-free

$$D\Sigma = 0. \tag{9.244}$$

We now want to impose the nearly parallel condition on the 3-form C given by (9.241), and show that it holds provided $F = \Sigma$, i.e., provided the 4D metric is ASD Einstein. This generalises the previous twistor space of S^4 construction to an arbitrary gravitational instanton. The calculations we have to do are similar to those previously encountered, except that we are no longer allowed to use arguments based on the explicit form of A.

9.7.2 Verification of $da = 2\omega$

The real 2-form ω in (9.243) is defined so that $da = 2\omega$, which in particular implies that ω is closed. Let us verify this relation explicitly. It is easier to do this calculation by changing to spinor index notations. We have

$$a = \frac{\hat\eta^{A'}D\eta_{A'} - D\hat\eta^{A'}\eta_{A'}}{2\mathrm{i}[\hat\eta\eta]}, \tag{9.245}$$

where $[\hat\eta\eta] := \hat\eta^{A'}\eta_{A'}$. Therefore,

$$\begin{aligned} da = {}& \frac{D\hat\eta^{A'}D\eta_{A'} + \hat\eta^{A'}F_{A'}{}^{B'}\eta_{B'}}{\mathrm{i}[\hat\eta\eta]} \\ & - \frac{1}{2\mathrm{i}[\hat\eta\eta]^2}(D\hat\eta^{A'}\eta_{A'} + \hat\eta^{A'}D\eta_{A'})(\hat\eta^{B'}D\eta_{B'} - D\hat\eta^{B'}\eta_{B'}). \end{aligned} \tag{9.246}$$

The second term is simplified by noticing that the terms containing $D\hat{\eta}^{A'} D\hat{\eta}^{B'}$ and $D\eta_{A'} D\eta_{B'}$ are $A'B'$ antisymmetric and hence vanish as causing the contraction of a spinor with itself. The remaining terms give

$$da = \frac{1}{\mathrm{i}[\hat{\eta}\eta]^2}(D\hat{\eta}^{A'}\hat{\eta}^{B'}(D\eta_{A'}\eta_{B'} - \eta_{A'}D\eta_{B'})) + \frac{\hat{\eta}^{A'}F_{A'}{}^{B'}\eta_{B'}}{\mathrm{i}[\hat{\eta}\eta]} \tag{9.247}$$
$$= \frac{1}{\mathrm{i}[\hat{\eta}\eta]^2}\hat{\eta}^{A'}D\hat{\eta}_{A'}\eta^{B'}D\eta_{B'} + \frac{\hat{\eta}^{A'}F_{A'}{}^{B'}\eta_{B'}}{\mathrm{i}[\hat{\eta}\eta]} = 2\omega,$$

where we used $\tau = -\eta^{A'}D\eta_{A'}, \bar{\tau} = -\hat{\eta}^{A'}D\hat{\eta}_{A'}$ to recognise the 2-form ω.

9.7.3 Computation of dC

The calculation of the exterior derivative of the remaining terms in C proceeds as in 9.6.9, with some changes related to inability to use the explicit form of A. We carry it out using the spinor notations. We have

$$d(\eta^T\epsilon D\eta\,\eta^T\epsilon\Sigma\eta) = d(\eta^{A'}D\eta_{A'}\eta^{B'}\Sigma_{B'}{}^{C'}\eta_{C'})$$
$$= D\eta^{A'}D\eta_{A'}\eta^{B'}\Sigma_{B'}{}^{C'}\eta_{C'} + \eta^{A'}F_{A'}{}^{B'}\eta_{B'}\eta^{C'}\Sigma_{C'}{}^{D'}\eta_{D'}$$
$$- 2\eta^{A'}D\eta_{A'}D\eta^{B'}\Sigma_{B'}{}^{C'}\eta_{C'},$$

where we have used (9.244). As before, the first and the third term here are actually equal, and so we get

$$2D\eta^{A'}D\eta_{A'}\eta^{B'}\Sigma_{B'}{}^{C'}\eta_{C'} + \eta^{A'}F_{A'}{}^{B'}\eta_{B'}\eta^{C'}\Sigma_{C'}{}^{D'}\eta_{D'}.$$

We then get

$$d\frac{1}{2(\eta^\dagger\eta)^2}\mathrm{Re}\,(e^{4\mathrm{i}\psi}\eta^T\epsilon D\eta\,\eta^T\epsilon\Sigma\eta)$$
$$= \frac{1}{[\hat{\eta}\eta]^2}\mathrm{Re}\left[e^{4\mathrm{i}\psi}\left(2\mathrm{i}d\psi - \frac{D\hat{\eta}^{E'}\eta_{E'} + \hat{\eta}^{E'}D\eta_{E'}}{\hat{\eta}^{F'}\eta_{F'}}\right)\eta^{A'}D\eta_{A'}\eta^{B'}\Sigma_{B'}{}^{C'}\eta_{C'}\right.$$
$$\left. + e^{4\mathrm{i}\psi}D\eta^{A'}D\eta_{A'}\eta^{B'}\Sigma_{B'}{}^{C'}\eta_{C'} + \frac{1}{2}e^{4\mathrm{i}\psi}\eta^{A'}F_{A'}{}^{B'}\eta_{B'}\eta^{C'}\Sigma_{C'}{}^{D'}\eta_{D'}\right].$$

The terms in brackets in the second line can be simplified by using antisymmetry in $E'A'$ in the last term. We then have

$$\hat{\eta}^{E'}D\eta_{E'}\eta^{A'}D\eta_{A'} = \frac{1}{2}D\eta^{A'}D\eta_{A'}\hat{\eta}^{B'}\eta_{B'}. \tag{9.248}$$

This makes the last term in the second line a multiple of the first term in the third line. This allows to rewrite the terms in brackets in the second line in terms of the connection (9.245). It is clear that to get this combination we just need to flip the sign in front of the last term in brackets in the second line. But adding the term in the third line does exactly that. So, we have

$$dRe(\Omega) = -4(d\psi + a)\text{Im}(\Omega) + \text{Re}\left(\frac{e^{4i\psi}}{2(\eta^\dagger\eta)^2}\eta^T\epsilon F\eta\,\eta^T\epsilon\Sigma\eta\right).$$

9.7.4 Imposing the Nearly Parallel Condition

Combining the results of previous calculations, we get

$$dC = 4\left(-\frac{1}{2}\omega\omega - (d\psi + a)\text{Im}(\Omega)\right) + \text{Re}\left(\frac{e^{4i\psi}}{2(\eta^\dagger\eta)^2}\eta^T\epsilon F\eta\,\eta^T\epsilon\Sigma\eta\right). \quad (9.249)$$

It is clear that the only chance of having the nearly parallel condition $dC = 4{}^*C$ satisfied is if $\eta^T\epsilon F\eta\,\eta^T\epsilon\Sigma\eta = 0$ for any η. This will be the case if the 4D metric is a gravitational instanton $F = \Sigma$. Then the metric condition on Σ reads $\Sigma^{A'B'}\Sigma^{C'D'} \sim \epsilon^{A'(C'}\epsilon^{D')B'}$, which then makes at least one pair of three copies of the unhatted spinor η contract.

We also note that the instanton condition $F = \Sigma$ also guarantees

$$\omega \wedge \Omega = 0, \quad (9.250)$$

thus allowing to interpret ω as a $(1,1)$ form for the almost complex structure defined by the decomposable 3-form Ω as a $(3,0)$ form. Indeed, the term in ω containing $\bar{\tau}\tau$ gives zero when wedged with Ω because the latter contains a factor of τ. The $\eta^\dagger F\eta$ term in ω gives zero when wedged with Ω in view of $F = \Sigma$.

Once F is set to be Σ the calculation of the dual form ${}^*C$ is simple, because the form (9.241) is given in the canonical form (9.157). The dual 4-form is then given by

$$C^* = -\frac{1}{2}\omega\omega - (d\psi + a)\text{Im}(\Omega), \quad (9.251)$$

which shows that the nearly parallel condition is satisfied for the 3-form (9.241) on the circle bundle over the twistor space of an arbitrary gravitational instanton.

It is possible, and in fact suggested by the result in Herfray et al. (2016a), that the previous construction can be made even more general and that the nearly parallel condition imposed on an appropriate generalisation of (9.241) can be equivalent to Einstein-like relation between F and Σ on M. We will, however, refrain from attempting to demonstrate this here.

9.7.5 Integrability

Let us also show why the nearly parallel condition imposed on the 3-form (9.241) implies integrability of the almost complex structure defined by declaring the decomposable 3-form Ω to be the $(3,0)$ form. With the dual 4-form ${}^*C$ being given by (9.251), the nearly parallel condition for C given by (9.241) is equivalent to two differential equations

$$da = 2\omega, \qquad d\Omega = 4ia \wedge \Omega, \quad (9.252)$$

where we indicated the wedge product explicitly. We note that these two equations together imply $\omega \wedge \Omega = 0$. Indeed, this follows by taking the exterior derivative of the second equation and using the first to replace da with ω.

We now shows that the second of the equations in (9.252) is sufficient to guarantee that the almost complex structure defined by Ω is integrable. Indeed, a 1-form θ is $(1,0)$ if and only if

$$\theta \wedge \Omega = 0. \tag{9.253}$$

Let us now take the exterior derivative of this equation, using the second equation in (9.252)

$$d\theta \wedge \Omega + 4i\theta \wedge a \wedge \Omega = 0. \tag{9.254}$$

But then by assumption θ is $(1,0)$, and so the second term vanishes and we get

$$d\theta \wedge \Omega = 0. \tag{9.255}$$

This implies that $d\theta$ does not have a $(0,2)$ part, which in turn implies that the almost complex structure is integrable, by Newlander–Nirenberg theorem.

The fact that $\omega \wedge \Omega = 0$ then implies that ω is a $(1,1)$ form, and the first equation in (9.252) means that it is closed. The data ω, Ω satisfying (9.252) then define a Kähler metric. The metric arises as $g(X,Y) = \omega(X,JY)$, where J is the complex structure defined by Ω. This metric can be shown to coincide with the one induced by (9.146) on the $\psi = const$ slices. Moreover, this metric can be shown to be Einstein with positive scalar curvature. These aspects of the geometry under discussion are explained in, e.g., Sparks (2011).

10
Concluding Remarks

Our journey took us from the usual formalism that views general relativity (GR) as a dynamical theory of Riemannian geometry of metrics through a sequence of formalisms based on connections and differential forms to more exotic 6D and 7D constructions. It is now time to attempt to summarise what has been learned.

In all formalisms related to Cartan's tetrads, gravity becomes very similar to the Yang–Mills gauge theory. The geometric structures that make this possible are essentially invisible in the usual metric formulation. But gravity is different from the Yang–Mills theory. From the geometric point of view the main difference is presence in gravity of an object that solders the geometry of the manifold to the geometry of whatever abstract bundle that is used. This geometric object is different in different formalisms, see Table 10.1.

Thus, in all these descriptions there is a geometric object that ties the geometry of an abstract fibre bundle over a manifold to the geometry of the tangent bundle. The metric is then constructed from this object. There is no such soldering in the Yang–Mills theory. We can therefore say that

Gravity Is Gauge Theory with Soldering

We have also seen that formalisms based on differential forms allow the equations of gravity to be rewritten in index-free notations. In 2D this is achieved by introducing a complex linear combination of the pair of frame 1-forms, see (3.40). In 3D this is achieved by constructing 1-forms with values in the Lie algebra of the appropriate 'Lorentz' group, concretely 1-forms with values in 2×2 tracefree matrices, both for the frame field as well as for the connection, see (4.11) and (4.13). Finally, in 4D the closest one gets to an index-free formalism is via the chiral Plebański setup. For instance, the index-free relation (9.138) is the Einstein equation describing the four-sphere. In general, however, when there is also Weyl curvature present, 4D Einstein equations can't be naturally written in a completely index-free notation due to the presence of the matrix

Table 10.1. *Table of formalisms with objects that implement soldering*

Formalism	Soldering object
Cartan formalism	Frame field or tetrad
BF formalism	2-form field valued in the Lie algebra of the Lorentz group
MacDowell–Mansouri formalism	De Sitter/Anti-De Sitter connection
Pure spin connection formalism	Curvature of the spin connection
Plebański formalism	Triple of self-dual 2-forms
Chiral pure connection formalism	Curvature of the chiral part of the spin connection

Ψ^{ij} representing the chiral part of the Weyl curvature on the right-hand side of Plebański equations (5.162). Thus, field equations of 4D gravity are like those of the Yang–Mills theory in the sense that they can't be written solely in terms of wedge products of Lie algebra–valued differential forms. In the case of the Yang–Mills theory, one needs the operation of the Hodge dual to write $d^*F = 0$. In the case of gravity, the analogous operation is the one required to form the right-hand side of the Plebański second equation in (5.162) from Σ^i. Schematically, the Plebański equations are $d_A\Sigma = 0$, which is written solely in terms of wedge product of forms, as well as $F =$ '$^*\Sigma$', where the 'Hodge star' in quotes is the operation that produces the Lie algebra–valued 2-form $(\Psi^{ij} + (\Lambda/3)\delta^{ij})\Sigma^j$ from the Lie algebra–valued 2-form Σ^i.

The analogy with the Yang–Mills theory becomes even more pronounced in the pure connection formalism, where the field equations take the form d_A'$^*F$'$= 0$. Now the 'Hodge star' is the operation (6.14) that is necessary to produce the Lie algebra–valued 2-form Σ^i_F from the curvature 2-forms. In both Yang–Mills and GR it is the presence of these 'Hodge stars' that prevents the equations to be writable solely in terms of wedge products of differential forms.

In terms of the computational efficiency, we have seen that 4D *chiral* formalisms are clearly superior in terms of their economy. In these formalisms, the connection components necessary for the computation of the curvature are stored very compactly and computations required to write Einstein equations are done with minimal effort. This is true both in the case of the original Plebański description that works with 2-forms Σ^i and connection A^i, as well as for the pure connection formalisms that work with either solely A^i or A^i and the auxiliary matrix M^{ij}.

We have also seen that the description of the linearised gravity and the gravitational perturbation theory simplifies greatly by the use of the chiral formalisms. First, the usage of chiral objects brings with it completely new types of differential operators; see Figure 8.1. This allows us to write the familiar spin one and spin two kinetic terms in a completely new way, see, e.g., (8.158) for how the usual linearised Lagrangian for the spin two perturbation $h_{\mu\nu}$ gets compactly rewritten by the use of the chiral 2-form fields $\Sigma^i_{\mu\nu}$.

The propagators and interaction vertices also get simplified by the chiral formalism. The gravitational action becomes polynomial in the fields in any first-order formalism. However, all such formalisms apart from the chiral ones introduce 'too many' auxiliary fields. This is manifested by the fact that the two-point function of the auxiliary field with itself is nonzero in all but the chiral formalisms. This is the case in the chiral description of Yang–Mills, see (8.98), as compared to the non-chiral version, see (8.158), as well as in the chiral description of GR as compared to standard GR, as we have verified in Section 8.5. The chiral perturbation theory for GR that we have developed in this book may well hold a lot of potential. It would be interesting to try to use it to simplify computations ranging from quantum loops to the perturbative calculations that are necessary to extract the gravitational wave signals.

In the last chapter we have developed an even more exotic viewpoint on 4D gravity, one that puts at the forefront the total space of the bundle of two-component spinors over the four-dimensional manifold M in question. The projective version of this bundle is known as the twistor space of M. The usual twistor story emphasises the complex analytic aspects of the twistor space. This, however, only works when the geometry of M is chiral in the sense that only one of the two chiral halves of the Weyl curvature is nonzero.

We have seen that there exists a version of the twistor story that works in the circle bundle over twistor space instead. This is a 7D manifold, and the geometric data on M define a certain natural 3-form C on it. There is then a natural first-order differential equation that can be imposed on M, namely $dC = \lambda {}^*C$, where λ is a constant. Such 3-forms are called nearly parallel and define a 7D metric via (9.146). Moreover, this metric is automatically Einstein with nonzero scalar curvature. Requiring that this equation is satisfied for the 3-form that is defined by the 4D data imposes Einstein-like equations on these data. We have then seen that the usual twistor story with its integrable almost complex structures lifts naturally to this 7D description. In particular, the first-order Cauchy–Riemann equation guaranteeing integrability of the almost complex structure on the twistor space follows from the first-order nearly parallel condition $dC = \lambda {}^*C$ satisfied by the 3-form.

Importantly, the described 6D and 7D viewpoint on 4D gravity is crucially based precisely on its chiral version, to which we devoted so much attention in this book. This is manifested particularly strongly by the example of the quaternionic Hopf fibration in Section 9.3. This example shows the chiral 4D description of the four-sphere with its chiral 2-forms Σ and the chiral connection $\mathbf{A}$ arising from the geometry of the total space of the Hopf three-sphere bundle over S^4. A related point is the fact that the Urbantke formula (5.37) that appears somewhat mysteriously in the chiral 4D descriptions gets explained by the observation that it is the dimensionally reduced to 4D version of the formula (9.146) for the metric defined by a generic 3-form in 7D; see (9.148).

At the same time, the higher-dimensional descriptions that we developed suffer from a very serious defficiency – they only work for the Euclidean version of the 4D gravity. This is the case for both the usual twistor description, which is only capable of describing the half-flat Euclidean gravitational instantons, as well as for the 7D description in terms of 3-forms that we developed. It is clear that if there is any truth in the higher-dimensional perspective of the type described, it should be possible to find also the version appropriate for the Lorentzian signature.

Let us end this discussion by listing questions that, in the opinion of this author, hold greatest potential to lead to a breakthrough in our understanding of gravity. The first question was already mentioned in Chapter 3 introducing formalisms based on differential forms. It is 'Why nonzero metric?' To expand on this, we now know that if there is a nonzero metric filling the universe, then its low-energy dynamics can only be described by GR, at least in 4D. At the same time, GR is unable to answer the question as to why such a nonzero metric exists. The same is true about any of its reformulations described in this book, even though re-formulations based on differential forms seem to come closer to an eventual answer, because in these formulations one can at least talk about the zero field configurations. So, it is clear that answering the 'Why nonzero metric?' question will require radically new ideas. It is possible that the puzzle of gravity can only be solved by answering this question.

The second question that we believe is also of fundamental importance is more well-posed, and so can probably be answered in the near future. This is the question of interpretation of the Lorentzian signature Urbantke formula (5.47). In our discussion following (9.148) we have seen that the Euclidean signature Urbantke formula can be understood as being a consequence of (9.146) defining a 7D metric from a stable 3-form. Thus, we have seen that assuming that the 7D manifold is fibred by three-dimensional submanifolds on which the 3-form is nonzero does exhibit the 4D Urbantke metric as the one induced on the 4D slices transverse to the fibres. The same interpretation exists for the split signature metrics in 4D. This also follows from the 7D formula (9.146) except for C lying in the orbit of the different sign; thus the one for which the metric defined by C is of signature $(3, 4)$. However, there is no such interpretation to the Lorentzian signature Urbantke that works with complex-valued 2-forms but still produces a real-valued metric. It is clear that if there is an interpretation that is related to 3-forms in seven dimensions, it must involve complex-valued forms in some way. We believe that finding such an interpretation, if it exists, holds potential for a breakthrough in understanding of 4D Lorentzian signature gravity, as it would point to a deeper geometric structure behind it.

We end this book by a provocative remark. GR is the unique low-energy theory of interacting massless spin two particles. This statement holds independently of any Lagrangian formulation that may be used to describe it. The usual metric formalism is by far the most explored one. But, in this book, we have seen that, surprisingly, GR admits many non-obviously equivalent formulations. In fact, GR appears to be the theory that admits by far many more reformulations than any other known theory. This is one 'empirical fact' about GR that is rarely emphasised, and that we believe becomes strikingly apparent from the developments we have followed. We don't know the significance of this fact, if any, but it may be that gravity is trying to tell us something. It is possible that the message is: 'I am more than just an effective low-energy theory of massless spin two particles; I hold the key to the puzzle of why the universe can be so successfully described in geometric terms'.

References

Agricola, Ilka. 2008. Old and new on the exceptional group G_2. *Notices Amer. Math. Soc.*, **55**(8), 922–929.

Aldrovandi, Ruben, and Pereira, José Geraldo. 2013. *Teleparallel Gravity*. Vol. 173. Dordrecht: Springer.

Alexander, S., Marciano, A., and Smolin, L. 2014. Gravitational origin of the weak interaction's chirality. *Phys. Rev. D*, **89**(6), 065017.

Alexandrov, S., and Krasnov, K. 2009. Hamiltonian analysis of non-chiral Plebanski theory and its generalizations. *Class. Quant. Grav*, **26**, 055005.

Arnowitt, R. L., Deser, S., and Misner, C. W. 1960. Canonical variables for general relativity. *Phys. Rev*, **117**, 1595.

Ashtekar, A. 1987. New Hamiltonian formulation of general relativity. *Phys. Rev. D*, **36**, 1587.

Atiyah, M. F., Hitchin, Nigel J., and Singer, I. M. 1978. Self-duality in four-dimensional Riemannian geometry. *Proc. Roy. Soc. Lond.*, **A362**, 425–461.

Atiyah, Michael, Dunajski, Maciej, and Mason, Lionel. 2017. Twistor theory at fifty: From contour integrals to twistor strings. *Proc. Roy. Soc. Lond.*, **A473**(2206), 20170530.

Basile, T., Bekaert, X., and Boulanger, N. 2016. Note about a pure spin-connection formulation of general relativity and spin-2 duality in (A)dS. *Phys. Rev. D*, **93**(12), 124047.

Beke, D. 2011. Scalar-Tensor theory as a singular subsector of $\Lambda(\phi)$ Plebanski gravity. e-Print: **arXiv:1111.1139 [gr-qc]**

Beke, D., Palmisano, G., and Speziale, S. 2012. Pauli-Fierz mass term in modified Plebanski gravity. *JHEP*, **1203**, 069.

Besse, A. L. 2008. *Einstein Manifolds*. Berlin: Springer (original edition 1987).

Birmingham, Danny, Blau, Matthias, Rakowski, Mark, and Thompson, George. 1991. Topological field theory. *Phys. Rept.*, **209**, 129–340.

Blau, Matthias, and Thompson, George. 2016. *Chern-Simons Theory with Complex Gauge Group on Seifert Fibred 3-Manifolds*. e-Print: **arXiv:1812.10966 [hep-th]**

Bott, R. and Tu, L. W. 1982. *Differential Forms in Algebraic Topology*. Berlin: Springer.

Brans, C. H. 1974. Complex structures and representations of the einstein equations. *J. Math. Phys.*, **15**, 1559–1566.

Cahen, M., Debever, R., and Defrise, L. 1967. A complex vectorial formalism in general relativity. *J. Math. Mech.*, **16**, 761–785.

Capovilla, R., Jacobson, T., Dell, J., and Mason, L. J. 1991. Self-dual two forms and gravity. *Class. Quant. Grav*, **8**, 41.

Capovilla, R., Montesinos, M., Prieto, V. A., and Rojas, E. 2001. BF gravity and the Immirzi parameter. *Class. Quant. Grav*, **18**, L49.

Carroll, S. 2019. *An Introduction to General Relativity: Spacetime and Geometry*. New York: Cambridge University Press.

Cheung, Clifford, and Remmen, Grant N. 2017. Hidden simplicity of the gravity action. *JHEP*, **09**, 002.

de Haro, S., Solodukhin, S. N., and Skenderis, K. 2001. Holographic reconstruction of space-time and renormalization in the AdS / CFT correspondence. *Commun. Math. Phys*, **217**, 595.

Deser, S. 1970. Selfinteraction and gauge invariance. *Gen. Rel. Grav*, **1**, 9.

Dray, T., Kulkarni, R., and J., Samuel. 1989. Duality and conformal structure. *J. of Math. Phys.*, **30**, 1306–1309.

Dubrovin, B. A., Novikov, A. T., and Fomenko, S. P. 1985. *Modern Geometry Methods and Applications: Part II: The Geometry and Topology of Manifolds*. Berlin: Springer.

Fine, Joel, Krasnov, Kirill, and Panov, Dmitri. 2014. A gauge theoretic approach to Einstein 4-manifolds. *New York J. Math.*, **20**, 293–323.

Frankel, T. 2012. *The Geometry of Physics: An Introduction*, 3rd ed. Boston, MA: Cambridge.

Freidel, L., and Starodubtsev, A. 2005. *Quantum Gravity in Terms of Topological Observables*. e-Print: **hep-th/0501191**

Freidel, L., Krasnov, K., and Puzio, R. 1999. BF description of higher dimensional gravity theories. *Adv. Theor. Math. Phys*, **3**, 1289.

Freidel, Laurent. 2008. *Modified gravity without new degrees of freedom*. e-Print: **arXiv:0812.3200 [gr-qc]**

Friedrich, Th., Kath, I., Moroianu, A., and Semmelmann, U. 1997. On nearly parallel G_2-structures. *J. Geom. Phys.*, **23**(3–4), 259–286.

Goenner, Hubert F. M. 2014. On the history of unified field theories. Part II. (ca. 1930–ca. 1965). *Living Rev. Rel.*, **17**, 5.

Goroff, Marc H., and Sagnotti, Augusto. 1986. The ultraviolet behavior of Einstein gravity. *Nucl. Phys.*, **B266**, 709–736.

Gray, Alfred. 1969. Vector cross products on manifolds. *Trans. Amer. Math. Soc.*, **141**, 465–504.

Haggett, S. A. and Tod, K. P. 1994. *An Introduction to Twistor Theory*, 2nd ed. New York: Cambridge.

Hehl, F. W. 2012. Gauge theory of gravity and spacetime. *Einstein Stud.* 13, 145–169.

Heisenberg, Lavinia. 2018. A systematic approach to generalisations of general relativity and their cosmological implications. *Phys. Rept.* 796, 1–113.

Hejna, M. 2006. *Symmetric Affine Theories and Nonlinear Einstein-Proca System*. e-Print: **gr-qc/0611118**

Herfray, Yannick, and Krasnov, Kirill. 2015. *New First Order Lagrangian for General Relativity*. e-Print: **arXiv:1503.08640 [gr-qc]**

Herfray, Yannick, Krasnov, Kirill, Scarinci, Carlos, and Shtanov, Yuri. 2016a. A 4D gravity theory and G2-holonomy manifolds. *Adv. Theor. Math. Phys.* 22, 2001–2034.

Herfray, Yannick, Krasnov, Kirill, and Shtanov, Yuri. 2016b. Anisotropic singularities in chiral modified gravity. *Class. Quant. Grav.*, **33**, 235001.

Herfray, Yannick, Krasnov, Kirill, and Scarinci, Carlos. 2017. 6D interpretation of 3D gravity. *Class. Quant. Grav.*, **34**(4), 045007.

Hitchin, Nigel J. 2000. The geometry of three-forms in six dimensions. *J. Diff. Geom.*, **55**(3), 547–576.

Hitchin, N. 2014. Differentiable manifolds. Lectures notes available from: https://people.maths.ox.ac.uk/hitchin/

Holst, S. 1996. Barbero's Hamiltonian derived from a generalized Hilbert-Palatini action. *Phys. Rev. D*, **53**, 5966.

Israel, W. 1970. Differential forms in general relativity. *Comm. Dublin Inst. Adv. Studies, Ser. A.*, **19**.

Jacobson, Ted, and Smolin, Lee. 1988. Covariant action for Ashtekar's form of canonical gravity. *Class. Quant. Grav.*, **5**, 583.

Jordan, P., Ehlers, J., and Kundt, W. 2009. Strenge Lösungen der Feldgleichungen der Allgemeinen Relativitätstheorie. *Gen. Rel. Grav.*, **41**, 2191–2280.

Krasnov, K. 2011. Pure connection action principle for general relativity. *Phys. Rev. Lett*, **106**, 251103.

Krasnov, Kirill. 2017a. Dynamics of 3-forms in seven dimensions. *Class. Quant. Grav.*, **34**(22), 225007.

Krasnov, Kirill. 2017b. Self-dual gravity. *Class. Quant. Grav.*, **34**(9), 095001.

Krasnov, Kirill. 2018. Field redefinitions and Plebanski formalism for GR. *Class. Quant. Grav.*, **35**(14), 147001.

Landau, L. D. and Lifshitz, E. M. 1987. *The Classical Theory of Fields*, 4th ed. New York: Elsevier.

Lisi, A. G., Smolin, L., and Speziale, S. 2010. Unification of gravity, gauge fields, and Higgs bosons. *J. Phys. A*, **43**, 445401.

MacDowell, S. W., and Mansouri, F. 1977. Unified geometric theory of gravity and supergravity. *Phys. Rev. Lett*, **38**, 739.

Mitsou, Ermis. 2019. Spin connection formulations of real Lorentzian general relativity. *Class. Quant. Grav.*, **36**, 045008.

Nakahara, M. 2003. *Geometry, Topology and Physics*, 2nd ed. New York: Taylor and Francis.

Newman, Ezra, and Penrose, Roger. 1962. An approach to gravitational radiation by a method of spin coefficients. *J. Math. Phys.*, **3**, 566–578.

Peldan, P. 1992. Connection formulation of (2+1)-dimensional Einstein gravity and topologically massive gravity. *Class. Quant. Grav*, **9**, 2079.

Peldan, Peter. 1994. Actions for gravity, with generalizations: A review. *Class. Quant. Grav.*, **11**, 1087–1132.

Penrose, R. 1960. A spinor approach to general relativity. *Annals Phys.*, **10**, 171–201.

Penrose, R. and Rindler, W. 1999. *Spinors and Space-Time I: Two-Spinor Calculus and Relativistic Fields*. New York: Cambridge (original edition 1984).

Perez, A. 2013. The spin foam approach to quantum gravity. *Living Rev. Rel*, **16**, 3.

Petrov, A. Z. 2000. The classification of spaces defining gravitational fields. *Gen. Rel. Grav.*, **32**, 1661–1663.

Pietri, R. De, and Freidel, L. 1999. so(4) Plebanski action and relativistic spin foam model. *Class. Quant. Grav*, **16**, 2187.

Plebanski, J. F. 1977. On the separation of Einsteinian substructures. *J. Math. Phys.*, **18**, 2511.

Pope, C. 2010. Kaluza-Klein theory. Lecture notes available from: http://people.physics.tamu.edu/pope/ihplec.pdf

Reese Harvey, F. 1990. *Spinors and Calibrations*. New York: Academic Press.

Reshetikhin, N., and Turaev, V. G. 1991. Invariants of three manifolds via link polynomials and quantum groups. *Invent. Math.*, **103**, 547–597.

Roberts, Justin. 1997. Refined state-sum invariants of 3- and 4-manifolds. Pages 217–234 of: *Geometric topology (Athens, GA, 1993)*. AMS/IP Stud. Adv. Math., vol. 2. Amer. Math. Soc., Providence, RI.

Smolin, L. 2009. The Plebanski action extended to a unification of gravity and Yang-Mills theory. *Phys. Rev. D*, **80**, 124017.

Sparks, James. 2011. Sasaki-Einstein Manifolds. *Surveys Diff. Geom.*, **16**, 265–324.

Speziale, S. 2010. Bi-metric theory of gravity from the non-chiral Plebanski action. *Phys. Rev. D*, **82**, 064003.

Stelle, K. S., and West, P. C. 1980. Spontaneously Broken De Sitter symmetry and the gravitational holonomy group. *Phys. Rev. D*, **21**, 1466.

Taubes, A.H. 1966. The Riemann-Christoffel tensor and tetrad and self-dual formalisms. Pages 360–368 of: *Perspectives in Geometry (Essays in honor of V. Hlavaty)*, ed. Hoffmann, Indianapolis, IN: Indiana University Press.

Taubes, C. H. 2011. *Differential Geometry: Bundles, Connections, Metrics and Curvature*. New York: Oxford.

Urbantke, H. 1984. On integrability properties of SU(2) Yang-Mills fields. I. Infinitesimal part. *J. Math. Phys.*, **25**(7), 2321–2324.

Ward, R. S. 1980. Self-dual space-times with cosmological constant. *Commun. Math. Phys.*, **78**, 1–17.

Weinberg, S. 1972. *Gravitation and Cosmology: Principles and Applications of the General Theory of Relativity*. Hoboken, NJ: Wiley.

West, P. C. 1978. A geometric gravity Lagrangian. *Phys. Lett. B*, **76**, 569.
Wise, Derek K. 2010. MacDowell-Mansouri gravity and Cartan geometry. *Class. Quant. Grav.*, **27**, 155010.
Witten, E. 1988. (2+1)-Dimensional gravity as an exactly soluble system. *Nucl. Phys. B*, **311**, 46.
Zee, A. 2013. *Einstein Gravity in a Nutshell.* Princeton, NJ: Princeton University Press.
Zinoviev, Y. M. 2005. *On Dual Formulation of Gravity.* e-Print: **hep-th/0504210**

Index